Inside the LaunchPad for *Practical Strategies for Technical Communication*

Missing something? Instructions may assign the online materials that accompany this text. For access to them, visit **launchpadworks.com.**

CASES

Document-based cases are presented online, where you can familiarize yourself with each scenario, download and work with related documents, and access assignment questions in a single space.

CASE 1: Using the Measures of Excellence in Evaluating a Résumé

CASE 2: The Ethics of Requiring Students To Subsidize a Plagiarism-Detection Service

CASE 3: Accommodating a Team Member's Scheduling Problems

CASE 4: Focusing on an Audience's Needs and Interests

CASE 5: Revising a Questionnaire

CASE 6: Emphasizing Important Information in a Technical Description

CASE 7: Designing a Flyer

CASE 8: Creating Appropriate Graphics To Accompany a Report

CASE 9: Setting Up and Maintaining a Professional Microblog Account

CASE 10: Identifying the Best-of-the-Best Job-Search Sites

CASE 11: Revising a Brief Proposal

CASE 12: Writing a Directive

CASE 13: Analyzing Decision Matrices

CASE 14: Choosing a Medium for Presenting Instructions

CASE 15: Understanding the Claim-and-Support Structure for Presentation Graphics

SUPPLEMENTAL E-BOOKS

Document-Based Cases for Technical Communication, Second Edition, by Roger Munger, features seven realistic scenarios in which you can practice workplace writing skills.

Team Writing, by Joanna Wolfe, focuses on the role of written communication in teamwork. Built around five short videos of real team interactions, *Team Writing* teaches you how to use written documentation to manage a team and provides models for working on large collaborative documents.

DOCUMENT ANALYSIS ACTIVITIES

Explore real multimedia documents that harness digital technologies in exciting new ways, and respond to prompts that will help you analyze them.

Mechanism Description Using Interactive Graphics: Hybridcenter.org and Union of Concerned Scientists, *Hybrids Under the Hood (Part 2)*

Online Portfolio: Blane C. Holden's Online Portfolio

Proposal Delivered as a Prezi Presentation: Andrew Washuta, *Marketing Project Proposal*

Report Presented as a Website: United States Geological Survey, *High Plains Water-Level Monitoring Study*

Informational Report Presented Through an Interactive Graphic: Matthew C. Hansen et al., University of Maryland, Google, USGS, and NASA, *"Global Forest Change" Interactive Map*

Recommendations Presented in a Video: One & Only Campaign, *Check Your Steps! Make Every Injection Safe*, Centers for Disease Control, *Influenza 2010–2011, ACIP Vaccination Recommendations*

Mechanism Description Using Interactive Graphics: Hybridcenter.org and Union of Concerned Scientists, *Hybrids Under the Hood (Part 2)*

Process Description Using Video Animation: North Carolina Department of Transportation (NCDOT), *Diverging Diamond Interchange Visualization*

Instructions Using a Combination of Video Demonstration ar Screen Capture: Texas Tech University, Multiple Literacy L (MuLL), *Recording Audio with iPod + iTalk*

Definition Using Video Animation: ABC News, *What Is the Cloud?*

DOWNLOADABLE FORMS

Download and work with a variety of helpful forms discussed throughout the text.

Work-Schedule Form (Chapter 3)
Team-Member Evaluation Form (Chapter 3)
Self-Evaluation Form (Chapter 3)
Audience Profile Sheet (Chapter 4)
Oral Presentation Evaluation Form (Chapter 15)

LEARNINGCURVE

Master the material in the text with midpoint and final exams powered by LearningCurve, an adaptive quizzing program that meets you where you are and gives you the extra support you need when you need it. Additional activities on grammar and concerns of multilingual writers are also available.

TEAM WRITING MODULES

These modules, built around five short videos of real team interactions, focus on the role of written communication in teamwork. They'll teach you how to use written documentation to manage a team by producing task schedules, minutes, charters, and other materials and also provide models for working on large collaborative documents.

TEST BANK

A test bank offers multiple-choice, true/false, and short-answer questions for key textbook content.

TUTORIALS

Engaging tutorials show you helpful tools and tips for creating your projects along with guidance on how to best use them, as well as the documentation process for citing the sources you use in MLA and APA style.

DIGITAL WRITING TUTORIALS

Cross-Platform Word Processing with CloudOn, Quip, and More (Chapter 3)
Tracking Sources with Evernote and Zotero (Chapter 5)
Photo Editing Basics with GIMP (Chapter 8)
Building Your Professional Brand with LinkedIn, Twitter, and More (Chapter 10)
Creating Presentations with PowerPoint and Prezi (Chapter 15)
Audio Recording and Editing with Audacity (Chapter 15)

DIGITAL TIPS TUTORIALS

Creating Styles and Templates (Chapter 3)
Scheduling Meetings Online (Chapter 3)
Conducting Online Meetings (Chapter 3)
Using Wikis for Collaborative Work (Chapter 3)
Incorporating Tracked Changes (Chapter 3)
Using Collaborative Software (Chapter 3)
Reviewing Collaborative Documents (Chapter 3)
Proofreading for Format Consistency (Chapter 7)

DOCUMENTATION TUTORIALS

How To Cite a Website in APA Style (Appendix, Part A: Documenting Your Sources)
How To Cite a Database in APA Style (Appendix, Part A: Documenting Your Sources)
How To Cite an Article in MLA Style (Appendix, Part A: Documenting Your Sources)
How To Cite a Book in MLA Style (Appendix, Part A: Documenting Your Sources)
How To Cite a Website in MLA Style (Appendix, Part A: Documenting Your Sources)
How To Cite a Database in MLA Style (Appendix, Part A: Documenting Your Sources)

Practical Strategies

FOR TECHNICAL COMMUNICATION

Practical Strategies

FOR TECHNICAL COMMUNICATION

A BRIEF GUIDE

THIRD EDITION

Yuri_Arcurs / Getty Images

Mike Markel
Boise State University

Stuart A. Selber
Penn State University

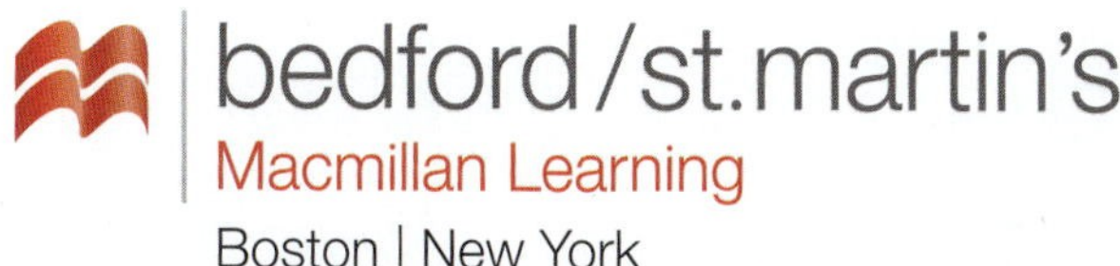
bedford/st.martin's
Macmillan Learning
Boston | New York

For Bedford/St. Martin's

Vice President, Editorial, Macmillan Learning Humanities: Edwin Hill
Executive Program Director for English: Leasa Burton
Senior Program Manager: Laura Arcari
Marketing Manager: Lauren Arrant
Director of Content Development, Humanities: Jane Knetzger
Development Manager: Susan McLaughlin
Developmental Editor: Jesse Hassenger
Senior Workflow Project Supervisor: Joe Ford
Production Supervisor: Robin Besofsky
Media Project Manager: Rand Thomas
Media Editor: Angela Beckett
Media Developmental Editor: Sherry Mooney
Manager of Publishing Services: Andrea Cava
Project Management: Lumina Datamatics, Inc.
Composition: Lumina Datamatics, Inc.
Text Permissions Editor: Arthur Johnson, Lumina Datamatics, Inc.
Photo Researchers: Julie Tesser and Krystyna Borgen, Lumina Datamatics, Inc.
Permissions Editor: Angela Boehler
Permissions Manager: Kalina Ingham
Director of Design, Content Management: Diana Blume
Text Design: Maureen McCutcheon Design; Lumina Datamatics, Inc.
Cover Design: William Boardman
Cover Image: Yuri_Arcurs/Getty Images
Printing and Binding: LSC Communications

Manufactured in the United States of America.

1 2 3 4 5 6 23 22 21 20

For information, write: Bedford/St. Martin's, 75 Arlington Street, Boston, MA 02116

ISBN 978-1-319-36229-4 [Student Edition]
ISBN 978-1-319-36231-7 [Loose-leaf Edition]

Acknowledgments

Acknowledgments and copyrights appear on the same page as the text and art selections they cover; these acknowledgments and copyrights constitute an extension of the copyright page.

Preface for Instructors

AS A LONGTIME USER and admirer of Mike Markel's work, I'm excited to take on the role of co-author in this new edition of *Practical Strategies*, a shorter version of *Technical Communication. Practical Strategies* focuses on the essential topics, writing strategies, and skills students need to succeed in the course and in their professional lives, and makes that material as accessible as possible with streamlined and reorganized chapters. Its goal is to teach students how to become successful technical communicators who produce documents that are both useful and professionally responsible. My hope as co-author of *Practical Strategies* is to continue advancing the fundamental goal of helping students understand both their professional and ethical responsibilities as communicators, and making that process easier and more manageable and understandable.

This goal can be challenging as technical communication continues to change and expand, with social media leading to increased audience interactions, evolving technologies offering new formats for traditional documents, and user experiences becoming more important than ever. The types of documents that technical communicators produce are also changing: microblog posts, infographic résumés, discussion forums, and status updates on LinkedIn are now accepted parts of the technical-communication landscape.

Despite these changes, the fundamentals of technical communication are at least as important as they always have been. An inaccuracy in an online status update communicating a project update is every bit as big a problem as an inaccuracy in a traditional progress report. And even though we live and work in an era that values brevity and quick turnaround, some information can be properly communicated only through the longer, detailed documents that have always been at the center of technical communication.

We have revised this new edition of *Practical Strategies for Technical Communication* to help students learn how to communicate in ways both old and new. Employers have never valued communication skills as much as they value them today, and for good reason. Today's professionals need to communicate more frequently, more rapidly, more accurately, and with more individuals than ever before. This book will help prepare students to do so—in their courses and in their careers.

Organization and Features of the Text

Practical Strategies for Technical Communication is organized into five parts.

- Part 1, "Working in the Technical-Communication Environment," orients students to the practice of technical communication, introducing important topics such as the roles of technical communicators, basic processes for writing technical documents, ethical and legal considerations, effective collaboration, and uses for social media in collaboration.
- Part 2, "Planning and Drafting the Document," focuses on rhetorical and stylistic concerns: considering audience and purpose, gathering information through primary and secondary research, and writing coherent, clear documents.
- Part 3, "Designing User-Friendly Documents and Websites," introduces students to design principles and techniques and to the creation and use of graphics in technical documents and websites.
- Part 4, "Learning Important Applications," offers practical advice for preparing the types of technical communication that students are most likely to encounter in their professional lives: letters, memos, emails, and microblogs; job-application materials; proposals; informational reports, such as progress and status reports; recommendation reports; definitions, descriptions, and instructions; and oral presentations.
- The appendix, "Reference Handbook," provides help with paraphrasing, quoting, and summarizing sources; documenting sources in the APA, IEEE, and MLA styles; and editing and proofreading documents.

Help with the writing process is integrated throughout the book in the form of two prominent features.

- Choices and Strategies charts (see page 78, for example) are designed to help students at decision points in their writing. These charts summarize various writing and design strategies and help students choose the one that best suits their specific audience and purpose.
- Focus on Process boxes in each of the applications chapters (see page 291, for example) highlight aspects of the writing process that require special consideration when writing specific types of technical communication. Each Focus on Process box in Part 4 relates back to a complete overview of the writing process in Chapter 1 (see page 11).

New to This Edition

The Third Edition has been revised to help students keep pace with a changing technical-communication environment. Throughout the book, every Tech Tip has been revised and updated to reflect not just *how* to create different parts of documents using different kinds of software, but *why* to use these tools. Chapter 10, which discusses job applications, now includes coverage of nontraditional résumé formats, including infographic-based résumés and video résumés, giving students tips on how to create them and also, crucially, when to consider using them. Chapter 14 adds new material addressing usability testing for instructions, a step that has gained importance as technology has become more complex and sophisticated. All of these changes, along with other updates throughout the book, serve to make *Practical Strategies* better reflect the technical-communication world of today's workers.

Acknowledgments

All of the examples in this book—from single sentences to complete documents—are real. Some were written by students at our schools. Some were written by engineers, scientists, health-care providers, and businesspeople with whom we have worked. Because much of the information in these documents is proprietary, we have silently changed brand names and other identifying information. We thank the dozens of individuals—students and professionals alike—who have graciously allowed us to reprint their writing. They have been our best teachers.

The Third Edition of *Practical Strategies for Technical Communication* has benefited greatly from the perceptive observations and helpful suggestions of our fellow instructors throughout the country. We thank Lauren Birdsong, Atlantic Cape Community College; Melissa Dugan, Clemson University; Julie Groesch, San Jacinto College; Sandra Johnston, University of Maryland Eastern Shore; Sacheen Mobley, Farmingdale State College; Clarise Nixon, Southern Adventist University; Beata Peterson, Fayetteville Technical Community College; Rebecca Ruggiero, University of Maine; Kathie Russell, Santa Fe College; Ember Smith, Spartanburg Community College; and Dzmitry Yuran, Florida Institute of Technology.

We also thank reviewers of past editions: Lisa Angius, Farmingdale State College; Katie Arosteguy, University of California, Davis; Monique Babin, Clackamas Community College; Jenny Billings Beaver, Rowan Cabarrus Community College; Sheri Benton, University of Toledo; Charles Bevis, University of Massachusetts Lowell; Olin Bjork, University of Houston–Downtown; An Cheng, Oklahoma State University; Elijah Coleman, Washington State University; Crystal Colombini, University of Texas at San Antonio; Teresa Cook, University of Cincinnati; Matthew Cox, East Carolina University; Ed Cuoco, Wentworth Institute of Technology; Jerry DeNuccio, Graceland University; Charlsye Smith Diaz, University of Maine; Carolyn Dunn, East Carolina University; Tomie Gowdy-Burke, Washington State University; Amber Kinonen, Bay College; Tamara Kuzmenkov, Tacoma Community College; Jodie Marion, Mt. Hood Community College; Donna Miguel, Bellevue College; Bonni Miller, University of Maryland Eastern Shore; Mary Ellen Muesing, University of

North Carolina at Charlotte; Ervin Nieves, Kirkwood Community College; Sabrina Peters-Whitehead, University of Toledo; Ehren Pflugfelder, Oregon State University; Neil Plakcy, Broward College; Kathleen Robinson, Eckerd College; Paula Sebastian, Bellevue College; Stella Setka, Loyola Marymount University; Terry Smith, University of Maryland Eastern Shore; Russel Stolins, Institute of American Indian Arts; Virginia Tucker, Old Dominion University; Gabriela Vlahovici-Jones, University of Maryland Eastern Shore; Lynne Walker, Bellevue College; Beverly Army Williams, Westfield State University; and several anonymous reviewers.

We have been fortunate, too, to work with a terrific team at Bedford/St. Martin's. We want to express our appreciation to Leasa Burton, Laura Arcari, Lauren Arrant, Jesse Hassenger, Andrea Cava, Sherry Mooney, Suzanne Chouljian, Rand Thomas, and Angela Beckett. For us, Bedford/St. Martin's continues to exemplify the highest standards of professionalism in publishing. The people there have been endlessly encouraging and helpful. We hope they realize the value of their contributions to this book.

Mike would like to give special thanks to his wife, Rita, who, over the course of many years, has helped him say what he means. Stuart would like to thank his family, Kate Latterell and Griffin and Avery Selber, for their ongoing support and encouragement.

A Final Word

We are more aware than ever before of how much we learn from our students, our fellow instructors, and our colleagues in industry and academia. If you have comments or suggestions for making this a better book, please send an email to **techcomm@macmillan.com**. We hope to hear from you.

Mike Markel and Stuart A. Selber

We're all in. As always.

Bedford/St. Martin's is as passionately committed to the discipline of English as ever, working hard to provide support and services that make it easier for you to teach your course your way.

Find **community support** at the Bedford/St. Martin's English Community (community.macmillan.com), where you can follow our Bits blog for new teaching ideas, download titles from our professional resource series, and review projects in the pipeline.

Choose **curriculum solutions** that offer flexible custom options, combining our carefully developed print and digital resources, acclaimed works from Macmillan's trade imprints, and your own course or program materials to provide the exact resources your students need. Our approach to customization makes it possible to create a tailor-made project uniquely suited for your students, and based on your enrollment size, return money to your department and raise your institutional profile with a high-impact author visit through the Macmillan Author Program (MAP).

Rely on **outstanding service** from your Bedford/St. Martin's sales representative and editorial team. Contact us or visit macmillanlearning.com to learn more about any of the options below.

LaunchPad for *Practical Strategies for Technical Communication*: Where Students Learn

LaunchPad provides engaging content and new ways to get the most out of your book. Get an interactive e-book combined with assessment tools in a fully customizable course space; then assign and mix our resources with yours.

- **Cases and Document Analysis Activities** for every chapter give students the opportunity to practice their skills in context. Students can familiarize themselves with case scenarios, then download and work with related documents to complete their assignment. They also have access to online equivalents of the Document Analysis Activities included in the print book, which introduce students to the kinds of multimedia documents that can exist only online, accompanied by a set of assessment questions to guide students in their analysis.
- **Diagnostics** provide opportunities to assess areas for improvement and assign additional exercises based on students' needs. Visual reports show performance by topic, class, and student as well as improvement over time.
- **Prebuilt units**—including readings, videos, quizzes, and more—are easy to adapt and assign by adding your own materials and mixing them with our high-quality multimedia content and ready-made assessment options, such as **LearningCurve** adaptive quizzing and Exercise Central.

- Use LaunchPad on its own or **integrate it** with your school's learning management system so that your class is always on the same page.

LaunchPad for *Practical Strategies for Technical Communication* can be purchased on its own or packaged with the print book at a significant discount. An activation code is required. To order LaunchPad for *Practical Strategies for Technical Communication* with the print book, use ISBN 978-1-319-22438-7. For more information, go to **launchpadworks.com**.

Choose from Alternative Formats of *Practical Strategies for Technical Communication*

Bedford/St. Martin's offers a range of formats. Choose what works best for you and your students:

- *Paperback* To order the paperback edition, use ISBN 978-1-319-10432-0.
- *Popular e-book formats* For details of our e-book partners, visit **macmillanlearning.com/ebooks**.

Select Value Packages

Add value to your text by packaging one of the following resources with *Practical Strategies*.

Document-Based Cases for Technical Communication, Second Edition, by Roger Munger (Boise State University), offers realistic writing tasks based on 7 context-rich scenarios, with more than 50 examples of documents that students are likely to encounter in the workplace. To order the print book packaged with *Document-Based Cases for Technical Communication*, contact your sales representative.

Team Writing, by Joanna Wolfe (Carnegie Mellon University), is a print supplement with online videos that provides guidelines and examples of collaborating to manage written projects by documenting tasks, deadlines, and team goals. Two- to five-minute videos corresponding with the chapters in *Team Writing* give students the opportunity to analyze team interactions and learn about communication styles. Practical troubleshooting tips show students how best to handle various types of conflicts within peer groups. To order the print book packaged with *Team Writing*, contact your sales representative.

Instructor Resources

You have a lot to do in your course. We want to make it easy for you to find the support you need—and to get it quickly.

Instructor's Resource Manual for Practical Strategies for Technical Communication, Third Edition, is available as a PDF that can be downloaded from macmillanlearning.com. Visit the instructor resources tab for *Practical Strategies*. In addition to chapter overviews and teaching tips, the instructor's manual includes sample syllabi, essays on teaching the technical-communication course, and suggested responses to all of the Document Analysis Activities, Exercises, and Cases.

Computerized Test Bank for Practical Strategies for Technical Communication, Third Edition, offers a convenient way to provide additional assessment to students and is available to download from macmillanlearning.com. Instructors using LaunchPad will find the test bank material there, where they can add prebuilt quizzes to any unit or build their own tests from the test bank questions.

Lecture Slides are available to download and adapt for each chapter.

Introduction for Writers

THE THIRD EDITION of *Practical Strategies for Technical Communication* offers a wealth of support to help you complete your technical-communication projects.

Annotated Examples make it easier for you to learn from the many model documents, illustrations, and screen shots throughout the text.

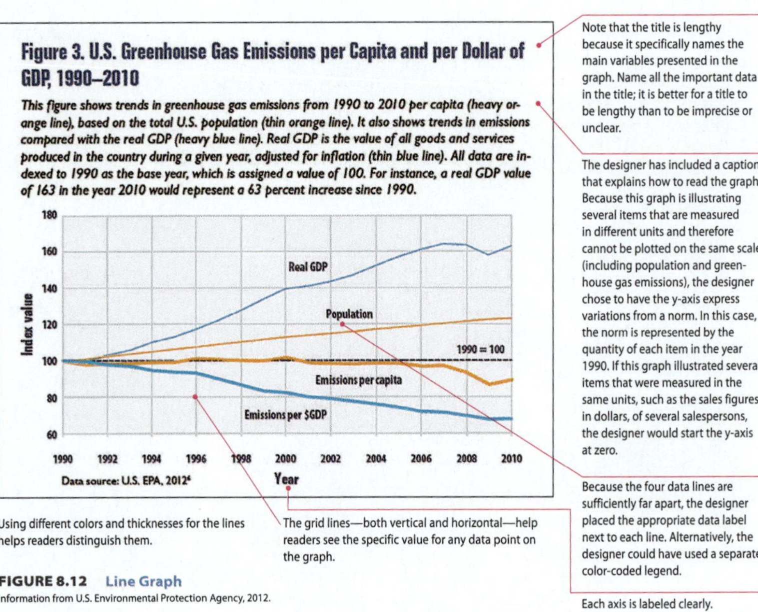

Figure 3. U.S. Greenhouse Gas Emissions per Capita and per Dollar of GDP, 1990–2010
This figure shows trends in greenhouse gas emissions from 1990 to 2010 per capita (heavy orange line), based on the total U.S. population (thin orange line). It also shows trends in emissions compared with the real GDP (heavy blue line). Real GDP is the value of all goods and services produced in the country during a given year, adjusted for inflation (thin blue line). All data are indexed to 1990 as the base year, which is assigned a value of 100. For instance, a real GDP value of 163 in the year 2010 would represent a 63 percent increase since 1990.
180
160
140
120
100
80
60
Index value
Real GDP
Population
1990 = 100
Emissions per capita
Emissions per $GDP
1990
1992
1994
1996
1998
2000
2002
2004
2006
2008
2010
Year
Data source: U.S. EPA, 2012[4]
Note that the title is lengthy because it specifically names the main variables presented in the graph. Name all the important data in the title; it is better for a title to be lengthy than to be imprecise or unclear.
The designer has included a caption that explains how to read the graph. Because this graph is illustrating several items that are measured in different units and therefore cannot be plotted on the same scale (including population and greenhouse gas emissions), the designer chose to have the y-axis express variations from a norm. In this case, the norm is represented by the quantity of each item in the year 1990. If this graph illustrated several items that were measured in the same units, such as the sales figures, in dollars, of several salespersons, the designer would start the y-axis at zero.
Because the four data lines are sufficiently far apart, the designer placed the appropriate data label next to each line. Alternatively, the designer could have used a separate color-coded legend.
Each axis is labeled clearly.
Using different colors and thicknesses for the lines helps readers distinguish them.
The grid lines—both vertical and horizontal—help readers see the specific value for any data point on the graph.

FIGURE 8.12 Line Graph

Information from U.S. Environmental Protection Agency, 2012.

GUIDELINES Planning a Nontraditional Résumé

The following five planning tasks will help you prepare a nontraditional résumé.

- **Analyze the job-search context.** Determine whether a nontraditional résumé is appropriate to the workplace culture for the job. Does the job posting invite applicants to submit a nontraditional résumé? If not, what is your reasoning for submitting one?
- **Consider your résumé content.** Think about which aspects of your content lend themselves to being redesigned for a nontraditional résumé, especially whether there are words and numbers that can be represented visually. Ask yourself what content might be well-suited to a video, and why.
- **Research software programs.** What software programs will you need to access in order to produce a nontraditional résumé? Will your current programs work, or will you need to learn new programs? Keep in mind that if you are using free programs, there may be limits to what you can do with them.
- **Research delivery options.** Consider whether you will host a video résumé on your own website, through a streaming service such as YouTube or Vimeo, or by using a job search portal that allows candidates to upload videos. What are the advantages and disadvantages of the different options?
- **Draft an element.** Test your ideas by drafting an element of a nontraditional résumé. What's a good starting point? Why? You may need to change your plans depending on the results of your draft.

Guidelines boxes throughout the book summarize crucial information and provide strategies related to key topics.

ETHICS NOTE
EUPHEMISMS AND TRUTH TELLING

There is nothing wrong with using the euphemism *restroom*, even though few people visit one to rest. The British use the phrase *go to the toilet* in polite company, and nobody seems to mind. In this case, if you want to use a euphemism, no harm done.

But it is unethical to use a euphemism to gloss over an issue that has important implications for people or the environment. People get uncomfortable when discussing layoffs—and they should. It's an uncomfortable issue. But calling a layoff a *redundancy elimination initiative* ought to make you even more uncomfortable. Don't use language to cloud reality. It's an ethical issue.

Ethics Notes in every chapter remind you to think about the ethical implications of your writing and oral presentations.

DOCUMENT ANALYSIS ACTIVITY
Integrating Graphics and Text on a Presentation Slide

The following slide is part of a presentation about the Human Genome Project. The questions in the margin ask you to think about the discussion of preparing presentation graphics.

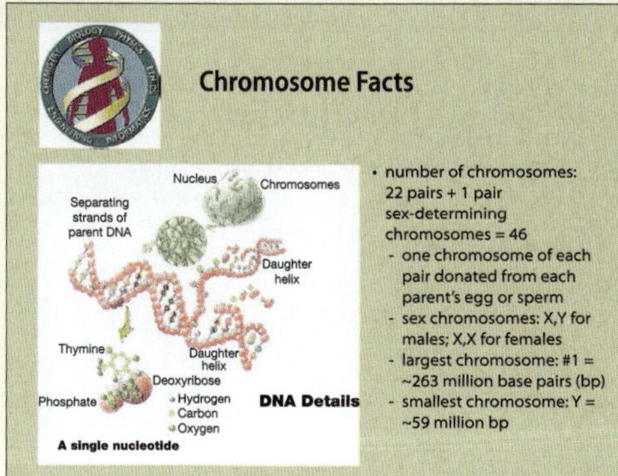

1. How effective is the Human Genome Project logo in the upper left-hand corner of the slide?
2. How well does the graphic of DNA support the accompanying text on chromosome facts?
3. Overall, how effective is the presentation graphic?

Document Analysis Activities, located both in print and online, allow you to apply what you have just read as you analyze a real business or technical document.

Tech Tips for using basic software tools give you step-by-step, illustrated instructions on topics such as tracking changes, creating graphics, and modifying templates. Each Tech Tip has also been updated to include explanation of not just how to perform these actions, but why. This enhances your understanding of your technical communication goals.

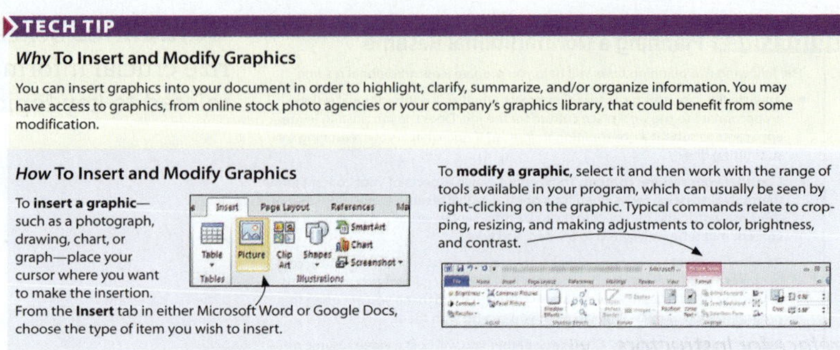

TECH TIP

***Why* To Insert and Modify Graphics**

You can insert graphics into your document in order to highlight, clarify, summarize, and/or organize information. You may have access to graphics, from online stock photo agencies or your company's graphics library, that could benefit from some modification.

***How* To Insert and Modify Graphics**

To **insert a graphic**—such as a photograph, drawing, chart, or graph—place your cursor where you want to make the insertion. From the **Insert** tab in either Microsoft Word or Google Docs, choose the type of item you wish to insert.

To **modify a graphic**, select it and then work with the range of tools available in your program, which can usually be seen by right-clicking on the graphic. Typical commands relate to cropping, resizing, and making adjustments to color, brightness, and contrast.

Writer's Checklists summarize important concepts and act as handy reminders as you draft and revise your work.

WRITER'S CHECKLIST

- ☐ Did you determine the questions you need to answer for your document? *(p. 75)*

Did you choose appropriate secondary-research tools to answer those questions, including, if appropriate,

- ☐ online catalogs? *(p. 80)*
- ☐ reference works? *(p. 80)*
- ☐ periodical indexes? *(p. 81)*
- ☐ newspaper indexes? *(p. 82)*
- ☐ abstract services? *(p. 82)*
- ☐ government information? *(p. 83)*
- ☐ social media and other interactive resources? *(p. 83)*

Did you choose appropriate primary-research methods to answer your questions, including, if appropriate,

- ☐ social-media data analysis? *(p. 88)*
- ☐ observations and demonstrations? *(p. 90)*
- ☐ inspections? *(p. 91)*
- ☐ experiments? *(p. 91)*
- ☐ field research? *(p. 92)*
- ☐ interviews? *(p. 93)*
- ☐ inquiries? *(p. 95)*
- ☐ questionnaires? *(p. 96)*
- ☐ Did you report and analyze the data honestly? *(p. 99)*

Cases in every chapter present real-world writing scenarios built around common workplace documents that you can critique, download, and revise.

CASE 9: Setting Up and Maintaining a Professional Microblog Account

As the editor-in-chief of your college newspaper, you have recently been granted permission to create a Twitter account. The newspaper's faculty advisor has requested that, before you set up the account, you develop a statement of audience and purpose based on your school's own social-media policy statement and statements from other schools, newspapers, and organizations. To begin putting together a bibliography to guide your research and craft your statement, go to LaunchPad.

For quick reference, many of these features are indexed on the last book page and the inside back cover of this book.

Brief Contents

Contents

Part 2 Planning and Drafting the Document 49

4 Analyzing Your Audience and Purpose 50

5 Researching Your Subject 74

13 Writing Recommendation Reports 338

14 Writing Definitions, Descriptions, and Instructions 383

Practical Strategies

FOR TECHNICAL COMMUNICATION

Part 1

Yuri_Arcurs/Getty Images

Working in the Technical-Communication Environment

CHAPTER 1 Introduction to Technical Communication

Palsur/Shutterstock

THIS TEXTBOOK EXPLORES how people in the working world find, create, and deliver technical information. Even if you do not plan on becoming a *technical communicator* (a person whose main job is to produce documents such as manuals, reports, and websites), you will often find yourself writing documents on your own, participating in teams that write them, and contributing technical information for others who read and write them. The purpose of *Practical Strategies for Technical Communication* is to help you learn the skills you need to communicate more effectively and more efficiently in your professional life.

What Is Technical Communication?

Technical information is frequently communicated through documents, such as proposals, emails, reports, podcasts, computer help files, blogs, and wikis. Although these documents are a key component of technical communication, so too is the *process*: writing and reading tweets and text messages, for example, or participating in videoconference exchanges with colleagues. Technical communication encompasses a set of *activities* that people do to discover, shape, and transmit information.

Technical communication uses the four basic communication modes—*listening, speaking, reading,* and *writing*—to analyze a problem, find and evaluate evidence, and draw conclusions. These are the same skills and processes you use when you write in college, and the principles you have studied in your earlier writing courses apply to technical communication. The biggest difference between technical communication and the other kinds of writing you have done is that technical communication has a somewhat different focus on *audience* and *purpose*.

UNDERSTANDING AUDIENCE AND PURPOSE

In most of your previous academic writing, your audience has been your instructor, and your purpose has been to show your instructor that you have mastered some body of information or skill. Typically, you have not tried to create new knowledge or motivate the reader to take a particular action—except to give you an A for that assignment.

By contrast, in technical communication, your audience will likely include peers and supervisors in your company, as well as people outside your company. Your purpose will likely fall into one of two categories:

- **To help others learn about a subject, carry out a task, or make a decision.** For instance, the president of a manufacturing company might write an article in the company newsletter to explain to employees why management decided to phase out production of one of the company's products. Administrators with the Department of Health and Human Services might hire a media-production company to make a video that explains to citizens how to manage their Medicare benefits. The board of directors of a community-service organization might produce a grant proposal to submit to a philanthropic organization in hopes of being awarded a grant.
- **To reinforce or change attitudes and motivate readers to take action.** A wind-energy company might create a website with videos and text intended to show that building windmills off the coast of a tourist destination would have many benefits and few risks. A property owners' association might create a website to make the opposite argument: that the windmills would have few benefits but many risks. In each of these two cases, the purpose of communicating the information is to persuade people to accept a point of view and encourage them to act—perhaps to contact their elected representatives and present their views about this public-policy issue.

These types of communication have a clearly defined audience—one or more people who are going to read the document, attend the oral presentation, visit the website, or view the video you produce. It's also possible, even likely, that a piece of technical communication will have multiple audiences with different purposes.

For example, suppose you are a public-health scientist working for a federal agency. You and your colleagues just completed a study showing that, for most adults, moderate exercise provides as much health benefit as strenuous exercise. After participating in numerous meetings with your colleagues and after writing, critiquing, and revising many drafts, you produce four different documents:

- a journal article for other scientists
- a press release to distribute to popular print and online publications
- a blog post and podcast for your agency's website

In each of these documents, you present the key information in a different way to meet the needs of a particular audience.

Why Technical Communication Skills Are Important in Your Career

Many college students believe that the most important courses they take are those in their major. Some biology majors think, for example, that if they just take that advanced course in genetic analysis, employers will conclude that they are prepared to do more-advanced projects and therefore will hire them.

But knowledge in a particular field is not the only thing employers are looking for. It's not even the most important skill or ability. Surveys over the past three or four decades have shown consistently that employers want people who can communicate. Look at it this way: When employers hire a biologist, they want a person who can communicate effectively about biology. When they hire a civil engineer, they want a person who can communicate about civil engineering.

A 2013 survey of 500 business executives found that almost half—44 percent—think that recent hires are weak in soft skills (including communication and collaboration), whereas only 22 percent think recent hires are weak in technical skills (Adecco Staffing, 2013). According to another 2013 survey, by the Workforce Solutions Group at St. Louis Community College, more than 60 percent of employers believe that job seekers are weak in communication and interpersonal skills. This figure is up 10 percentage points from 2011 (*Time*, 2013).

Job Outlook 2014, a report produced by the National Association of Colleges and Employers, found that communication skills were second only to problem-solving skills among the abilities employers seek (National Association, 2014, p. 8). On a 5-point scale, communication skills scored a 4.6, as did the ability to obtain and process information. Also scoring above 4 were the ability to plan, organize, and prioritize work (4.5); the ability to analyze quantitative data (4.4); technical knowledge related to the job (4.2); and proficiency with computer software programs (4.1). The ability to create and/or edit written reports and the ability to sell or influence others scored a 3.6 and a 3.7, respectively. Most of these skills relate to the previous discussion about the importance of process in technical communication.

A 2014 study of more than 400 freelancer profiles conducted by the online grammar-checking service Grammarly found a direct correlation between the number of errors in a freelancer's client service profile on the website Elance and that freelancer's client rating (Grammarly, 2014). This pattern held across eight industries. Grammarly also found that in most skill-driven jobs, better writers tended to earn more money from clients. This was especially true in the fields of engineering and manufacturing, finance and management, the law, and sales and marketing.

You're going to be producing and contributing to a lot of technical documents. The facts of life in the working world are simple: the better you communicate, the more valuable you are. This textbook can help you learn and practice the skills that will make you a better communicator.

The Challenges of Producing Technical Communication

One of the most challenging activities you will engage in as a professional is communicating your ideas to audiences. Why? Because communication is a higher-order skill that involves many complex factors.

The good news is that there are ways to think about these complex factors, to think *through* them, that will help you communicate better. No matter what document you produce or contribute to, you need to begin by considering five sets of factors:

- **Audience-related factors.** What problem (or problems) is your audience trying to solve? Does your audience know enough about your subject to understand a detailed discussion, or do you need to limit the scope, the amount of technical detail, or the type of graphics you use? Does your audience already have certain attitudes or expectations about your subject that you wish to reinforce or change? Does your audience speak English well, or should you present the information in more than one language? Does your audience share your cultural assumptions about such matters as how to organize and interpret documents, or do you need to adjust your writing approach to match a different set of assumptions? Does your audience include people with disabilities (of vision, hearing, movement, or cognitive ability) who have needs you want to meet?
- **Purpose-related factors.** Before you can write, you need to determine your purpose: What do you want your audience to know or believe or do after having read your document? Do you have multiple purposes? If so, is one more important than the others? Although much technical communication is intended to help people perform tasks, such as configuring privacy settings in a social-media environment, many organizations large and small devote significant communication resources to the increasingly vital purpose of *branding*: creating an image that helps

customers distinguish the company from competitors. Most companies now employ community specialists as technical communicators to coordinate the organization's day-to-day online presence and its social-media campaigns. These specialists publicize new products and initiatives and respond to questions and new developments. They also manage the organization's documents, including tweets, blog posts, Facebook pages, and company-sponsored discussion forums.

- **Setting-related factors.** What is the situation surrounding the problem you are trying to solve? Is there a lot at stake in this situation, such as the budget for a project, or is your document a more routine communication, such as technical notes for a software update? What is the context in which your audience will use your document? Will the ways in which they use it—or the physical or digital environment in which they use it—affect how you write? Will the document be used in a socially or politically charged setting? Does the setting include established norms of ethical behavior? Is the setting formal or informal? Settings can have a great deal of influence over how audiences think about and use technical communication.
- **Document-related factors.** What type of content will the document include? How will the content aid problem solving? Does the subject dictate what kind of document (such as a report or a blog post) you choose to write? Does your subject dictate what medium (print or digital) you choose for your documents? Do you need to provide audiences with content in more than one medium? If you're using a document template, how should you modify it for your audiences and purposes? Does the application call for a particular writing style or level of formality? (For the sake of convenience, we will use the word *document* throughout this book to refer to all forms of technical communication, from written documents to oral presentations and online forms, such as podcasts and wikis.)
- **Process-related factors.** What process will you use to produce the document? Is there an established process to support the work, or do you need to create a new one? Do you have sufficient time for planning tasks, choosing writing tools, and researching background information? Does your budget limit the number of people you can enlist to help you or limit the size or shape of the document? Does your schedule limit how much information you can include in the document? Does your schedule limit the type or amount of document testing you can do? Will the document require updating or maintenance?

Because all these factors interact in complicated ways, every technical document you create involves a compromise. If you are writing a set of instructions for installing a water heater and you want those instructions to be easily understood by people who speak only Spanish, you will need more time and a bigger budget to have the document translated, and it will be longer and thus a little bit harder to use, for both English and Spanish speakers.

You might need to save money by using smaller type, smaller pages, and cheaper paper, and you might not be able to afford to print it in full color. In technical communication, you do the best you can with your resources of time, information, and money. The more carefully you think through your options, the better able you will be to use your resources wisely and make a document that will get the job done.

Skills and Qualities Shared by Successful Workplace Communicators

People who are good at communicating in the workplace share a number of skills and qualities. Three of them relate to the skills you have been honing in school and will continue to develop in your career:

- **Ability to perform research.** Successful communicators know how to perform primary research (discovering new information through experiments, observations, interviews, surveys, and calculations) and secondary research (finding existing information by reading what others have written or said). Successful communicators seek out information from people who *use* the products and services, not just from the manufacturers and service providers.
- **Ability to analyze information.** Successful communicators know how to identify the best information—the most accurate, relevant, recent, and unbiased—and then figure out how it helps in understanding a problem and finding ways to solve it. Successful communicators know how to sift through mountains of data, identifying relationships among apparently unrelated facts. They know how to evaluate a situation, look at it from other people's perspectives, and zero in on the most important issues.
- **Ability to speak and write clearly.** Successful communicators know how to express themselves clearly and simply, both to audiences that know a lot about the subject and to audiences that do not. They take care to revise, edit, and proofread their documents so that the documents present accurate information, are easy to read, and make a professional impression. And they know how to produce different types of documents, from tweets to memos to presentations.

In addition to the skills just described, successful workplace communicators have seven qualities that you should model:

- **Be honest.** Successful communicators tell the truth. They don't promise what they know they can't deliver, and they don't bend facts. When they make mistakes, they admit them and work harder to solve the problem.
- **Be willing to learn.** Successful communicators know that they don't know everything—not about what they studied in college, what their company does, or how to write and speak. Every professional is a lifelong learner.

- **Display emotional intelligence.** Because technical communication usually calls for collaboration, successful communicators understand their own emotions and those of others. Because they can read people—through body language, facial expressions, gestures, and words—they can work effectively in teams.
- **Be generous.** Successful communicators share information willingly. (Of course, they don't share confidential information, such as trade secrets, information about new products being developed, or personal information about colleagues.)
- **Monitor the best information.** Successful communicators seek out opinions from others in their organization and in their industry. They monitor the best blogs, discussion forums, and podcasts for new approaches that can spark their own ideas. They know how to use social media and can represent their organization online.
- **Be self-disciplined.** Successful communicators are well organized and diligent. They know, for instance, that proofreading an important document might not be fun but is always essential. They know that when a colleague asks a simple technical question, answering the question as soon as possible is more helpful than answering it in a couple of weeks. They finish what they start, and they always do their best on any document, from the least important text message to the most important report.
- **Prioritize and respond quickly.** Successful communicators know that the world doesn't always conform to their own schedules. Because social media never sleep, communicators sometimes need to put their current projects aside in order to respond immediately when a stakeholder reports a problem that needs prompt action or publishes inaccurate information that can hurt the organization.

A Process for Writing Technical Documents

Although every technical document is unique, in most of your writing you will likely carry out the tasks described in the Focus on Process box on page 11.

This writing process consists of five steps: planning, drafting, revising, editing, and proofreading. The frustrating part of writing, however, is that these five steps are not linear. That is, you don't plan the document, then check off a box and go on to drafting. At any step, you might double back to do more planning, drafting, or revising. Even when you think you're almost done—when you're proofreading—you still might think of something that would improve the planning. That means you'll need to go back and rethink all five steps.

As you backtrack, you will have one eye on the clock, because the deadline is sneaking up on you. That's the way it is for all writers. A technical writer stops working on a user manual because she has to get it off to the print shop. An engineer stops working on a set of slides for a conference presentation because it's time to head for the airport.

THINKING VISUALLY

Characteristics of a Technical Document

Almost every technical document that gets the job done has six major characteristics.

It addresses particular READERS.

Knowing who your readers are, what they understand about the subject, how well they speak English, and how they will use the document will help you determine the type of document to create.

It helps readers solve PROBLEMS.

You might produce a video that shows your company's employees how to select their benefits, or a text document that explains the company's social-media policy.

It is produced COLLABORATIVELY.

No one person has all the skills, information, or time to create a large document. You will work with a variety of technical professionals inside and outside your organization to obtain the information you need.

It uses DESIGN to increase readability.

Design features such as typography, spacing, and color help make a document attractive, navigable, and understandable.

It reflects the organization's GOALS and CULTURE.

Organizations produce documents that help them further their goals and that demonstrate their values and culture to the outside world.

It consists of WORDS or IMAGES or both.

Images—both static and moving—can communicate difficult concepts, instructions, descriptions of objects and processes, and large amounts of data. They can also communicate information to nonnative speakers.

Clockwise from top left: (1) Colorlife/Shutterstock; (2) Palsur/Shutterstock; (4) puruan/Shutterstock; (5) Brothers Good/Shutterstock; (6) PureSolution/Shutterstock.

THINKING VISUALLY

Measures of Excellence in Technical Documents

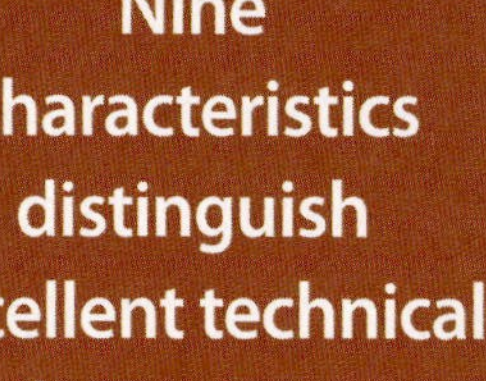
Nine characteristics distinguish excellent technical documents.

HONESTY

The most-important measure of excellence is honesty. Not only are dishonest documents unethical, but they can hurt readers and can result in serious legal repercussions.

CLARITY

An unclear technical document can be dangerous and incur additional expenses.

ACCESSIBILITY

A good technical document can be accessed, explored, and used by people with varying physical abilities.

COMPREHENSIVENESS

A comprehensive document provides readers with a complete, self-contained discussion.

PROFESSIONAL APPEARANCE

If the document looks professional, readers will form a positive impression of it and of you. Your document should adhere to the format standards of your organization or your field, and it should be well designed.

USABILITY

In technical communication, usability measures how successfully a document achieves its purposes and meets its audience's needs.

ACCURACY

A slight inaccuracy can confuse and annoy your readers; a major inaccuracy can be dangerous and expensive.

CONCISENESS

You can shorten most writing by 10 to 20 percent by eliminating unnecessary phrases, choosing shorter words, and using economical grammatical forms.

CORRECTNESS

A correct document adheres to the conventions of grammar, punctuation, spelling, mechanics, and usage. Incorrect writing can confuse readers, make your document inaccurate, and make you look unprofessional.

Clockwise from top left: (1) a Sk/Shutterstock; (2) RedKoala/Shutterstock; (3) VoodooDot/Shutterstock; (4) Armita/Shutterstock; (5) ProStockStudio/Shutterstock; (6) Nobelus/Shutterstock; (7) Brothers Good/Shutterstock; (8) graphixmania/Shutterstock; (9) Artram/Shutterstock.

FOCUS ON PROCESS Writing Technical Documents

PLANNING

- **Analyze your audience.** Who are your readers? What are their attitudes and expectations? How will they use the document? See Chapter 4 for advice about analyzing your audience.
- **Analyze your purpose.** After they have read the document, what do you want your readers to *know* or to *do*? See Chapter 4 (p. 69) for advice about determining your purpose.
- **Generate ideas about your subject.** Ask journalistic questions (*who*, *what*, *when*, *where*, *why*, and *how*), brainstorm, freewrite, talk with someone, or make clustering or branching diagrams.
- **Research additional information.** See Chapter 5 for advice about researching your subject.
- **Organize and outline your document.** See Chapter 6 (p. 103) for information about common organizational patterns.
- **Select an application, a design, and a delivery method.** See Chapter 7 for advice about designing your document.
- **Devise a schedule and a budget.** How much time will you need to complete each task of the project? Will you incur expenses for travel, research, or usability testing?

DRAFTING

- **Draft effectively.** Get comfortable. Start with the easiest topics, and don't stop writing to revise.
- **Use templates—carefully.** Check that their design is appropriate and that they help you communicate your information effectively to your readers.
- **Use styles.** Styles are like small templates that apply to the design of elements such as headings and bullet lists. They help you present the elements of your document clearly and consistently.

REVISING

Look again at your draft to see if it works. Revising by yourself and with the help of others, focus on three questions:

- Has your understanding of your audience changed?
- Has your understanding of your purpose changed?
- Has your understanding of your subject changed?

If the answer to any of these questions is yes, what changes should you make to the content and style of your document? See the Writer's Checklist in each chapter for information about what to look for when revising.

EDITING

Check your revised draft to improve six aspects of your writing: grammar, punctuation, style, usage, diction (word choice), and mechanics (matters such as use of numbers and abbreviations). See Appendix, Part B (p. 482) for more information about these topics.

PROOFREADING

Check to make sure you have typed what you meant to type. Don't rely on the spell-checker or the grammar-checker. They will miss some errors and flag correct words and phrases. See Appendix, Part B (p. 482) for more information about proofreading.

So, when you read about how to write, remember that you are reading about a messy process that goes backward as often as it goes forward and that, most likely, ends only when you run out of time.

Later chapters will discuss how to vary this basic process in writing various applications such as proposals, reports, and descriptions. The Focus on Process boxes at the beginning of various chapters will highlight important steps in this process for each application.

Should you use the process described here? If you don't already have a process that works for you, yes. But your goal should be to devise a process that enables you to write *effective* documents (that is, documents that accomplish what you want them to do) *efficiently* (without taking more time than necessary).

A Look at Three Technical Documents

Figures 1.1, 1.2, and 1.3 (p. 14) present excerpts from technical documents. Together, they illustrate a number of the ideas about technical communication discussed in this chapter.

This screen is from a video produced by the Department of Energy and is intended to educate the general public about the basics of solar energy. Because the document includes narration, still images, video, and animation, creating it required the efforts of many professionals.

The video is meant to be easy to share on social media.

The video takes advantage of our cultural assumptions about color: red suggests heat, blue suggests cold.

The video was designed to accommodate people with disabilities: the viewer can listen to the narration or turn on the subtitles.

The document includes a text-only version that provides a complete transcript of the narration and describes the images.

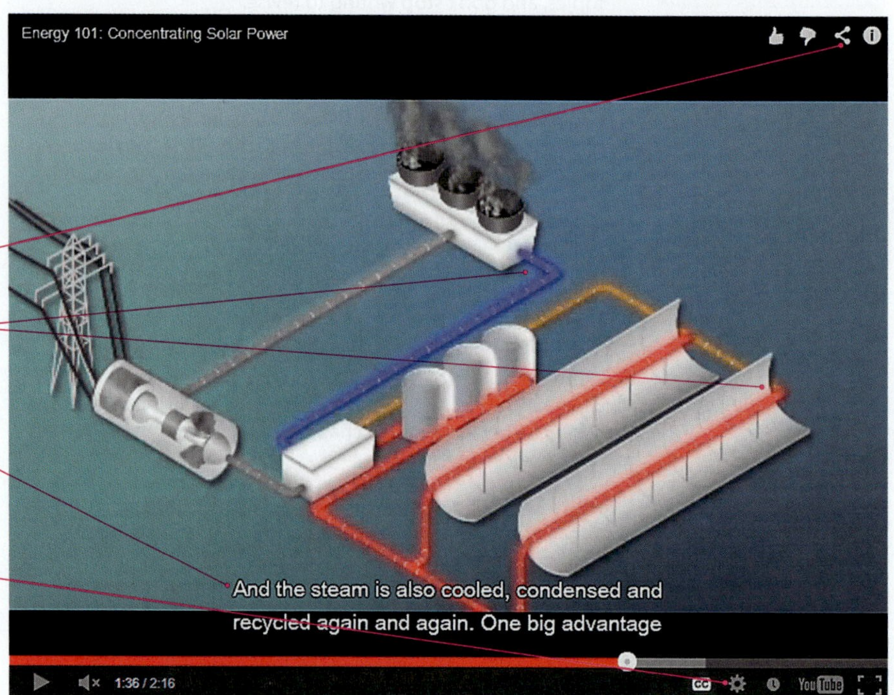

FIGURE 1.1 **A Video That Educates the Public About a Technical Subject**
Information from U.S. Department of Energy, 2012: http://energy.gov/articles/energy-101-concentrating-solar-power.

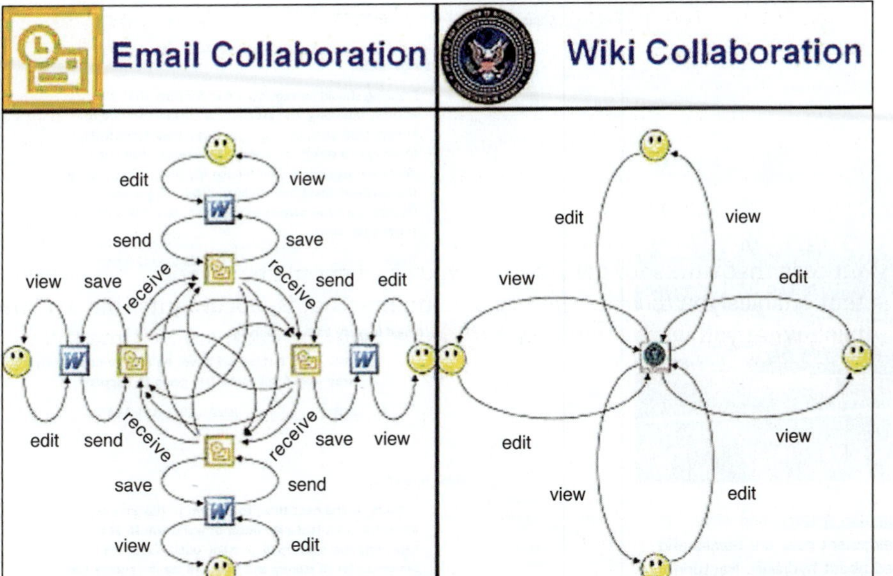

One characteristic that distinguishes technical communication from many other kinds of writing is its heavy use of graphics to clarify concepts and present data. This graphic, from a PowerPoint presentation, compares two technologies used for collaborative writing. The image on the left represents how a writer creates a document and then distributes it via email to others for editing. The image on the right represents how a writer creates a document in a wiki (an online writing and editing space), to which others come to view and edit the document.

The history of this graphic says something about how information flows in the digital age. The graphic was originally created by one person, Manny Wilson of U.S. Central Command, who shared it with a colleague at another U.S. government agency. Eventually, it made its way to another person, Anthony D. Williams, who incorporated it into a presentation he delivered at a corporation. From there, it went viral.

The writer who created this image doesn't need to say that a wiki is a better tool than email for editing a document. The complexity of the image on the left, compared with the simplicity of the image on the right, *shows* why the wiki is the better tool for this job.

FIGURE 1.2 A Graphic Comparing Two Communication Media

Information from Williams, 2008: www.wikinomics.com/blog/index.php/2008/03/26/wiki-collaboration-leads-to-happiness/.

Fracking In Our Backyard

Through our current campaign, Our Common Waters, and with exposure to increased oil and gas development near our homes and communities, we have grown concerned about hydraulic fracturing (commonly called "fracking") and its impact on water, air, soil, wildlife habitat, and human health. Over 90% of oil and gas wells in the U.S. use fracking to aid in extraction, and many fracking fluids and chemicals are known toxins for humans and wildlife.

For decades, natural gas (methane) deposits were tapped by single wells drilled vertically over large, free-flowing pockets of gas. Then came fracking, a water- and chemical-intensive method that promised the profitable extraction of natural gas trapped in shale.

Patagonia, the manufacturer of outdoor clothing, hosts a blog called The Cleanest Line. In one post, "Fracking In Our Backyard," the company sought to educate its readers about the controversy surrounding hydraulic fracturing. The post included links to many online sources about the controversy and presented the company's perspective: "Because of fracking's wide-ranging risks and impacts, we support each community's right to educate itself and regulate and/or ban fracking, and we support local, state and federal government efforts to monitor and regulate fracking."

Comments

John Jennings said...

Fracking should be regulated but consider that it's actually less risky and produces a cleaner method of energy than coal. Until we get away from fossil fuels, fracking is a smart technology that allows America to decrease dependence on foreign oil, while also produces the cleanest fossil fuel we know. Natural gas and fracking isn't the permanent solution, but it is a better in between step.

Reply July 11, 2013 at 12:12 AM

R said in reply to John Jennings...

Sorry John, but we will never be able to completely get away from fossil fuels. Not going to happen.

Reply July 11, 2013 at 05:31 AM

Andrew H. said...

Fracking is the best thing to happen to this country since the manufacturing boom of world war II. It's improving this country in so many good ways. It's bringing a lot of money and jobs...so maybe people can finally afford to buy patagonia clothes.

Reply July 11, 2013 at 05:56 PM

The post generated many comments, of which the first three are presented here. Notice that the third comment ends with a swipe at the company. Blogs are a popular way for organizations to interact with their stakeholders, and even though blog posts routinely elicit negative comments, most organizations believe that the occasionally embarrassing or critical comment is a reasonable price to pay for the opportunity to generate honest discussions about issues—and thereby learn what is on the minds of their stakeholders.

FIGURE 1.3 A Corporate Blog Post Presenting a Public-Policy Viewpoint

Left: Information from Patagonia, 2013: www.thecleanestline.com/2013/07/fracking-in-our-backyard.html#more.
Right: Information from Topher Donahue.

EXERCISES

For more about memos, see Chapter 9, page 248.

1. Form small groups and study the home page of your college or university's website. Focus on three measures of excellence in technical communication: clarity, accessibility, and professional appearance. How effectively does the home page meet each of these measures of excellence? Be prepared to share your findings with the class.
2. Locate an owner's manual for a consumer product, such as a coffee maker, bicycle, or hair dryer. In a memo to your instructor, discuss two or three decisions the writers and designers of the manual appear to have made to address audience-related factors, purpose-related factors, setting-related factors, document-related factors, or process-related factors. For instance, if the manual is available only in English, the writers and designers presumably decided that they didn't have the resources to create versions in other languages.
3. Using a job site such as Indeed.com or Monster.com, locate three job ads for people in your academic major. In each ad, identify references to writing and communication skills, and then identify references to professional attitudes and work habits. Be prepared to share your findings with the class.

CASE 1: Using the Measures of Excellence in Evaluating a Résumé

Your technical-communication instructor is planning to invite guest speakers to deliver presentations to the class on various topics throughout the semester, and she has asked you to work with one of them to tailor his job-application presentation to the "Measures of Excellence" illustrated in this chapter. To access relevant documents and get started on your project, go to LaunchPad.

CHAPTER 2 Understanding Ethical and Legal Obligations

Fejas/Shutterstock

ETHICAL AND LEGAL ISSUES are all around you in your work life. If you look at the website of any bike manufacturer, for example, you will see that bicyclists are always shown wearing helmets. Is this because bike manufacturers care about safety? Certainly. But bike makers also care about product liability. If a company website showed cyclists without helmets, an injured cyclist might sue, claiming that the company was suggesting it is safe to ride without a helmet.

Ethical and legal pitfalls lurk in the words and graphics of many kinds of formal documents. In producing a proposal, you might be tempted to exaggerate or lie about your organization's accomplishments, pad the résumés of the project personnel, list as project personnel some workers who will not be contributing to the project, or present an unrealistically short schedule. In drafting product information, you might feel pressured to exaggerate the quality of the products shown in catalogs or manuals or to downplay their hazards. In creating graphics, you might be asked to hide a product's weaknesses by manipulating a photo of it.

One thing is certain: there are many serious ethical and legal issues related to technical communication, and all professionals need a basic understanding of them.

A Brief Introduction to Ethics

Ethics is the study of the principles of conduct that apply to an individual or a group. For some people, ethics is a matter of intuition—what their gut feelings tell them about the rightness or wrongness of an act. Others see ethics in terms of their own religion or the Golden Rule: treat others as you would like them to treat you. Ethicist Manuel G. Velasquez (2011) outlines four moral standards that are useful in thinking about ethical dilemmas:

- **Rights.** This standard concerns individuals' basic needs and welfare. Everyone agrees, for example, that people have a right to a reasonably safe workplace. When we buy a product, we have a right to expect that the information that accompanies it is honest and clear. However, not everything that is desirable is necessarily a right. For example, in some countries, high-quality health care is considered a right that the government is required to provide; in others, it does not receive this designation.
- **Justice.** This standard concerns how the costs and benefits of an action or a policy are distributed among a group. For example, the cost of maintaining a high-speed broadband infrastructure should be borne, in part, by people who use it. However, because everyone benefits from the infrastructure, the standard of justice suggests that general funds can also be used to pay for it. Another example: justice requires that people doing the same job receive the same pay, regardless of sex or color.
- **Utility.** This standard concerns the positive and negative effects that an action or a policy has, will have, or might have on others. For example, if a company is considering closing a plant, the company's leaders should consider not only the money they would save but also the financial hardship of laid-off workers and the economic effects on the community. One tricky issue in thinking about utility is figuring out the time frame to examine. An action such as laying off employees can have one effect in the short run—improving the company's quarterly balance sheet—and a very different effect in the long run—hurting the company's productivity or the quality of its products.
- **Care.** This standard concerns the relationships we have with other individuals. We owe care and consideration to all people, but we have greater responsibilities to people in our families, our workplaces, and our communities. The closer a person is to us, the greater care we owe that person. Therefore, we have greater obligations to members of our family than we do to others in our community.

Although these standards provide a vocabulary for thinking about how to resolve ethical conflicts, they are imprecise and often conflict with each other. Therefore, they cannot provide a foolproof method of resolving ethical conflicts. Take the case of a job opportunity in your company. You are a member of the committee that will recommend which of six applicants to hire to redesign a customer portal that hosts tutorials and documentation. One of the six is a friend of yours who has been unable to secure a professional job since graduating from college two years ago. She therefore does not have as much website design experience as the other five candidates. However, she is enthusiastic about gaining experience in this particular field—and eager to start paying off her student loans.

How can the four standards help you think through the situation? According to the *rights* standard, lobbying for your friend or against the other applicants would be wrong because all applicants have an ethical right to an evaluation process that considers only their qualifications to do the job. Looking at the situation from the perspective of *justice* yields the same conclusion: it would be wrong to favor your friend. From the perspective of *utility*, lobbying for your friend would probably not be in the best interests of the organization, although it might be in your friend's best interests. Only according to the *care* standard does lobbying for your friend seem reasonable.

As you think about this case, you have to consider a related question: should you tell the other people on the hiring committee that one of the applicants is your friend? Yes, because they have a right to know about your personal relationship so that they can better evaluate your contributions to the discussion. You might also offer to recuse yourself (that is, not participate in the discussion of this position), leaving it to the other committee members to decide whether your friendship represents a conflict of interest.

One more complication in thinking about this case: Let's say your friend is one of the top two candidates for the job. In your committee, which is made up of seven members, three vote for your friend, but four vote for the other candidate, who already has a very good job. She is a young, highly skilled employee with degrees from prestigious universities. In other words, she is likely to be very successful in the working world, regardless of whether she is offered this particular job. Should the fact that your friend has yet to start her own career affect your thinking about this problem? Some people would say no: the job should be offered to the most qualified applicant. Others would say yes: society does not adequately provide for its less-fortunate members, and because your friend needs the job more and is almost as qualified as the other top applicant, she should get the offer. In other words, some people would focus on the narrow, technical question of determining the best candidate for the job, whereas others would see a much broader social question involving human rights.

Because ethical questions can be complex, ethicists have described a general set of principles that can help people organize their thinking about the role of ethics within an organizational context. These principles form a web of rights and obligations that connect an employee, an organization, and the world in which the organization is situated.

Your Ethical and Legal Obligations

In addition to enjoying rights, an employee assumes obligations, which can form a clear and reasonable framework for discussing the ethics of technical communication. The following discussion outlines four sets of obligations that you have as an employee: to your employer, to the public, to the environment, and to copyright holders.

OBLIGATIONS TO YOUR EMPLOYER

You are hired to further your employer's legitimate aims and to refrain from any activities that run counter to those aims. Specifically, you have five obligations:

- **Competence and diligence.** *Competence* refers to your skills; you should have the training and experience to do the job adequately. *Diligence* simply means hard work. Unfortunately, a recent survey of over 1,000 workers revealed that more than half of employees waste up to one hour of their eight-hour day surfing the web, socializing with co-workers, and doing other tasks unrelated to their jobs (Salary.com, 2013).
- **Generosity.** Although *generosity* might sound like an unusual obligation, you are obligated to help your co-workers and stakeholders outside your organization by sharing your knowledge and expertise. What this means is that if you are asked to respond to appropriate questions or provide recommendations on some aspect of your organization's work, you should do so. If a customer or supplier contacts you, make the time to respond helpfully.
- **Honesty and candor.** You should not steal from your employer. Stealing includes such practices as embezzlement, "borrowing" office supplies, and padding expense accounts. *Candor* means truthfulness; you should report to your employer problems that might threaten the quality or safety of the organization's product or service.

  Issues of honesty and candor include what Sigma Xi, the Scientific Research Society, calls trimming, cooking, and forging (Sigma Xi, 2000, p. 11). *Trimming* is the smoothing of irregularities to make research data look extremely accurate and precise. *Cooking* is retaining only those results that fit the theory and discarding the others. And *forging* is inventing some or all of the data or even reporting experiments that were never performed. In carrying out research, employees must resist any pressure to report only positive findings.
- **Confidentiality.** You should not divulge company business outside of the company. If a competitor finds out that your company is planning to introduce a new product, it might introduce its own version of that product, robbing your company of its competitive advantage. Many other kinds of privileged information—such as information on quality-control

problems, personnel matters, relocation or expansion plans, and financial restructuring—also could be used against the company. A well-known confidentiality problem involves *insider information*: an employee who knows about a development that will increase (or decrease) the value of the company's stock, for example, buys (or sells) the stock before the information is made public, thus unfairly—and illegally—reaping a profit (or avoiding a loss).

- **Loyalty.** You should act in the employer's interest, not in your own. Therefore, it is unethical to invest heavily in a competitor's stock, because that could jeopardize your objectivity and judgment. For the same reason, it is unethical (and illegal) to accept bribes or kickbacks. It is unethical to devote considerable time to moonlighting (performing an outside job, such as private consulting), because the outside job could lead to a conflict of interest and because the heavy workload could make you less productive in your primary position. However, you do not owe your employer absolute loyalty; if your employer is acting unethically, you have an obligation to try to change that behavior—even, if necessary, by blowing the whistle.

For more about whistle-blowing, see page 27.

OBLIGATIONS TO THE PUBLIC

Every organization that offers products or provides services is obligated to treat its customers fairly. As a representative of an organization, and especially as an employee communicating technical information, you will frequently confront ethical questions.

In general, an organization is acting ethically if its product or service is both safe and effective. The product or service must not injure or harm the consumer, and it must fulfill its promised function. However, these commonsense principles provide little guidance in dealing with the complicated ethical problems that arise routinely.

The U.S. Consumer Product Safety Commission (2015) estimates that more than 3,700 deaths and 15 million injuries occurred in the United States in 2015 because of consumer products—not counting automobiles and medications. Even more common, of course, are product and service failures: products or services don't do what they are supposed to do, products are difficult to assemble or operate, they break down, or they require more expensive maintenance than the product information indicates.

Although in some cases it is possible to blame either the company or the consumer for the injury or product failure, in many cases it is not. Today, most court rulings are based on the premise that the manufacturer knows more about its products than the consumer does and therefore has a greater responsibility to make sure the products comply with all of the manufacturer's claims and are safe. Therefore, in designing, manufacturing, testing, and communicating about a product, the manufacturer has to make sure the product will be safe and effective when used according to the instructions.

However, the manufacturer is not liable when something goes wrong that it could not have foreseen or prevented.

OBLIGATIONS TO THE ENVIRONMENT

One of the most important lessons we have learned in recent decades is that we are polluting and depleting our limited natural resources at a high rate. Our excessive use of fossil fuels not only deprives future generations of them but also creates pollution problems. Everyone—government, businesses, and individuals—must work to preserve the environment to ensure the survival not only of our own species but also of the other species with which we share the planet.

But what does this have to do with you? In your daily work, you probably do not cause pollution or deplete the environment in any extraordinary way. Yet you will often know how your organization's actions affect the environment. For example, if you work for a manufacturing company, you might be aware of the environmental effects of making or using your company's products. Or you might help write an environmental impact statement.

As communicators, we should treat every actual or potential occurrence of environmental damage seriously. We should alert our supervisors to the situation and work with them to try to reduce the damage. The difficulty, of course, is that protecting the environment can be expensive. Clean fuels often cost more than dirty ones. Disposing of hazardous waste properly costs more (in the short run) than merely dumping it. Organizations that want to reduce costs may be tempted to cut corners on environmental protection.

OBLIGATIONS TO COPYRIGHT HOLDERS

As a student, you are often reminded to avoid plagiarism. A student caught plagiarizing would likely fail the assignment or the course or even be expelled from school. A medical researcher or a reporter caught plagiarizing would likely be fired, or at least find it difficult to publish in the future. But plagiarism is an ethical, not a legal, issue. Although a plagiarist might be expelled from school or be fired, he or she will not be fined or sent to prison.

By contrast, copyright is a legal issue. *Copyright law* is the body of law that relates to the appropriate use of a person's intellectual property: written documents, pictures, musical compositions, and the like. Copyright literally refers to a person's *right* to *copy* the work that he or she has created.

The most important concept in copyright law is that only the copyright holder—the person or organization that owns the work—can copy it. For instance, if you work for Zipcar, you can legally copy information from the Zipcar website and use it in other Zipcar documents. This reuse of information is routine because it helps ensure that the information a company distributes is both consistent and accurate.

However, if you work for Zipcar, you cannot simply copy information that you find on the Car2Go website and put it in Zipcar publications. Unless you obtained written permission from Car2Go to use its intellectual property, you would be infringing on Car2Go's copyright.

Why doesn't the Zipcar employee who wrote the information for Zipcar own the copyright to that information? The answer lies in a legal concept known as *work made for hire*. Anything written or revised by an employee on the job is the company's property, not the employee's.

Although copyright gives the owner of the intellectual property some rights, it doesn't give the owner all rights. You can place small portions of copyrighted text in your own document without getting formal permission from the copyright holder. When you quote a few lines from an article, for example, you are taking advantage of an aspect of copyright law called *fair use*. Under fair-use guidelines, you have the right to use material, without getting permission, for purposes such as criticism, commentary, news reporting, teaching, scholarship, or research. Unfortunately, *fair use* is based on a set of general guidelines that are meant to be interpreted on a case-by-case basis. Keep in mind that you should still cite the source accurately to avoid plagiarism.

GUIDELINES Determining Fair Use

Courts consider four factors in disputes over fair use:

- **The purpose and character of the use, especially whether the use is for profit.** Profit-making organizations are scrutinized more carefully than nonprofits.
- **The nature and purpose of the copyrighted work.** When the information is essential to the public — for example, medical information — the fair-use principle is applied more liberally.
- **The amount and substantiality of the portion of the work used.** A 200-word passage would be a small portion of a book but a large portion of a 500-word brochure.
- **The effect of the use on the potential market for the copyrighted work.** Any use of the work that is likely to hurt the author's potential to profit from the original work would probably not be considered fair use.

A new trend is for copyright owners to stipulate which rights they wish to retain and which they wish to give up. You might see references to Creative Commons, a not-for-profit organization that provides symbols for copyright owners to use to communicate their preferences. A benefit of using materials with Creative Commons licenses is that you don't have to track down the copyright holder to ask for permission. Note that many Internet search engines allow you to search for Creative Commons materials.

GUIDELINES Dealing with Copyright Questions

Consider the following advice when using material from another source.

- **Abide by the fair-use concept.** Do not rely on excessive amounts of another source's work (unless the information is your company's own boilerplate).
- **Seek permission.** Write to the source, stating what portion of the work you wish to use and the publication you wish to use it in. The source is likely to charge you for permission.
- **Cite your sources accurately.** Citing sources fulfills your ethical obligation and strengthens your writing by showing the reader the range of your research.
- **Consult legal counsel if you have questions.** Copyright law is complex. Don't rely on instinct or common sense.

For more about documenting your sources, see Appendix, Part A.

ETHICS NOTE

DISTINGUISHING PLAGIARISM FROM ACCEPTABLE REUSE OF INFORMATION

Plagiarism is the act of using someone else's words or ideas without giving credit to the original author. It doesn't matter whether the user of the material intended to plagiarize. Obviously, it is plagiarism to borrow or steal graphics, video or audio media, written passages, or entire documents and then use them without attribution. Web-based sources are particularly vulnerable to plagiarism, partly because people mistakenly think that if information is on the web it is free to borrow and partly because this material is so easy to copy, paste, and reformat.

However, writers within a company often reuse one another's information without giving credit — and that is completely ethical. For instance, companies publish press releases when they wish to publicize news. These press releases typically conclude with descriptions of the company and how to get in touch with an employee who can answer questions about the company's products or services. These descriptions, sometimes called *boilerplate*, are simply copied and pasted from previous press releases. Because these descriptions are legally the intellectual property of the company, reusing them in this way is completely honest. Similarly, companies often *repurpose* their writing. That is, they copy a description of the company from a press release and paste it into a proposal or an annual report. This reuse also is acceptable.

When you are writing a document and need a passage that you suspect someone in your organization might already have written, ask a more-experienced co-worker whether the culture of your organization permits reusing someone else's writing. If the answer is yes, check with your supervisor to see whether he or she approves of what you plan to do.

The Role of Corporate Culture in Ethical and Legal Conduct

Most employees work within organizations, such as corporations and government agencies. We know that organizations exert a powerful influence on their employees' actions. According to a study by the Ethics Resource Center of more than 6,500 employees in various businesses (2014), organizations that value ethics and build strong cultures experience fewer ethical problems than organizations with weak ethical cultures.

Companies can take specific steps to improve their ethical culture:

- The organization's leaders can set the right tone by living up to their commitment to ethical conduct.
- Supervisors can set good examples and encourage ethical conduct.
- Peers can support those employees who act ethically.
- The organization can use informal communication to reinforce the formal policies, such as those presented in a company code of conduct.

In other words, it is not enough for an organization to issue a statement that ethical and legal behavior is important. The organization has to create a culture that values and rewards ethical and legal behavior. That culture starts at the top and extends to all employees, and it permeates the day-to-day operations of the organization.

An important element of a culture of ethical and legal conduct is a formal code of conduct. Most large corporations in the United States have one, as do almost all professional societies. (U.S. companies that are traded publicly are required to state whether they have a code of conduct—and if not, why not.) Codes of conduct vary greatly from organization to organization, but most of them address such issues as the following:

- adhering to local laws and regulations, including those intended to protect the environment
- avoiding discrimination
- maintaining a safe and healthy workplace
- respecting privacy
- avoiding conflicts of interest
- protecting the company's intellectual property
- avoiding bribery and kickbacks in working with suppliers and customers

A code of conduct focuses on behavior, including such topics as adhering to the law. Many codes of conduct are only a few paragraphs long; others are lengthy and detailed, some consisting of several volumes.

An effective code of conduct has three major characteristics:

- **It protects the public rather than members of the organization or profession.** For instance, the code should condemn unsafe building practices but not advertising, which increases competition and thus lowers prices.
- **It is specific and comprehensive.** A code is ineffective if it merely states that people must not steal or if it does not address typical ethical offenses such as bribery in companies that do business in other countries.

- **It is enforceable.** A code is ineffective if it does not stipulate penalties, including dismissal from the company or expulsion from the profession.

Although many codes are too vague to be useful in determining whether a person has violated one of their principles, writing and implementing a code can be valuable because it forces an organization to clarify its own values and fosters an increased awareness of ethical issues.

If you think there is a serious ethical problem in your organization, find out what resources your organization offers to deal with it. If there are no resources, work with your supervisor to solve the problem.

What do you do if the ethical problem persists even after you have exhausted all the resources at your organization and, if appropriate, the professional organization in your field? The next step will likely involve *whistle-blowing*—the practice of going public with information about serious unethical conduct within an organization. For example, an engineer is blowing the whistle when she tells a regulatory agency or a newspaper that quality-control tests on a company product were faked.

Ethicists such as Velasquez (2011) argue that whistle-blowing is justified if you have tried to resolve the problem through internal channels, if you have strong evidence that the problem is hurting or will hurt other parties, and if the whistle-blowing is reasonably certain to prevent or stop the wrongdoing. But Velasquez also points out that whistle-blowing is likely to hurt the employee, his or her family, and other parties. Whistle-blowers can be penalized through negative performance appraisals, transfers to undesirable locations, or isolation within the company. The Ethics Resource Center reports in its 2013 survey that 21 percent of whistle-blowers experienced retaliation (2013, p. 13).

Understanding Ethical and Legal Issues Related to Social Media

There is probably some truth to social-media consultant Peter Shankman's comment "For the majority of us, social media is nothing more than a faster way to screw up in front of a larger number of people in a shorter amount of time" (Trillos-Decarie, 2012). User-generated content, whether it is posted to Facebook, Twitter, LinkedIn, YouTube, Google Groups, Yelp, Pinterest, or any of the many other online services, presents significant new ethical and legal issues. Just as employers are trying to produce social-media policies that promote the interests of the organization without infringing on employees' rights of free expression, all of us need to understand the basics of ethical and legal principles related to these new media.

A 2014 report by the law firm Proskauer Rose LLP, "Social Media in the Workplace Around the World 3.0," surveyed some 150 companies from the

United States and many other countries. Here are some of the survey findings (Proskauer Rose LLP, 2014, p. 2):

- Almost 80 percent of employers have social-media policies.
- One-third of employers block employee access to social media.
- More than half of the employers reported problems caused by misuse of social media by employees. Over 70 percent of businesses have had to take disciplinary action against an employee for misuse of social media.

Over the next few years, organizations will revise their policies about how employees may use social media in the workplace, just as courts will clarify some of the more complicated issues related to social media and the law. For these reasons, what we now see as permissible and ethical is likely to

GUIDELINES Using Social Media Ethically and Legally

These nine guidelines can help you use social media to your advantage in your career.

- **Keep your private social-media accounts separate from your company-sponsored accounts.** After you leave a company, you don't want to get into a dispute over who "owns" an account. Companies can argue, for example, that your collection of Twitter followers on a company-sponsored account is in fact a customer list and therefore the company's intellectual property. Regardless of whether you post from the workplace or at home, post only about business on your company-sponsored accounts.
- **Read the terms of service of every service to which you post.** Although you retain the copyright on original content that you post, most social-media services state that they can repost your content wherever and whenever they want, without informing you, getting your permission, or paying you. Many employers would consider this policy unacceptable.
- **Avoid revealing unauthorized news about your own company.** A company that wishes to apply for a patent has, according to the law, only one year to do so after the product or process is first mentioned or illustrated in a "printed publication." Because courts have found that a photo on Facebook or a blog or even in a tweet is equivalent to a printed publication (Bettinger, 2010), you could inadvertently start the clock ticking. Even worse, some other company could use the information to apply for a patent for the product or process that your company is developing. Or suppose that on your personal blog, you reveal that your company's profits will dip in the next quarter. This information could prompt investors to sell shares of your company's stock, thereby hurting everyone who owns shares.
- **Avoid self-plagiarism.** Self-plagiarizing is the act of publishing something you have already published. If you write an article for your company newsletter and later publish it on a personal blog, you are violating your company's copyright, because your newsletter article was a work made for hire and therefore the company's intellectual property.

(continued)

- **Avoid defaming anyone.** Defamation is the legal term for making false statements of fact about a person that could harm that person. Defamation includes libel (making such statements in writing, as in a blog post) and slander (making them in speech, as in a video posted online). In addition, you should not repost libelous or slanderous content that someone else has created.
- **Don't live stream or quote from a speech or meeting without permission.** Although you may describe a speech or meeting online, you may not stream video or post quotations without permission.
- **Avoid false endorsements.** The Federal Trade Commission has clear rules defining false advertising. The most common type of false advertising involves posting a positive review of a product or company in exchange for some compensation. Also common is endorsing your own company's products without stating your relationship with the company (U.S. Federal Trade Commission, 2009).
- **Avoid impersonating someone else online.** If that person is real (whether alive or dead), you could be violating his or her right of publicity (the right to control his or her name, image, or likeness). If that person is a fictional character, such as a character on a TV show or in a movie, you could be infringing on the copyright of whoever created that character.
- **Avoid infringing on trademarks by using protected logos or names.** Don't include copyrighted or trademarked names, slogans, or logos in your posts unless you have received permission to do so. Even if the trademark owner likes your content, you probably will be asked to stop posting it. If the trademark owner dislikes your content, you are likely to face a more aggressive legal response.

change. Still, it is possible to identify a list of best practices that can help you use social media wisely—and legally—in your career.

Finally, although defamation laws forbid making untrue factual statements about your employer, you are in fact permitted to criticize your employer, online or offline. The National Labor Relations Board has ruled that doing so is legal because it is protected discussion about "working conditions." Our advice: if you're angry, first move away from the keyboard. Once you post something, you've lost control of it.

However, if you think your employer is acting illegally or unethically, start by investigating the company's own resources for addressing such problems. Then, if you are still dissatisfied, consider whistle-blowing as a first step.

Communicating Ethically Across Cultures

Every year, the United States exports more than $2.2 trillion worth of goods and services to the rest of the world (U.S. Census Bureau, 2016). U.S. companies do not necessarily have the same ethical and legal obligations when they export as when they sell in the United States. For this reason, communicators should understand the basics of two aspects of writing for people in other countries: communicating with cultures with different ethical beliefs and communicating in countries with different laws.

COMMUNICATING WITH CULTURES WITH DIFFERENT ETHICAL BELIEFS

Companies face special challenges when they market their products and services to people in other countries (and to people in their home countries who come from other cultures). Companies need to decide how to deal with situations in which the target culture's ethical beliefs clash with those of their own culture. For instance, in many countries, sexual discrimination makes it difficult for women to assume responsible positions in the workplace. If a U.S. company that sells cell phones, for example, wishes to present product information in such a country, should it reinforce this discrimination by excluding women from photographs of its products? Ethicist Thomas Donaldson argues that doing so is wrong (1991). According to the principle he calls the *moral minimum*, companies are ethically obligated not to reinforce patterns of discrimination in product information.

However, Donaldson argues, companies are not obligated to challenge the prevailing prejudice directly. A company is not obligated, for example, to include photographs that show women performing roles they do not normally perform within a particular culture, nor is it obligated to portray women wearing clothing, makeup, or jewelry that is likely to offend local standards. But there is nothing to prevent an organization from adopting a more activist stance. Organizations that actively oppose discrimination are acting admirably.

COMMUNICATING IN COUNTRIES WITH DIFFERENT LAWS

When U.S. companies export goods and services to other countries, they need to adhere to those countries' federal and regional laws. For instance, a company that wishes to export to Montreal must abide by the laws of Quebec Province and of Canada. A company that wishes to export to Germany must abide by the laws of Germany and of the European Union, of which it is a part. In many cases, the target region will not allow the importation of goods and services that do not conform to local laws. The hazardous-product laws of the European Union, in particular, are typically more stringent than those of the United States.

Because exporting goods to countries with different laws is such a complex topic, companies that export devote considerable resources to finding out what they need to do, not only in designing and manufacturing products but also in writing the product information. For a good introduction to this topic, see Lipus (2006).

DOCUMENT ANALYSIS ACTIVITY

Presenting Guidelines for Using Social Media

1 *Overview*

In today's world, just about everything we do online can be traced back to us and can have an impact (for better or worse) on a company. Paragon wants to remind you that the company policies on anti-harassment, ethics, and company loyalty extend to all media. There is a certain etiquette you should abide by when you participate online. This document is not intended to be restrictive, but to provide some guidelines on proper social-networking etiquette.

2 *What Are Social Media?*

Social media are the tools and content that enable people to connect online, share their interests, and engage in conversations.

Guidelines

These policies apply to individuals who want to participate in social-media conversations on behalf of Paragon. Please be mindful that your behavior at all times reflects on Paragon as a whole. Do not write or post anything that might reflect negatively on Paragon.

3
- Always use your best judgment and be honest.
- Be respectful of confidential information (such as clients, financials).
- Always be professional, especially when accepting criticism.
- Participate, don't promote. Bring value. Give to get.
- Write only about what you know.
- When in doubt, ask for help or clarification.
- Seek approval before commenting on any articles that portray Paragon negatively.

This excerpt is from a corporate social-media policy statement. The questions below ask you to think about how to make the policy statement clearer and more useful.

1. The "Overview" section discusses the company's social-media policy guidelines in terms of etiquette. In what way is "etiquette" an appropriate word to describe the policy? In what way is it inappropriate?
2. The "What Are Social Media?" section provides little useful information. What other information might it include to make the document more useful to Paragon employees?
3. The bulleted guidelines are vague. Revise any two of them to include more specific information.

THINKING VISUALLY

Principles for Ethical Communication

The following ten principles provide a starting point for communicating ethically in the workplace.

Abide by COPYRIGHT LAWS.
Get written permission when you wish to include copyrighted material in a document.

Abide by your organization's professional CODE OF CONDUCT.
Your field's professional organization is likely to have a code that expresses ethical principles.

Abide by your organization's POLICY ON SOCIAL MEDIA.
If there is no written policy, check with Human Resources or your supervisor for advice.

Take advantage of your employer's ETHICS RESOURCES, such as its Ethics Office. Your employer will likely have a mechanism for registering complaints anonymously.

Tell the TRUTH.
Resist pressure to lie, going over your supervisor's head if necessary.

Don't MISLEAD your readers.
Avoid false implications about products, euphemisms, exaggerations about product specifications, and legalistic constructions.

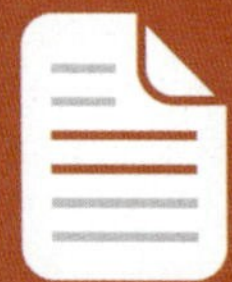

Use DESIGN to highlight important ethical and legal information.
Don't bury this information or downplay it using very small type.

Be CLEAR, using tables of contents, indexes, headings, and other accessing devices to help your readers find what they need.

Avoid LANGUAGE that DISCRIMINATES against people because of their sex, religion, ethnicity, race, sexual orientation, or physical or mental abilities.

Cite YOUR SOURCES and YOUR COLLABORATORS accurately and graciously.

From left to right: (2) artizarus/Shutterstock; (3) Fejas/Shutterstock; (4) Sk/Shutterstock; (5) Sk/Shutterstock; (6) alexandrovskyi/Shutterstock; (7) ThisSTOCK/Shutterstock; (8) VoodooDot/Shutterstock; (9) punsayaporn/Shutterstock.

WRITER'S CHECKLIST

- ☐ Did you abide by copyright laws? *(p. 21)*
- ☐ Did you abide by your organization's professional code of conduct? *(p. 24)*
- ☐ Did you abide by your organization's policy on social media? *(p. 25)*
- ☐ Did you take advantage of your employer's ethics resources? *(p. 23)*
- ☐ Did you tell the truth? *(p. 19)*
- ☐ Did you avoid using misleading language? *(p. 20)*
- ☐ Did you use design to highlight important ethical and legal information? *(p. 30)*
- ☐ Were you clear? *(p. 30)*
- ☐ Did you avoid language that discriminates? *(p. 28)*
- ☐ Did you cite your sources and your collaborators? *(p. 30)*

EXERCISES

For more about memos, see Ch. 9, page 248.

1. It is late April, and you need a summer job. On your town's news website, you see an ad for a potential job. The only problem is that the ad specifically mentions that the job is "a continuing, full-time position." You know that you will be returning to college in the fall. Is it ethical for you to apply for the job without mentioning this fact? Why or why not? If you believe it is unethical to withhold that information, is there any ethical way you can apply? Be prepared to share your ideas with the class.
2. You serve on the Advisory Committee of your college's bookstore, which is a private business that leases space on campus and donates 10 percent of its profits to student scholarships. The head of the bookstore wishes to stock Simple Study Guides, a popular series of plot summaries and character analyses of classic literary works. In similar bookstores, the sale of Simple Study Guides yields annual profits of over $10,000. Six academic departments have signed a statement condemning the idea. Should you support the bookstore head or the academic departments? Be prepared to discuss your answer with the class.
3. Using the search term "social-media policy examples," find a corporate policy statement on employee use of social media. In a 500-word memo to your instructor, explain whether the policy statement is clear, specific, and comprehensive. Does the statement include a persuasive explanation of why the policy is necessary? Is the tone of the statement positive or negative? How would you feel if you were required to abide by this policy? If appropriate, include a copy of the policy statement (or a portion of it) so that you can refer to it in your memo.
4. **TEAM EXERCISE** Form small groups. Study the website of a company or other organization that has a prominent role in your community or your academic field. Find information about the organization's commitment to ethical and legal conduct. Often, organizations present this information in sections with titles such as "Information for Investors," "About the Company," or "Values and Principles of Conduct."
   - One group member could identify the section that states the organization's values. How effective is this section in presenting information that goes beyond general statements about the importance of ethical behavior?
   - A second group member could identify the section that describes the organization's code of conduct. Does the organization seem to take principles of ethical and legal behavior seriously? Can you get a clear idea from the description of whether the organization has a specific, well-defined set of policies, procedures, and resources available for employees who wish to discuss ethical and legal issues?
   - A third group member could identify any information related to the organization's commitment to the environment. What does the organization do, in its normal operations, to limit its carbon footprint and otherwise encourage responsible use of natural resources and limit damage to the environment?
   - As a team, write a memo to your instructor presenting your findings. Attach the organization's code to your memo.

CASE 2: The Ethics of Requiring Students To Subsidize a Plagiarism-Detection Service

The provost of your university has sent a letter to you and other members of the Student Government proposing that the university subscribe to a plagiarism-detection service, the cost of which would be subsidized by students' tuition. You and other Student Government members have some serious concerns about the proposal and decide to respond with a letter that analyzes the ethical implications of requiring students to subsidize such a program. To read the provost's letter and begin drafting your response, go to LaunchPad.

CHAPTER 3

Writing Collaboratively

Faberr Ink/Shutterstock

THE EXPLOSIVE GROWTH of social media over the last decade has greatly expanded the scope of workplace collaboration, reducing former barriers of time and space. Today, people routinely collaborate not only with members of their project teams but also with others within and outside their organization.

Workplace collaboration takes numerous forms. For example, you and other members of your project team might use social media primarily to gather information that you will use in your research. You bring this information back to your team, and then you work exclusively with your team in drafting, revising, and proofreading your document. In a more complex collaboration pattern, you and other members of your team might use social media to gather information from sources around the globe and then reach out to others in your organization to see what they think of your new ideas. Later in the process, you create the outline of your document, in the form of a wiki, and authorize everyone in your organization to draft sections, pose questions and comments, and even edit what others have written. In short, you can collaborate with any number of people at one or at several stages of the writing process.

Every document is unique and will therefore call for a unique kind of collaboration. Your challenge is to think creatively about how you can work effectively with others to make your document as good as it can be. Being aware of the strengths and limitations of collaborative tools can prompt you to find people in your building and around the world who can help you think about your subject and write about it compellingly and persuasively.

To complete a series of interactive team writing modules, go to LaunchPad.

THINKING VISUALLY

Advantages and Disadvantages of Collaboration

ADVANTAGES	DISADVANTAGES
Collaboration...	Collaboration...
 draws on a **WIDER KNOWLEDGE BASE.**	 takes **MORE TIME** than individual writing.
 draws on a **WIDER SKILLS BASE.**	 can lead to **GROUPTHINK**, which values harmony over critical thinking.
 provides a **BETTER IDEA OF HOW THE AUDIENCE WILL READ THE DOCUMENT.**	 can yield a **DISJOINTED DOCUMENT.** Sections can contradict or repeat each other or be written in different styles.
 improves **EMPLOYEE COMMUNICATION.**	 can lead to **INEQUITABLE WORKLOADS.**
 helps **ACCLIMATE NEW EMPLOYEES** to an organization.	 can **REDUCE A PERSON'S MOTIVATION** if he or she feels alienated from the group.
 **MOTIVATES EMPLOYEES TO SHARE** skills and knowledge.	 can lead to **DISAGREEMENT AND INTERPERSONAL CONFLICT**, leaving a lasting impact on working relationships.

From left to right: (1) WonderfulPixel/Shutterstock; (2) PureSolution/Shutterstock; (3)Ecelop/Shutterstock; (4) milo827/Shutterstock; (5) Colorlife/Shutterstock; (6) NEGOVURA/Shutterstock; (7) Jiripravda/Shutterstock; (8) iDesign/Shutterstock; (9) Faberr Ink/Shutterstock; (10) Jiw Ingka/Shutterstock; (11) Introwiz1/Shutterstock; (12) Macrovector/Shutterstock.

Managing Projects

At some point in your career, you will likely collaborate on a project that is just too big, too technical, too complex, and too difficult for your team to complete successfully without some advance planning and careful oversight. Often, collaborative projects last several weeks or months, and the efforts of several people are required at scheduled times for the project to proceed. For this reason, collaborators need to spend time managing the project to ensure that it not only meets the needs of the audience but also is completed on time and, if relevant, within budget.

GUIDELINES Managing Your Project

These seven suggestions can help you keep your project on track.

- **Break down a large project into several smaller tasks.** Working backward from what you must deliver to your client or manager, partition your project into its component parts, making a list of what steps your team must take to complete the project. This task is not only the foundation of project management but also a good strategy for determining the resources you will need to complete the project successfully and on time. Once you have a list of tasks to complete, you can begin to plan your project, assign responsibilities, and set deadlines.
- **Plan your project.** Planning allows collaborators to develop an effective approach and reach agreement before investing a lot of time and resources. Planning prevents small problems from becoming big problems when a deadline looms. Effective project managers use planning documents such as *needs analyses*, *information plans*, *specifications*, and *project plans*.
- **Create and maintain an accurate schedule.** An accurate schedule helps collaborators plan ahead, allocate their time, and meet deadlines. Update your schedule when changes are made, and either place the up-to-date schedule in an easily accessible location (for example, on a project website) or send the schedule to each team member. If the team misses a deadline, immediately create a new deadline. Team members should always know when tasks must be completed.
- **Put your decisions in writing.** Writing down your decisions, and communicating them to all collaborators, helps the team remember what happened. In addition, if questions arise, the team can refer easily to the document and, if necessary, update it.
- **Monitor the project.** By regularly tracking the progress of the project, the team can learn what it has accomplished, whether the project is on schedule, and if any unexpected challenges exist.
- **Distribute and act on information quickly.** Acting fast to get collaborators the information they need helps ensure that the team makes effective decisions and steady progress toward completing the project.
- **Be flexible regarding schedule and responsibilities.** Adjust your plan and methods when new information becomes available or problems arise. When tasks are held up because earlier tasks have been delayed or need reworking, the team should consider revising responsibilities to keep the project moving forward.

To watch a tutorial on using online tools to schedule meetings, go to LaunchPad.

Conducting Meetings

Collaboration involves meetings. Whether you are meeting face-to-face or using videoconferencing tools, the five aspects of meetings discussed in this section can help you use your time productively and produce the best possible document.

LISTENING EFFECTIVELY

Participating in a meeting involves listening and speaking. If you listen carefully to other people, you will understand what they are thinking and you will be able to speak knowledgeably and constructively. Unlike hearing, which involves receiving and processing sound waves, listening involves understanding what the speaker is saying and interpreting the information.

GUIDELINES Listening Effectively

Follow these five steps to improve your effectiveness as a listener.

- **Pay attention to the speaker.** Look at the speaker, and don't let your mind wander.
- **Listen for main ideas.** Pay attention to phrases that signal important information, such as "What I'm saying is . . ." or "The point I'm trying to make is. . . ."
- **Don't get emotionally involved with the speaker's ideas.** Even if you disagree, continue to listen. Keep an open mind. Don't stop listening in order to plan what you are going to say next.
- **Ask questions to clarify what the speaker said.** After the speaker finishes, ask questions to make sure you understand. For instance, "When you said that each journal recommends different protocols, did you mean that each journal recommends several protocols or that each journal recommends a different protocol?"
- **Provide appropriate feedback.** The most important feedback is to look into the speaker's eyes. You can nod your approval to signal that you understand what he or she is saying. Appropriate feedback helps assure the speaker that he or she is communicating effectively.

SETTING YOUR TEAM'S AGENDA

It's important to get your team off to a smooth start. In the first meeting, start to define your team's agenda.

GUIDELINES Setting Your Team's Agenda

Carrying out these eight tasks will help your team work effectively and efficiently.

- **Define the team's task.** Every team member has to agree on the task, the deadline, and the approximate length of the document. You also need to agree on more conceptual points, including the document's audience, purpose, and scope.
- **Choose a team leader.** This person serves as the link between the team and management. (In an academic setting, the team leader represents the team in communicating with the instructor.) The team leader also keeps the team on track, leads the meetings, and coordinates communication among team members.
- **Define tasks for each team member.** There are three main ways to divide the tasks: according to technical expertise (for example, one team member, an engineer, is responsible for the information about engineering); according to stages of the writing process (one team member contributes to all stages, whereas another participates only during the planning stage); or according to sections of the document (several team members work on the whole document but others work only on, say, the appendixes). People will likely assume informal roles, too. One person might be good at clarifying what others have said, another at preventing arguments, and another at asking questions that force the team to reevaluate its decisions.
- **Establish working procedures.** Before starting to work, collaborators need answers—in writing, if possible—to the following questions:
  - — When and where will we meet?
  - — What procedures will we follow in the meetings?
  - — What tools will we use to communicate with other team members, including the leader, and how often will we communicate?
- **Establish a procedure for resolving conflict productively.** Disagreements about the project can lead to a better product. Give collaborators a chance to express ideas fully and find areas of agreement, and then resolve the conflict with a vote.
- **Create a style sheet.** A style sheet defines the characteristics of the document's writing style. For instance, a style sheet states how many levels of headings the document will have, whether it will have lists, whether it will have an informal tone (for example, using "you" and contractions), and so forth. If all collaborators draft using a similar writing style, the document will need less revision. And be sure to use styles to ensure a consistent design for headings and other textual features.

To watch a tutorial on creating styles and templates, go to LaunchPad.

- **Establish a work schedule.** For example, for a proposal to be submitted on February 10, you might aim to complete the outline by January 25, the draft by February 1, and the revision by February 8. These dates are called *milestones.*

To download a work-schedule form, a team-member evaluation form, and a self-evaluation form, go to LaunchPad.

- **Create evaluation materials.** Team members have a right to know how their work will be evaluated. In college, students often evaluate themselves and other team members. In the working world, managers are more likely to do the evaluations.

ETHICS NOTE

PULLING YOUR WEIGHT ON COLLABORATIVE PROJECTS

Collaboration involves an ethical dimension. If you work hard and well, you help the other members of the team. If you don't, you hurt them.

You can't be expected to know and do everything, and sometimes unanticipated problems arise in other courses or in your private life that prevent you from participating as actively and effectively as you otherwise could. When problems occur, inform the other team members as soon as possible. For instance, call the team leader as soon as you realize you will have to miss a meeting. Be honest about what happened. Suggest ways you might make up for missing a task. If you communicate clearly, the other team members are likely to cooperate with you.

If you are a member of a team that includes someone who is not participating fully, keep records of your attempts to get in touch with that person. When you do make contact, you owe it to that person to try to find out what the problem is and suggest ways to resolve it. Your goal is to treat that person fairly and to help him or her do better work, so that the team will function more smoothly and more effectively.

CONDUCTING EFFICIENT MEETINGS

Human communication has nonverbal elements. Although people communicate through words and through the tone, rate, and volume of their speech, they also communicate through body language. For this reason, meetings provide the most information about what a person is thinking and feeling—and the best opportunity for team members to understand one another.

To help make meetings effective and efficient, team members should arrive on time and stick to the agenda. One team member should serve as secretary, recording the important decisions made at the meeting. At the end of the meeting, the team leader should summarize the team's accomplishments and state the tasks each team member is to perform before the next meeting. If possible, the secretary should give each team member this informal set of meeting minutes.

For a discussion of meeting minutes, see Ch. 12, "Writing Meeting Minutes."

COMMUNICATING DIPLOMATICALLY

Because collaborating can be stressful, it can lead to interpersonal conflict. People can become frustrated and angry with one another because of personality clashes or because of disputes about the project. If the project is to succeed, however, team members have to work together productively. When you speak in a team meeting, you want to appear helpful, not critical or overbearing.

CRITIQUING A TEAM MEMBER'S WORK

In your college classes, you probably have critiqued other students' writing. In the workplace, you will do the same sort of critiquing of notes and drafts written by other team members. Knowing how to do it without offending the writer is a valuable skill.

GUIDELINES Communicating Diplomatically

These seven suggestions for communicating diplomatically will help you communicate effectively.

- **Listen carefully, without interrupting.** See the Guidelines box on page 36.
- **Give everyone a chance to speak.** Don't dominate the discussion.
- **Avoid personal remarks and insults.** Be tolerant and respectful of other people's views and working methods. Doing so is right—and smart: if you anger people, they will go out of their way to oppose you.
- **Don't overstate your position.** A modest qualifier such as "I think" or "it seems to me" is an effective signal to your listeners that you realize that everyone might not share your views.

  OVERBEARING My plan is a sure thing; there's no way we're not going to kill Allied next quarter.

  DIPLOMATIC I think this plan has a good chance of success: we're playing off our strengths and Allied's weaknesses.

  Note that in the diplomatic version, the speaker says, "this plan," not "my plan."
- **Don't get emotionally attached to your own ideas.** When people oppose you, try to understand why. Digging in is usually unwise—unless it's a matter of principle—because, although it's possible that you are right and everyone else is wrong, it's not likely.
- **Ask pertinent questions.** Bright people ask questions to understand what they hear and to connect it to other ideas. Asking questions also encourages other team members to examine what they hear.
- **Pay attention to nonverbal communication.** Bob might *say* that he understands a point, but his facial expression might suggest that he doesn't. If a team member looks confused, ask him or her about it. A direct question is likely to elicit a statement that will help the team clarify its discussion.

GUIDELINES Critiquing a Colleague's Work

Most people are very sensitive about their writing. Following these three suggestions for critiquing writing will increase the chances that your colleague will consider your ideas positively.

- **Start with a positive comment.** Even if the work is weak, say, "You've obviously put a lot of work into this, Joanne. Thanks." Or, "This is a really good start. Thanks, Joanne."
- **Discuss the larger issues first.** Begin with the big issues, such as organization, development, logic, design, and graphics. Then work on smaller issues, such as paragraph development, sentence-level matters, and word choice. Leave editing and proofreading until the end of the process.

(*continued*)

- **Talk about the document, not the writer.**

  RUDE You don't explain clearly why this criterion is relevant.

  BETTER I'm having trouble understanding how this criterion relates to the topic.

  Your goal is to improve the quality of the document you will submit, not to evaluate the writer or the draft. Offer constructive suggestions.

  RUDE Why didn't you include the price comparisons here, as you said you would?

  BETTER I wonder if the report would be stronger if we included the price comparisons here.

  In the better version, the speaker focuses on the goal (to create an effective report) rather than on the writer's draft. Also, the speaker qualifies his recommendation by saying, "I wonder if . . ." This approach sounds constructive rather than boastful or annoyed.

Using Social Media and Other Electronic Tools in Collaboration

Professionals use many types of electronic tools to exchange information and ideas as they collaborate. The following discussion highlights the major technologies that enable collaboration: word-processing tools, messaging technologies, videoconferencing, and wikis and shared document workspaces.

WORD-PROCESSING TOOLS

To watch tutorials on using the commenting and track-changes features in Word, Adobe Acrobat, and Google Drive, go to LaunchPad.

Most word processors offer three powerful features that you will find useful in collaborative work:

- The *comment feature* lets readers add electronic comments to a file.
- The *revision feature* lets readers mark up a text by deleting, revising, and adding words and graphics and indicates who made which suggested changes.
- The *highlighting feature* lets readers use one of about a dozen "highlighting pens" to call the writer's attention to a particular passage.

MESSAGING TECHNOLOGIES

Two messaging technologies have been around for decades: instant messaging and email. *Instant messaging (IM)* is real-time communication between two or more people. In the working world, IM enables people in different locations to communicate information at the same time. *Email* is an asynchronous medium for sending brief messages and for transferring files such as documents, spreadsheets, images, and videos. On mobile devices such as phones, the two most popular technologies are text messaging and microblogging.

For more about writing emails, see Ch. 9, "Writing Emails."

Text messaging enables people to use mobile devices to send messages that can include text, audio, images, and video. Texting is the fastest-growing

DOCUMENT ANALYSIS ACTIVITY

Critiquing a Draft Clearly and Diplomatically

This is an excerpt from the methods section of a report about computer servers. In this section, the writer is explaining the tasks he performed in analyzing different servers. In a later section, he explains what he learned from the analysis. The comments in the balloons were inserted into the document by the author's colleague.

The questions in the margin ask you to think about techniques for critiquing (as outlined on page 39).

The first task of the on-site evaluations was to set up and configure each server. We noted the relative complexity of setting up each system to our network.

Comment: Huh? What exactly does this mean?

After we had the system configured, we performed a set of routine maintenance tasks: add a new memory module, swap a hard drive, swap a power supply, and perform system diagnostics.

Comment: Okay, good. Maybe we should explain why we chose these tests.

We recorded the time and relative difficulty of each task. Also, we tried to gather a qualitative feeling for how much effort would be involved in the day-to-day maintenance of the systems.

Comment: What kind of scale are you using? If we don't explain it, it's basically useless.

Comment: Same question as above.

After each system was set up, we completed the maintenance evaluations and began the benchmark testing. We ran the complete WinBench and NetBench test suites on each system. We chose several of the key factors from these tests for comparison.

Comment: Will readers know these are the right tests? Should we explain?

1. What is the tone of the comments? How can they be improved?
2. How well does the collaborator address the larger issues?
3. How well does the collaborator address the writing, not the writer?
4. How well do the collaborator's comments focus on the goal of the document, rather than judging the quality of the writing?

technology for exchanging messages electronically because most people keep their phones nearby. Organizations use text messaging for such purposes as sending a quick update or alerting people that an item has been delivered or a task completed. On your campus, the administration might use a texting system to alert people about a campus emergency.

Microblogging is a way of sending very brief textual messages to your personal network. You might use the world's most popular microblog, Twitter, which now has more than 300 million users (SocialTimes, 2016). Although some organizations use Twitter, many use Twitter-like microblogs such as Yammer, which can be administered from within an organization.

VIDEOCONFERENCING

Videoconferencing technology allows two or more people at different locations to simultaneously see and hear one another as well as exchange documents, share data on computer displays, and use electronic whiteboards. Systems such as Skype are simple and inexpensive, requiring only a webcam, free software, and an Internet connection. However, there are also organizational systems that involve more extensive components and provide more features.

To watch a tutorial on using videoconferencing software to conduct online meetings, go to LaunchPad.

GUIDELINES Participating in a Videoconference

Follow these six suggestions for participating effectively in a videoconference.

- **Practice using the technology.** For many people, being on camera is uncomfortable, especially the first time. Before participating in a high-stakes videoconference, become accustomed to the camera by participating in a few informal videoconferences.
- **Arrange for tech support at each site.** Participants can quickly become impatient or lose interest when someone is fumbling to make the technology work. Each site should have a person who can set up the equipment and troubleshoot if problems arise.
- **Organize the room to encourage participation.** If there is more than one person at the site, arrange the chairs so that they face the monitor and camera. Each person should be near a microphone. Before beginning the conference, check that each location has adequate audio and video as well as access to other relevant technology such as computer monitors. Finally, remember to introduce everyone in the room, even those off camera, to everyone participating in the conference.
- **Make eye contact with the camera.** Eye contact is an important element of establishing your professional persona. The physical setup of some videoconferencing systems means you will likely spend most of your time looking at your monitor and not directly into the camera. However, this might give your viewers the impression that you are avoiding eye contact. Make a conscious effort periodically to look directly into the camera when speaking.
- **Dress as you would for a face-to-face meeting.** Wearing inappropriate clothing can distract participants and damage your credibility.
- **Minimize distracting noises and movements.** Sensitive microphones can magnify the sound of shuffling papers, fingers tapping on tables, and whispering. Likewise, depending on your position in the picture frame, excessive movements can be distracting.

WIKIS AND SHARED DOCUMENT WORKSPACES

Not all that long ago, people would collaborate on a document by using email to send it from one person to another. One person would write or assemble the document and then send it to another person, who would revise it and send it along to the next person, and so forth. Although the process was effective, it was inefficient: only one person could work on the document at any given moment. Today, two new technologies—wikis and shared document workspaces—make collaborating on a document much simpler and more convenient.

To watch a tutorial on using wikis for collaborative work, go to LaunchPad.

A *wiki* is a web-based document that authorized users can write and edit. The best-known wiki is Wikipedia, an online encyclopedia that contains millions of articles written and edited by people around the world. In the working world, people use software such as Jive and Socialtext to host wikis used for creating many kinds of documents, such as instructions, manuals, lists of frequently asked questions, and policy documents. The concept is that a wiki draws on the expertise and insights of people throughout the organization and, sometimes, outside the organization.

A *shared document workspace* makes it convenient for a team of users to edit a file, such as a Prezi or PowerPoint slide set or a Word document. A shared document workspace such as Microsoft SharePoint or Google Drive archives all the revisions made by each of the team members, so that the team can create a single document that incorporates selected revisions. Some shared document workspaces enable a user to download the document, revise it on his or her computer, and then upload it again. This feature is extremely convenient because the user does not need to be connected to the Internet to work on the document.

To watch a tutorial on cross-platform word processing, go to LaunchPad.

To watch a tutorial on using collaboration software, go to LaunchPad.

TASER, a company that manufactures law-enforcement products, uses the shared document workspace Quip to collaborate on press releases and other documents. Figure 3.1 shows one of those press releases in development. In Quip, team members can edit a single version of a document simultaneously. Those edits are recorded in a chat thread, where team members can also add comments and questions. TASER PR Director Sydney Siegmeth notes that Quip helps the company overcome one of collaboration's biggest disadvantages: inefficiency.

FIGURE 3.1 A Quip Document and Chat Thread

TASER employees collaborate in Quip on final edits to a press release announcing the purchases of products by several municipal police forces. The chat thread, which is located directly to the left of the document, serves as a single space where all communication about a project is recorded, such as approvals from other departments, as well as communication about the document itself. Here, team members discuss how to format information about purchases from multiple departments within a single police force, and they quickly resolve the questions. See the following page for the press release under discussion.

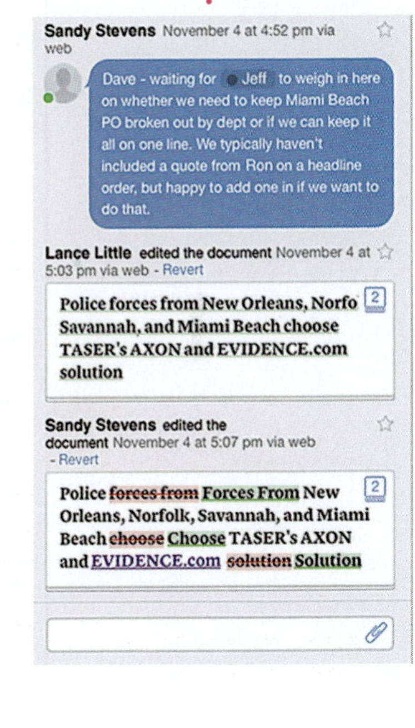

(continued)

NOV Headline Order: 11/5

SHARE   

FOR RELEASE ON: November 5, 2014 at 0730 ET

CONTACT:
Sandy Stevens
PR Director
EVIDENCE.com
Media ONLY Hotline: (480) 444-4000
Press@TASER.com

Police Forces From New Orleans, Norfolk, Savannah, and Miami Beach Choose TASER's AXON and EVIDENCE.com Solution

Miami Beach First City to Outfit AXON Cameras on All Public Facing Employees Including Code, Fire and Parking Departments

SEATTLE, WA, November 5, 2014 – TASER International (NASDAQ: TASR) today announced multiple large orders of it's AXON body-worn video cameras and www.EVIDENCE.com solution, a back-end digital evidence management system. These orders were received in the fourth quarter of 2014 and are expected to ship in the fourth quarter of 2014.

Significant orders were received from the following domestic agencies:

- Miami Beach Police Department (FL): 556 AXON body cameras for all public facing employees including Building, Code, Fire and Police Departments with five years of EVIDENCE.com and TASER Assurance Plan
- Savannah-Chatham Metro Police Department (GA): 360 AXON flex cameras with five years of EVIDENCE.com and TASER Assurance Plan
- Norfolk Police Department (VA): 310 AXON flex cameras with five years of EVIDENCE.com and TASER Assurance Plan
- Okaloosa County Sheriff's Office (FL): 279 AXON flex cameras with five years of EVIDENCE.com and TASER Assurance Plan
- Ontario Police Department (CA): 227 AXON body cameras with five years of EVIDENCE.com and TASER Assurance Plan
- New Orleans Police Department (LA): 103 AXON body cameras with five years of EVIDENCE.com and TASER Assurance Plan
- Gadsden Police Department (AL): 83 AXON body and flex cameras with five years of EVIDENCE.com and TASER Assurance Plan
- Shawnee County Sheriff's Office (KS): 73 AXON flex cameras
- Oxford Police Department (AL): 60 AXON flex cameras with five years of EVIDENCE.com and TASER Assurance Plan
- Colonial Heights Police Department (VA): 42 AXON flex cameras with five years of EVIDENCE.com
- Laurel Police Department (MS): 42 AXON body and flex cameras with five years of EVIDENCE.com and TASER Assurance Plan
- Talladega Police Department (AL): 41 AXON flex cameras with five years of EVIDENCE.com and TASER Assurance Plan
- Cullman Police Department (AL): 35 AXON flex cameras with five years of EVIDENCE.com and TASER Assurance Plan
- Cambridge Police Department (MD): 31 AXON flex cameras with five years of EVIDENCE.com and TASER Assurance Plan

TASER's AXON cameras are small, yet highly visible, and can be attached securely to sunglasses, a cap, a shirt collar, or a head mount. They are powered by a pocketsize battery pack, which ensures recording capability during an entire shift. When recording, the cameras capture a wide-angle, full-color view of what an officer is facing. The video automatically uploads via a docking station to EVIDENCE.com, a cloud-based storage and management system, where it can be easily accessed for review. The video files stored online or on the AXON video camera are secure and cannot be tampered with.

EVIDENCE.com helps police capture, manage, and share their digital evidence without the complexity or cost of installing in-house servers. It enables greater transparency through seamless integration with the industry-leading AXON body-worn video cameras. EVIDENCE.com is the most secure, scalable, and cost-effective solution for managing all types of digital evidence. EVIDENCE.com automates the upload process to ensure security and integrity while keeping officers in the field rather than sitting at computers.

A year-long Cambridge University study conducted at the Rialto, CA Police Department investigated whether officers' use of AXON flex cameras could bring measurable benefits to relations between police and civilians. The results showed an 88% reduction in citizen complaints and a 60% reduction in uses of force after implementation of TASER's AXON flex cameras.

Follow EVIDENCE.com

- Facebook: https://www.facebook.com/EVIDENCEcom
- Twitter: http://www.twitter.com/EVIDENCEcom

About TASER International, Inc.

TASER International makes communities safer with innovative public safety technologies. Founded in 1993, TASER first transformed law enforcement with its electrical weapons. TASER continues to define smarter policing with its growing suite of technology solutions, including AXON body-worn video cameras and EVIDENCE.com, a secure digital evidence management platform. More than 133,000 lives and countless dollars have been saved with TASER's products and services.

Learn more at www.TASER.com and www.EVIDENCE.com or by calling (800) 978-2737.
TASER® is a registered trademark of TASER International, Inc., registered in the U.S. All rights reserved. TASER logo and AXON are trademarks of TASER International, Inc.

FIGURE 3.1 **A Quip Document and Chat Thread** *(continued)*

TASER's press releases are developed by many teams, including some from outside the company. Before using Quip, Siegmeth's team had to consolidate edits made to various versions of the document by other teams and make sure all the contributors received updates about the progress of the press release. Quip's chat thread streamlines all communication regarding the document in a single place, ensuring that all parties are kept in the loop. TASER has also found that Quip motivates employees to make their contributions more quickly than they did when working with email and a word processor. "I've found that once someone goes in to make an edit or a comment, others will jump in there too and offer their approval or other edits," Siegmeth says.

ETHICS NOTE

MAINTAINING A PROFESSIONAL PRESENCE ONLINE

According to a report from Cisco Systems (2010), half of the surveyed employees claim to routinely ignore company guidelines that prohibit the use of social media for non-work-related activities during company time. If you use your organization's social media at work, be sure to act professionally so that your actions reflect positively on you and your organization. Be aware of several important legal and ethical issues related to social media.

For more about maintaining a professional presence online, see Ch. 2, "Understanding Ethical and Legal Issues Related to Social Media."

Although the law has not always kept pace with recent technological innovations, a few things are clear. You and your organization can be held liable if you make defamatory statements (statements that are untrue and damaging) about people or organizations, publish private information (such as trade secrets) or something that publicly places an individual "in a false light," publish personnel information, harass others, or participate in criminal activity.

In addition, follow these guidelines to avoid important ethical pitfalls:

- Don't waste company time using social media for nonbusiness purposes. You owe your employer diligence (hard work).
- Don't divulge secure information, such as a login and password that expose your organization to unauthorized access, and don't reveal information about products that have not yet been released.
- Don't divulge private information about anyone. Private information relates to such issues as religion, politics, and sexual orientation.
- Don't make racist or sexist comments or post pictures of people drinking.

If your organization has a written policy on the use of social media, study it carefully. Ask questions if anything in it is unclear. If the policy is incomplete, work to make it complete. If there is no policy, work to create one.

For an excellent discussion of legal and ethical aspects of using your organization's social media, see Kaupins and Park (2010).

Although this section has discussed various collaboration tools as separate technologies, software companies are bundling programs in commercial products such as IBM Sametime, Adobe Creative Cloud, and Microsoft Office 365, which are suites of voice, data, and video services. These services usually share four characteristics:

- **They are cloud based.** That is, organizations lease the services and access them over the Internet. They do not have to acquire and maintain special hardware. This model is sometimes called *software as a service.*

- **They are integrated across desktop and mobile devices.** Because employees can access these services from their desktops or mobile devices, they are free to collaborate in real time even if they are not at their desks. Some services provide *presence awareness*, the ability to determine a person's online status, availability, and geographic location.
- **They are customizable.** Organizations can choose whichever services they wish and then customize the services to work effectively with the rest of their electronic infrastructure, such as computer software and telephone systems.
- **They are secure.** Organizations store the software behind a firewall, providing security: only authorized employees have access to the services.

Gender and Collaboration

Effective collaboration involves two related challenges: maintaining the team as a productive, friendly working unit and accomplishing the task. Scholars of gender and collaboration see these two challenges as representing feminine and masculine perspectives.

This discussion should begin with a qualifier: in discussing gender, we are generalizing. The differences in behavior between two men or between two women could very well be greater than the differences between men and women in general.

Differences in how men and women communicate and work in teams have been traced to traditional family structures. Because women were traditionally the primary caregivers in American culture, they learned to value nurturing, connection, growth, and cooperation; because men were the primary breadwinners, they learned to value separateness, competition, debate, and even conflict (Karten, 2002). In collaborative teams, women appear to value consensus and relationships more than men do, to show more empathy, and to demonstrate superior listening skills. Women talk more about topics unrelated to the task (Duin, Jorn, & DeBower, 1991), but this talk is central to maintaining team coherence. Men appear to be more competitive than women and more likely to assume leadership roles. Scholars of gender recommend that all professionals strive to achieve an androgynous mix of the skills and aptitudes commonly associated with both women and men.

Culture and Collaboration

For more about multicultural issues, see Ch. 4, "Communicating Across Cultures."

Most collaborative teams in industry and in the classroom include people from other cultures. The challenge for all team members is to understand the ways in which cultural differences can affect team behavior. People from other cultures

- might find it difficult to assert themselves in collaborative teams
- might be unwilling to respond with a definite "no"
- might be reluctant to admit when they are confused or to ask for clarification
- might avoid criticizing others
- might avoid initiating new tasks or performing creatively

Even the most benign gesture of friendship on the part of a U.S. student can cause confusion. If a U.S. student casually asks a Japanese student about her major and the courses she is taking, the Japanese student might find the question too personal—yet she might consider it perfectly appropriate to talk about her family and her religious beliefs (Lustig & Koester, 2012). Therefore, you should remain open to encounters with people from other cultures without jumping to conclusions about what their actions might or might not mean.

WRITER'S CHECKLIST

In managing your project, did you

- ☐ break it down into several smaller tasks if it was large? *(p. 35)*
- ☐ create a plan? *(p. 35)*
- ☐ create and maintain an accurate schedule? *(p. 35)*
- ☐ put your decisions in writing? *(p. 35)*
- ☐ monitor progress? *(p. 35)*
- ☐ distribute and act on information quickly? *(p. 35)*
- ☐ act flexibly regarding schedule and responsibilities? *(p. 35)*

At your first team meeting, did you

- ☐ define the team's task? *(p. 37)*
- ☐ choose a team leader? *(p. 37)*
- ☐ define tasks for each team member? *(p. 37)*
- ☐ establish working procedures? *(p. 37)*
- ☐ establish a procedure for resolving conflict productively? *(p. 37)*
- ☐ create a style sheet? *(p. 37)*
- ☐ establish a work schedule? *(p. 37)*
- ☐ create evaluation materials? *(p. 37)*

To help make meetings efficient, do you

- ☐ arrive on time? *(p. 38)*
- ☐ stick to the agenda? *(p. 38)*
- ☐ make sure that a team member records important decisions made at the meeting? *(p. 38)*
- ☐ make sure that the leader summarizes the team's accomplishments and that every member understands what his or her tasks are? *(p. 38)*

To communicate diplomatically, do you

- ☐ listen carefully, without interrupting? *(p. 39)*
- ☐ give everyone a chance to speak? *(p. 39)*
- ☐ avoid personal remarks and insults? *(p. 39)*
- ☐ avoid overstating your position? *(p. 39)*
- ☐ avoid getting emotionally attached to your own ideas? *(p. 39)*
- ☐ ask pertinent questions? *(p. 39)*
- ☐ pay attention to nonverbal communication? *(p. 39)*

In critiquing a team member's work, do you

- ☐ start with a positive comment? *(p. 39)*
- ☐ discuss the larger issues first? *(p. 39)*
- ☐ talk about the document, not the writer? *(p. 40)*
- ☐ use the comment, revision, and highlighting features of your word processor, if appropriate? *(p. 40)*

When you participate in a videoconference, do you

- ☐ first practice using videoconferencing technology? *(p. 42)*
- ☐ arrange for tech support at each site? *(p. 42)*
- ☐ organize the room to encourage participation? *(p. 42)*
- ☐ make eye contact with the camera? *(p. 42)*
- ☐ dress as you would for a face-to-face meeting? *(p. 42)*
- ☐ minimize distracting noises and movements? *(p. 42)*

EXERCISES

For more about memos, see Ch. 9, "Writing Memos."

1. Experiment with the comment, revision, and highlighting features of your word processor. Using online help if necessary, learn how to make, revise, and delete comments; make, undo, and accept revisions; and add and delete highlights.
2. Locate free videoconferencing software on the Internet. Download the software, and install it on your computer at home. Learn how to use the feature that lets you send attached files.
3. Using a wiki site such as wikiHow.com, find a set of instructions on a technical process that interests you. Study one of the revisions to the instructions, noting the types of changes made. Do the changes relate to the content of the instructions, to the use of graphics, or to the correctness of the writing? Be prepared to share your findings with the class.
4. **TEAM EXERCISE** If you are enrolled in a technical-communication course that calls for you to do a large collaborative project, such as a recommendation report or an oral presentation, meet with your team members. Study the assignment for the project, and then fill out the work-schedule form. (You can download the form in LaunchPad.) Be prepared to share your completed form with the class.
5. You have probably had a lot of experience working in collaborative teams in previous courses or on the job. Brainstorm for five minutes, listing some of your best and worst experiences participating in collaborative teams. Choose one positive experience and one negative experience. Think about why the positive experience went well. Was there a technique that a team member used that accounted for the positive experience? Think about why the negative experience went wrong. Was there a technique or action that accounted for the negative experience? How might the negative experience have been prevented—or fixed? Be prepared to share your responses with the class.
6. **TEAM EXERCISE** Your college or university wishes to update its website to include a section called "For Prospective International Students." Along with members of your team, first determine whether your school's website already has information of particular interest to prospective international students. If it does, write a memo to your instructor describing and evaluating the information. Is it accurate? Comprehensive? Clear? Useful? What kind of information should be added to the site to make it more effective?

   If the school's site does not have this information, perform the following two tasks:

   - *Plan.* What kind of information should this new section include? Does some of this information already exist elsewhere on the web, or does it all have to be created from scratch? For example, can you create a link to an external site with information on how to obtain a student visa? Write an outline of the main topics that should be covered.
   - *Draft.* Write the following sections: "Where To Live on or near Campus," "Social Activities on or near Campus," and "If English Is Not Your Native Language." What graphics could you include? Are they already available? What other sites should you link to from these three sections?

   In a memo, present your suggestions to your instructor.

For more practice with the concepts covered in Chapters 1–3, complete the LearningCurve activity "Working in the Technical Communication Environment" in LaunchPad.

CASE 3: Accommodating a Team Member's Scheduling Problems

Your technical-communication instructor has organized you into groups of three. Your group will collaborate on a series of projects throughout the semester. Before your first assignment is due, you learn that one team member must deal with a family emergency that will interfere with his ability to participate in the project for some time. Now, you and your other teammate must devise a plan to proceed with the project. You also decide to propose a class-wide policy for communicating with teammates when problems arise. To get started on your assignment, go to LaunchPad.

Part 2

Yuri_Arcurs/Getty Images

Planning and Drafting the Document

CHAPTER 4

Analyzing Your Audience and Purpose

Domofon/Shutterstock

Nobelus/Shutterstock

DIGITAL STRATEGIST JASON FALLS writes frequently about how companies can use social media to create relationships with customers. What does he say is the key to using social media for business? Knowing your audience. The analytics report shown in Figure 4.1 provides basic information about the people who have visited a specific Facebook page in the previous four weeks: their gender, age, and geographic location. If you want more detailed information, you can purchase sophisticated tools and software that allow you to analyze the data on your own, or you can hire an expert to do it for you. Understanding your followers—your audience—can help you tailor your page to the people who are already visiting and broaden or adjust your content to reach more people.

Organizations of all sorts, not just businesses, analyze their audiences. Government agencies that want to appeal to the general public—to urge them to eat better, get vaccinated, or sign up for health insurance, to name just a few activities—start by analyzing their audiences to learn how to motivate them. Political campaigns analyze voters to determine the issues they want to see addressed. Charities such as the March of Dimes analyze their audiences to improve the effectiveness of their communications.

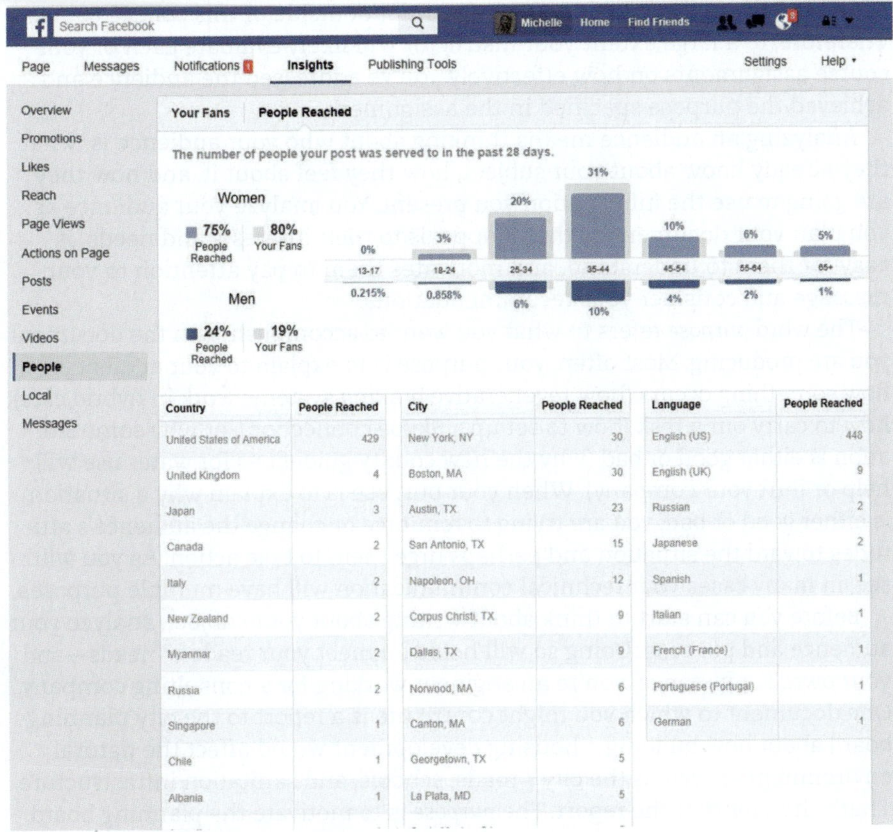

FIGURE 4.1 **Analysis of Facebook Users**
Information from Facebook, 2016: www.facebook.com.

Understanding Audience and Purpose

Projects of all sizes and types succeed only if they are based on an accurate understanding of the needs and desires of their audiences and have a clear, focused purpose. Because the documents and other communication you produce in the workplace will, more often than not, form the foundations of these projects, they too will succeed only if they are based on an accurate understanding of your audience and have a clear purpose.

Although you might not realize it, you probably consider audience in your day-to-day communication. For example, when you tell your parents about a new job you've landed, you keep the discussion general and focus on the job details you know they care most about: its location, its salary and benefits, and your start date. But when you email a former internship supervisor with the news, you discuss your upcoming duties and projects in more detail.

As you produce documents for this technical-communication course, you will of course consider your instructor's expectations, just as you do when you write anything for any other course. But keep in mind that your instructor in this course is also playing the role of the audience that you would be addressing if you had produced the document outside of this college course. Therefore, to a large extent your instructor will likely evaluate each of your course assignments on how effectively you've addressed the audience and achieved the purpose specified in the assignment.

Analyzing an audience means thinking about who your audience is, what they already know about your subject, how they feel about it, and how they are going to use the information you present. You analyze your audience as you plan your document so that it appeals to their interests and needs, is easy for them to understand, and motivates them to pay attention to your message and consider your recommendations.

The word *purpose* refers to what you want to accomplish with the document you are producing. Most often, your purpose is to explain to your audience how something occurs (how regenerative braking systems work in hybrid cars), how to carry out a task (how to set up a Skype connection), or why some situation is either good or bad (why the new county guidelines for water use will help or hurt your company). When your purpose is to explain why a situation is either good or bad, you are trying to reinforce or change the audience's attitudes toward the situation and perhaps urge them to take action. As you will see, in many cases your technical communication will have multiple purposes.

Before you can start to think about writing about your subject, analyze your audience and purpose. Doing so will help you meet your readers' needs—and your own. For instance, you're an engineer working for a consulting company. One document to which you might contribute is a report to the city planning board about how building a housing development would affect the natural environment as well as the city's roads, schools, and sanitation infrastructure. That's the *subject* of the report. The *purpose* is to motivate the planning board to approve the project so that it can begin. How does the audience affect how you analyze your purpose? You think about who the board members are. If most of them are not engineers, you don't want to use specialized vocabulary and advanced engineering graphics and concepts. You don't want to dwell on the technical details. Rather, you want to use general vocabulary, graphics, and concepts. You want to focus on the issues the board members are concerned about. Would the development affect the environment negatively? If so, is the developer including a plan to offset that negative effect? Can the roads handle the extra traffic? Can the schools handle the extra kids? Will the city have to expand its police force? Its fire department? Its sewer system?

In other words, when you write to the planning board, you focus on topics they are most interested in, and you write the document so that it is easy for them to read and understand. If the project is approved and you need to communicate with other audiences, such as architects and contractors, you will have different purposes, and you will adjust your writing to meet each audience's needs.

What can go wrong when you don't analyze your audience? McDonald's Corporation found out when it printed takeout bags decorated with flags from around the world. Among them was the flag of Saudi Arabia, which contains scripture from the Koran. This was extremely offensive to Muslims, who consider it sacrilegious to throw out items bearing sacred scripture. As a result, McDonald's lost public credibility.

Throughout this chapter, the text will refer to your *reader* and your *document*. But all of the information refers as well to oral presentations, which are the subject of Chapter 15, as well as to nonprint documents, such as podcasts or videos.

Using an Audience Profile Sheet

As you read the discussions in this chapter about audience characteristics and techniques for learning about your audience, you might think about using an audience profile sheet: a form that prompts you to consider various audience characteristics as you plan your document. For example, the profile sheet can help you realize that you do not know much about your primary reader's work history and what that history can tell you about how to shape your document.

For an audience profile sheet, which will help you think about audience characteristics, go to LaunchPad.

CHOICES AND STRATEGIES Responding to Readers' Attitudes

IF . . .	TRY THIS . . .
Your reader is neutral or positively inclined toward your subject	Write the document so that it responds to the reader's needs; make sure that vocabulary, level of detail, organization, and style are appropriate.
Your reader is hostile to the subject or to your approach to it	• Find out what the objections are, and then answer them directly. Explain why the objections are not valid or are less important than the benefits. For example, you want to hire an online-community manager to coordinate your company's social-media efforts, but you know that one of your primary readers won't like this idea. Try to find out why. Does this person think social media are a fad? That they are irrelevant and can't help your company? If you understand the objections, you can explain your position more effectively. • Organize the document so that your recommendation follows your explanation of the benefits. This strategy encourages the hostile reader to understand your argument rather than to reject it out of hand. • Avoid describing the subject as a dispute. Seek areas of agreement and concede points. Avoid trying to persuade readers overtly; people don't like to be persuaded, because it threatens their ego. Instead, suggest that there are new facts that need to be considered. People are more likely to change their minds when they realize this.
Your reader was instrumental in creating the policy or procedure that you are arguing is ineffective	In discussing the present system's shortcomings, be especially careful if you risk offending one of your readers. When you address such an audience, don't write, "The present system for logging customer orders is completely ineffective." Instead, write, "While the present system has worked well for many years, new developments in electronic processing of orders might enable us to improve logging speed and reduce errors substantially."

For tips on critiquing a team member's draft diplomatically, see Ch. 3, "Critiquing a Colleague's Work."

If your document has several readers, you must decide whether to fill out only one sheet (for your most important reader) or several sheets. One technique is to fill out sheets for one or two of your most important readers and one for each major category of other readers. For instance, you could fill out one sheet for your primary reader, Harry Becker; one for managers in other areas of your company; and one for readers from outside your company.

When do you fill out an audience profile sheet? Although some writers like to do so at the start of the process as a way to prompt themselves to consider audience characteristics, others prefer to do so at the end of the process as a way to help themselves summarize what they have learned about their audience. Of course, you can start to fill out the sheet before you begin and then complete it or revise it at the end.

Techniques for Learning About Your Audience

To learn about your audience, you figure out what you do and do not already know, interview people, read about them, and read documents they have written. Of course, you cannot perform extensive research about every possible reader of every document you write, but you should learn what you can about your most important readers of your most important documents.

DETERMINING WHAT YOU ALREADY KNOW ABOUT YOUR AUDIENCE

Start by asking yourself what you already know about your most important readers: their demographics (such as age, education, and job responsibilities); their expectations and attitudes toward you and the subject; and the ways they will use your document. Then list the important factors you don't know. That is where you will concentrate your energies. The audience profile sheet available in LaunchPad can help you identify gaps in your knowledge about your readers.

INTERVIEWING PEOPLE

For your most important readers, make a list of people who you think have known those readers and their work the longest or who are closest to them on the job. These people might include those who joined the organization at about the same time your readers did, people who work in the same department as your readers, and people at other organizations who have collaborated with your readers.

For a discussion of interviewing, see Ch. 5, "Interviews."

Prepare a few interview questions that are likely to elicit information about your readers and their preferences and needs. For instance, you are writing a proposal for a new project at work. You want to present return-on-investment calculations to show how long it will take the company to recoup what it invested, but you're not sure how much detail to present because you don't know whether an important primary reader has a background in this aspect of accounting. Several of this reader's colleagues will know. Interview them in person, on the phone, or by email.

READING ABOUT YOUR AUDIENCE ONLINE

If you are writing for people in your own organization, start your research there. If your primary reader is a high-level manager or executive, search the organization's website or internal social network. Sections such as "About Us," "About the Company," and "Information for Investors" often contain a wealth of biographical information, as well as links to other sources.

In addition, use a search engine to look for information on the Internet. You are likely to find newspaper and magazine articles, industry directories, websites, and blog posts about your audience.

SEARCHING SOCIAL MEDIA FOR DOCUMENTS YOUR AUDIENCE HAS WRITTEN

Documents your readers have written can tell you a lot about what they like to see with respect to design, level of detail, organization and development, style, and vocabulary. If your primary audience consists of those within your organization, start searching for documents they've produced within the company. Then broaden the search to the Internet.

Although some of your readers might have written books or articles, many or even most of them might be active users of social media, such as Facebook. Pay particular attention to LinkedIn, a networking site for professionals. LinkedIn profiles are particularly useful because they include a person's current and former positions and education, as well as recommendations from other professionals.

A typical LinkedIn entry directs you to a person's websites and blogs and to the LinkedIn groups to which the person belongs. You can also see the person's connections (his or her personal network). And if you are a LinkedIn member, you can see whether you and the person share any connections.

For an employee's LinkedIn profile, see Ch. 10, page 267.

In addition, the person you are researching might have a social-media account on which he or she posts about matters related to his or her job. Reading a person's recent posts will give you a good idea of his or her job responsibilities and professionalism.

ANALYZING SOCIAL-MEDIA DATA

Private companies and public agencies alike analyze social media to better understand their audiences. Private companies use these data primarily to determine who their customers are, how they feel about various marketing messages, and how these messages influence their buying behavior. Public agencies use these data to help them refine their own messages.

For instance, the Centers for Disease Control and Prevention (CDC), a U.S. federal agency, analyzes social media to improve the quality and effectiveness of its public health information. The agency starts by classifying people into various categories by age (such as tweens, teens, baby boomers) and determining which media each group uses most. On the basis of these data, the agency designs and implements health campaigns on such topics as cancer screening, HIV/AIDS prevention and treatment, vaccines, and smoking cessation.

THINKING VISUALLY

Determining the Important Characteristics of Your Audience

WHO ARE YOUR READERS?

For each of your most important readers, consider his or her

EDUCATIONAL BACKGROUND.

CULTURAL CHARACTERISTICS.

PROFESSIONAL EXPERIENCE, including areas of competence or expertise.

PERSONAL CHARACTERISTICS, such as age or physical impairments.

JOB RESPONSIBILITY and how your document will help that person accomplish it.

READING, SPEAKING, AND LISTENING PREFERENCES.

WHY IS YOUR AUDIENCE READING YOUR DOCUMENT?

Consider why each of your most important readers is reading your document. Some writers find it helpful to classify readers into categories that identify each reader's distance from the writer.

A **PRIMARY** audience consists of people to whom the communication is directed.

A **SECONDARY** audience consists of people who will not directly act on or respond to the document but who need to be aware of it.

A **TERTIARY** audience consists of people who might take an interest in the document, such as interest groups, government officials, and the general public.

From left to right: (1, 3, 5) Rungrote/Shutterstock; (2, 6) Domofon/Shutterstock; (4) Redberry/Shutterstock; (7, 8, 9) Nobelus/Shutterstock.

WHAT ARE YOUR READERS' ATTITUDES AND EXPECTATIONS?

In thinking about the attitudes and expectations of each of your most important readers, consider these three factors:

Your reader's attitude toward **YOU.** If a reader's animosity toward you is irrational or unrelated to the current project, try discussing other, less volatile projects or some shared interest.

Your reader's attitude toward the **SUBJECT.** If possible, gauge your primary readers' attitudes beforehand.

Your reader's expectations about the **DOCUMENT,** including scope, organizational pattern, amount of detail, and application.

HOW WILL YOUR READERS USE YOUR DOCUMENT?

In thinking about how your reader will use your document, consider the following four factors:

The **WAY** your reader will read your document. Will he or she file it, skim it, study it carefully, modify it, try to implement its recommendations, use it to perform a test, or use it as a source document?

The **PHYSICAL ENVIRONMENT** in which your reader will read your document, including light levels and exposure to the elements.

Your reader's **READING SKILL.** Consider whether you should be writing at all and at what level, or whether you should use another medium, such as a video, an oral presentation, or a podcast.

The **DIGITAL PLATFORMS** on which your reader will read your document. How can you design the document so that it is easy to access—easy to get to, to see, to navigate, and to use—in these environments?

From left to right: (1) Colorlife/Shutterstock; (2) Colorlife/Shutterstock, (cover inset) Aleutie/Shutterstock; (3) Colorlife/Shutterstock; (4) Eman Design/Shutterstock; (5) vasabii/Shutterstock; (6) Vector.design/Shutterstock; (7) Domofon/Shutterstock.

Then the CDC monitors social media to determine how many people are seeing the agency's information, how they are engaging with the information (whether they share the information or follow links to other sites), and whether the information is changing their behavior (Centers for Disease Control, 2013). Among the data the CDC analyzes each month are the following:

- the number of visitors to each of the CDC web pages
- the most popular keywords searched on CDC pages as well as on selected other sites and popular search engines such as Google
- the numbers of Facebook fans and Twitter followers
- the number of click-throughs to CDC web pages from Facebook and Twitter

On the basis of these data, the CDC adjusts its social-media campaigns to use its campaign resources most effectively.

Communicating Across Cultures

Our society and our workforce are becoming increasingly diverse, both culturally and linguistically, and businesses are exporting more goods and services. As a result, professionals often communicate with individuals from different cultural backgrounds, many of whom are nonnative speakers of English, both in the United States and abroad, and with speakers of other languages who read texts translated from English into their own languages.

The economy of the United States depends on international trade. In 2012, according to the Bureau of Economic Analysis (2014b), the United States exported over $3 trillion of goods and services. In that year, direct investment abroad by U.S. companies totaled more than $4.6 trillion (Bureau of Economic Analysis, 2014a). In addition, the population of the United States itself is truly multicultural. Each year, the United States admits a million immigrants. In 2012, 13 percent of the U.S. population was foreign born; of those foreign born, more than a third had entered the country since 2000 (U.S. Census Bureau, 2014).

Effective communication requires an understanding of culture: the beliefs, attitudes, and values that motivate people's behavior.

UNDERSTANDING THE CULTURAL VARIABLES "ON THE SURFACE"

Communicating effectively with people from another culture requires understanding a number of cultural variables that lie on the surface. You need to know, first, what language or languages to use. You also need to be aware of political, social, religious, and economic factors that can affect how readers will interpret your documents. Understanding these factors is not an exact science, but it does require that you learn as much as you can about the culture of those you are addressing.

A brief example: Microsoft's search engine had trouble catching on in China, in part because of its name—*Bing*—which means "sickness" in Chinese (Yan, 2015).

In *International Technical Communication*, Nancy L. Hoft (1995) describes seven major categories of cultural variables that lie on the surface:

- **Political.** This category relates to trade issues and legal issues (for example, some countries forbid imports of certain foods or chemicals) and laws about intellectual property, product safety, and liability.
- **Economic.** A country's level of economic development is a critical factor. In many developing countries, most people cannot afford devices for accessing the Internet.
- **Social.** This category covers many issues, including gender and business customs. In most Western cultures, women play a much greater role in the workplace than they do in many Middle Eastern and Asian cultures. Business customs—including forms of greeting, business dress, and gift giving—vary from culture to culture.
- **Religious.** Religious differences can affect diet, attitudes toward particular colors, styles of dress, holidays, and hours of business.
- **Educational.** In the United States, 40 million people are only marginally literate. In other cultures, the rate can be much higher or much lower. In some cultures, classroom learning with a teacher is considered the most acceptable way to study; in others, people tend to study on their own.
- **Technological.** If you sell high-tech products, you need to know whether your readers have the hardware, the software, and the technological infrastructure to use them.
- **Linguistic.** In some countries, English is taught to all children starting in grade school; in other countries, English is seen as a threat to the national language. In many cultures, the orientation of text on a page and in a book is not from left to right.

In addition to these basic differences, you need to understand dozens of other factors. For instance, the United States is the only major country that has not adopted the metric system. Whereas Americans use periods to separate whole numbers from decimals, and commas to separate thousands from hundreds, much of the rest of the world reverses this usage.

UNITED STATES	3,425.6
EUROPE	3.425,6

Also, in the United States, the format for writing out and abbreviating dates is different from that of most other cultures:

UNITED STATES	March 2, 2019	3/2/19
EUROPE	2 March 2019	2/3/19
JAPAN	2019 March 2	19/3/2

These cultural variables are important in obvious ways: for example, you can't send a file to a person who doesn't have access to the Internet.

However, there is another set of cultural characteristics—those beneath the surface—that you also need to understand.

UNDERSTANDING THE CULTURAL VARIABLES "BENEATH THE SURFACE"

Scholars of multicultural communication have identified cultural variables that are less obvious than those discussed in the previous section but just as important. Writing scholars Elizabeth Tebeaux and Linda Driskill (1999) explain five key variables and how they are reflected in technical communication.

- **Focus on individuals or groups.** Some cultures, especially in the West, value individuals more than groups. The typical Western employee doesn't see his or her identity as being defined by the organization for which he or she works. Other cultures, particularly those in Asia, value groups more than individuals. The typical employee in such cultures sees himself or herself more as a representative of the organization than as an individual who happens to work there.

  Communication in individualistic cultures focuses on the writer's and reader's needs rather than on those of their organizations. Writers use the pronoun *I* rather than *we*. Communication in group-oriented cultures focuses on the organization's needs by emphasizing the benefits to be gained through a cooperative relationship between organizations. Writers emphasize the relationship between the writer and the reader rather than the specific technical details of the message. Writers use *we* rather than *I*.

- **Distance between business life and private life.** In some cultures, especially in the West, many people separate their business lives from their private lives. When the workday ends, they are free to go home and spend their time as they wish. Although many employees are increasingly expected to be available by email or phone outside official working hours, those in the West still usually think of themselves primarily as individuals rather than as part of an organizational body. In other cultures, particularly in Asia, people see a much smaller distance between their business lives and their private lives. Even after the day ends, they still see themselves as employees of their organization.

  Cultures that value individualism tend to see a great distance between business and personal lives. In these cultures, communication focuses on technical details, with relatively little reference to personal information about the writer or the reader.

  Cultures that are group oriented tend to see a smaller distance between business life and private life. In these cultures, communication contains much more personal information—about the reader's family and health—and more information about general topics—for example, the weather and the seasons. The goal is to build a formal relationship between the two organizations. Both the writer and the reader are, in effect, on call after business hours and are likely to transact business during long social activities such as elaborate dinners or golf games.

- **Distance between ranks.** In some cultures, the distance in power and authority between workers within an organization is small. Supervisors work closely with their subordinates. In other cultures, the distance in power and authority between workers within an organization is great. Supervisors do not consult with their subordinates. Subordinates use formal names and titles—"Mr. Smith," "Dr. Perez"—when addressing people of higher rank.

  Individualistic cultures that separate business and private lives tend to have a smaller distance between ranks. In these cultures, communication is generally less formal. Informal documents (emails and memos) are appropriate, and writers often sign their documents with their first names only. Keep in mind, however, that many people in these cultures resent inappropriate informality, such as letters or emails addressed "Dear Jim" when they have never met the writer.

  In cultures with a great distance between ranks, communication is generally formal. Writers tend to use their full professional titles and to prefer formal documents (such as letters) to informal ones (such as memos and emails). Writers make sure their documents are addressed to the appropriate person and contain the formal design elements (such as title pages and letters of transmittal) that signal their respect for their readers.

- **Need for details to be spelled out.** Some cultures value full, complete communication. The written text must be comprehensive, containing all the information a reader needs to understand it. These cultures are called *low-context cultures*. Other cultures value documents in which some of the details are merely implied. This implicit information is transmitted by other forms of communication that draw on the personal relationship between the reader and the writer, as well as social and business norms of the culture. These cultures are called *high-context cultures*.

- **Attitudes toward uncertainty.** In some cultures, people are comfortable with uncertainty. They communicate less formally and rely less on written policies. In many cases, they rely more on a clear set of guiding principles, as communicated in a code of conduct or a mission statement. In other cultures, people are uncomfortable with uncertainty. Businesses are structured formally, and they use written procedures for communicating.

  In cultures that tolerate uncertainty, written communication tends to be less detailed. Oral communication is used to convey more of the information that is vital to the relationship between the writer and the readers. In cultures that value certainty, communication tends to be detailed. Policies are lengthy and specific, and forms are used extensively. Roles are firmly defined, and there is a wide distance between ranks.

As you consider this set of cultural variables, keep four points in mind:

- **Each variable represents a spectrum of attitudes.** Terms such as *high-context* and *low-context*, for instance, represent the opposite end points on a scale. Most cultures occupy a middle ground.

- **The variables do not line up in a clear pattern.** Although the variables sometimes correlate—for example, low-context cultures tend to be individualistic—in any one culture, the variables do not form a consistent pattern. For example, the dominant culture in the United States is highly individualistic rather than group oriented but only about midway along the scale in terms of tolerance of uncertainty.
- **Different organizations within the same culture can vary greatly.** For example, one software company in Germany might have a management style that does not tolerate uncertainty, whereas another software company in that country might tolerate a lot of uncertainty.
- **An organization's cultural attitudes are fluid, not static.** How an organization operates is determined not only by the dominant culture but also by its own people. As new people join an organization, its culture changes. The IBM of 2020 is not the IBM of 2000.

For you as a communicator, this set of variables therefore offers no answers. Instead, it offers a set of questions. You cannot know in advance the attitudes of the people in an organization. You have to interact with them for a long time before you can reach even tentative conclusions. The value of being aware of the variables is that they can help you study the communication from people in that organization and become more aware of underlying values that affect how they will interpret your documents.

CONSIDERING CULTURAL VARIABLES AS YOU WRITE

The challenge of communicating effectively with a person from another culture is that you are communicating with a person, not a culture. You cannot be sure which cultures have influenced that person (Lovitt, 1999). For example, a 50-year-old Japanese-born manager at the computer manufacturer Fujitsu in Japan has been shaped by the Japanese culture, but she also has been influenced by the culture of her company and of the Japanese industry in general. Because she works on an export product, it is also likely that she has traveled extensively outside of Japan and has absorbed influences from other cultures.

A further complication is that when you communicate with a person from another culture, to that person *you* are from another culture, and you cannot know how much that person is trying to accommodate your cultural patterns. As writing scholar Arthur H. Bell (1992) points out, the communication between the two of you is carried out in a third, hybrid culture. When you write to a large audience, the complications increase.

No brief discussion of cultural variables can answer questions about how to write for a particular multicultural audience. You need to study your readers' culture and, as you plan your document, seek assistance from someone native to the culture who can help you avoid problems that might confuse or offend your readers.

For books and other resources about writing to people from other cultures, see the Selected Bibliography, located in LaunchPad.

Start by reading some of the basic guides to communicating with people from other cultures, and then study guides to the particular culture you are investigating. In addition, numerous sites on the Internet provide useful guidelines that can help you write to people from another culture. If possible, study documents written by people in your audience. If you don't have access to these resources, try to locate documents written in English by people from the culture you are interested in.

Figure 4.2 provides a useful glimpse into cultural variables. The excerpt, from a training manual used by Indian Railways, describes a medical exam that prospective applicants are required to take.

> **501. Introduction:** (1) The standards of physical fitness to be adopted should make due allowance for the age and length of service, if any, of the candidate concerned.
>
> (2) No person will be deemed qualified for admission to the public service who shall not satisfy the Government, or the appointing authority, as the case may be, that he has no disease, constitutional affliction or bodily infirmity unfitting him, or likely to unfit him for that service.
>
> (3) It should be understood that the question of fitness involves the future as well as the present and that one of the main objectives of medical examination is to secure continuous effective service, and in the case of candidates for permanent appointment, to prevent early pension or payment in case of premature death.
>
> . . .
>
> Note: The minimum height prescribed can be relaxed in case of candidates belonging to races such as Gorkhas, Garhwalis, Assamese, Nagaland tribal, whose average height is distinctly lower.

The passage sounds as if it was written a hundred years ago, full of complicated sentences and formal vocabulary. The writing style is closer to that of the British (who colonized India) than that of the United States.

However, the explanation of why the exam is used is particularly candid: to save the government from having to support employees who become ill and therefore cannot perform the tasks for which they were hired.

The wording of this note, which follows a table showing the minimum height requirements for male and female applicants, would likely be considered offensive in most cultures. In India, a culture made of many ethnic groups and with a rigid caste system, most readers would not be offended.

FIGURE 4.2 **Statement from an Indian Railways Training Manual**
Information from Indian Railways, 2000.

GUIDELINES Writing for Readers from Other Cultures

The following eight suggestions will help you communicate more effectively with multicultural readers.

- **Limit your vocabulary.** Every word should have a narrow range of meaning, as called for in Simplified English and in other basic-English languages.
- **Keep sentences short.** There is no magic number, but try for an average sentence length of no more than 20 words.
- **Define abbreviations and acronyms in a glossary.** Don't assume that your readers know what a GFI (ground fault interrupter) is, because the abbreviation is derived from English vocabulary and word order.
- **Avoid jargon unless you know your readers are familiar with it.** For instance, your readers might not know what a graphical user interface is.

(continued)

- **Avoid idioms and slang.** These terms are culture specific. If you tell your Japanese readers that your company plans to put on a "full-court press," most likely they will be confused.
- **Use the active voice whenever possible.** The active voice is easier for nonnative speakers of English to understand than the passive voice.
- **Be careful with graphics.** The question-mark icon for "information" does not translate well, because outside the United States a lower-case *i* is often used to represent "information."
- **Be sure someone from the target culture reviews your document.** Even if you have had help in planning the document, have it reviewed before you publish and distribute it.

For more about voice, see Ch. 6 (pp. 134–36).

For more about graphics, see Ch. 8 (pp. 193–233).

USING GRAPHICS AND DESIGN FOR MULTICULTURAL READERS

One of the challenges of writing to people from another culture is that they are likely to be nonnative speakers of English. One way to overcome the language barrier is to use effective graphics and appropriate document design.

However, the most appropriate graphics and design can differ from culture to culture. Business letters written in Australia use a different size paper and a different format from those in the United States. An icon for a file folder in a software program created in the United States could confuse European readers, who use file folders of a different size and shape (Bosley, 1999). A series of graphics arranged left to right could confuse readers from the Middle East, who read from right to left. For this reason, you should study samples of documents written by people from the culture you are addressing to learn the important differences.

For more about design for multicultural readers, see Ch. 7 (p. 153). For more about graphics for international readers, see Ch. 8 (pp. 229–30).

Applying What You Have Learned About Your Audience

You want to use what you know about your audience to tailor your communication to their needs and preferences. Obviously, if your most important reader does not understand the details of DRAM technology, you cannot use the concepts, vocabulary, and types of graphics used in that field. If she uses one-page summaries at the beginning of her documents, decide whether one will work for your document. If your primary reader's paragraphs always start with clear topic sentences, yours should, too.

The samples of technical communication shown in Figure 4.3 (p. 67) illustrate some of the ways writers have applied what they know about their audiences in text and graphics.

DOCUMENT ANALYSIS ACTIVITY

Examining Cultural Variables in a Business Letter

Server Solutions
Cincinnati, OH 46539

July 3, 2017

Nadine Meyer
Director of Marketing

Mr. Philip Henryson, Director of Purchasing
Allied Manufacturing
1321 Industrial Boulevard
Boise, ID 83756

1 Dear Mr. Henryson:

Thank you for your inquiry about our PowerServer servers. I'm happy to answer your questions.

The most popular configuration is our PowerServer 3000. This model is based on the Intel® Xeon ES-4600 processor, ServerSure High-End UltraLite chipset with quadpeer PCI architecture, and embedded RAID. The system comes with our InstallIt system-management CD, which lets you install the server and monitor and manage your network with a simple graphical interface. With six PCI slots, the PowerServer 3000 is equipped with redundant cooling as well as redundant power, and storage expandability to 1.0TB. I'm taking the liberty of enclosing the brochure for this system to fill you in on the technical details.

The PowerServer 3000 has performed extremely well on a number of industry benchmark tests. I'm including with this letter copies of feature articles on the system from *PC World*, *CIO*, and *DigiTimes*.

It would be a pleasure for me to arrange for an on-site demo at your convenience. I will phone you on Monday to see what dates would be best for you. In the meantime, please do not hesitate to get in touch with me directly if you have any questions about the PowerServer line.

I look forward to talking with you next week.

Sincerely,

Nadine Meyer

Nadine Meyer
Director of Marketing

Attachments:
"PowerServer 3000 Facts at a Glance"
"Another Winner from Server Solutions"
"Mid-Range Servers for 2016"
"Four New Dual-Processor Workhorses"

(*continued*)

These two versions of the same business letter were written by a sales manager for an American computer company. The first letter was addressed to a potential customer in the United States; the second version was addressed to a potential customer in Japan. The questions in the margin ask you to think about how cultural variables affect the nature of the evidence, the structure of the letters, and their tone (see pp. 58–64).

Examining Cultural Variables in a Business Letter *(continued)*

1. How does the difference in the salutations (the "Dear . . ." part of the letter) reflect a cultural difference?
2. Does the first paragraph of the second letter have any function beyond delaying the discussion of business?
3. What is the point of telling Mr. Kirisawa about his own company? How does this paragraph help the writer introduce her own company's products?
4. To a reader from the United States, the third paragraph of the second letter would probably seem thin. What aspect of Japanese culture makes it effective in the context of this letter?
5. Why doesn't the writer make a more explicit sales pitch at the end of the second letter?

Server Solutions
Cincinnati, OH 46539

Nadine Meyer
Director of Marketing

Mr. Kato Kirisawa, Director of Purchasing
Allied Manufacturing
3-7-32 Kita Urawa
Saitama City, Saitama Pref. 336-0002
Japan

(1) Dear Sir:

(2) It is my sincere hope that you and your loved ones are healthy and enjoying the pleasures of summer. Here in the American Midwest, the warm rays of the summer sun are accompanied by the sounds of happy children playing in the neighborhood swimming pools. I trust that the same pleasant sounds greet you in Saitama City.

(3) Your inquiry about our PowerServer 3000 suggests that your company is growing. Allied Manufacturing has earned a reputation in Japan and all of Asia for a wide range of products manufactured to the most demanding standards of quality. We are not surprised that your company requires new servers that can be expanded to provide fast service for more and more clients.

(4) For more than 20 years, Server Solutions has had the great honor of manufacturing the finest computer servers to meet the needs of our valued customers all over the world. We use only the finest materials and most innovative techniques to ensure that our customers receive the highest-quality, uninterrupted service that they have come to expect from us.

(5) One of my great pleasures is to talk with esteemed representatives such as yourself about how Server Solutions can help them meet their needs for the most advanced servers. I would be most gratified if our two companies could enter into an agreement that would be of mutual benefit.

Sincerely,

Nadine Meyer

Nadine Meyer
Director of Marketing

Attachments:
"PowerServer 3000 Facts at a Glance"
"Another Winner from Server Solutions"
"Mid-Range Servers for 2016"
"Four New Dual-Processor Workhorses"

2016 July 3

What is XSLT?

XSL Transformations (XSLT 2.0) is a language for transforming XML documents into other XML documents, text documents or HTML documents. You might want to format a chapter of a book using XSL-FO, or you might want to take a database query and format it as HTML.

With XSLT 2.0, processors can operate not only on XML but on anything that can be made to look like XML: relational database tables, geographical information systems, file systems, anything from which your XSLT processor can build an XDM instance. In some cases an XSLT 2.0 processor might also be able to work directly from a database of XDM instances. This ability to operate on multiple input files in multiple formats, and to treat them all as if they were XML files, is very powerful. It is shared with XQuery, and with anything else using XPath 2.0:

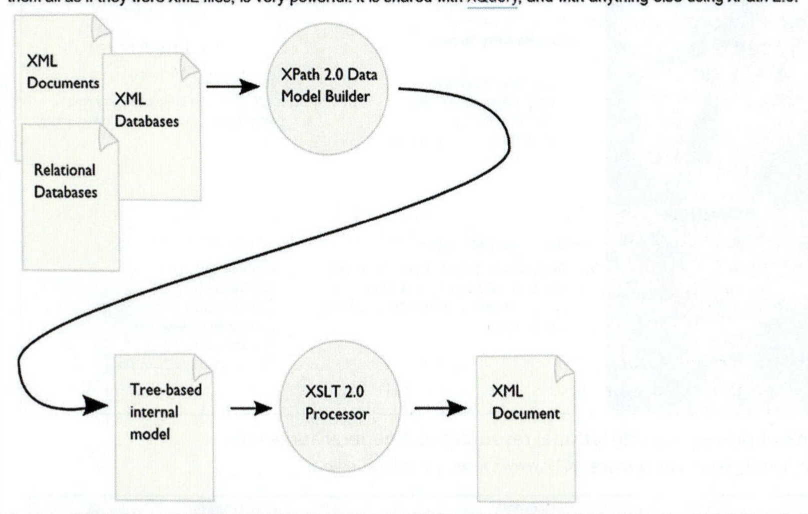

a. Document presenting technical information to an expert audience
Information from World Wide Web Consortium, 2013: www.w3.org/standards/xml/transformation.html.

This excerpt from a technical description of a web coding language appears on the site of the World Wide Web Consortium (W3C).

Because the readers are coding experts, the writers use highly technical language and refer to advanced topics. Note, however, that the nontechnical information is written simply and directly.

Notice that the graphic is based on a simple flowchart and basic icons. Why? Because the readers are interested only in understanding the logic of the process illustrated in the flowchart.

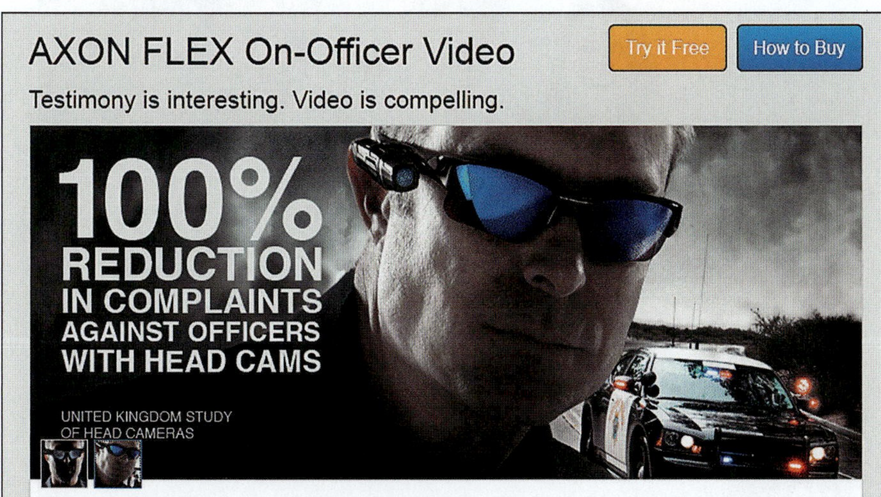

b. Document motivating decision makers to learn about a product
Information from *TASER International, 2013.*

This passage, from TASER, describes a videocam to be worn by police officers. It is addressed to high-level officers responsible for ordering equipment.

The writer makes a simple argument: that an unbiased study showed how the product eliminated complaints by providing a video record of an incident. (The web page also includes text and a video highlighting the argument about the product.)

The argument consists of two elements: the photograph and the brief text. The photograph shows that the device is easy to wear. The text makes two points: (1) video of the incident will be even more compelling than an officer's court testimony, and (2) therefore the product can greatly reduce the number of incidents in which officers are disciplined. Together, these points appeal to the interests of police supervisors, who don't want their officers to be unfairly charged. The links at the top of the page encourage the audience to take action, either by learning how to try out the device or by buying it.

FIGURE 4.3 **Using Text and Graphics to Appeal to Readers' Needs and Interests**
(continued)

This excerpt from the "Stay Healthy" section of the American Cancer Society website shows several techniques for providing information to a general audience.

The page begins with a short video and brief text intended to motivate readers to find out more about how to stay healthy. The tone throughout—from the words to the images of the smiling man and woman—is encouraging: it says, "You can do this."

The page includes a set of seven links to detailed information on more-specific topics about preventing cancer.

The Tools and Calculators section gives readers opportunities to learn more about how to improve their health. This section is consistent with the main point of this page: you can take steps to improve your health.

Notice that the writers use simple, direct language, as well as the second person ("you"), to maintain an informal tone.

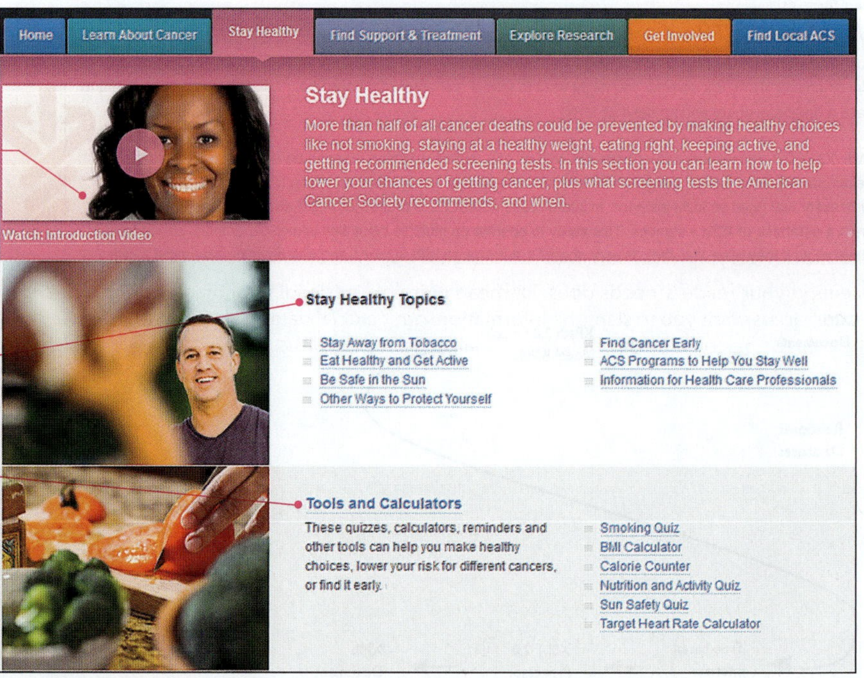

c. Document presenting educational resources to a general audience
Information from *American Cancer Society, 2013:* www.cancer.org/healthy/index.

Hewlett-Packard, one of the world's most innovative technology companies, recently lost much of its luster because of management miscalculations and scandals. This page from the HP site is part of the company's attempt to project a new, positive message.

The smiling faces of company President and CEO Meg Whitman and of the other people suggest that bad times are over.

In the quotation, the CEO is forthright about the company's mistakes but asserts that the company has the people and resources necessary to "turn HP around." Note that five sentences in a row begin with "we," thereby including customers in the HP story.

Whitman uses simple, strong words and sentences to project honesty and determination. And she signs her name "Meg." The point is that she is like HP's customers: hard working and down-to-earth.

d. Document reinforcing a brand
Information from *Hewlett-Packard, 2013:* www8.hp.com/us/en/hp-information/index.html.

FIGURE 4.3 **Using Text and Graphics** *(continued)*

ETHICS NOTE

MEETING YOUR READERS' NEEDS RESPONSIBLY

A major theme of this chapter is that effective technical communication meets your readers' needs. What this theme means is that as you plan, draft, revise, and edit, you should always be thinking of who your readers are, why they will read your document, and how they will read the document. For example, if your readers include many nonnative speakers of English, you will adjust your vocabulary, sentence structure, and other textual elements so that those readers can understand your document easily. If your readers will access the document on a mobile device, you will ensure that the design is optimized for their screen.

Meeting your readers' needs does *not* mean writing a misleading or inaccurate document. If your readers want you to slant the information, omit crucial data, or downplay bad news, they are asking you to act unethically. You should not do so. For more information on ethics, see Chapter 2.

Writing for Multiple Audiences

Many documents of more than a few pages are addressed to more than one reader. Often, an audience consists of people with widely different backgrounds, needs, and attitudes.

If you think your document will have a number of readers, consider making it *modular*: break it up into components addressed to different readers. A modular report might contain an executive summary for managers who don't have the time, knowledge, or desire to read the whole report. It might also contain a full technical discussion for expert readers, an implementation schedule for technicians, and a financial plan in an appendix for budget officers. Figure 4.4 (p. 70) shows the table of contents for a modular report.

Determining Your Purpose

Once you have identified and analyzed your audience, it is time to examine your purpose. Ask yourself this: "What do I want this document to accomplish?" When your readers have finished reading what you have written, what do you want them to *know* or *believe*? What do you want them to do? Your writing should help your readers understand a concept, adopt a particular belief, or carry out a task.

In defining your purpose, think of a verb that represents it. (Sometimes, of course, you have several purposes.) The following list presents verbs in two categories: those used to communicate information to your readers and those used to convince them to accept a particular point of view.

Communicating verbs	**Convincing verbs**
authorize	assess
define	evaluate
describe	forecast

Communicating verbs	**Convincing verbs**
explain	propose
illustrate	recommend
inform	request
outline	
present	
review	
summarize	

This classification is not absolute. For example, *review* could in some cases be a *convincing verb* rather than a *communicating verb*: one writer's review of a complicated situation might be very different from another's.

This table of contents shows the organization of a modular document.

Few readers will want to read the whole document—it's almost 1,000 pages long.

Most readers will want to read the 18-page summary for policymakers.

Some readers will want to read selected sections of the technical summary or "annexes" (appendixes).

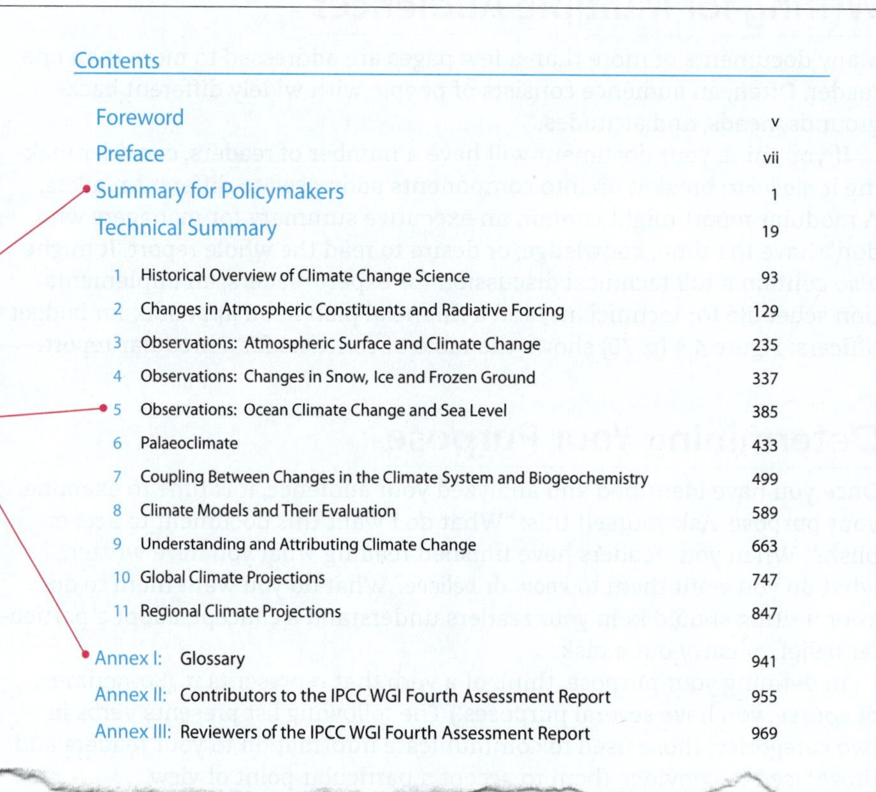

Contents

Foreword		v
Preface		vii
Summary for Policymakers		1
Technical Summary		19
1	Historical Overview of Climate Change Science	93
2	Changes in Atmospheric Constituents and Radiative Forcing	129
3	Observations: Atmospheric Surface and Climate Change	235
4	Observations: Changes in Snow, Ice and Frozen Ground	337
5	Observations: Ocean Climate Change and Sea Level	385
6	Palaeoclimate	433
7	Coupling Between Changes in the Climate System and Biogeochemistry	499
8	Climate Models and Their Evaluation	589
9	Understanding and Attributing Climate Change	663
10	Global Climate Projections	747
11	Regional Climate Projections	847
Annex I:	Glossary	941
Annex II:	Contributors to the IPCC WGI Fourth Assessment Report	955
Annex III:	Reviewers of the IPCC WGI Fourth Assessment Report	969

FIGURE 4.4 **Table of Contents for a Modular Report**

Information from Solomon et al., *Climate Change 2007: The Physical Science Basis.* Working Group I Contribution to the Fourth Assessment Report of the Intergovernmental Panel on Climate Change, p. xix. Cambridge University Press.

Here are a few examples of how you can use these verbs to clarify the purpose of your document (the verbs are italicized).

- This wiki *presents* the draft of our policies on professional use of social media within the organization.
- This letter *authorizes* the purchase of six new tablets for the Jenkintown facility.
- This report *recommends* that we revise the website as soon as possible.

Sometimes your real purpose differs from your expressed purpose. For instance, if you want to persuade your reader to lease a new computer system rather than purchase it, you might phrase the purpose this way: *to explain the advantages of leasing over purchasing*. As mentioned earlier, many readers don't want to be persuaded but are willing to learn new facts or ideas.

In situations like this, stick to the facts. No matter how much you want to convince your readers, it is unacceptable to exaggerate or to omit important information. Trust that the strength and accuracy of your writing will enable you to achieve your intended purpose.

WRITER'S CHECKLIST

Following is a checklist for analyzing your audience and purpose. Remember that your document might be read by one person, several people, a large group, or several groups with various needs.

- ☐ Did you fill out an audience profile sheet for your primary and secondary audiences? *(p. 53)*

In analyzing your audience, did you consider the following questions about each of your most important readers:

- ☐ What is your reader's educational background? *(p. 56)*
- ☐ What is your reader's professional experience? *(p. 56)*
- ☐ What is your reader's job responsibility? *(p. 56)*
- ☐ What are your reader's cultural characteristics? *(p. 56)*
- ☐ What are your reader's personal characteristics? *(p. 56)*
- ☐ What are your reader's personal preferences? *(p. 56)*
- ☐ Why will the reader read your document? *(p. 56)*
- ☐ What is your reader's attitude toward you? *(p. 57)*
- ☐ What is your reader's attitude toward the subject? *(p. 57)*
- ☐ What are your reader's expectations about the document? *(p. 57)*
- ☐ How will your reader read your document? *(p. 57)*
- ☐ What is your reader's reading skill? *(p. 57)*
- ☐ In what physical environment and on what digital platforms will your reader read your document? *(p. 57)*

In learning about your readers, did you

- ☐ determine what you already know about them? *(p. 54)*
- ☐ interview people? *(p. 54)*
- ☐ read about your audience online? *(p. 55)*
- ☐ search social media for documents your audience has written? *(p. 55)*
- ☐ analyze social-media data, if available? *(p. 55)*

In planning to write for an audience from another culture, did you consider the following cultural variables:

- ☐ political? *(p. 59)*
- ☐ economic? *(p. 59)*
- ☐ social? *(p. 59)*
- ☐ religious? *(p. 59)*
- ☐ educational? *(p. 59)*
- ☐ technological? *(p. 59)*
- ☐ linguistic? *(p. 59)*

In planning to write for an audience from another culture, did you consider other cultural variables:

- ☐ focus on individuals or groups? *(p. 60)*
- ☐ distance between business life and private life? *(p. 60)*
- ☐ distance between ranks? *(p. 61)*
- ☐ need for details to be spelled out? *(p. 61)*
- ☐ attitudes toward uncertainty? *(p. 61)*

In writing for a multicultural audience, did you

- ☐ limit your vocabulary? *(p. 63)*
- ☐ keep sentences short? *(p. 63)*
- ☐ define abbreviations and acronyms in a glossary? *(p. 63)*
- ☐ avoid jargon unless you knew that your readers were familiar with it? *(p. 63)*
- ☐ avoid idioms and slang? *(p. 64)*
- ☐ use the active voice whenever possible? *(p. 64)*
- ☐ use graphics carefully? *(p. 64)*
- ☐ have the document reviewed by someone from the reader's culture? *(p. 64)*

☐ In writing for multiple audiences, did you consider creating a modular document? *(p. 69)*

☐ Did you state your purpose in writing and express it in the form of a verb or verbs? *(p. 69)*

EXERCISES

For more about memos, see Ch. 9, page 248.

1. Choose a 200-word passage from a technical article related to your major course of study and addressed to an expert audience. (You can find a technical article on the web by using Google Scholar or the Directory of Open Access Journals. In addition, many federal government agencies publish technical articles and reports on the web.) Rewrite the passage so that it will be clear and interesting to a general reader. Submit the original passage to your instructor along with your revision.

2. The following passage is an advertisement for a translation service. Revise the passage to make it more appropriate for a multicultural audience. Submit the revision to your instructor.

   If your technical documents have to meet the needs of a global market but you find that most translation houses are swamped by the huge volume, fail to accommodate the various languages you require, or fail to make your deadlines, where do you turn?

   Well, your search is over. Translations, Inc. provides comprehensive translations in addition to full-service documentation publishing.

   We utilize ultrasophisticated translation programs that can translate a page in a blink of an eye. Then our crack linguists comb each document to give it that personalized touch.

   No job too large! No schedule too tight! Give us a call today!

3. Study the website of a large manufacturer of computer products, such as Hewlett-Packard, Acer, Dell, or Lenovo. Identify three different pages that address different audiences and fulfill different purposes. Here is an example:

   Name of the page: Lenovo Group Fact Page

   Audience: prospective investors

   Purpose: persuade the prospective investor to invest in the company

   Be prepared to share your findings with the class.

4. **TEAM EXERCISE** Form small groups and study two websites that advertise competing products. For instance, you might choose the websites of two car makers, two television shows, or two music producers. Have each person in the group, working *alone*, compare and contrast the two sites according to these three criteria:

   a. the kind of information they provide: hard, technical information or more emotional information

   b. the use of multimedia such as animation, sound, or video

   c. the amount of interactivity they invite—that is, the extent to which you can participate in activities while you visit the site

After each person has separately studied the sites and taken notes about the three criteria, come together as a group. After each person shares his or her findings, discuss the differences as a group. Which aspects of these sites caused the most difference in group members' reactions? Which aspects seemed to elicit the most consistent reactions? In a brief memo to your instructor, describe and analyze how the two sites were perceived by the different members of the group.

For more practice with the concepts covered in this chapter, complete the LearningCurve activity "Analyzing Your Audience and Purpose" in LaunchPad.

CASE 4: Focusing on an Audience's Needs and Interests

You're interning in the marketing department of a cell-phone service provider, and your supervisor has asked you to perform research into a competing provider's products and services for the over-65 market, paying special attention to the ways in which the company successfully appeals to the needs and interests of its audience. She then asks you to prepare an oral presentation about your findings. To begin this project, go to LaunchPad.

CHAPTER

5 Researching Your Subject

Analyzing the data

IN THE WORKPLACE, you will conduct research all the time. As a buyer for a clothing retailer, for example, you might need to conduct research to help you determine whether a new line of products would be successful in your store. As a civil engineer, you might need to perform research to determine whether to replace your company's traditional surveying equipment with 3D-equipped stations. And as a pharmacist, you might need to research whether a prescribed medication might have a harmful interaction with another medication a patient is already taking.

In the workplace, you will also conduct research using a variety of methods. You will consult websites, blogs, and discussion boards, and you might listen to podcasts or watch videos. Sometimes you will interview people, and you will likely distribute surveys electronically to acquire information from customers and suppliers. Regardless of which technique you use, your challenge will be to sort the relevant information from the irrelevant, and the accurate from the bogus.

This chapter focuses on conducting primary research and secondary research. *Primary research* involves discovering or creating technical information yourself. *Secondary research* involves finding information that other people

have already discovered or created. This chapter presents secondary research first. Why? Because you will probably do secondary research first. To design the experiments or the field research that constitutes primary research, you need a thorough understanding of the information that already exists about your subject.

Understanding the Differences Between Academic and Workplace Research

Although academic research and workplace research can overlap, in most cases they differ in their goals.

In *academic research*, your goal is to find information that will help answer a scholarly question: "What would be the effect on the trade balance between the United States and China if China lowered the value of its currency by 10 percent?" or "At what age do babies learn to focus on people's eyes?" Academic research questions are often more abstract than applied. That is, they get at the underlying principles of a phenomenon. Academic research usually requires extensive secondary research: reading scholarly literature in academic journals and books. If you do primary research, as scientists do in labs, you do so only after extensive secondary research.

In *workplace research*, your goal is to find information to help you answer a practical question: "Should we replace our sales staff's notebook computers with tablets?" or "What would be the advantages and disadvantages to our company of adopting a European-style privacy policy for customer information?" Workplace research questions frequently focus on improving a situation at a particular organization. These questions call for considerable primary research because they require that you learn about your own organization's processes and how the people in your organization would respond to your ideas. Sometimes, workplace research questions address the needs of customers or other stakeholders. You will need a thorough understanding of your organization's external community in order to effectively align your products or services with their needs.

Regardless of whether you are conducting academic or workplace research, the basic research methods—primary and secondary research—are fundamentally the same, as is the goal: to help you answer questions.

Understanding the Research Process

When you perform research, you want the process to be effective and efficient. That is, you want to find information that answers the questions you need to answer. And you don't want to spend any more time than necessary getting that information. To meet these goals, you have to think about how the research relates to the other aspects of the overall project. The Focus on Process box provides an overview of the research process. Although all these tasks are described as part of the planning stage, remember that you might also need to perform additional research during the drafting, revising, editing, and proofreading stages. Whenever you need additional information to help you make your argument clear and persuasive, do more research.

FOCUS ON PROCESS Researching a Topic

PLANNING

- **Analyze your audience.** Who are your most important readers? What are their personal characteristics, their attitudes toward your subject, their motivations for reading? If you are writing to an expert audience that might be skeptical about your message, you need to do a lot of research to gather the evidence for a convincing argument. See Chapter 4.
- **Analyze your purpose.** Why are you writing? Understanding your purpose helps you understand the types of information readers will expect. Think in terms of what you want your readers to know or believe or do after they finish reading your document. See Chapter 4.
- **Analyze your subject.** What do you already know about your subject? What do you still need to find out? Using techniques such as freewriting and brainstorming, you can determine those aspects of the subject you need to investigate.
- **Visualize the deliverable.** What application will you need to deliver: a proposal, a report, a website? What kind of oral presentation will you need to deliver?
- **Work out a schedule and a budget for the project.** When is the deliverable due? Do you have a budget for database searches or travel to libraries or other sites?
- **Determine what information will need to be part of the deliverable.** Draft an outline of the contents, focusing on the kinds of information that readers will expect to see in each part.
- **Determine what information you still need to acquire.** Make a list of the pieces of information you don't yet have.
- **Create questions you need to answer in your deliverable.** Writing the questions in a list forces you to think carefully about your topic. One question suggests another, and soon you have a lengthy list that you need to answer.
- **Conduct secondary research.** Study journal articles and web-based sources such as online journals, discussion boards, blogs, and podcasts.

(continued)

- **Conduct primary research.** You can answer some of your questions by consulting company records, by interviewing experts in your organization, by distributing questionnaires, and by interviewing other people in your organization and industry. Other questions call for using social media to gather information from your customers, suppliers, and other stakeholders.
- **Evaluate your information.** Once you have your information, you need to evaluate its quality: is it accurate, comprehensive, unbiased, and current?
- **Do more research.** If the information you have acquired doesn't sufficiently answer your questions, do more research. And if you have thought of additional questions that need to be answered, do more research. When do you stop doing research? You will stop only when you think you have enough high-quality information to create the deliverable.

Choosing Appropriate Research Methods

Different research questions require different research methods. Once you have determined the questions you need to answer, think about the various research techniques you could use to answer them.

For example, your research methods for finding out how a current situation is expected to change would differ from your research methods for finding out how well a product might work for your organization. That is, if you want to know how outsourcing will change the computer-support industry over the next 10 to 20 years, you might search for long-range predictions in journal and magazine articles and on reputable websites and blogs. By contrast, if you want to figure out whether a specific scanner will produce the quality of scan that you need and will function reliably, you might do the same kind of secondary research and then observe the operation of the scanner at a vendor's site; schedule product demos at your site; follow up by interviewing others in your company; and perform an experiment in which you try two different scanners and analyze the results.

The Choices and Strategies feature provides a good starting point for thinking about how to acquire the information you need. You are likely to find that your research plan changes as you conduct your research. You might find, for instance, that you need more than one method to get the information you need or that the one method you thought would work doesn't. Still, having a plan can help you discover the most appropriate methods more quickly and efficiently.

CHOICES AND STRATEGIES Choosing Appropriate Research Techniques

TYPE OF QUESTION	EXAMPLE OF QUESTION	APPROPRIATE RESEARCH TECHNIQUE
What is the theory behind this process or technique?	How do greenhouse gases contribute to global warming?	**Encyclopedias**, **handbooks**, and **journal articles** present theory. Also, you can find theoretical information on **websites** from reputable professional organizations and universities. Search using keywords such as "greenhouse gases" and "global warming."
What is the history of this phenomenon?	When and how did engineers first try to extract shale oil?	**Encyclopedias** and **handbooks** present history. Also, you can find historical information on **websites** from reputable professional organizations and universities. Search using keywords such as "shale oil" and "petroleum history."
What techniques are being used now to solve this problem?	How are companies responding to the federal government's new laws on health-insurance portability?	If the topic is recent, you will have better luck using digital resources such as **websites** and **social media** than using traditional print media. Search using keywords and tags such as "health-insurance portability." Your search will be most effective if you use standard terminology in your search, such as "HIPAA" for the health-insurance law.
How is a current situation expected to change?	What changes will outsourcing cause in the computer-support industry over the next 10 to 20 years?	For long-range predictions, you can find information in **journal articles** and **magazine articles** and on reputable **websites**. Experts might write forecasts on **discussion boards** and **blogs**.
What products are available to perform a task or provide a service?	Which vendors are available to upgrade and maintain our company's website?	For current products and services, search **websites**, **discussion boards**, and **blogs**. Reputable vendors—manufacturers and service providers—have sites describing their offerings. But be careful not to assume vendors' claims are accurate. Even the specifications they provide might be exaggerated.
What are the strengths and weaknesses of competing products and services?	Which portable GPS system is the lightest?	Search for benchmarking articles from experts in the field, such as a **journal article**—either in print or on the web—about camping and outfitting that compares the available GPS systems according to reasonable criteria. Also check **discussion boards** for reviews and **blogs** for opinions. If appropriate, do **field research** to answer your questions.
Which product or service do experts recommend?	Which four-wheel-drive SUV offers the best combination of features and quality for our needs?	Experts write **journal articles**, **magazine articles**, and sometimes **blogs**. Often, they participate in **discussion boards**. Sometimes, you can **interview** them, in person or on the phone, or write **inquiries**.
What are the facts about how we do our jobs at this company?	Do our chemists use gas chromatography in their analyses?	Sometimes, you can **interview** someone, in person or on the phone, to answer a simple question. To determine whether your chemists use a particular technique, start by asking someone in that department.

(*continued*)

TYPE OF QUESTION	EXAMPLE OF QUESTION	APPROPRIATE RESEARCH TECHNIQUE
What can we learn about what caused a problem in our organization?	What caused the contamination in the clean room?	You can **interview** personnel who were closest to the problem and **inspect** the scene to determine the cause of the problem.
What do our personnel think we should do about a situation?	Do our quality-control analysts think we need to revise our sampling quotient?	If there are only a few personnel, **interview** them. If there are many, use **questionnaires** to get the information more quickly.
How well would this product or service work in our organization?	Would this scanner produce the quality of scan that we need and interface well with our computer equipment?	Read product reviews on reputable **websites**. Study **discussion boards. Observe** the use of the product or service at a vendor's site. Schedule product **demos** at your site. Follow up by **interviewing** others in your company to get their thinking. Do an **experiment** in which you try two different solutions to a problem, then analyze the results.

GUIDELINES Researching a Topic

Follow these three guidelines as you gather information to use in your document.

- **Be persistent.** Don't be discouraged if a research method doesn't yield useful information. Even experienced researchers fail at least as often as they succeed. Be prepared to rethink how you might find the information. Don't hesitate to ask reference librarians for help or to post questions on discussion boards.
- **Record your data carefully.** Prepare the materials you will need. Write information down, on paper or electronically. Record interviews (with the respondents' permission). Paste the URLs of the sites you visit into your notes. Bookmark sites so that you can return to them easily.
- **Triangulate your research methods.** *Triangulating* your research methods means using more than one or two methods. If a manufacturer's website says a printer produces 17 pages per minute, an independent review in a reputable journal also says 17, and you get 17 in a demo at your office with your documents, the printer probably will produce 17 pages per minute. When you need to answer important questions, don't settle for only one or two sources.

If you are doing research for a document that will be read by people from other cultures, think about what kinds of evidence your readers will consider appropriate. In many non-Western cultures, tradition or the authority of the person making the claim can be extremely important, in some cases more important than the kind of scientific evidence that is favored in Western cultures.

And don't forget that all people pay particular attention to information that comes from their own culture. If you are writing to European readers about telemedicine, for instance, try to find information from European authorities and about European telemedicine. This information will interest your readers and will likely reflect their cultural values and expectations.

Conducting Secondary Research

When you conduct secondary research, you are trying to learn what experts have to say about a topic. Whether that expert is a world-famous scientist revising an earlier computer model about the effects of climate change on agriculture in Europe or the head of your Human Resources Department checking company records to see how the Affordable Care Act changed the way your company hired part-time workers, your goal is the same: to acquire the best available information—the most accurate, most unbiased, most comprehensive, and most current.

Sometimes you will do research in a library, particularly if you need specialized handbooks or access to online subscription services that are not freely available on the Internet. Sometimes you will do your research on the web. As a working professional, you might find much of the information you need in your organization's information center. An *information center* is an organization's library, a resource that collects different kinds of information critical to the organization's operations. Many large organizations have specialists who can answer research questions or who can get articles or other kinds of data for you.

USING TRADITIONAL RESEARCH TOOLS

There is a tremendous amount of information in the different media. The trick is to learn how to find what you want. This section discusses six basic research tools.

Online Catalogs An online catalog is a database of books, microform materials, films, compact discs, phonograph records, tapes, and other materials. In most cases, an online catalog lists and describes the holdings at one particular library or a group of libraries. Your college library has an online catalog of its holdings. To search for an item, consult the instructions, which explain how to limit your search by characteristics such as types of media, date of publication, and language. The instructions also explain how to use punctuation and words such as *and*, *or*, and *not* to focus your search effectively.

Reference Works Reference works include general dictionaries and encyclopedias, biographical dictionaries, almanacs, atlases, and dozens of other research tools. These print and online works are especially useful when you are beginning a research project because they provide an overview of the subject and often list the major works in the field.

How do you know if there is a dictionary of the terms used in a given field? The following reference books—the guides to the guides—list some of the many resources available:

Hacker, D., and Fister, B. (2015). *Research and documentation in the digital age* (6th ed.). Boston: Bedford/St. Martin's.

Kennedy, X. J., Kennedy, D. M., and Muth, M. F. (2014). *The Bedford guide for college writers with reader, research manual, and handbook* (10th ed.). Boston: Bedford/St. Martin's.

Lester, R. (Ed.). (2008). *The new Walford guide to reference resources* (Vol. 1: Science, Technology and Medicine; Vol. 2: Social Sciences). London: Neal-Schuman.

To find information on the web, go to the "reference" section of a library website or search engine. There you will find links to excellent collections of reference works online, such as Infomine and ipl2.

Periodical Indexes Periodicals are excellent sources of information because they offer recent, authoritative discussions of specific subjects. The biggest challenge in using periodicals is identifying and locating the dozens of articles relevant to any particular subject that are published each month. Although only half a dozen major journals might concentrate on your field, a useful article could appear in one of hundreds of other publications. A periodical index, which is a list of articles classified according to title, subject, and author, can help you determine which journals you want to locate.

There are periodical indexes in all fields. The following brief list will give you a sense of the diversity of titles:

- *Applied Science & Technology Index*
- *Business Source Premier*
- *Engineering Village*
- *Readers' Guide to Periodical Literature*

You can also use a directory search engine. Many directory categories include a subcategory called "journals" or "periodicals" that lists online and printed sources.

Once you have created a bibliography of printed articles you want to study, you have to find them. Check your library's online catalog, which includes all the journals your library receives. If your library does not have an article you want, you can use one of two techniques for securing it:

- **Interlibrary loan.** Your library finds a library that has the article. That library scans the article and sends it to your library. This service can take more than a few days.
- **Document-delivery service.** If you are in a hurry, you can log on to a document-delivery service, such as IngentaConnect, a free database of 5 million articles in 10,000 periodicals. There are also fee-based document-delivery services.

Newspaper Indexes Many major newspapers around the world are indexed by subject. The three most important indexed U.S. newspapers are

- the *New York Times*, perhaps the most reputable U.S. newspaper for national and international news
- the *Christian Science Monitor*, another highly regarded general newspaper
- the *Wall Street Journal*, an authoritative news source on business, finance, and the economy

Many newspapers available on the web can be searched electronically, although sometimes there is a charge for archived articles. Keep in mind that the print version and the electronic version of a newspaper can vary greatly. If you wish to quote from an article in a newspaper, be sure that your citation format references the medium you're using.

For more about abstracts, see Ch. 13, p. 349.

Abstract Services Abstract services are like indexes but also provide abstracts: brief technical summaries of the articles. In most cases, reading the abstract will enable you to decide whether to seek out the full article. The title of an article alone can often mislead you about its contents.

Some abstract services, such as *Chemical Abstracts Service*, cover a broad field, but many are specialized rather than general. Figure 5.1 shows an abstract from *AnthroSource*, an abstract service covering anthropology journals. Note that it provides statistical information about the full article that would not be evident from just reading the title.

PLOT AND IRONY IN CHILDBIRTH NARRATIVES OF MIDDLE-CLASS BRAZILIAN WOMEN

Brazil's rate of cesarean deliveries is among the highest in the world and constitutes the majority of childbirths in private hospitals. This study examines ways middle-class Brazilian women are exercising agency in this context. It draws from sociolinguistics to examine narrative structure and dramatic properties of 120 childbirth narratives of 68 low- to high-income women. Surgical delivery constituted 62% of the total. I focus on 20 young middle-class women, of whom 17 had C-sections. Doctors determined mode of childbirth pre-emptively or appeared to accommodate women's wishes, while framing the scenario as necessitating surgical delivery. The women strove to imbue C-section deliveries with value and meaning through staging, filming, familial presence, attempting induced labor, or humanized childbirth. Their stories indicate that class privilege does not lead to choice over childbirth mode. The women nonetheless struggle over the significance of their agency in childbirth.

FIGURE 5.1 An Abstract from *AnthroSource*

Information from O'Dougherty, Maureen. "Plot and Irony in Childbirth Narratives of Middle-Class Brazilian Women," from *Medical Anthropology Quarterly*, Volume 27, Issue 1, pages 43–62, March 2013. Republished with permission of University of California Press; permission conveyed through Copyright Clearance Center, Inc.

Government Information The U.S. government is the world's biggest publisher. In researching any field of science, engineering, or business, you are likely to find that a federal agency or department has produced a relevant brochure, report, or book.

Government publications are cataloged and shelved separately from other kinds of materials. They are classified according to the Superintendent of Documents system, not the Library of Congress system. A reference librarian or a government-documents specialist at your library can help you use government publications.

You can also access various government sites and databases on the Internet. For example, if your company wishes to respond to a request for proposals (RFP) published by a federal government agency, you will find that RFP on a government site. The major entry point for federal government sites is USA.gov (usa.gov), which links to hundreds of millions of pages of government information and services. It also features tutorials, a topical index, online transactions, and links to state and local government sites.

For more about RFPs, see Ch. 11, p. 293.

USING SOCIAL MEDIA AND OTHER INTERACTIVE RESOURCES

Social media and other interactive resources enable people to collaborate, share, link, and generate content in ways that traditional websites offering static content cannot. The result is an Internet that can harness the collective intelligence of people around the globe—and do so quickly. However, the ease and speed with which new content can be posted, as well as the lack of formal review of the content, creates challenges for people who do research on the Internet. Everyone using social-media resources must be extra cautious in evaluating and documenting sources.

To watch a tutorial on using online tools to organize your research, go to LaunchPad.

This discussion covers three categories of social media and web-based resources used by researchers—discussion forums, wikis, and blogs—as well as two techniques for streamlining the process of using these resources: tagged content and RSS.

Discussion Forums Online discussion forums sponsored by professional organizations, private companies, and others enable researchers to tap a community's information. Discussion forums are especially useful for presenting quick, practical advice. However, the advice might or might not be authoritative. Figure 5.2 shows one interchange related to starting a business as a foreign national.

Wikis A wiki is a website that makes it easy for members of a community, company, or organization to create and edit content collaboratively. Often, a wiki contains articles, information about student and professional conferences, reading lists, annotated sets of links, book reviews, and documents used by members of the community. You might have participated in creating

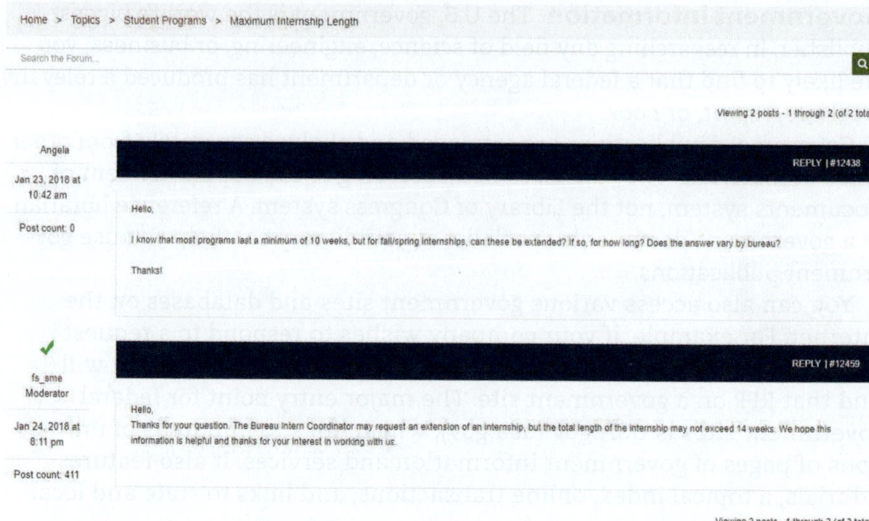

FIGURE 5.2 A Discussion Board Exchange
Information from Small Business Administration, 2014: http://www.sba.gov/community/discussion-boards/starting-business-us-foreign-national.

If you use a search engine to find this interchange, you are performing secondary research: discovering what has already been written or said about a topic. If you post a question to a discussion forum (or comment on a blog post) and someone responds, you are performing primary research, just as if you were interviewing that person. For more on primary research, see p. 88. But don't worry too much about whether you are doing primary or secondary research; worry about whether the information is accurate and useful.

and maintaining a wiki in one of your courses or as a member of a community group outside of your college.

Wikis are popular with researchers because they contain information that can change from day to day, on topics in fields such as medicine or business. In addition, because wikis rely on information contributed voluntarily by members of a community, they represent a much broader spectrum of viewpoints than media that publish only information that has been approved by editors. For this reason, however, you should be especially careful when you use wikis; the information they contain might not be trustworthy. It's a good idea to corroborate any information you find on a wiki by consulting other sources. You can search wikis by using any search engine and adding the word "wiki" to the search. Or you can use a specialized search engine such as Wiki.com.

For more about blogs, see Ch. 9, "Writing Microblogs."

Blogs Many technical and scientific organizations, universities, and private companies sponsor blogs that offer useful information for researchers.

Keep in mind that bloggers are not always independent voices. A Hewlett-Packard employee writing on a company-sponsored blog will likely be presenting the company's viewpoint on the topic. Don't count on that blogger to offer objective views about products.

Figure 5.3, a screen shot of a portion of NASA's My Big Fat Planet blog, offers information that is likely to be credible, accurate, and timely.

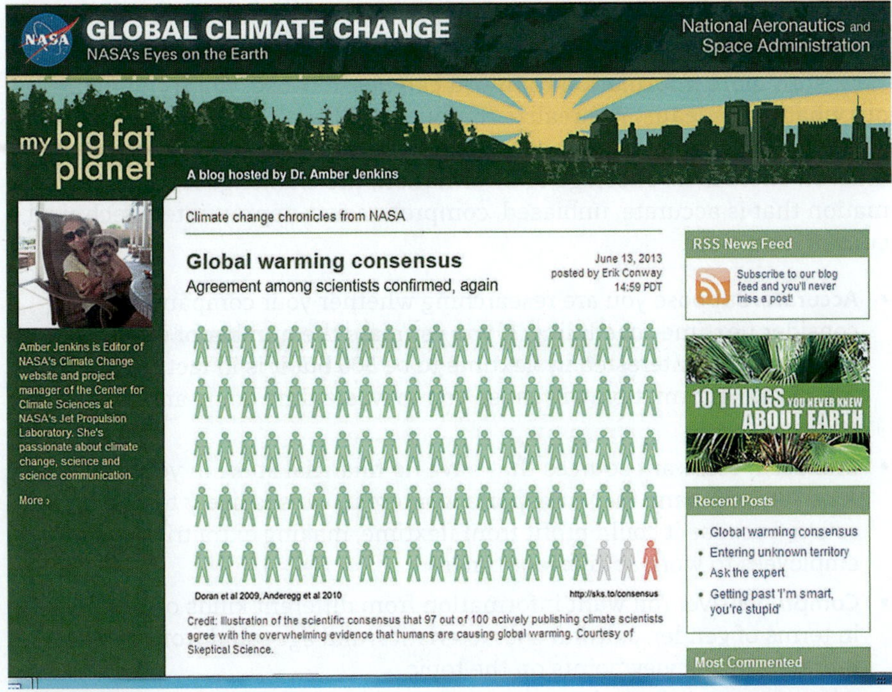

FIGURE 5.3 **A Blog**

In the first 10 days after it went online, this post was liked 522 times on Facebook, was tweeted 86 times on Twitter, and elicited 11 comments from readers. Almost all blogs invite readers to post comments or questions.
Information from NASA, 2013: http://climate.nasa.gov/blog/938.

Tagged Content Tags are descriptive keywords people use to categorize and label content such as blog entries, videos, podcasts, and images they post to the Internet or bookmarks they post to social-bookmarking sites. Tags can be one-word descriptors without spaces or punctuation (such as "sandiegozoo") or multiword descriptors (such as "San Diego Zoo"). More and more social-media platforms, including Facebook and Twitter, have adopted the hashtag (#) as a way to tag an item to make it easier to find by searching.

RSS Feeds Repeatedly checking for new content on many different websites can be a time-consuming and haphazard way to research a topic. RSS (short for *rich site summary* or *really simple syndication*) *technology* allows readers to check just one place (such as a software program running on their computer or an email program) for alerts to new content posted on selected websites. Readers use a special type of software program called an *RSS aggregator* to be alerted by *RSS feeds* (notifications of new or changed content from sites of interest to them).

EVALUATING THE INFORMATION

For more about taking notes, paraphrasing, and quoting, see Appendix, Part A, pp. 449–51.

You've taken notes, paraphrased, and quoted from your secondary research. Now, with more information than you can possibly use, you try to figure out what it all means. You realize that you still have some questions—that some of the information is incomplete, some contradictory, and some unclear. There is no shortage of information; the challenge is to find information that is accurate, unbiased, comprehensive, appropriately technical, current, and clear.

- **Accurate.** Suppose you are researching whether your company should consider flextime scheduling. If you estimate the number of employees who would be interested in flextime to be 500 but it is in fact closer to 50, inaccurate information will cause you to waste time doing an unnecessary study.
- **Unbiased.** You want sources that have no financial stake in your project. A private company that transports workers in vans is likely to be a biased source because it could profit from flextime, making extra trips to bring employees to work at different times.
- **Comprehensive.** You want information from different kinds of people—in terms of gender, cultural characteristics, and age—and from people representing all viewpoints on the topic.
- **Appropriately technical.** Good information is sufficiently detailed to meet the needs of your readers, but not so detailed that they cannot understand it or do not need it. For the flextime study, you need to find out whether opening your building an hour earlier and closing it an hour later would significantly affect your utility costs. You can get this information by interviewing people in the Operations Department; you do not need to do a detailed inspection of all the utility records of the company.
- **Current.** If your information is 10 or even 5 years old, it might not accurately reflect today's situation.
- **Clear.** You want information that is easy to understand. Otherwise, you'll waste time figuring it out, and you might misinterpret it.

The most difficult kind of material to evaluate is user-generated content from the Internet—such as information on discussion forums or in blogs—because it rarely undergoes the formal review procedure used for books and professional journals. A general principle for using any information you find on the Internet is to be extremely careful. Because content is unlikely to have been reviewed before being published on a social-media site, use one or more trusted sources to confirm the information you locate. Some instructors do not allow their students to use blogs or wikis, including Wikipedia, for their research. Check with your instructor to learn his or her policies.

GUIDELINES Evaluating Print and Online Sources

FOR PRINTED SOURCES	FOR ONLINE SOURCES
Authorship Do you recognize the name of the author? Does the source describe the author's credentials and current position? If not, can you find this information in a "who's who" or by searching for other books or other journal articles by the author?	If you do not recognize the author's name, is the site mentioned on another reputable site? Does the site contain links to other reputable sites? Does it contain biographical information—the author's current position and credentials? Can you use a search engine to find other references to the author's credentials? Be especially careful with unedited sources such as Wikipedia; some articles in it are authoritative, others are not. Be careful, too, with blogs, some of which are written by disgruntled former employees with a score to settle.
Publisher What is the publisher's reputation? A reliable book is published by a reputable trade, academic, or scholarly publisher; a reliable journal is sponsored by a professional association or university. Are the editorial board members well known? Trade publications—magazines about a particular industry or group—often promote the interests of that industry or group. For example, information in a trade publication for either loggers or environmentalists might be biased. If you doubt the authority of a book or journal, ask a reference librarian or a professor.	Can you determine the publisher's identity from headers or footers? Is the publisher reputable? If the site comes from a personal account, the information it offers might be outside the author's field of expertise. Many Internet sites exist largely for public relations or advertising. For instance, websites of corporations and other organizations are unlikely to contain self-critical information. For blogs, examine the blogroll, a list of links to other blogs and websites. Credible blogs are likely to link to blogs already known to be credible. If a blog links only to the author's friends, blogs hosted by the same corporation, or blogs that express the same beliefs, be very cautious.
Knowledge of the literature Does the author appear to be knowledgeable about the major literature on the topic? Is there a bibliography? Are there notes throughout the document?	Analyze the Internet source as you would any other source. Often, references to other sources will take the form of links.

(continued)

FOR PRINTED SOURCES	FOR ONLINE SOURCES
Accuracy and verifiability of the information	
Is the information based on reasonable assumptions? Does the author clearly describe the methods and theories used in producing the information, and are they appropriate to the subject? Has the author used sound reasoning? Has the author explained the limitations of the information?	Is the site well constructed? Is the information well written? Is it based on reasonable assumptions? Are the claims supported by appropriate evidence? Has the author used sound reasoning? Has the author explained the limitations of the information? Are sources cited? Online services such as BlogPulse help you evaluate how active a blog is, how the blog ranks compared to other blogs, and who is citing the blog. Active, influential blogs that are frequently linked to and cited by others are more likely to contain accurate, verifiable information.
Timeliness	
Does the document rely on recent data? Was the document published recently?	Was the document created recently? Was it updated recently? If a site is not yet complete, be wary.

Conducting Primary Research

Although the library and the Internet offer a wealth of authoritative information, in the workplace you will often need to conduct primary research because you need new information. There are eight major categories of primary research: analysis of social-media data, observations and demonstrations, inspections, experiments, field research, interviews, inquiries, and questionnaires.

ANALYSIS OF SOCIAL-MEDIA DATA

Every hour, tens of millions of posts are made on social media. A torrent of information is continuously coming online, and many organizations are working hard to sift through it to find useful insights.

Businesses are spending the most time on social-media research, trying to figure out what customers like and dislike about their products and services, learn what customers want, and reinforce brand loyalty. Take the case of Nielsen, which for fifty years has been monitoring the TV viewing habits of Americans by distributing questionnaires and attaching devices to their TVs, and then selling the data it collects to TV networks and producers, who use the information to determine how much to charge advertisers. The problem for Nielsen is that many people don't watch TV on TV or they don't watch

DOCUMENT ANALYSIS ACTIVITY

Evaluating Information from Internet Sources

How can you ignore thousands of scientists who say manmade global warming is a serious threat?

The idea that there is a "scientific consensus" does not hold up. Scientists who are skeptical about "dangerous manmade climate change" have been speaking out for years. Just this year, two prominent former believers in man-made global warming announced they were reconsidering the science.

"Gaia" scientist James Lovelock had been "alarmist" about climate change for years. Now he says "The problem is we don't know what the climate is doing. We thought we knew 20 years ago."

German meteorologist Klaus-Eckart Puls also reversed his belief in man-made global warming in 2012 and called the idea CO_2 can regulate climate "sheer absurdity." "Ten years ago I simply parroted what the IPCC told us," he said. "One day I started checking the facts and data. First I started with a sense of doubt, but then I became outraged when I discovered that much of what the IPCC and media were telling us was sheer nonsense and was not even supported by any scientific facts and measurements. To this day, I still feel shame that as a scientist I made presentations of their science without first checking it."

In 2010, a report documented that More Than 1000 International Scientists Dissented Over Man-Made Global Warming Claims. Many of them were former IPCC scientists. Climate scientist Mike Hulme dismantled the "thousands of scientists agree" claim put forth by the United Nations and news media. Claims that "2,500 of the world's leading scientists have reached a consensus that human activities are having a significant influence on the climate" are disingenuous, Hulme noted. The key scientific case for CO_2 driving global warming, like many others in the IPCC reports, "is reached by only a few dozen experts in the specific field of detection and attribution studies; other IPCC authors are experts in other fields." Other scientists are excluded or not consulted.

Dr. William Schlesinger agrees with the UN climate view but has admitted that only 20% of UN IPCC scientists deal with climate. In other words, 80% of the UN's IPCC membership are experts in other fields and have no dealing with or expertise in climate change as part of their academic studies.

Information from Committee for a Constructive Tomorrow, 2013: www.cfact.org/issues/climate-change/climate-change-truth-file/.

This blog post appears in the FAQ section of the "Climate Change Truth File" section of the website of the Committee for a Constructive Tomorrow. (For another view on climate change, see Figure 5.3 on p. 85.) The questions below ask you to consider the guidelines for evaluating Internet sources (pp. 87–88).

1. This blog post comes from a portion of the site called "Climate Change Truth File." Does this title make you more likely or less likely to consider the information authoritative?
2. If you were considering using this source in a document you were writing, what information would you want to discover about the site and the organization that publishes it? How would you locate it?
3. The bulk of this passage is devoted to two prominent scientists who have changed their minds on the question of whether human-caused global warming is a serious threat. If the claim about the two scientists is true, does the case for human-caused global warming collapse?

shows when they are broadcast. Now Nielsen also uses social-media analysis: gathering data by monitoring social media to listen in on what people are saying on Twitter, Facebook, and other services about different TV programs (DeVault, 2013).

But organizations other than businesses are analyzing social-media data, too. For instance, the U.S. Geological Survey created the Twitter Earthquake Detector (TED), a program to monitor Twitter for the use of the word *earthquake*. Why? Because they realized that when people experience earthquakes, a lot of them tweet about it. The Centers for Disease Control, a U.S. federal agency, analyzes keywords on social media to monitor the spread of diseases, such as the H7N9 flu virus, in the United States and around the world. According to one scientist, "The world is equipped with human sensors—more than 7 billion and counting. It's by far the most extensive sensor network on the planet. What can we learn by paying attention?" (McCaney, 2013).

How do you perform social-media data analysis? There are many software programs that can help you devise searches. Among the most popular is HootSuite, which includes tools for listening in on what people are saying about your company on social media such as Twitter, Facebook, LinkedIn, and many other services. In addition, HootSuite helps you monitor and manage your company's social-media presence and provides analytics: demographic data about who is following your company, their attitudes, and their behaviors.

OBSERVATIONS AND DEMONSTRATIONS

Observation and demonstration are two common forms of primary research. When you *observe*, you simply watch some activity to understand some aspect of it. For instance, if you were trying to determine whether the location of the break room was interfering with work on the factory floor, you could observe the situation, preferably at different times of the day and on different days of the week. If you saw workers distracted by people moving in and out of the room or by sounds made in the room, you would record your observations by taking notes, taking photos, or shooting video of events. An observation might lead to other forms of primary research. You might, for example, follow up by interviewing some employees who could help you understand what you observed.

When you witness a *demonstration* (or *demo*), you are watching someone carry out a process. For instance, if your company was considering buying a mail-sorting machine, you could arrange to visit a manufacturer's facility, where technicians would show how the machine works. If your company was considering a portable machine, such as a laptop computer, manufacturers or dealers could demo their products at your facility.

When you plan to observe a situation or witness a demo, prepare beforehand. Write down the questions you need answered or the factors you want to investigate. Prepare interview questions in case you have a chance to speak with someone. Think about how you are going to incorporate the information you acquire into the document you will write. Finally, bring whatever equipment you will need (pen and paper, computer, camera, etc.) to the site of the observation or demo.

INSPECTIONS

Inspections are like observations, but you participate more actively. For example, a civil engineer can determine what caused a crack in a foundation by inspecting the site: walking around, looking at the crack, photographing it and the surrounding scene, examining the soil. An accountant can determine the financial health of an organization by inspecting its financial records, perhaps performing calculations and comparing the data she finds with other data.

These professionals are applying their knowledge and professional judgment as they inspect a site, an object, or a document. Sometimes inspection techniques are more complicated. A civil engineer inspecting foundation cracking might want to test his hunches by bringing soil samples back to the lab for analysis.

When you carry out an inspection, do your homework beforehand. Think about how you will use the data in your document: will you need photographs or video files or computer data? Then prepare the materials and equipment you'll need to capture the data.

EXPERIMENTS

Learning to conduct the many kinds of experiments used in a particular field can take months or even years. This discussion is a brief introduction. In many cases, conducting an experiment involves four phases.

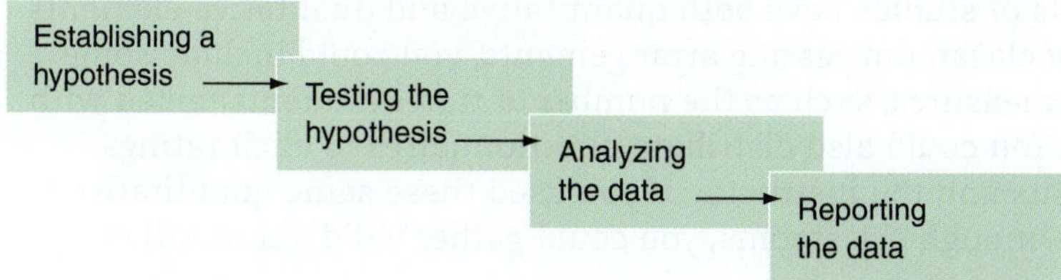

- **Establishing a hypothesis.** A hypothesis is an informed guess about the relationship between two factors. In a study relating gasoline octane and miles per gallon, a hypothesis might be that a car will get 5 percent better mileage with 89-octane gas than with 87-octane gas.

- **Testing the hypothesis.** Usually, you need an experimental group and a control group. These two groups should be identical except for the condition you are studying: in the above example, the gasoline. The control group would be a car running on 87 octane. The experimental group would be an identical car running on 89 octane. The experiment would consist of driving the two cars over an identical course at the same speed—preferably in some sort of controlled environment—over a given distance, such as 1,000 miles. Then you would calculate the miles per gallon. The results would either support or refute your original hypothesis.
- **Analyzing the data.** Do your data show a correlation—one factor changing along with another—or a causal relationship? For example, we know that sports cars are involved in more fatal accidents than sedans (there is a stronger correlation for sports cars), but we don't know what the causal relationship is—whether the car or the way it is driven is the important factor.

For more about reports, see Chs. 12 and 13.

- **Reporting the data.** When researchers report their findings, they explain what they did, why they did it, what they saw, what it means, and what ought to be done next.

FIELD RESEARCH

Whereas an experiment yields quantitative data that typically can be measured precisely, most field research is qualitative; that is, it yields data that typically cannot be measured precisely. Often in field research, you seek to understand the quality of an experience. For instance, you might want to understand how a new seating arrangement affects group dynamics in a classroom. You could design a study in which you observed and shot video of classes and interviewed the students and the instructor about their reactions to the new arrangement. Then you could do the same in a traditional classroom and compare the results.

Some kinds of studies have both quantitative and qualitative elements. In the case of classroom seating arrangements, you could include some quantitative measures, such as the number of times students talked with one another. You could also distribute questionnaires to elicit ratings by the students and the instructor. If you used these same quantitative measures on enough classrooms, you could gather valid quantitative information.

When you are doing quantitative or qualitative studies on the behavior of animals—from rats to monkeys to people—try to minimize two common problems:

- **The effect of the experiment on the behavior you are studying.** In studying the effects of the classroom seating arrangement, minimize the effects of your own presence. For instance, if you observe in person, avoid drawing attention to yourself. Also, make sure that the video camera is placed unobtrusively and that it is set up before the students arrive, so they don't see the process. But be aware that anytime you bring in a camera, you cannot be sure that what you witness is typical.

- **Bias in the recording and analysis of the data.** Bias can occur because researchers want to confirm their hypotheses. In an experiment to determine whether students write differently on physical keyboards than on touch screens, a researcher might see differences where other people don't. For this reason, the experiment should be designed so that it is *double blind*. That is, the students shouldn't know what the experiment is about so that they don't change their behavior to support or negate the hypothesis, and the data being analyzed should be disguised so that researchers don't know whether they are examining the results from the control group or the experimental group. For example, the documents produced on keyboards and touch screens should be printed out the same way.

Conducting an experiment or field research is relatively simple; the hard part is designing your study so that it accurately measures what you want it to measure.

INTERVIEWS

Interviews are extremely useful when you need information on subjects that are too new to have been discussed in the professional literature or are too narrow for widespread publication (such as local political questions).

In choosing a respondent—a person to interview—answer three questions:

- **What questions do you want to answer?** Only when you know this can you begin to search for a person who can provide the information.
- **Who could provide this information?** The ideal respondent is an expert willing to talk. Unless there is an obvious choice, such as the professor carrying out the research you are studying, use directories, such as local industrial guides, to locate potential respondents. You may also be able to find experts through your social-media network.
- **Is the person willing to be interviewed?** Contact the potential respondent by phone or in writing and state what you want to ask about. If the person is not able to help you, he or she might be willing to refer you to someone who can. Explain why you have decided to ask him or her. (A compliment works better than admitting that the person you really wanted to interview is out of town.) Explain what you plan to do with the information, such as write a report or present a talk. Then, if the person is willing to be interviewed, set up an appointment at his or her convenience.

GUIDELINES Conducting an Interview

PREPARING FOR THE INTERVIEW

Follow these suggestions for preparing for and conducting an interview—and for following up after the interview.

- **Do your homework.** If you ask questions that have already been answered in the professional literature, the respondent might become annoyed and uncooperative.
- **Prepare good questions.** Good questions are clear, focused, and open.
  - **Be clear.** The respondent should be able to understand what you are asking.

    UNCLEAR Why do you sell Trane products?

    CLEAR What are the characteristics of Trane products that led you to include them in your product line?

    The unclear question can be answered in a number of unhelpful ways: "Because they're too expensive to give away" or "Because I'm a Trane dealer."
  - **Be focused.** The question must be narrow enough to be answered briefly. If you want more information, you can ask a follow-up question.

    UNFOCUSED What is the future of the computer industry?

    FOCUSED What will the American chip industry look like in 10 years?
  - **Ask open questions.** Your purpose is to get the respondent to talk. Don't ask a lot of questions that have yes or no answers.

    CLOSED Do you think the federal government should create industrial partnerships?

    OPEN What are the advantages and disadvantages of the federal government's creating industrial partnerships?
- **Check your equipment.** If you will be recording the interview, test your voice recorder or video camera to make sure it is operating properly.

BEGINNING THE INTERVIEW

- **Arrive on time.**
- **Thank the respondent for taking the time to talk with you.**
- **State the subject and purpose of the interview and what you plan to do with the information.**
- **If you wish to record the interview, ask permission.**

CONDUCTING THE INTERVIEW

- **Take notes.** Write down important concepts, facts, and numbers, but don't take such copious notes that you can't make eye contact with the respondent or that you are still writing when the respondent finishes an answer.
- **Start with prepared questions.** Because you are likely to be nervous at the start, you might forget important questions. Have your first few questions ready.

(continued)

- **Be prepared to ask follow-up questions.** Listen carefully to the respondent's answer and be ready to ask a follow-up question or request a clarification. Have your other prepared questions ready, but be willing to deviate from them if the respondent leads you in unexpected directions.
- **Be prepared to get the interview back on track.** Gently return to the point if the respondent begins straying unproductively, but don't interrupt rudely or show annoyance. Do not say, "Whoa! I asked about layoffs in this company, not in the whole industry." Rather, say, "On the question of layoffs at this company, do you anticipate . . . ?"

CONCLUDING THE INTERVIEW

- **Thank the respondent.**
- **Ask for a follow-up interview.** If a second meeting would be useful, ask to arrange one.
- **Ask for permission to quote the respondent.** If you think you might want to quote the respondent by name, ask for permission now.

AFTER THE INTERVIEW

- **Write down the important information while the interview is fresh in your mind.** (This step is unnecessary, of course, if you have recorded the interview.) If you will be printing a transcript of the interview, make the transcript now.
- **Send a brief thank-you note.** Within a day or two, send a note showing that you appreciate the respondent's courtesy and that you value what you have learned. In the note, confirm any previous offers you have made, such as to send the respondent a copy of your final document.

When you wish to present the data from an interview in a document you are preparing, include a transcript of the interview (or an excerpt from the interview). You will probably present the transcript as an appendix so that readers can refer to it but are not slowed down when reading the body of the document. You might decide to present brief excerpts from the transcript in the body of the document as evidence for points you make.

INQUIRIES

A useful alternative to a personal interview is to send an inquiry. This inquiry can take the form of a letter, an email, or a message sent through an organization's website.

For more about inquiry letters, see Ch. 9, p. 242.

If you are lucky, your respondent will provide detailed and helpful answers. However, the respondent might not clearly understand what you want to know or might choose not to help you. Although the strategy of the inquiry is essentially that of a personal interview, inquiries can be less successful because the recipient has not already agreed to provide information and might not respond. Also, an inquiry, unlike an interview, gives you little opportunity to follow up by asking for clarification.

QUESTIONNAIRES

To find software for conducting surveys, search for "survey software."

Questionnaires enable you to solicit information from a large group of people. You can send questionnaires through the mail, email them, present them as forms on a website, or use survey software (such as SurveyMonkey).

Unfortunately, questionnaires rarely yield completely satisfactory results, for three reasons:

- **Some of the questions will misfire.** Respondents will misinterpret some of your questions or supply useless answers.
- **You won't obtain as many responses as you want.** The response rate will almost never exceed 50 percent. In most cases, it will be closer to 10 to 20 percent.
- **You cannot be sure the respondents are representative.** People who feel strongly about an issue are much more likely to respond to questionnaires than are those who do not. For this reason, you need to be careful in drawing conclusions based on a small number of responses to a questionnaire.

When you send a questionnaire, you are asking the recipient to do you a favor. Your goal should be to construct questions that will elicit the information you need as simply and efficiently as possible.

Asking Effective Questions

To ask effective questions, follow two suggestions:

- **Use unbiased language.** Don't ask, "Should U.S. clothing manufacturers protect themselves from unfair foreign competition?" Instead, ask, "Are you in favor of imposing tariffs on men's clothing?"
- **Be specific.** If you ask, "Do you favor improving the safety of automobiles?" only an eccentric would answer no. Instead, ask, "Do you favor requiring automobile manufacturers to equip new cars with electronic stability control, which would raise the price by an average of $300 per car?"

The Choices and Strategies box explains common types of questions used in questionnaires.

Include an introductory explanation with the questionnaire. This explanation should clearly indicate who you are, why you are writing, what you plan to do with the information from the questionnaire, and when you will need it.

CHOICES AND STRATEGIES Choosing Types of Questions for Questionnaires

IF YOU WANT TO . . .	CONSIDER USING THIS QUESTION TYPE	EXAMPLE
Have respondents choose among alternatives	**Multiple choice.** The respondent selects one answer from a list.	Would you consider joining a company-sponsored sports team? Yes_____ No _____
Measure respondents' feelings about an idea or concept	**Likert scale.** The respondent ranks the degree to which he or she agrees or disagrees with the statement. Using an even number of possible responses (six, in this case) increases your chances of obtaining useful data. With an odd number, many respondents will choose the middle response.	The flextime program has been a success in its first year. strongly agree __ __ __ __ __ __ strongly disagree
Measure respondents' feelings about a task, an experience, or an object	**Semantic differentials.** The respondent registers a response along a continuum between a pair of opposing adjectives. As with Likert scales, an even number of possible responses yields better data.	Logging on to the system is simple __ __ __ __ __ __ difficult The description of the new desalinization process is interesting __ __ __ __ __ __ boring
Have respondents indicate a priority among a number of alternatives	**Ranking.** When using ranking questions, be sure to give instructions about what method (in this case, numbering) the respondents should use to rank the items.	Please rank the following work schedules in order of preference. Put a 1 next to the schedule you would most like to have, a 2 next to your second choice, and so on. 8:00–4:30_____ 9:00–5:30_____ 8:30–5:00_____ flexible_____
Ask open-ended questions	**Short answer.** The respondent writes a brief answer using phrases or sentences.	What do you feel are the major advantages of the new parts-requisitioning policy? 1. ____________ 2. ____________ 3. ____________
Give respondents an opportunity to present fuller responses	**Short essay.** Although essay questions can yield information you never would have found using closed-ended questions, you will receive fewer responses because they require more effort. Also, essays cannot be quantified precisely, as data from other types of questions can.	The new parts-requisitioning policy has been in effect for a year. How well do you think it is working? ____________ ____________ ____________ ____________ ____________

Testing the Questionnaire Before you send out *any* questionnaire, show it and its accompanying explanation to a few people who can help you identify any problems. After you have revised the materials, test them on people whose backgrounds are similar to those of your intended respondents. Revise the materials a second time, and, if possible, test them again. Once you have sent the questionnaire, you cannot revise it and resend it to the same people.

Administering the Questionnaire Determining who should receive the questionnaire can be simple or difficult. If you want to know what the residents of a particular street think about a proposed construction project, your job is easy. But if you want to know what mechanical-engineering students in colleges across the country think about their curricula, you will need a background in sampling techniques to identify a representative sample.

Make it easy for respondents to present their information. For mailed questionnaires, include a self-addressed, stamped envelope.

Presenting Questionnaire Data in Your Document To decide where and how to present the data that you acquire from your questionnaire, think about your audience and purpose. Start with this principle: important information is presented and analyzed in the body of a document, whereas less-important information is presented in an appendix (a section at the end that only some of your audience will bother to read). Most often, different versions of the same information appear in both places.

Typically, the full questionnaire data are presented in an appendix. If you can, present the respondents' data—the answers they provided—in the questionnaire itself, as shown here:

If you think your reader will benefit from analyses of the data, present such analyses. For instance, you could calculate the percentage for each response: for question 1, "12 people—17 percent—say they do not eat in the cafeteria at all." Or you could present the percentage in parentheses after each number: "12 (17%)."

1. Approximately how many days per week do you eat lunch in the lunchroom?

   0 **12** 1 **16** 2 **18** 3 **12** 4 **9** 5 **4**

2. At approximately what time do you eat in the lunchroom?

   11:30–12:30 **3** 12:00–1:00 **26** 12:30–1:30 **7** varies **23**

Selected data might then be interpreted in the body of the document. For instance, you might devote a few sentences or paragraphs to the data for one of the questions. The following example shows how a writer might discuss the data from question 2.

> Question 2 shows that 26 people say that they use the cafeteria between noon and 1:00. Only 10 people selected the two other times: 11:30–12:30 or 12:30–1:30. Of the 23 people who said they use the cafeteria at various times, we can conclude that at least a third—8 people—use it between noon and 1:00. If this assumption is correct, at least 34 people (26 + 8) use the cafeteria between noon and 1:00. This would explain why people routinely cannot find a table in the noon hour, especially between 12:15 and 12:30. To alleviate this problem, we might consider asking department heads not

to schedule meetings between 11:30 and 1:30, to make it easier for their people to choose one of the less-popular times.

The body of a document is also a good place to discuss important nonquantitative data. For example, you might wish to discuss and interpret several representative textual answers to open-ended questions.

ETHICS NOTE

REPORTING AND ANALYZING DATA HONESTLY

When you put a lot of time and effort into a research project, it's frustrating if you can't find the information you need or if the information you find doesn't help you say what you want to say. As discussed in Chapter 2, your responsibility as a professional is to tell the truth.

If the evidence suggests that the course of action you propose won't work, don't omit that evidence or change it. Rather, try to figure out why the evidence does not support your proposal. Present your explanation honestly.

If you can't find reputable evidence to support your claim that one device works better than another, don't just keep silent and hope your readers won't notice. Explain why you think the evidence is missing and how you propose to follow up by continuing your research.

If you make an honest mistake, you are a person. If you cover up a mistake, you're a dishonest person. If you get caught fudging the data, you could be an unemployed dishonest person. If you don't get caught, you're still a smaller person.

WRITER'S CHECKLIST

- ☐ Did you determine the questions you need to answer for your document? *(p. 75)*

Did you choose appropriate secondary-research tools to answer those questions, including, if appropriate,

- ☐ online catalogs? *(p. 80)*
- ☐ reference works? *(p. 80)*
- ☐ periodical indexes? *(p. 81)*
- ☐ newspaper indexes? *(p. 82)*
- ☐ abstract services? *(p. 82)*
- ☐ government information? *(p. 83)*
- ☐ social media and other interactive resources? *(p. 83)*

In evaluating information, did you carefully assess

- ☐ the author's credentials? *(p. 87)*
- ☐ the publisher? *(p. 87)*
- ☐ the author's knowledge of literature in the field? *(p. 87)*
- ☐ the accuracy and verifiability of the information? *(p. 88)*
- ☐ the timeliness of the information? *(p. 88)*

Did you choose appropriate primary-research methods to answer your questions, including, if appropriate,

- ☐ social-media data analysis? *(p. 88)*
- ☐ observations and demonstrations? *(p. 90)*
- ☐ inspections? *(p. 91)*
- ☐ experiments? *(p. 91)*
- ☐ field research? *(p. 92)*
- ☐ interviews? *(p. 93)*
- ☐ inquiries? *(p. 95)*
- ☐ questionnaires? *(p. 96)*
- ☐ Did you report and analyze the data honestly? *(p. 99)*

EXERCISES

For more about memos, see Ch. 9, "Writing Memos."

1. Imagine you are an executive working for a company that distributes books to bookstores in the Seattle, Washington, area. Your company, with a 20,000-square-foot warehouse and a fleet of 15 small delivery vans, employs 75 people. The following are three questions that an academic researcher specializing in energy issues might focus on in her research. Translate each of these academic questions into a workplace question that your company might need to answer.
   a. What are the principal problems that need to be resolved before biomass (such as switchgrass) can become a viable energy source for cars and trucks?
   b. How much money will need to be invested in the transmission grid before windmills can become a major part of the energy solution for business and residential customers in the Western United States?
   c. Would a federal program that enables companies to buy and sell carbon offsets help or hurt industry in the United States?
2. For each of the following questions, select a research technique that is likely to yield a useful answer. For instance, if the question is "Which companies within a 20-mile radius of our company headquarters sell recycled paper?" a search of the web is likely to provide a useful answer.
   a. Does the Honda CR-V include traction control as a standard feature?
   b. How much money has our company's philanthropic foundation donated to colleges and universities in each of the last three years?
   c. How does a 3D printer work?
   d. Could our Building 3 support a rooftop green space?
   e. How can we determine whether we would save more money by switching to LED lighting in our corporate offices?
3. Using a search engine, answer the following questions. Provide the URL of each site that provides information for your answer. If your instructor requests it, submit your answers in an email to him or her.
   a. What are the three largest or most-important professional organizations in your field? (For example, if you are a construction management major, your field is construction management, civil engineering, or industrial engineering.)
   b. What are three important journals read by people in your field?
   c. What are three important online discussion forums read by people in your field?
   d. What are the date and location of an upcoming national or international professional meeting for people in your field?
   e. Name and describe, in one paragraph for each, three major issues being discussed by practitioners or academics in your field. For instance, nurses might be discussing the effect of managed care on the quality of medical care delivered to patients.
4. Revise the following interview questions to make them more effective. In a brief paragraph for each, explain why you have revised the question as you have.
   a. What is the role of communication in your daily job?
   b. Do you think it is better to relocate your warehouse or go to just-in-time manufacturing?
   c. Isn't it true that it's almost impossible to train an engineer to write well?
   d. Where are your company's headquarters?
   e. Is there anything else you think I should know?
5. Revise the following questions from questionnaires to make them more effective. In a brief paragraph for each, explain why you have revised the question as you have.
   a. Does your company provide tuition reimbursement for its employees? Yes_____ No_____
   b. What do you see as the future of bioengineering?
   c. How satisfied are you with the computer support you receive?
   d. How many employees work at your company? 5–10____ 10–15____ 15 or more____
   e. What kinds of documents do you write most often? memos_____ letters_____ reports_____

6. **TEAM EXERCISE** Form small groups, and describe and evaluate your college or university's website. A different member of the group might carry out each of the following tasks:
   - In an email to the site's webmaster, ask questions about the process of creating the site. For example, how involved was the webmaster with the content and design of the site? What is the webmaster's role in maintaining the site?
   - Analyze the kinds of information the site contains, and determine whether the site is intended primarily for faculty, students, alumni, legislators, or prospective students.
   - Determine the overlap between information on the site and information in printed documents published by the school. In those cases in which there is overlap, is the information on the site merely a duplication of the printed information, or has it been revised to take advantage of the unique capabilities of the web?

   In a memo to your instructor, present your findings and recommend ways to improve the site.

For more practice with the concepts covered in this chapter, complete the LearningCurve activity "Researching Your Subject" in LaunchPad.

CASE 5: Revising a Questionnaire

You're a marketing director at a real-estate company, and you are trying to determine whether it would be cost-effective to have the company's agents take property photos instead of having the photos taken by the professional photographers from the supplier with which you currently contract. You ask one of your agents to develop a questionnaire to gauge agents' reactions to and opinions about the possibility of adding photography to their responsibilities, but you find that her questionnaire needs considerable revising before it will be an effective tool. To access the questionnaire and begin assessing it, go to LaunchPad.

CHAPTER 6

Writing for Your Readers

WRITING FOR YOUR READERS means writing documents that are easy for readers to use and understand. It starts with making sure you present yourself effectively, as a professional whose writing is worth reading. In addition, writing for your readers involves creating a usable document—one in which readers can easily find the information they need—that presents ideas and data clearly and emphasizes the most important information. Finally, writing for your readers means choosing words carefully and crafting accurate, clear, concise, and forceful sentences. If a sentence doesn't say what you intended, misunderstandings can occur, and misunderstandings can cause harm and cost money. More important, the ability to write for your readers—word by word and sentence by sentence—reflects positively on you and your organization.

Presenting Yourself Effectively

A big part of presenting yourself effectively is showing that you know the appropriate information about your subject. However, you also need to come across as a professional.

GUIDELINES Creating a Professional Persona

Your *persona* is how you appear to your readers. Demonstrating the following four characteristics will help you establish an attractive professional persona.

- **Cooperativeness.** Make clear that your goal is to solve a problem, not to advance your own interests.
- **Moderation.** Be moderate in your judgments. The problem you are describing will not likely spell doom for your organization, and the solution you propose will not solve all the company's problems.
- **Fair-mindedness.** Acknowledge the strengths of opposing points of view, even as you offer counterarguments.
- **Modesty.** If you fail to acknowledge that you don't know everything, someone else will be sure to volunteer that insight.

The following paragraph shows how a writer can demonstrate the qualities of cooperativeness, moderation, fair-mindedness, and modesty:

> This plan is certainly not perfect. For one thing, it calls for a greater up-front investment than we had anticipated. And the return on investment through the first three quarters is likely to fall short of our initial goals. However, I think this plan is the best of the three alternatives for the following reasons. . . . Therefore, I recommend that we begin implementing the plan. I am confident that this plan will enable us to enter the flat-screen market successfully, building on our fine reputation for high-quality advanced electronics.

In the first three sentences, the writer acknowledges the problems with the plan.

The use of "I think" adds an attractive modesty; the recommendation might be unwise.

The recommendation itself is moderate; the writer does not claim that the plan will save the world.

In the last sentence, the writer shows a spirit of cooperativeness by focusing on the company's goals.

Using Basic Organizational Patterns

Every document calls for its own organizational pattern. You should begin by asking yourself whether a conventional pattern for presenting your information already exists. A conventional pattern makes things easier for you as a writer because it serves as a template or checklist, helping you remember which information to include in your document and where to put it. For your audience, a conventional pattern makes your document easier to read and understand because the organization of information meets readers' expectations. When you write a proposal, for example, readers who are familiar with proposals can find the information they want in your document if you put it where others have put similar information. The Choices and Strategies box beginning on page 104 explains the relationship between organizational patterns and the kinds of information you want to present.

Does this mean that technical communication is merely the process of filling in the blanks? No. You need to assess the writing situation continuously as you work. If you think you can communicate your ideas better by modifying a conventional pattern or by devising a new pattern, do so. Long, complex arguments often require several organizational patterns. For instance, one part of a document might be a causal analysis of a problem, and another might be a comparison and contrast of two options for solving that problem.

CHOICES AND STRATEGIES Choosing Effective Organizational Patterns

IF YOU WANT TO . . .	CONSIDER USING THIS ORGANIZATIONAL PATTERN	EXAMPLE
Explain events that occurred or might occur, or tasks the reader is to carry out in sequence	**Chronological.** Most of the time, you present information in chronological order. Sometimes, however, you use reverse chronology.	You describe the process you used to diagnose the problem with the accounting software. In a job résumé, you describe your more-recent jobs before your less-recent ones.
Describe a physical object or scene, such as a device or a location	**Spatial.** You choose an organizing principle such as top-to-bottom, east-to-west, or inside-to-outside.	You describe the three buildings that will make up the new production facility.
Explain a complex situation or idea, such as the factors that led to a problem or the theory that underlies a process	**General to specific.** You present general information first, then specific information. Understanding the big picture helps readers understand the details.	You explain the major changes in and the details of the law mandating the use of a new refrigerant in cooling systems.
Present a set of factors	**More important to less important.** You discuss the most-important issue first, then the next-most-important issue, and so forth. In technical communication, you don't want to create suspense. You want to present the most-important information first.	When you launch a new product, you discuss market niche, competition, and then pricing.
Present similarities and differences between two or more items	**Comparison and contrast.** You choose from one of two patterns: discuss all the factors related to one item, then all the factors related to the next item, and so forth; or discuss one factor as it relates to all the items, then another factor as it relates to all the items, and so forth.	You discuss the strengths and weaknesses of three companies bidding on a contract your company is offering.
Assign items to logical categories, or discuss the elements that make up a single item	**Classification or partition.** Classification involves placing items into categories according to some basis. Partition involves breaking a single item into its major elements.	You group the motors your company manufactures according to the fuel they burn: gasoline or diesel. Or you explain the operation of each major component of one of your motors.

(continued)

IF YOU WANT TO . . .	CONSIDER USING THIS ORGANIZATIONAL PATTERN	EXAMPLE
Discuss a problem you encountered, the steps you took to address the problem, and the outcome or solution	**Problem-methods-solution.** You can use this pattern in discussing the past, the present, or the future. Readers understand this organizational pattern because they use it in their everyday lives.	In describing how your company is responding to a new competitor, you discuss the problem (the recent loss in sales), the methods (how you plan to examine your product line and business practices), and the solution (which changes will help your company remain competitive).
Discuss the factors that led to (or will lead to) a given situation, or the effects that a situation led to or will lead to	**Cause and effect.** You can start from causes and speculate about effects, or start with the effect and work backward to determine the causes.	You discuss factors that you think contributed to a recent sales dip for one of your products. Or you explain how you think changes to an existing product will affect its sales.

Writing Clear, Informative Titles and Headings

The title of a document is crucial because it is your first chance to define your subject and purpose for your readers, giving them their first clue to whether the document contains the information they need. The title is an implicit promise to readers: "This document is about Subject A, and it was written to achieve Purpose B." Everything that follows has to relate clearly to the subject and purpose defined in the title; if it doesn't, either the title is misleading or the document has failed to make good on the title's promise.

You might want to put off giving a final title to your document until you have completed the document, because you cannot be sure that the subject and purpose you established during the planning stages will not change. However, you should jot down a working title before you start drafting; you can revise it later. To give yourself a strong sense of direction, make sure the working title defines not only the subject of the document but also its purpose. The working title "Snowboarding Injuries" states the subject but not the purpose. "How to Prevent Snowboarding Injuries" is better because it helps keep you focused on your purpose.

An effective title is precise. For example, if you are writing a feasibility study on the subject of offering free cholesterol screening at your company, the title should contain the key terms *free cholesterol screening* and *feasibility*. The following title would be effective:

Offering Free Cholesterol Screening at Thrall Associates: A Feasibility Study

If your document is an internal report discussing company business, you might not need to identify the company. In that case, the following would be clear:

Offering Free Cholesterol Screening: A Feasibility Study

Or you could present the purpose before the subject:

A Feasibility Study of Offering Free Cholesterol Screening

Avoid substituting general terms, such as *health screening* for *cholesterol screening* or *study* for *feasibility study*; the more precise your terms, the more useful your readers will find the title. An added benefit of using precise terms is that your document can be more accurately and effectively indexed in databases and online libraries, increasing the chances that someone researching your subject will be able to find the document.

You'll notice that clear, comprehensive titles can be long. If you need eight or ten words to say what you want to say about your subject and purpose, use them.

Headings, which are lower-level titles for the sections and subsections in a document, do more than announce the subject that will be discussed in the document. Collectively, they create a *hierarchy of information*, dividing the document into major sections and subdividing those sections into subsections. In this way, coherent headings communicate the relative importance and generality of the information that follows, helping readers recognize major sections as *primary* (likely to contain more-important or more-general information) and subsections as *secondary* or *subordinate* (likely to contain less-important or more-specific information).

Clear, informative headings communicate this relationship not only through their content but also through their design. For this reason, make sure that the design of a primary heading (sometimes referred to as a *level 1 heading*, *1 heading*, or *A heading*) clearly distinguishes it from a subordinate heading (a *level 2 heading*, *2 heading*, or *B heading*), and that the design of that subordinate heading clearly distinguishes it from yet a lower level of subordinate heading (a *level 3 heading*, *3 heading*, or *C heading*).

The headings used in this book illustrate this principle, as does the example below. Notice that the example uses both typography and indentation to distinguish one heading from another and to communicate visually how information at one level logically relates to information at other levels.

Level 1 Heading

Level 2 Heading

Level 3 Heading

Effective headings help both reader and writer by forecasting not only the subject and purpose of the discussion that follows but also its scope and organization. When readers encounter the heading "Three Health Benefits of Yoga: Improved Muscle Tone, Enhanced Flexibility, Better Posture," they can reasonably assume that the discussion will consist of three parts (not two or four) and that it will begin with a discussion of muscle tone, followed by a discussion of flexibility and then posture.

Because headings introduce text that discusses or otherwise elaborates on the subject defined by the heading, avoid back-to-back headings. In other words, avoid following one heading directly with another heading:

> **3. Approaches to Neighborhood Policing**
>
> **3.1 Community Policing**
> According to the COPS Agency (a component of the U.S. Department of Justice), "Community policing focuses on crime and social disorder." . . .

What's wrong with back-to-back headings? First, they're illogical. If your document contains a level 1 heading, you have to say something at that level before jumping to the discussion at level 2. Second, back-to-back headings distract and confuse readers. The heading "3. Approaches to Neighborhood Policing" announces to readers that you have something to say about neighborhood policing—but you don't say anything. Instead, another, subordinate heading appears, announcing to readers that you now have something to say about community policing.

To avoid confusing and frustrating readers, separate the headings with text, as in this example:

> **3. Approaches to Neighborhood Policing**
> Over the past decade, the scholarly community has concluded that community policing offers significant advantages over the traditional approach based on patrolling in police cars. However, the traditional approach has some distinct strengths. In the following discussion, we define each approach and then explain its advantages and disadvantages. Finally, we profile three departments that have successfully made the transition to community policing while preserving the major strengths of the traditional approach.
>
> **3.1 Community Policing**
> According to the COPS Agency (a component of the U.S. Department of Justice), "Community policing focuses on crime and social disorder." . . .

The text after the heading "3. Approaches to Neighborhood Policing" is called an *advance organizer*. It indicates the background, purpose, scope, and organization of the discussion that follows it. Advance organizers give readers an overview of the discussion's key points before they encounter the details in the discussion itself.

GUIDELINES Revising Headings

Follow these four suggestions to make your headings more effective.

- **Avoid long noun strings.** The following example is ambiguous and hard to understand:

  Proposed Production Enhancement Strategies Analysis Techniques

For more about noun strings, see p. 138.

(continued)

Is the heading introducing a discussion of techniques for analyzing strategies that have been proposed? Or is it introducing a discussion that proposes using certain techniques to analyze strategies? Readers shouldn't have to ask such questions. Adding prepositions makes the heading clearer:

Techniques for Analyzing the Proposed Strategies for Enhancing Production

This heading announces more clearly that the discussion describes techniques for analyzing strategies, that those strategies have been proposed, and that the strategies are aimed at enhancing production. It's a longer heading than the original, but that's okay. It's also much clearer.

- **Be informative.** In the preceding example, you could add information about how many techniques will be described:

  Three Techniques for Analyzing the Proposed Strategies for Enhancing Production

  You can go one step further by indicating what you wish to say about the three techniques:

  Advantages and Disadvantages of the Three Techniques for Analyzing the Proposed Strategies for Enhancing Production

  Again, don't worry if the heading seems long; clarity is more important than conciseness.

- **Use a grammatical form appropriate to your audience.** The question form works well for readers who are not knowledgeable about the subject (Benson, 1985) and for nonnative speakers:

  What Are the Three Techniques for Analyzing the Proposed Strategies for Enhancing Production?

  The "how-to" form is best for instructional material, such as manuals:

  How to Analyze the Proposed Strategies for Enhancing Production

  The gerund form (*-ing*) works well for discussions and descriptions of processes:

  Analyzing the Proposed Strategies for Enhancing Production

- **Avoid back-to-back headings.** Use advance organizers to separate the headings.

For more about how to format headings, see Ch. 7, p. 170.

Writing Clear, Informative Paragraphs

There are two kinds of paragraphs—body paragraphs and transitional paragraphs—both of which play an important role in helping you emphasize important information.

A *body paragraph*, the basic unit for communicating information, is a group of sentences (or sometimes a single sentence) that is complete and self-sufficient and that contributes to a larger discussion. In an effective paragraph, all the sentences clearly and directly articulate one main point, either by introducing the point or by providing support for it. In addition, the whole paragraph follows logically from the material that precedes it.

A *transitional paragraph* helps readers move from one major point to another. Like a body paragraph, it can consist of a group of sentences or be a single sentence. Usually it summarizes the previous point, introduces the next point, and helps readers understand how the two are related.

The following example of a transitional paragraph appeared in a discussion of how a company plans to use this year's net proceeds.

> Our best estimate of how we will use these net proceeds, then, is to develop a second data center and increase our marketing efforts. We base this estimate on our current plans and on projections of anticipated expenditures. However, at this time we cannot precisely determine the exact cost of these activities. Our actual expenditures may exceed what we've predicted, making it necessary or advisable to reallocate the net proceeds within the two uses (data center and marketing) or to use portions of the net proceeds for other purposes. The most likely uses appear to be reducing short-term debt and addressing salary inequities among software developers; each of these uses is discussed below, including their respective advantages and disadvantages.

The first sentence contains the word "then" to signal that it introduces a summary.

The final sentence clearly indicates the relationship between what precedes it and what follows it.

STRUCTURE PARAGRAPHS CLEARLY

Most paragraphs consist of a topic sentence and supporting information.

The Topic Sentence Because a topic sentence states, summarizes, or forecasts the main point of the paragraph, put it up front. Technical communication should be clear and easy to read, not suspenseful. If a paragraph describes a test you performed, include the result of the test in your first sentence:

> The point-to-point continuity test on Cabinet 3 revealed an intermittent open circuit in the Phase 1 wiring.

Then go on to explain the details. If the paragraph describes a complicated idea, start with an overview. In other words, put the "bottom line" on top:

> Mitosis is the usual method of cell division, occurring in four stages: (1) prophase, (2) metaphase, (3) anaphase, and (4) telophase.

Putting the bottom line on top makes the paragraph much easier to read, as shown in Figure 6.1.

Make sure each of your topic sentences relates clearly to the organizational pattern you are using. In a discussion of the physical condition of a building, for example, you might use a spatial pattern and start a paragraph with the following topic sentence:

> On the north side of Building B, water damage to about 75 percent of the roof insulation and insulation in some areas in the north wall indicates that the roof has been leaking for some time. The leaking has contributed to . . .

Your next paragraph should begin with a topic sentence that continues the spatial organizational pattern:

> On the east side of the building, a downspout has eroded the lawn and has caused a small silt deposit to form on the neighboring property directly to the east. Riprap should be placed under the spout to . . .

The topic sentences are italicized for emphasis in this figure.

Notice that placing the topic sentence at the start gives a focus to the paragraph, helping readers understand the information in the rest of the paragraph.

Topic sentence at the end of the paragraph	Topic sentence at the start of the paragraph
A solar panel affixed to a satellite in distant geosynchronous orbit receives about 1400 watts of sunlight per square meter. On Earth, cut this number in half, due to the day/night cycle. Cut it in half again because sunlight hits the Earth obliquely (except exactly on the equator). Cut it in half again due to clouds and dust in the atmosphere. *The result: eight times the amount of sunlight falls on a solar panel in sun-synchronous orbit as falls on the same size area on Earth.*	*Eight times the amount of sunlight falls on a solar panel in distant geosynchronous orbit as falls on the same size area on Earth.* A solar panel affixed to a satellite in sun-synchronous orbit receives about 1400 watts of sunlight per square meter. On Earth, cut this number in half, due to the day/night cycle. Cut it in half again because sunlight hits the Earth obliquely (except exactly on the equator). Cut it in half again due to clouds and dust in the atmosphere.

FIGURE 6.1 A Topic Sentence Works Better at the Start of the Paragraph

Note that the phrases "on the north side" and "on the east side" signal that the discussion is following the points of the compass in a clockwise direction, further emphasizing the spatial pattern. Readers can reasonably assume that the next two parts of the discussion will be about the south side of the building and the west side, in that order.

Similarly, if your first topic sentence is "First, we need to . . . ," your next topic sentence should refer to the chronological pattern: "Second, we should . . ." (Of course, sometimes well-written headings can make such references to the organizational pattern unnecessary, as when headings are numbered to emphasize that the material is arranged in a chronological pattern.)

The Supporting Information The supporting information makes the topic sentence clear and convincing. Sometimes a few explanatory details provide all the support you need. At other times, however, you need a lot of information to clarify a difficult thought or to defend a controversial idea. How much supporting information to provide also depends on your audience and purpose. Readers knowledgeable about your subject may require little supporting information; less-knowledgeable readers might require a lot. Likewise, you may need to provide little supporting information if your purpose is merely to *state* a controversial point of view rather than *persuade* your reader to agree with it. In deciding such matters, your best bet is to be generous with your supporting information. Paragraphs with too little support are far more common than paragraphs with too much.

Supporting information, which is most often developed using the basic patterns of organization discussed earlier in this chapter, usually fulfills one of these five roles:

- It defines a key term or idea included in the topic sentence.
- It provides examples or illustrations of the situation described in the topic sentence.

- It identifies causes: factors that led to the situation.
- It defines effects: implications of the situation.
- It supports the claim made in the topic sentence.

A topic sentence is like a promise to readers. At the very least, when you write a topic sentence that says "Within five years, the City of McCall will need to upgrade its wastewater-treatment facilities because of increased demands from a rapidly rising population," you are implicitly promising readers that the paragraph not only will be about wastewater-treatment facilities but also will explain that the rapidly rising population is the reason the facilities need to be upgraded. If your paragraph fails to discuss these things, it has failed to deliver on the promise you made. If the paragraph discusses these things but also goes on to speculate about the price of concrete over the next five years, it is delivering on promises that the topic sentence never made. In both situations, the paragraph has gone astray.

Paragraph Length How long should a paragraph be? In general, 75 to 125 words are enough for a topic sentence and four or five supporting sentences. Long paragraphs are more difficult to read than short paragraphs because they require more focused concentration. They can also intimidate some readers, who might skip over them.

ETHICS NOTE

AVOIDING BURYING BAD NEWS IN PARAGRAPHS

The most-emphatic location in a paragraph is the topic sentence, usually the first sentence in a paragraph. The second-most-emphatic location is the end of the paragraph. Do not bury bad news in the middle of the paragraph, hoping readers won't see it. It would be misleading to structure a paragraph like this:

> In our proposal, we stated that the project would be completed by May. In making this projection, we used the same algorithms that we have used successfully for more than 14 years. In this case, however, the projection was not realized, due to several factors beyond our control. . . . We have since completed the project satisfactorily and believe strongly that this missed deadline was an anomaly that is unlikely to be repeated. In fact, we have beaten every other deadline for projects this fiscal year.

The writer has buried the bad news in a paragraph beginning with a topic sentence that appears to suggest good news. The last sentence, too, suggests good news.

A more forthright approach would be as follows:

> We missed our May deadline for completing the project. Although we derived this schedule using the same algorithms that we have used successfully for more than 14 years, several factors, including especially bad weather at the site, delayed the construction. . . .
>
> However, we have since completed the project satisfactorily and believe strongly that this missed deadline was an anomaly that is unlikely to be repeated. . . . In fact, we have beaten every other deadline for projects this fiscal year.

Here the writer forthrightly presents the bad news in a topic sentence. Then he creates a separate paragraph with the good news.

GUIDELINES Dividing Long Paragraphs

Here are three techniques for dividing long paragraphs.

- **Break the discussion at a logical place.** The most logical place to divide this material is at the introduction of the second factor. Because the paragraphs are still relatively long and cues are minimal, this strategy should be reserved for skilled readers.

  High-tech companies have been moving their operations to the suburbs for two main reasons: cheaper, more modern space and a better labor pool. A new office complex in the suburbs will charge from one-half to two-thirds of the rent charged for the same square footage in the city. And that money goes a lot further, too. The new office complexes are bright and airy; new office space is already wired for computers; and exercise clubs, shopping centers, and even libraries are often on-site.

  The second major factor attracting high-tech companies to the suburbs is the availability of experienced labor. Office workers and middle managers are abundant. In addition, the engineers and executives, who tend to live in the suburbs anyway, are happy to forgo the commuting, the city wage taxes, and the noise and stress of city life.

- **Make the topic sentence a separate paragraph and break up the supporting information.** This version is easier to understand than the one above because the brief paragraph at the start clearly introduces the information. In addition, each of the two main paragraphs now has a clear topic sentence.

  High-tech companies have been moving their operations to the suburbs for two main reasons: cheaper, more modern space and a better labor pool.

  First, office space is a bargain in the suburbs. A new office complex in the suburbs will charge from one-half to two-thirds of the rent charged for the same square footage in the city. And that money goes a lot further, too. The new office complexes are bright and airy; new office space is already wired for computers; and exercise clubs, shopping centers, and even libraries are often on-site.

  Second, experienced labor is plentiful. Office workers and middle managers are abundant. In addition, the engineers and executives, who tend to live in the suburbs anyway, are happy to forgo the commuting, the city wage taxes, and the noise and stress of city life.

- **Use a list.** This is the easiest of the three versions for all readers because of the extra visual cues provided by the list format.

  High-tech companies have been moving their operations to the suburbs for two main reasons:

  - *Cheaper, more modern space.* Office space is a bargain in the suburbs. A new office complex in the suburbs will charge anywhere from one-half to two-thirds of the rent charged for the same square footage in the city. And that money goes a lot further, too. The new office complexes are bright and airy; new office space is already wired for computers; and exercise clubs, shopping centers, and even libraries are often on-site.
  - *A better labor pool.* Office workers and middle managers are abundant. In addition, the engineers and executives, who tend to live in the suburbs anyway, are happy to forgo the commuting, the city wage taxes, and the noise and stress of city life.

But don't let arbitrary guidelines about length take precedence over your own analysis of the audience and purpose. You might need only one or two sentences to introduce a graphic, for example. Transitional paragraphs are also likely to be short. If a brief paragraph fulfills its function, let it be. Do not combine two ideas in one paragraph simply to achieve a minimum word count.

You may need to break up your discussion of one idea into two or more paragraphs. An idea that requires 200 or 300 words to develop should probably not be squeezed into one paragraph.

A note about one-sentence paragraphs: body paragraphs and transitional paragraphs alike can consist of a single sentence. However, many single-sentence paragraphs are likely to need revision. Sometimes the idea in that sentence belongs with the paragraph immediately before it or immediately after it or in another paragraph elsewhere in the document. Sometimes the idea needs to be developed into a paragraph of its own. And sometimes the idea doesn't belong in the document at all.

When you think about paragraph length, consider how the information will be printed or displayed. If the information will be presented in a narrow column, such as in a newsletter, short paragraphs are much easier to read. If the information will be presented in a wider column, readers will be able to handle a longer paragraph.

USE COHERENCE DEVICES WITHIN AND BETWEEN PARAGRAPHS

For the main idea in the topic sentence to be clear and memorable, you need to make the support—the rest of the paragraph—coherent. That is, you must link the ideas together clearly and logically, and you must express parallel ideas in parallel grammatical constructions. Even if the paragraph already moves smoothly from sentence to sentence, you can strengthen the coherence by adding transitional words and phrases, repeating key words, and using demonstrative pronouns followed by nouns.

Adding Transitional Words and Phrases Transitional words and phrases help the reader understand a discussion by explicitly stating the logical relationship between two ideas. Table 6.1 lists the most common logical relationships between two ideas and some of the common transitions that express those relationships.

Transitional words and phrases benefit both readers and writers. When a transitional word or phrase explicitly states the logical relationship between two ideas, readers don't have to guess at what that relationship might be. Using transitional words and phrases in your writing forces you to think more deeply about the logical relationships between ideas than you might otherwise.

TABLE 6.1 Transitional Words and Phrases

RELATIONSHIP	TRANSITION
addition	also, and, finally, first (second, etc.), furthermore, in addition, likewise, moreover, similarly
comparison	in the same way, likewise, similarly
contrast	although, but, however, in contrast, nevertheless, on the other hand, yet
illustration	for example, for instance, in other words, to illustrate
cause-effect	as a result, because, consequently, hence, so, therefore, thus
time or space	above, around, earlier, later, next, soon, then, to the right (left, west, etc.)
summary or conclusion	at last, finally, in conclusion, to conclude, to summarize

To better understand how transitional words and phrases benefit both reader and writer, consider the following pairs of examples:

WEAK Demand for flash-memory chips is down by 15 percent. We have laid off 12 production-line workers.

IMPROVED Demand for flash-memory chips is down by 15 percent; *as a result,* we have laid off 12 production-line workers.

WEAK The project was originally expected to cost $300,000. The final cost was $450,000.

IMPROVED The project was originally expected to cost $300,000. *However,* the final cost was $450,000.

The next sentence pair differs from the others in that the weak example *does* contain a transitional word, but it's a weak transitional word:

WEAK According to the report from Human Resources, the employee spoke rudely to a group of customers waiting to enter the store, *and* he repeatedly ignored requests from co-workers to unlock the door so the customers could enter.

IMPROVED According to the report from Human Resources, the employee spoke rudely to a group of customers waiting to enter the store; *moreover*, he repeatedly ignored requests from co-workers to unlock the door so the customers could enter.

In the weak version, *and* implies simple addition: the employee did this, and then he did that. The improved version is stronger, adding to simple addition the idea that refusing to unlock the door compounded the employee's rude behavior, elevating it to something more serious. By using *moreover*, the writer is saying that speaking rudely to customers was bad enough, but the employee *really* crossed the line when he refused to open the door.

Whichever transitional word or phrase you use, place it as close as possible to the beginning of the second idea. As shown in the examples above, the link between two ideas should be near the start of the second idea, to provide context for it. Consider the following example:

> The vendor assured us that the replacement parts would be delivered in time for the product release. The parts were delivered nearly two weeks after the product release, however.

The idea of Sentence 2 stands in contrast to the idea of Sentence 1, but the reader doesn't see the transition until the end of Sentence 2. Put the transition at the start of the second idea, where it will do the most good.

You should also use transitional words to maintain coherence *between* paragraphs, just as you use them to maintain coherence *within* paragraphs. The link between two paragraphs should be near the start of the second paragraph.

Repeating Key Words Repeating key words—usually nouns—helps readers follow the discussion. In the following example, the first version could be confusing:

UNCLEAR For months the project leaders carefully planned their research. The cost of the work was estimated to be over $200,000.

What is *the work*: the planning or the research?

CLEAR For months the project leaders carefully planned their research. The cost of the research was estimated to be over $200,000.

From a misguided desire to be interesting, some writers keep changing their important terms. *Plankton* becomes *miniature seaweed*, then *the ocean's fast food*. Avoid this kind of word game; it can confuse readers.

Of course, too much repetition can be boring. You can vary nonessential terms as long as you don't sacrifice clarity.

SLUGGISH The purpose of the new plan is to *reduce* the *problems* we are seeing in our accounting operations. We hope to see a *reduction* in the *problems* by early next quarter.

BETTER The purpose of the new plan is to *reduce* the *problems* we are seeing in our accounting operations. We hope to see an *improvement* by early next quarter.

Using Demonstrative Pronouns Followed by Nouns Demonstrative pronouns—*this*, *that*, *these*, and *those*—can help you maintain the coherence of a discussion by linking ideas securely. In almost all cases, demonstrative pronouns should be followed by nouns, rather than stand alone in the sentence. In the following examples, notice that a demonstrative pronoun by itself can be vague and confusing.

UNCLEAR New screening techniques are being developed to combat viral infections. *These* are the subject of a new research effort in California.

What is being studied in California: *new screening techniques* or *viral infections?*

CLEAR — New screening techniques are being developed to combat viral infections. *These techniques* are the subject of a new research effort in California.

UNCLEAR — The task force could not complete its study of the mine accident. *This* was the subject of a scathing editorial in the union newsletter.

What was the subject of the editorial: *the mine accident* or the task force's *inability to complete its study* of the accident?

CLEAR — The task force failed to complete its study of the mine accident. *This failure* was the subject of a scathing editorial in the union newsletter.

Even when the context is clear, a demonstrative pronoun used without a noun might interrupt readers' progress by forcing them to refer back to an earlier idea.

INTERRUPTIVE — The law firm advised that the company initiate proceedings. This caused the company to search for a second legal opinion.

FLUID — The law firm advised that the company initiate proceedings. *This advice* caused the company to search for a second legal opinion.

Writing Grammatically Correct Sentences

Grammar is the study of how words can be combined into sentences to make meaning. Why does being able to write grammatically correct sentences matter? One reason is that many grammar conventions are functional. If you write, "After sitting on a mildewed shelf in the garage for thirty years, my brother decided to throw out the old computer," you've said that your brother sat on a mildewed shelf in the garage for thirty years, which gave him plenty of time to decide what to do with the old computer. If you write, "Did Sean tell Liam when he was expected to report to work?" the reader might have a hard time figuring out whether *he* refers to Sean or Liam.

Even if a grammar mistake doesn't make you sound silly or confuse the reader, it can hurt you by making readers doubt your credibility. The logic is that if you are careless about grammar, you might also be careless about the quality of the technical information you communicate. Many readers will assume that documents that are unprofessional because of grammar problems might also be unprofessional in other ways.

This section will review nine principles for using clear, correct grammar:

- Avoid sentence fragments.
- Avoid comma splices.
- Avoid run-on sentences.
- Avoid ambiguous pronoun references.
- Compare items clearly.
- Use adjectives clearly.
- Maintain subject-verb agreement.
- Maintain pronoun-antecedent agreement.
- Use tenses correctly.

AVOID SENTENCE FRAGMENTS

A sentence fragment is an incomplete sentence. A sentence fragment occurs when a sentence is missing either a verb or an independent clause. To correct a sentence fragment, use one of the following two strategies:

1. **Introduce a verb.**

FRAGMENT The pressure loss caused by a worn gasket.

This example is a fragment because it lacks a verb. (The word *caused* does not function as a verb here; rather, it introduces a phrase that describes the pressure loss.)

COMPLETE The pressure loss was caused by a worn gasket.

Pressure loss now has a verb: *was caused*.

COMPLETE We identified the pressure loss caused by a worn gasket.

Pressure loss becomes the object in a new main clause: *We identified the pressure loss*.

FRAGMENT A plotting program with clipboard plotting, 3D animation, and FFTs.

COMPLETE It is a plotting program with clipboard plotting, 3D animation, and FFTs.

COMPLETE A plotting program with clipboard plotting, 3D animation, and FFTs will be released today.

2. **Link the fragment (a dependent element) to an independent clause.**

FRAGMENT The article was rejected for publication. Because the data could not be verified.

Because the data could not be verified is a fragment because it lacks an independent clause: a clause that has a subject and a verb and could stand alone as a sentence. To be complete, the clause needs more information.

COMPLETE The article was rejected for publication because the data could not be verified.

The dependent element is joined to the independent clause that precedes it.

COMPLETE Because the data could not be verified, the article was rejected for publication.

The dependent element is followed by the independent clause.

FRAGMENT Delivering over 150 horsepower. The two-passenger coupe will cost over $32,000.

COMPLETE Delivering over 150 horsepower, the two-passenger coupe will cost over $32,000.

COMPLETE The two-passenger coupe will deliver over 150 horsepower and cost over $32,000.

AVOID COMMA SPLICES

A comma splice is an error that occurs when two independent clauses are joined, or spliced together, by a comma. Independent clauses in a comma splice can be linked correctly in three ways:

1. **Use a comma and a coordinating conjunction (*and*, *or*, *nor*, *but*, *for*, *so*, or *yet*).**

SPLICE The 909 printer is our most popular model, it offers an unequaled blend of power and versatility.

CORRECT The 909 printer is our most popular model, for it offers an unequaled blend of power and versatility.

The coordinating conjunction *for* explicitly states the relationship between the two clauses.

2. **Use a semicolon.**

SPLICE The 909 printer is our most popular model, it offers an unequaled blend of power and versatility.

CORRECT The 909 printer is our most popular model; it offers an unequaled blend of power and versatility.

The semicolon creates a somewhat more distant relationship between the two clauses than the comma and coordinating conjunction do; the link remains implicit.

3. **Use a period or another form of terminal punctuation.**

SPLICE The 909 printer is our most popular model, it offers an unequaled blend of power and versatility.

CORRECT The 909 printer is our most popular model. It offers an unequaled blend of power and versatility.

The two independent clauses are separate sentences. Of the three ways to punctuate the two clauses correctly, this one suggests the most distant relationship between them.

AVOID RUN-ON SENTENCES

In a run-on sentence (sometimes called a *fused sentence*), two independent clauses appear together with no punctuation between them. A run-on sentence can be corrected in the same three ways as a comma splice:

1. **Use a comma and a coordinating conjunction (*and*, *or*, *nor*, *but*, *for*, *so*, or *yet*).**

RUN-ON The 909 printer is our most popular model it offers an unequaled blend of power and versatility.

CORRECT The 909 printer is our most popular model, for it offers an unequaled blend of power and versatility.

2. **Use a semicolon.**

RUN-ON The 909 printer is our most popular model it offers an unequaled blend of power and versatility.

CORRECT The 909 printer is our most popular model; it offers an unequaled blend of power and versatility.

3. **Use a period or another form of terminal punctuation.**

RUN-ON The 909 printer is our most popular model it offers an unequaled blend of power and versatility.

CORRECT The 909 printer is our most popular model. It offers an unequaled blend of power and versatility.

AVOID AMBIGUOUS PRONOUN REFERENCES

Pronouns must refer clearly to their antecedents—the words or phrases they replace. To correct ambiguous pronoun references, use one of these four strategies:

1. **Clarify the pronoun's antecedent.**

UNCLEAR Remove the cell cluster from the medium and analyze it.

Analyze what: *the cell cluster* or *the medium*?

CLEAR Analyze the cell cluster after removing it from the medium.

CLEAR Analyze the medium after removing the cell cluster from it.

CLEAR Remove the cell cluster from the medium. Then analyze the cell cluster.

CLEAR Remove the cell cluster from the medium. Then analyze the medium.

2. **Clarify the relative pronoun, such as *which*, introducing a dependent clause.**

UNCLEAR She decided to evaluate the program, which would take five months.

What would take five months: *the program* or *the evaluation*?

CLEAR She decided to evaluate the program, a process that would take five months.

By replacing *which* with *a process that*, the writer clearly indicates that it is the evaluation that will take five months.

CLEAR She decided to evaluate the five-month program.

By using the adjective *five-month*, the writer clearly indicates that it is the program that will take five months.

3. **Clarify the subordinating conjunction, such as *where*, introducing a dependent clause.**

UNCLEAR This procedure will increase the handling of toxic materials outside the plant, where adequate safety measures can be taken.

Where can adequate safety measures be taken: inside the plant or outside?

CLEAR This procedure will increase the handling of toxic materials outside the plant. Because adequate safety measures can be taken only in the plant, the procedure poses risks.

CLEAR This procedure will increase the handling of toxic materials outside the plant. Because adequate safety measures can be taken only outside the plant, the procedure will decrease safety risks.

Sometimes the best way to clarify an unclear reference is to split the sentence in two, drop the subordinating conjunction, and add clarifying information.

4. **Clarify the ambiguous pronoun that begins a sentence.**

UNCLEAR Allophanate linkages are among the most important structural components of polyurethane elastomers. They act as cross-linking sites.

What act as cross-linking sites: *allophanate linkages* or *polyurethane elastomers?*

CLEAR Allophanate linkages, which are among the most important structural components of polyurethane elastomers, act as cross-linking sites.

The writer has rewritten part of the first sentence to add a clear nonrestrictive modifier and has combined the rewritten phrase with the second sentence.

If you begin a sentence with a demonstrative pronoun that might be unclear to the reader, be sure to follow it immediately with a noun that clarifies the reference.

UNCLEAR The new parking regulations require that all employees pay for parking permits. These are on the agenda for the next senate meeting.

What are on the agenda: *the regulations* or *the permits?*

CLEAR The new parking regulations require that all employees pay for parking permits. These regulations are on the agenda for the next senate meeting.

COMPARE ITEMS CLEARLY

When comparing or contrasting items, make sure your sentence communicates their relationship clearly. A simple comparison between two items usually causes no problems: "The X3000 has more storage than the X2500." Simple comparisons, however, can sometimes result in ambiguous statements:

AMBIGUOUS Trout eat more than minnows.

Do trout eat minnows in addition to other food, or do trout eat more than minnows eat?

CLEAR Trout eat more than minnows do.

If you are introducing three items, make sure the reader can tell which two are being compared:

AMBIGUOUS Trout eat more algae than minnows.

CLEAR Trout eat more algae than they do minnows.

CLEAR Trout eat more algae than minnows do.

Beware of comparisons in which different aspects of the two items are compared:

ILLOGICAL The resistance of the copper wiring is lower than the tin wiring.

LOGICAL The resistance of the copper wiring is lower than that of the tin wiring.

Resistance cannot be logically compared with *tin wiring*. In the revision, the pronoun *that* substitutes for *resistance* in the second part of the comparison.

USE ADJECTIVES CLEARLY

In general, adjectives are placed before the nouns that they modify: *the plastic washer*. In technical communication, however, writers often need to use clusters of adjectives. To prevent confusion in technical communication, follow two guidelines:

1. **Use commas to separate coordinate adjectives.** Adjectives that describe different aspects of the same noun are known as coordinate adjectives.

   portable, programmable device

   adjustable, removable housings

   The comma is used instead of the word *and*.

   Sometimes an adjective is considered part of the noun it describes: *electric drill*. When one adjective modifies *electric drill*, no comma is required: *a reversible electric drill*. The addition of two or more adjectives, however, creates the traditional coordinate construction: *a two-speed, reversible electric drill*.

2. **Use hyphens to link compound adjectives.** A compound adjective is made up of two or more words. Use hyphens to link these elements when compound adjectives precede nouns.

   a variable-angle accessory

   increased cost-of-living raises

   The hyphens in the second example prevent *increased* from being read as an adjective modifying *cost*.

   A long string of compound adjectives can be confusing even if you use hyphens appropriately. To ensure clarity, turn the adjectives into a clause or a phrase following the noun.

   UNCLEAR an *operator-initiated default-prevention* technique

   CLEAR a technique *initiated by the operator to prevent default*

MAINTAIN SUBJECT-VERB AGREEMENT

The subject and verb of a sentence must agree in number, even when a prepositional phrase comes between them. The object of the preposition might be plural in a singular sentence.

INCORRECT The *result* of the tests *are* promising.

CORRECT The *result* of the tests *is* promising.

The object of the preposition might be singular in a plural sentence.

INCORRECT The *results* of the test *is* promising.

CORRECT The *results* of the test *are* promising.

Don't be misled by the fact that the object of the preposition and the verb don't sound natural together, as in *tests is* or *test are*. Here, the noun *test(s)* precedes the verb, but it is not the subject of the verb. As long as the subject and verb agree, the sentence is correct.

MAINTAIN PRONOUN-ANTECEDENT AGREEMENT

A pronoun and its antecedent (the word or phrase being replaced by the pronoun) must agree in number. Often an error occurs when the antecedent is a collective noun—one that can be interpreted as either singular or plural, depending on its usage.

INCORRECT The *company* is proud to announce a new stock option plan for *their* employees.

CORRECT The *company* is proud to announce a new stock option plan for *its* employees.

Company acts as a single unit; therefore, the singular pronoun is appropriate.

When the individual members of a collective noun are emphasized, however, a plural pronoun is appropriate.

CORRECT The inspection team have prepared their reports.

CORRECT The members of the inspection team have prepared their reports.

The use of *their* emphasizes that the team members have prepared their own reports.

USE TENSES CORRECTLY

Two verb tenses are commonly used in technical communication: the present tense and the past perfect tense. It is important to understand the specific purpose of each.

1. **The present tense is used to describe scientific principles and recurring events.**

   INCORRECT In 1992, McKay and his coauthors argued that the atmosphere of Mars *was* salmon pink.

   CORRECT In 1992, McKay and his coauthors argued that the atmosphere of Mars *is* salmon pink.

   Although the argument was made in the historical past—1992—the point is expressed in the present tense because the atmosphere of Mars continues to be salmon pink.

When the date of the argument is omitted, some writers express the entire sentence in the present tense.

CORRECT McKay and his coauthors *argue* that the atmosphere of Mars *is* salmon pink.

2. **The past perfect tense is used to describe the earlier of two events that occurred in the past.**

CORRECT We *had begun* excavation when the foreman *discovered* the burial remains.

Had begun is the past perfect tense. The excavation began before the burial remains were discovered.

CORRECT The seminar *had concluded* before I *got* a chance to talk with Dr. Tran.

Structuring Effective Sentences

Good technical communication consists of clear, graceful sentences that convey information economically. This section describes seven principles for structuring effective sentences:

- Use lists.
- Emphasize new and important information.
- Choose an appropriate sentence length.
- Focus on the "real" subject.
- Focus on the "real" verb.
- Use parallel structure.
- Use modifiers effectively.

USE LISTS

A sentence list is a list in which the bulleted or numbered items are words, phrases, or single sentences. Figure 6.2 on page 126 shows a traditional sentence and a list presenting the same information.

If you don't have enough space to list the items vertically or if you are not permitted to do so, number the items within the sentence:

We recommend that more work on heat-exchanger performance be done (1) with a larger variety of different fuels at the same temperature, (2) with similar fuels at different temperatures, and (3) with special fuels such as diesel fuel and shale-oil-derived fuels.

GUIDELINES Creating Effective Lists

These five suggestions will help you write clearer, more effective paragraph lists and sentence lists.

- **Set off each listed item with a number, a letter, or a symbol (usually a bullet).**
  - — Use numbered lists to suggest sequence (as in the steps in a set of instructions) or priority (the first item being the most important). Using numbers

(continued)

help readers see the total number of items in a list. For sublists, use lowercase letters:

1. Item
   a. subitem
   b. subitem
2. Item
   a. subitem
   b. subitem

— Use bullets to avoid suggesting either sequence or priority, such as for lists of people (otherwise, everyone except number 1 will be offended). For sublists, use dashes.

- Item
  — subitem
  — subitem

— Use an open (unshaded) box (☐) for checklists.

For more about designing checklists, see Ch. 8, p. 220.

- **Break up long lists.** Because most people can remember only 5 to 9 items easily, break up lists of 10 or more items.

ORIGINAL LIST	REVISED LIST
Tool kit:	Tool kit:
• handsaw	• Saws
• coping saw	– handsaw
• hacksaw	– coping saw
• compass saw	– hacksaw
• adjustable wrench	– compass saw
• box wrench	• Wrenches
• Stillson wrench	– adjustable wrench
• socket wrench	– box wrench
• open-end wrench	– Stillson wrench
• Allen wrench	– socket wrench
	– open-end wrench
	– Allen wrench

- **Present the items in a parallel structure.** A list is parallel if all the items have the same grammatical form. For instance, in the parallel list below, each item is a verb phrase.

NONPARALLEL	PARALLEL
Here is the sequence we plan to follow:	Here is the sequence we plan to follow:
1. writing of the preliminary proposal	1. write the preliminary proposal
2. do library research	2. do library research
3. interview with the Bemco vice president	3. interview the Bemco vice president
4. first draft	4. write the first draft
5. revision of the first draft	5. revise the first draft
6. preparing the final draft	6. prepare the final draft

- **Structure and punctuate the lead-in correctly.** The lead-in tells readers how the list relates to the discussion and how the items in the list relate to each other. Although standards vary from one organization to another, the most common

(continued)

lead-in consists of a grammatically complete clause followed by a colon, as shown in the following examples:

> Following are the three main assets:
>
> The three main assets are as follows:
>
> The three main assets are the following:

If you cannot use a grammatically complete lead-in, use a dash or no punctuation at all:

> The committee found that the employee
> - did not cause the accident
> - acted properly immediately after the accident
> - reported the accident according to procedures

- **Punctuate the list correctly.** Because rules for punctuating lists vary, you should find out whether people in your organization have a preference. If not, punctuate lists as follows:

— If the items are phrases, use a lowercase letter at the start. Do not use a period or a comma at the end. The space beneath the last item indicates the end of the list.

The new facility will offer three advantages:
- lower leasing costs
- shorter commuting distance
- a larger pool of potential workers

— If the items are complete sentences, use an uppercase letter at the start and a period at the end.

The new facility will offer three advantages:
- The leasing costs will be lower.
- The commuting distance for most employees will be shorter.
- The pool of potential workers will be larger.

— If the items are phrases followed by complete sentences, start each phrase with an uppercase letter and end it with a period. Begin the complete sentences with uppercase letters and end them with periods. Use italics to emphasize the phrases.

The new facility will offer three advantages:
- *Lower leasing costs.* The lease will cost $1,800 per month; currently we pay $2,300.
- *Shorter commuting distance.* Our workers' average commute of 18 minutes would drop to 14 minutes.
- *Larger pool of potential workers.* In the last decade, the population has shifted westward to the area near the new facility. As a result, we would increase our potential workforce in both the semiskilled and the managerial categories by relocating.

— If the list consists of two kinds of items—phrases and complete sentences—capitalize each item and end it with a period.

(*continued*)

The new facility will offer three advantages:

- Lower leasing costs.
- Shorter commuting distance. Our workers' average commute of 18 minutes would drop to 14 minutes.
- Larger pool of potential workers. In the last decade, the population has shifted westward to the area near the new facility. As a result, we would increase our potential workforce in both the semiskilled and the managerial categories by relocating.

In most lists, the second and subsequent lines, called *turnovers*, align under the first letter of the first line, highlighting the bullet or number to the left of the text. This *hanging indentation* helps the reader see and understand the organization of the passage.

Traditional sentence	Sentence list
We recommend that more work on heat-exchanger performance be done with a larger variety of different fuels at the same temperature, with similar fuels at different temperatures, and with special fuels such as diesel fuel and shale-oil-derived fuels.	We recommend that more work on heat-exchanger performance be done • with a larger variety of different fuels at the same temperature • with similar fuels at different temperatures • with special fuels such as diesel fuel and shale-oil-derived fuels

FIGURE 6.2 A Traditional Sentence and a Sentence List

EMPHASIZE NEW AND IMPORTANT INFORMATION

Sentences are often easier to understand and more emphatic if new information appears at the end. For instance, if your company has labor problems and you want to describe the possible results, structure the sentence like this:

> Because of labor problems, we anticipate a three-week delay.
>
> In this case, *three-week delay* is the new information.

If your readers already expect a three-week delay but don't know the reason for it, reverse the structure:

> We anticipate the three-week delay in production because of labor problems.
>
> Here, the new and important information is *labor problems.*

Try not to end the sentence with qualifying information that blunts the impact of the new information.

WEAK	The joint could fail under special circumstances.
IMPROVED	Under special circumstances, the joint could fail.

USE MODIFIERS EFFECTIVELY

Modifiers are words, phrases, and clauses that describe other elements in the sentence. To make your meaning clear, you must indicate whether a modifier provides necessary information about the word or phrase it refers to (its *antecedent*) or whether it simply provides additional information. You must also clearly identify the antecedent—the element in the sentence that the modifier is describing or otherwise referring to.

Distinguish Between Restrictive and Nonrestrictive Modifiers As the term implies, a *restrictive modifier* restricts the meaning of its antecedent; it provides information that the reader needs to identify the antecedent and is, therefore, crucial to understanding the sentence. Notice that restrictive modifiers—italicized in the following examples—are not set off by commas:

> The airplanes *used in the exhibitions* are slightly modified.
>
> The modifying phrase *used in the exhibitions* identifies which airplanes the writer is referring to. Presumably, there are at least two groups of airplanes: those that are used in the exhibitions and those that are not. The restrictive modifier tells readers which of the two groups is being discussed.
>
> Please disregard the notice *you recently received from us.*
>
> The modifying phrase *you recently received from us* identifies which notice. Without it, the sentence could be referring to one of any number of notices.

In most cases, the restrictive modifier doesn't require a relative pronoun, such as *that*, but you can choose to use the pronoun *that* (or *who*, for people):

> Please disregard the notice *that* you recently received from us.

A *nonrestrictive modifier* does not restrict the meaning of its antecedent: the reader does not need the information to identify what the modifier is describing or referring to. If you omit the nonrestrictive modifier, the basic sentence retains its primary meaning.

> The Hubble telescope, *intended to answer fundamental questions about the origin of the universe*, was last repaired in 2002.
>
> Here, the basic sentence is *The Hubble telescope was last repaired in 2002*. Removing the modifier doesn't change the meaning of the basic sentence.

If you use a relative pronoun with a nonrestrictive modifier, choose *which* (or *who* or *whom* for a person).

> Go to the Registration Area, *which is located on the second floor.*

Use commas to separate a nonrestrictive modifier from the rest of the sentence. In the example about the Hubble telescope, a pair of commas encloses the nonrestrictive modifier and separates it from the rest of the sentence. In that respect, the commas function much like parentheses, indicating that the modifying information is parenthetical. In the example about the Registration Area, the comma indicates that the modifying information is tacked on at the end of the sentence as additional information.

Avoid Misplaced Modifiers The placement of the modifier often determines the meaning of the sentence, as the placement of *only* in the following sentences illustrates:

Only Turner received a cost-of-living increase last year.
Meaning: Nobody else received one.

Turner received *only* a cost-of-living increase last year.
Meaning: He didn't receive a merit increase.

Turner received a cost-of-living increase *only* last year.
Meaning: He received a cost-of-living increase as recently as last year.

Turner received a cost-of-living increase last year *only*.
Meaning: He received a cost-of-living increase in no other year.

Misplaced modifiers—those that appear to modify the wrong antecedent—are a common problem. Usually, the best solution is to place the modifier as close as possible to its intended antecedent.

MISPLACED	The subject of the meeting is the future of geothermal energy *in the downtown Webster Hotel*.
CORRECT	The subject of the meeting *in the downtown Webster Hotel* is the future of geothermal energy.

A *squinting modifier* falls ambiguously between two possible antecedents, so the reader cannot tell which one is being modified:

UNCLEAR	We decided *immediately* to purchase the new system. Did we decide immediately, or did we decide to make the purchase immediately?
CLEAR	We *immediately* decided to purchase the new system.
CLEAR	We decided to purchase the new system *immediately*.

A subtle form of misplaced modification can also occur with *correlative constructions*, such as *either . . . or*, *neither . . . nor*, and *not only . . . but also*:

MISPLACED	The new refrigerant *not only decreases* energy costs *but also* spoilage losses.

Here, the writer is implying that the refrigerant does at least two things to energy costs: it decreases them and then does something else to them.

Unfortunately, that's not how the sentence unfolds. The second thing the refrigerant does to energy costs never appears.

CORRECT The new refrigerant *decreases not only* energy costs *but also* spoilage losses.

In the revised sentence, the phrase *decreases not only* implies that at least two things will be decreased, and as the sentence develops that turns out to be the case. *Decreases* applies to both *energy costs* and *spoilage losses*. Therefore, the first half of the correlative construction (*not only*) follows the verb (*decreases*). Note that if the sentence contains two different verbs, each half of the correlative construction precedes a verb:

The new refrigerant *not only decreases* energy costs *but also reduces* spoilage losses.

Avoid Dangling Modifiers A *dangling modifier* has no antecedent in the sentence and can therefore be unclear:

DANGLING Trying to solve the problem, the instructions seemed unclear.

This sentence says that the instructions are trying to solve the problem. To correct the sentence, rewrite it, adding the clarifying information either within the modifier or next to it:

CORRECT As I was trying to solve the problem, the instructions seemed unclear.

CORRECT Trying to solve the problem, I thought the instructions seemed unclear.

Sometimes you can correct a dangling modifier by switching from the *indicative mood* (a statement of fact) to the *imperative mood* (a request or command):

DANGLING To initiate the procedure, the BEGIN button should be pushed. (indicative mood)

CORRECT To initiate the procedure, push the BEGIN button. (imperative mood)

Choosing the Right Words and Phrases

This section discusses four principles that will help you use the right words and phrases in the right places: select an appropriate level of formality, be clear and specific, be concise, and use inoffensive language.

SELECT AN APPROPRIATE LEVEL OF FORMALITY

Although no standard definition of levels of formality exists, most experts would agree that there are three levels:

INFORMAL The Acorn 560 is a real screamer. With 5.5 GHz of pure computing power, it slashes through even the thickest spreadsheets before you can say 2 + 2 = 4.

MODERATELY FORMAL With its 5.5-GHz microprocessor, the Acorn 560 can handle even the most complicated spreadsheets quickly.

HIGHLY FORMAL With a 5.5-GHz microprocessor, the Acorn 560 is a high-speed personal computer appropriate for computation-intensive applications such as large, complex spreadsheets.

Technical communication usually requires a moderately formal or highly formal style.

To achieve the appropriate level and tone, think about your audience, your subject, and your purpose:

- **Audience.** You would probably write more formally to a group of retired executives than to a group of college students. You would likewise write more formally to the company vice president than to your co-workers, and you would probably write more formally to people from most other cultures than to people from your own.

For more about writing to a multicultural audience, see Ch. 4, p. 58.

- **Subject.** You would write more formally about a serious subject—safety regulations or important projects—than about plans for an office party.
- **Purpose.** You would write more formally in presenting quarterly economic results to shareholders than in responding to an email requesting sales figures on one of the company's products.

In general, it is better to err on the side of formality. Avoid an informal style in any writing you do at the office, for two reasons:

- **Informal writing tends to be imprecise.** In the example "The Acorn 560 is a real screamer," what exactly is a *screamer*?
- **Informal writing can be embarrassing.** If your boss spots your email to a colleague, you might wish it didn't begin, "'Sup, dawg?"

BE CLEAR AND SPECIFIC

Follow these seven guidelines to make your writing clear and specific:

- Use active and passive voice appropriately.
- Be specific.
- Avoid unnecessary jargon.
- Use positive constructions.
- Avoid long noun strings.
- Avoid clichés.
- Avoid euphemisms.

Use the Active and Passive Voices Appropriately In a sentence using the active voice, the subject performs the action expressed by the verb: the "doer" of the action is the grammatical subject. By contrast, in a sentence using the passive voice, the recipient of the action is the grammatical subject. Compare the following examples (the subjects are italicized):

ACTIVE *Dave Brushaw* drove the launch vehicle.

The doer of the action is the subject of the sentence.

PASSIVE The launch *vehicle* was driven by Dave Brushaw.

The recipient of the action is the subject of the sentence.

In most cases, the active voice works better than the passive voice because it emphasizes the *agent* (the doer of the action). An active-voice sentence also is shorter because it does not require a form of the verb *to be* and the past participle, as a passive-voice sentence does. In the active version of the example sentence, the verb is *drove* rather than *was driven*, and the word *by* does not appear.

The passive voice, however, is generally better in these four cases:

- When the agent is clear from the context:

  Students are required to take both writing courses.

  Here, the context makes it clear that the college sets the requirements.

- When the agent is unknown:

  The comet was first referred to in an ancient Egyptian text.

  We don't know who wrote this text.

- When the agent is less important than the action:

  The blueprints were hand-delivered this morning.

  It doesn't matter who the messenger was.

- When a reference to the agent is embarrassing, dangerous, or in some other way inappropriate:

  Incorrect figures were recorded for the flow rate.

  It might be unwise or tactless to specify who recorded the incorrect figures. Perhaps it was your boss. However, it is unethical to use the passive voice to avoid responsibility for an action.

For more about ethics, see Ch. 2.

The passive voice can also help you maintain the focus of your paragraph.

Cloud computing offers three major advantages. First, the need for server space is reduced. Second, security updates are installed automatically. . . .

Some people believe that the active voice is inappropriate in technical communication because it emphasizes the person who does the work rather than the work itself, making the writing less objective. In many cases, this objection is valid. Why write "I analyzed the sample for traces of iodine" if there is no ambiguity about who did the analysis or no need to identify who did it? The passive focuses on the action, not the actor: "The samples were analyzed for traces of iodine." But if in doubt, use the active voice.

Other people argue that the passive voice produces a double ambiguity. In the sentence "The samples were analyzed for traces of iodine," the reader is not quite sure who did the analysis (the writer or someone else) or when it was done (during the project or some time previously). Identifying the actor can often clarify both ambiguities.

The best approach is to recognize that the two voices differ and to use each one where it is most effective.

Many grammar-checkers can help you locate the passive voice. Some of them will advise you that the passive is undesirable, almost an error, but this advice is misleading. Use the passive voice when it works better than the active voice for your purposes.

Any word processor allows you to search for the forms of *to be* used most commonly in passive-voice expressions: is, *are*, *was*, and *were*. You can also search for *ed* to isolate past participles (for example, *purchased*, *implemented*, and *delivered*); such past participles appear in most passive-voice constructions.

Be Specific Being specific involves using precise words, providing adequate detail, and avoiding ambiguity.

- **Use precise words.** A Ford Focus is an automobile, but it is also a vehicle, a machine, and a thing. In describing the Focus, *automobile* is better than the less-specific *vehicle*, because *vehicle* can also refer to pickup trucks, trains, hot-air balloons, and other means of transport. As words become more abstract—from *machine* to *thing*, for instance—chances for misunderstanding increase.
- **Provide adequate detail.** Readers probably know less about your subject than you do. What might be perfectly clear to you might be too vague for them.

  VAGUE An engine on the plane experienced some difficulties.

  Which *engine*? What *plane*? What kinds of *difficulties*?

  CLEAR The left engine on the Cessna 310 temporarily lost power during flight.
- **Avoid ambiguity.** Don't let readers wonder which of two meanings you are trying to convey.

  AMBIGUOUS After stirring by hand for 10 seconds, add three drops of the iodine mixture to the solution.

  After stirring *the iodine mixture* or *the solution*?

  CLEAR Stir the iodine mixture by hand for 10 seconds. Then add three drops to the solution.

  CLEAR Stir the solution by hand for 10 seconds. Then add three drops of the iodine mixture.

If you don't have the specific data, you should approximate—and clearly tell readers you are doing so—or explain why the specific data are unavailable and indicate when they will be available:

The fuel leakage is much greater than we had anticipated; we estimate it to be at least 5 gallons per minute, not 2.

The fuel leakage is much greater than we had anticipated; we expect to have specific data by 4 P.M. today.

Avoid Unnecessary Jargon Jargon is shoptalk. To an audiophile, *LP* is a long-playing record; to an engineer, it is liquid propane; to a guitarist, it is a Gibson Les Paul model; to a physician, it is a lumbar puncture; to a drummer, it is Latin Percussion, a drum maker.

Jargon is often ridiculed; many dictionaries define it as "writing that one does not understand" or "nonsensical, incoherent, or meaningless talk." However, jargon is useful in its proper sphere. For one thing, jargon enables members of a particular profession to communicate clearly and economically with one another.

If you are addressing a technically knowledgeable audience, use jargon recognized in that field. However, keep in mind that technical documents often have many audiences in addition to the primary audience. When in doubt, avoid jargon; use more-common expressions or simpler terms.

Using jargon inappropriately is inadvisable for four reasons:

- **It can be imprecise.** If you ask a co-worker to review a document and provide *feedback*, are you asking for a facial expression, body language, a phone call, or a written evaluation?
- **It can be confusing.** If you ask a computer novice to *cold swap the drive*, he or she might have no idea what you're talking about.
- **It is often seen as condescending.** Many readers will react as if you were showing off—displaying a level of expertise that excludes them. If readers feel alienated, they will likely miss your message.
- **It is often intimidating.** People might feel inadequate or stupid because they do not know what you are talking about. Obviously, this reaction undermines communication.

Use Positive Constructions The term *positive construction* has nothing to do with being cheerful. It indicates that the writer is describing what something is instead of what it is not. In the sentence "I was sad to see this project completed," "sad" is a positive construction. The negative construction would be "not happy."

Here are a few more examples of positive and negative constructions:

Positive construction	Negative construction
most	not all
few	not many
on time	not late, not delayed
positive	not negative
inefficient	not efficient
reject	cannot accept
impossible	not possible

Readers understand positive constructions more quickly and more easily than negative constructions. Consider the following examples:

DIFFICULT	Because the team did not have sufficient time to complete the project, it was unable to produce a satisfactory report.
SIMPLER	Because the team had too little time to complete the project, it produced an unsatisfactory report.

Avoid Long Noun Strings A noun string contains a series of nouns (or nouns, adjectives, and adverbs), all of which together modify the last noun. For example, in the phrase *parking-garage regulations*, the first pair of words modifies *regulations*. Noun strings save time, and if your readers understand them, they are fine. It is easier to write *passive-restraint system* than *a system that uses passive restraints*.

For more about hyphens, see Appendix, Part B, p. 494.

Hyphens can clarify noun strings by linking words that go together. For example, in the phrase *flat-panel monitor*, the hyphen links *flat* and *panel*. Together they modify *monitor*. In other words, it is not a *flat panel* or a *panel monitor*, but a *flat-panel monitor*. However, noun strings are sometimes so long or so complex that hyphens can't ensure clarity. To clarify a long noun string, untangle the phrases and restore prepositions, as in the following example:

UNCLEAR	preregistration procedures instruction sheet update
CLEAR	an update of the instruction sheet for preregistration procedures

Noun strings can sometimes be ambiguous—they can have two or more plausible meanings, leaving readers to guess at which meaning you're trying to convey.

AMBIGUOUS	The building contains a special incoming materials storage area.
	What's special? Are the incoming materials special? Or is the area they're stored in special?
UNAMBIGUOUS	The building contains a special area for storing incoming materials.
UNAMBIGUOUS	The building contains an area for storing special incoming materials.

An additional danger is that noun strings can sometimes sound pompous. If you are writing about a simple smoke detector, there is no reason to call it a *smoke-detection device* or, worse, a *smoke-detection system*.

Avoid Clichés Good writing is original and fresh. Rather than use a cliché, say what you want to say in plain English. Current clichés include *pushing the envelope*; *game-changer*; *synergy*; *mission critical*; *bleeding edge*; *paradigm shift*; and *content is king*. The best advice is to avoid clichés: if you are used to hearing or reading a phrase, don't use it. Jimi Hendrix was a rock star; the Employee of the Month sitting in the next cubicle isn't. Don't think outside the box, pick low-hanging fruit, leverage your assets, bring your "A" game, be a change agent, raise the bar, throw anyone under a bus, be proactive, put lipstick on a pig, or give 110 percent. And you can assume that everyone already knows that it is what it is.

Avoid Euphemisms A euphemism is a polite way of saying something that makes people uncomfortable. For instance, a near miss between two airplanes is officially an "air proximity incident." The more uncomfortable the subject, the more often people resort to euphemisms. Dozens of euphemisms deal with drinking, bathrooms, sex, and death. Here are several euphemisms for firing someone:

personnel-surplus reduction
workforce-imbalance correction
rightsizing
indefinite idling
downsizing
administrative streamlining
synergy-related headcount restructuring
dehiring
decruiting
redundancy elimination
career-change-opportunity creation
permanent furloughing
personnel realignment

ETHICS NOTE

EUPHEMISMS AND TRUTH TELLING

There is nothing wrong with using the euphemism *restroom*, even though few people visit one to rest. The British use the phrase *go to the toilet* in polite company, and nobody seems to mind. In this case, if you want to use a euphemism, no harm done.

But it is unethical to use a euphemism to gloss over an issue that has important implications for people or the environment. People get uncomfortable when discussing layoffs—and they should. It's an uncomfortable issue. But calling a layoff a *redundancy elimination initiative* ought to make you even more uncomfortable. Don't use language to cloud reality. It's an ethical issue.

BE CONCISE

The following five principles can help you write concise technical documents:

- Avoid obvious statements.
- Avoid filler.
- Avoid unnecessary prepositional phrases.
- Avoid wordy phrases.
- Avoid fancy words.

Avoid Obvious Statements Writing can become sluggish if it overexplains. The italicized words in the following example are sluggish:

SLUGGISH The market for *the sale of* flash memory chips is dominated by *two chip manufacturers*: Intel and Advanced Micro Devices. These two *chip manufacturers* are responsible for 76 percent of the $1.3 billion market *in flash memory chips* last year.

IMPROVED The market for flash memory chips is dominated by Intel and Advanced Micro Systems, two companies that claimed 76 percent of the $1.3 billion industry last year.

Avoid Filler In our writing, we sometimes use filler, much of which is more suited to speech. Consider the following examples:

basically	kind of
certain	rather
essentially	sort of

Such words are common in oral communication, when we need to think fast, but they are meaningless in writing.

BLOATED *I think that, basically,* the board felt *sort of* betrayed, *in a sense,* by the *kind of* behavior the president displayed.

BETTER The board felt betrayed by the president's behavior.

But modifiers are not always meaningless. For instance, it might be wise to use *I think* or *it seems to me* to show that you are aware of other views.

BLUNT Next year we will face unprecedented challenges to our market dominance.

LESS BLUNT In my view, next year we will face unprecedented challenges to our market dominance.

Of course, a sentence that sounds blunt to one reader can sound self-confident to another. As you write, keep your audience's preferences and expectations in mind.

Other fillers include redundant expressions, such as *collaborate together, past history, end result, any and all, still remain, completely eliminate,* and *very unique*. Say it once.

REDUNDANT This project would not have succeeded if not for the *hard work and considerable effort* of *each and every one* of the auditors assigned to the project.

BETTER This project would not have succeeded if not for the *hard work* of *every one* of the auditors assigned to the project.

Avoid Unnecessary Prepositional Phrases A prepositional phrase consists of a preposition followed by a noun or a noun equivalent, such as *in the summary, on the engine,* and *under the heading.* Unnecessary prepositional phrases, often used along with abstract nouns and nominalizations, can make your writing long and boring.

LONG The increase *in* the number *of* students enrolled *in* the materials-engineering program *at* Lehigh University is suggestive *of* the regard *in* which that program is held *by* the university's new students.

SHORTER The increased enrollment in Lehigh University's materials-engineering program suggests that the university's new students consider it a good program.

Avoid Wordy Phrases Wordy phrases also make writing long and boring. For example, some people write *on a daily basis* rather than *daily*. The long phrase may sound more important, but *daily* says the same thing more concisely.

Table 6.2 lists common wordy phrases and their more concise equivalents.

TABLE 6.2 Wordy Phrases and Their Concise Equivalents

WORDY PHRASE	CONCISE PHRASE	WORDY PHRASE	CONCISE PHRASE
a majority of	most	in the event that	if
a number of	some, many	in view of the fact that	because
at an early date	soon	it is often the case that	often
at the conclusion of	after, following	it is our opinion that	we think that
at the present time	now	it is our recommendation that	we recommend that
at this point in time	now	it is our understanding that	we understand that
based on the fact that	because	make reference to	refer to
check out	check	of the opinion that	think that
despite the fact that	although	on a daily basis	daily
due to the fact that	because	on the grounds that	because
during the course of	during	prior to	before
during the time that	during, while	relative to	regarding, about
have the capability to	can	so as to	to
in connection with	about, concerning	subsequent to	after
in order to	to	take into consideration	consider
in regard to	regarding, about	until such time as	until

Compare the following wordy sentence and its concise translation:

WORDY I am of the opinion that, in regard to profit achievement, the statistics pertaining to this month will appear to indicate an upward tendency.

CONCISE I think this month's statistics will show an increase in profits.

Avoid Fancy Words Writers sometimes think they will impress their readers by using fancy words—*utilize* for *use*, *initiate* for *begin*, *perform* for *do*, *due to* for *because*, and *prioritize* for *rank*. In technical communication, plain talk is best. Compare the following fancy sentence with its plain-English version:

FANCY The purchase of a database program will enhance our record-maintenance capabilities.

PLAIN Buying a database program will help us maintain our records.

Table 6.3 lists commonly used fancy words and their plain equivalents.

TABLE 6.3 Fancy Words and Their Plain Equivalents

FANCY WORD	PLAIN WORD	FANCY WORD	PLAIN WORD
advise	tell	herein	here
ascertain	learn, find out	impact (verb)	affect
attempt (verb)	try	initiate	begin
commence	start, begin	manifest (verb)	show
demonstrate	show	parameters	variables, conditions
due to	because of	perform	do
employ (verb)	use	prioritize	rank
endeavor (verb)	try	procure	get, buy
eventuate	happen	quantify	measure
evidence (verb)	show	terminate	end, stop
finalize	end, settle, agree, finish	utilize	use
furnish	provide, give		

USE INOFFENSIVE LANGUAGE

Writing to avoid offense is not merely a matter of politeness; it is a matter of perception. Language reflects attitudes, but it also helps form attitudes. Writing inoffensively is one way to break down stereotypes.

For books about nonsexist writing, see the Selected Bibliography, located in LaunchPad. Many of the books in the "Usage and General Writing," "Handbooks for Grammar and Style," and "Style Manuals" sections address nonsexist writing.

Nonsexist Language You can use your word processor to search for *he*, *man*, and *men*, the words and parts of words most often associated with sexist writing. Some grammar-checkers identify common sexist terms and suggest alternatives.

GUIDELINES Avoiding Sexist Language

Follow these six suggestions for writing gender-neutral text.

- **Replace male-gender words with non-gender-specific words.** *Chairman,* for instance, can become *chairperson* or *chair. Firemen* are *firefighters; policemen* are *police officers.*

- **Switch to a different form of the verb.**

  SEXIST — The operator must pass rigorous tests before he is promoted.

  NONSEXIST — The operator must pass rigorous tests before being promoted.

- **Switch to the plural.**

  NONSEXIST — Operators must pass rigorous tests before they are promoted.

  Some organizations accept the use of plural pronouns with singular nouns, particularly in memos and other informal documents:

  > If an employee wishes to apply for tuition reimbursement, they should consult Section 14.5 of the Employee Manual.

  Careful writers and editors, however, resist this construction because it is grammatically incorrect (it switches from singular to plural). In addition, switching to the plural can make a sentence unclear:

  UNCLEAR — Operators are responsible for their operating manuals.

  Does each operator have one operating manual or more than one?

  CLEAR — Each operator is responsible for his or her operating manual.

- **Switch to *he or she, he/she, s/he,* or *his or her.*** *He or she, his or her,* and related constructions are awkward, especially if overused, but at least they are clear and inoffensive.

- **Address the reader directly.** Use *you* and *your* or the understood *you.*

  > [You] Enter the serial number in the first text box.

- **Alternate *he* and *she.*** Language scholar Joseph Williams (2007) and many other language authorities recommend alternating *he* and *she* from one paragraph or section to the next.

People-First Language for Referring to People with Disabilities

Around one in six Americans—some 56 million people—has a physical, sensory, emotional, or mental impairment that interferes with daily life (U.S. Census Bureau, 2012). In writing about people with disabilities, use the "people-first" approach: treat the person as someone with a disability, not as someone defined by that disability.

DOCUMENT ANALYSIS ACTIVITY

Revising for Conciseness and Simplicity

The following passage is from a request for proposals published by the National Science Foundation. (Sentence numbers have been added here.) The questions in the margin ask you to think about word choice (as discussed on pp. 139–43).

1. This passage contains many prepositional phrases. Identify two of them. For each one, determine whether its use is justified or whether the sentence would be easier to understand if it were eliminated.
2. Part of this passage is written in the passive voice. Select one sentence in the passive voice that would be clearer in the active voice, and rewrite it in the active voice.
3. This passage contains a number of fancy words. Identify two of them. How can they be translated into plain English?

(1) Grants.gov, part of the President's Management Agenda to improve government services to the public, provides a single Government-wide portal for finding and applying for Federal grants online.

(2) Proposals submitted via Grants.gov must be prepared and submitted in accordance with the *NSF Grants.gov Application Guide*, available through Grants.gov as well as on the NSF website at: www.nsf.gov/bfa/dias/policy/docs/grantsgovguide.pdf.

(3) The Grants.gov Application Guide contains important information on:

- general instructions for submission via Grants.gov, including the Grants.gov registration process and Grants.gov software requirements;
- NSF-specific instructions for submission via Grants.gov, including creation of PDF files;
- grant application package instructions;
- required SF 424 (R&R) forms and instructions; and
- NSF-specific forms and instructions.

(4) Upon successful insertion of the Grants.gov submitted proposal in the NSF FastLane system, no further interaction with Grants.gov is required.

(5) All further interaction is conducted via the NSF FastLane system.

Information from National Science Foundation, 2008: www.nsf.gov/pubs/policydocs/pappguide/nsf08_1/gpg_1.jsp#IA1.

GUIDELINES Using the People-First Approach

When writing about people with disabilities, follow these five guidelines, which are based on Snow (2009).

- **Refer to the person first, the disability second.** Write *people with cognitive impairments*, not *the cognitively impaired*.
- **Don't confuse *handicap* with *disability*.** *Disability* refers to the impairment or condition; *handicap* refers to the interaction between the person and his or her environment. A person can have a disability without being handicapped.
- **Don't refer to victimization.** Write *a person with AIDS*, not *an AIDS victim* or *an AIDS sufferer*.
- **Don't refer to a person as *wheelchair bound* or *confined to a wheelchair*.** People who use wheelchairs to get around are not confined.
- **Don't refer to people with disabilities as abnormal.** They are atypical, not abnormal.

WRITER'S CHECKLIST

- ☐ Did you consider using a conventional pattern of organization? (*p.* 103)

Titles and Headings

Did you revise the title of your document so that it

- ☐ clearly states the subject and purpose of your document? (*p.* 105)
- ☐ is precise and informative? (*p.* 105)

Did you revise the headings to

- ☐ avoid long noun strings? (*p.* 107)
- ☐ be informative? (*p.* 108)
- ☐ use a grammatical form appropriate to your audience? (*p.* 108)
- ☐ Did you avoid back-to-back headings by including an advance organizer? (*p.* 108)

Paragraphs

Did you revise your paragraphs so that each one

- ☐ begins with a clear topic sentence? (*p.* 109)
- ☐ has adequate and appropriate support? (*p.* 110)
- ☐ is not too long for readers? (*p.* 111)
- ☐ uses coherence devices such as transitional words and phrases, repetition of key words, and demonstrative pronouns followed by nouns? (*p.* 113)

Lists

- ☐ Is each list of the appropriate kind: numbered, lettered, bulleted, or checklist? (*p.* 123)
- ☐ Does each list contain an appropriate number of items? (*p.* 124)
- ☐ Are all the items in each list grammatically parallel? (*p.* 124)
- ☐ Is the lead-in to each list structured and punctuated properly? (*p.* 124)
- ☐ Are the items in each list punctuated properly? (*p.* 125)

Sentences

- ☐ Are the sentences structured with the new or important information near the end? (*p.* 126)
- ☐ Are the sentences the appropriate length? (*p.* 127)
- ☐ Does each sentence focus on the "real" subject? (*p.* 128)

- ☐ Does each sentence focus on the "real" verb, without weak nominalizations? *(p. 129)*
- ☐ Have you used parallel structure in your sentences? *(p. 130)*
- ☐ Have you used restrictive and nonrestrictive modifiers appropriately? *(p. 131)*
- ☐ Have you eliminated misplaced modifiers, squinting modifiers, and dangling modifiers? *(p. 132)*

Words and Phrases

Did you

- ☐ select an appropriate level of formality? *(p. 133)*
- ☐ use the active and passive voices appropriately? *(p. 134)*
- ☐ use precise words? *(p. 136)*
- ☐ provide adequate detail? *(p. 136)*
- ☐ avoid ambiguity? *(p. 136)*
- ☐ avoid unnecessary jargon? *(p. 136)*
- ☐ use positive rather than negative constructions? *(p. 137)*
- ☐ avoid long noun strings? *(p. 138)*
- ☐ avoid clichés? *(p. 138)*
- ☐ avoid euphemisms? *(p. 139)*
- ☐ avoid unnecessary prepositional phrases? *(p. 140)*
- ☐ use the most concise phrases? *(p. 141)*
- ☐ avoid fancy words? *(p. 142)*
- ☐ use nonsexist language? *(p. 142)*
- ☐ use the people-first approach in referring to people with disabilities? *(p. 143)*

EXERCISES

1. Identify the best organizational pattern for a discussion of each of the subjects that follow. For example, a discussion of distance education and on-campus courses could be organized using the comparison-and-contrast pattern. Write a brief explanation supporting your selection. (Use each of the organizational patterns discussed in this chapter at least once.)
   - **a.** how to register for courses at your college or university
   - **b.** how you propose to reduce the time required to register for classes or to change your schedule
   - **c.** your car's dashboard
   - **d.** the current price of gasoline
   - **e.** the reasons you chose your college or major
   - **f.** two music-streaming services
   - **g.** MP3 players
   - **h.** college courses
   - **i.** increased security in airports
   - **j.** how to prepare for a job interview
2. Write a one-paragraph evaluation of each of the following titles. How clearly does the title indicate the subject and purpose of the document? In what ways does it fall short of incorporating this chapter's advice about titles? On the basis of your analysis, rewrite each title.
   - **a.** Recommended Forecasting Techniques for Haldane Company
   - **b.** A Study of Digital Cameras
   - **c.** Agriculture in the West: A 10-Year View
3. Write a one-paragraph evaluation of each of the following headings. How clearly does the heading indicate the subject and purpose of the text that will follow it? In what ways does it fall short of incorporating this chapter's advice about headings? On the basis of your analysis, rewrite each heading to make it clearer and more informative. Invent any necessary details.
   - **a.** Multigroup Processing Technique Review Board Report Findings
   - **b.** The Great Depression of 1929
   - **c.** Intensive-Care Nursing
4. Revise the following list so that the lead-in is clear, easy to understand, and punctuated correctly. In addition, be sure the bullet items are grammatically parallel with one another.

   There are several goals being pursued by the Natural and Accelerated Bioremediation Research office;

   - the development of cost-effective *in situ* bio-remediation strategies for subsurface radionuclides and metals;

- an understanding of intrinsic bioremediation as well as accelerated bioremediation using nutrient amendments to immobilize contaminants;
- identifying societal issues associated with bioremediation research, and communication of bioremediation research findings to stakeholders.

5. Provide a topic sentence for this paragraph.

The reason for this difference is that a larger percentage of engineers working in small firms may be expected to hold high-level positions. In firms with fewer than 20 engineers, for example, the median income was $62,200. In firms of 20 to 200 engineers, the median income was $60,345. For the largest firms, the median was $58,600.

6. In the following paragraph, transitional words and phrases have been removed. Add an appropriate transition in each blank space. Where necessary, add punctuation.

One formula that appeared foolproof for selling computers was direct sales by the manufacturer to the consumer. Dell, __________, climbed to number two in PC sales by selling customized products directly on its website. __________, the recent success of Acer, now number three in sales, suggests that the older formula of distributing commodity items through retailers might be best for today's PC industry. Acer's success can be attributed to three decisions it made. First, it sold off its division that manufactured components for other PC brands. __________, it correctly concluded that consumers, who generally prefer preconfigured PCs, would outnumber business customers. And __________, it decided to expand its line of inexpensive netbooks (small PCs for surfing the web) just when the economic downturn increased the demand for cheaper PC products. These decisions appear to have paid off for Acer: last year, its market share rose 3 percentage points, from 8 to 11. __________, Dell rose only 0.1 point, from 14.8 to 14.9.

7. In each of the following exercises, the second sentence begins with a demonstrative pronoun. Add a noun after the demonstrative to enhance coherence.

a. The Zoning Commission has scheduled an open hearing for March 14. This __________ will enable concerned citizens to voice their opinions on the proposed construction.

b. The university has increased the number of parking spaces, instituted a shuttle system, and increased parking fees. These __________ are expected to ease the parking problems.

NOTE: In Exercises 8–29, pay close attention to what you are being asked to do, and do only as much revising as is necessary. Take special care to preserve the meaning of the original material. If necessary, invent reasonable details.

8. Refer to the advice on pages 123–26, and rewrite the following sentence in the form of a list.

The causes of burnout can be studied from three perspectives: physiological—the roles of sleep, diet, and physical fatigue; psychological—the roles of guilt, fear, jealousy, and frustration; and environmental—the role of physical surroundings at home and at work.

9. The following sentences might be too long for some readers. Refer to the advice on page 127, and break each sentence into two or more sentences.

a. If we get the contract, we must be ready by June 1 with the necessary personnel and equipment, so with this in mind a staff meeting, which all group managers are expected to attend, is scheduled for February 12.

b. Once we get the results of the stress tests on the 125-Z fiberglass mix, we will have a better idea of whether the project is on schedule, because if the mix isn't suitable we will really have to hurry to find and test a replacement by the Phase 1 deadline.

10. In the following sentences, the real subjects are buried in prepositional phrases or obscured by expletives. Refer to the advice on pages 128–29, and revise the sentences so that the real subjects appear prominently.

a. There has been a decrease in the number of students enrolled in our training sessions.

b. The use of in-store demonstrations has resulted in a dramatic increase in business.

11. In the following sentences, unnecessary nominalization obscures the real verb. Refer to the advice on pages 129–30, and revise the sentences to focus on the real verb.

a. Pollution constitutes a threat to the Matthews Wildlife Preserve.

b. Evaluation of the gumming tendency of the four tire types will be accomplished by comparing the amount of rubber that can be scraped from the tires.

12. Refer to the advice on page 130, and revise the following sentences to eliminate nonparallelism.

a. The next two sections of the manual discuss how to analyze the data, the conclusions that can be drawn from your analysis, and how to decide what further steps are needed before establishing a journal list.

b. In the box, we should include a copy of the documentation, the cables, and the docking station.

13. Refer to the advice on pages 131–33, and revise the following sentences to correct punctuation or pronoun errors related to modifiers.

a. Press the Greeting Record button to record the greeting that is stored on a microchip inside the machine.

b. This problem that has been traced to manufacturing delays, has resulted in our losing four major contracts.

14. Refer to the advice on pages 132–33, and revise the following sentences to eliminate the misplaced modifiers.

a. Information provided by this program is displayed at the close of the business day on the information board.

b. The computer provides a printout for the Director that shows the likely effects of the action.

15. Refer to the advice on page 133, and revise the following sentences to eliminate the dangling modifiers.

a. By following these instructions, your computer should provide good service for many years.

b. To examine the chemical homogeneity of the plaque sample, one plaque was cut into nine sections.

16. Refer to the advice on pages 133–34, and revise the following informal sentences to make them moderately formal.

a. The learning modules were put together by a couple of profs in the department.

b. If the University of Arizona can't figure out where to dump its low-level radioactive waste, Uncle Sam could pull the plug on millions of dollars of research grants.

17. Refer to the advice on pages 134–36, and rewrite the following sentences to remove inappropriate use of the passive voice.

a. Mistakes were made.

b. Come to the reception desk when you arrive. A packet with your name on it can be picked up there.

18. Refer to the advice on page 136, and revise the following sentences by replacing the vague elements with specific information. Make up any reasonable details.

a. The results won't be available for a while.

b. The chemical spill in the lab caused extensive damage.

19. Refer to the advice on pages 136–37, and revise the following sentences to remove unnecessary jargon.

a. We need to be prepared for blowback from the announcement.

b. The mission-critical data on the directory will be migrated to a new server on Tuesday.

20. Refer to the advice on pages 137–38, and revise the following sentences to convert the negative constructions to positive constructions.

a. Management accused Williams of filing trip reports that were not accurate.

b. We must make sure that all our representatives do not act unprofessionally toward potential clients.

21. General readers might find the following sentences awkward or difficult to understand. Refer to the advice on page 138, and rewrite the sentences to eliminate the long noun strings.

a. The corporate-relations committee meeting location has been changed.

b. The research team discovered a glycerin-initiated, alkylene-oxide-based, long-chain polyether.

22. Refer to the advice on page 138, and revise the following sentences to eliminate clichés.

a. If we are to survive this difficult period, we are going to have to keep our ears to the ground and our noses to the grindstone.

b. At the end of the day, if everyone is on the same page and it turns out to be the wrong page, you're really up a creek without a paddle.

23. Refer to the advice on page 139, and revise the following sentences to eliminate euphemisms.
    a. Downsizing our workforce will enable our division to achieve a more favorable cash-flow profile.
    b. Of course, accident statistics can be expected to show a moderate increase in response to a streamlining of the training schedule.
24. Refer to the advice on pages 139–41, and revise the following sentences to remove the redundancies.
    a. In grateful appreciation of your patronage, we are pleased to offer you this free gift as a small token gesture of our gratitude.
    b. An anticipated major breakthrough in storage technology will allow us to proceed ahead in the continuing evolution of our products.
25. Refer to the advice on page 140, and revise the following sentences to eliminate unnecessary prepositional phrases.
    a. The complexity of the module will hamper the ability of the operator in the diagnosis of problems in equipment configuration.
    b. The purpose of this test of your aptitudes is to help you with the question of the decision of which major to enroll in.
26. Refer to the advice on pages 139–41, and revise the following sentences to make them more concise.
    a. The instruction manual for the new copier is lacking in clarity and completeness.
    b. We remain in communication with our sales staff on a weekly basis.
27. Refer to the advice on page 142, and revise the following sentences to eliminate fancy words.
    a. This state-of-the-art soda-dispensing module is to be utilized by Marketing Department personnel.
    b. We have failed to furnish the proposal to the proper agency by the mandated date by which such proposals must be in receipt.
28. Refer to the advice on page 143, and revise the following sentences to eliminate sexist language.
    a. Each doctor is asked to make sure he follows the standard procedure for handling Medicare forms.
    b. Policemen are required to live in the city in which they work.
29. Refer to the advice on pages 143–45, and revise the following sentences to eliminate the offensive language.
    a. This year, the number of female lung-cancer victims is expected to rise because of increased smoking.
    b. Mentally retarded people are finding greater opportunities in the service sector of the economy.

For more practice with the concepts covered in this chapter, complete the LearningCurve activities "Organizing and Emphasizing Information" and "Writing Correct and Effective Sentences" in LaunchPad.

CASE 6: Emphasizing Important Information in a Technical Description

You and two classmates have been asked to write a technical description of a new 3D printer purchased by the engineering college at your school. Your professor, however, has concerns about the draft you have submitted, and he has outlined those concerns in an email. To read the email, identify passages that warrant improvement, and begin revising the technical description for emphasis and coherence, go to LaunchPad.

Part 3

Yuri_Arcurs/Getty Images

Designing User-Friendly Documents and Websites

CHAPTER

7 Designing Print and Online Documents

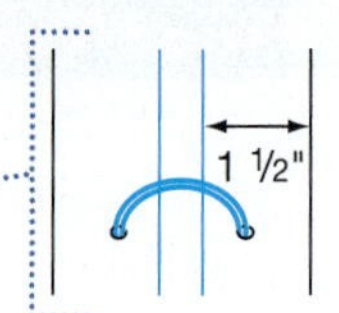

THE DESIGN OF a print or online document can help a writer achieve many goals: to entertain, to amaze, to intrigue, to sell. In technical communication, the goal is typically to help the reader learn something, perform a task, or accept a point of view. When you look at a well-designed page or screen, you understand how to use it.

Design refers to the physical appearance of print and online documents. For print documents, design features include binding, page size, typography, and color. For online documents, many of the same design elements apply, but there are unique elements, too. On a web page, for instance, there are navigation bars, site maps, and search engines.

The effectiveness of a document depends largely on how well it is designed, because readers *see* the document before they actually *read* it. In a matter of seconds, the document makes an impression on them, one that might determine how well they read it—or even whether they decide to read it at all.

Goals of Document Design

In designing a document, you have five major goals:

- **To make a good impression on readers.** Your document should reflect your own professional standards and those of your organization.

FIGURE 7.1 **Effective Use of Proximity**
Information from U.S. Department of State, 2011: http://future.state.gov.

ALIGNMENT

The principle of alignment says that you should consciously line up text and graphics along a real or imaginary vertical axis so that the reader can understand the relationships among elements. Figure 7.2 shows how alignment works to help organize information.

REPETITION

The principle of repetition says that you should format the same kind of information in the same way so that readers can recognize consistent patterns. For example, all first-level headings should have the same typeface, type size, spacing above and below, and so forth. This repetition signals a connection between headings, making the content easier to understand. Other elements that are used to create consistent visual patterns are colors, icons, rules, and screens. Figure 7.3 shows an effective use of repetition.

This panel from an FAQ section of a website uses alignment to help organize the information.

The writer is using three levels of importance, each signaled by a different alignment.

Writers often use more than one technique at a time to help organize information. In this case, text size and color also indicate levels of importance.

V. Completing the SF424 (R&R) Application

A. About the SF424 (R&R) Application Form

1. **Which form should be used to submit electronic applications to NIH via Grants.gov?**

   Applicants should use the Standard Form (SF) 424 Research & Related (R&R) family of forms. SF424 consolidates grant applications, related data and forms currently used by Federal grant-making agencies to enable applicants to use familiar forms regardless of the program or agency to which they are applying. The SF424 Research & Related (R&R) will become the government-wide data set for research grant applications. The SF424 (R&R) will replace the Public Health Service (PHS) 398 form at NIH.

2. **Are SF424 components portable? Can components be reused for other applications?**

   Currently there is no way to reuse the forms from one opportunity to another. Grants.gov hopes to have the functionality next year to import and export data for reuse with other applications.

3. **Where is the budget justification located?**

   In the SF424 (R&R) detailed budget component, the budget justification is item K--a PDF upload. In the PHS398 Modular budget component, budget justifications for Personnel, Consortium and Additional Narrative are requested as separate PDF uploads as part of the Cumulative Budget Information.

B. Application Instructions

1. **Where will an applicant need to look to find application instructions?**

   Application instructions are available in two places: the SF424 (R&R) Application Guide and within each Funding Opportunity Announcement (FOA). The Application Guide includes all general instructions and a

FIGURE 7.2 Effective Use of Alignment

Information from National Institutes of Health, 2013a: http://grants.nih.gov/grants/ElectronicReceipt/faq_full.htm#application.

This page shows repetition used effectively as a design element.

Different colors, typefaces, and type sizes are used for the headings, instructions, and text.

The two lists make use of stylized bullets and oversize numbers, both in one of the two main colors.

Identify Key Terms

Identify and explain the significance of each item below.

open-field system (p. 289)
merchant guild (p. 299)
craft guild (p. 300)
Hanseatic League (p. 303)
commercial revolution (p. 304)
sumptuary laws (p. 307)
Scholastics (p. 311)
vernacular literature (p. 314)
troubadours (p. 315)
cathedral (p. 317)
Romanesque (p. 317)
Gothic (p. 317)

Review the Main Ideas

Answer the focus questions from each section of the chapter.

- What was village life like in medieval Europe? (p. 288)
- How did religion shape everyday life in the High Middle Ages? (p. 293)
- What led to Europe's economic growth and reurbanization? (p. 298)
- What was life like in medieval cities? (p. 305)
- How did universities serve the needs of medieval society? (p. 309)
- How did literature and architecture express medieval values? (p. 313)

Make Connections

Think about the larger developments and continuities within and across chapters.

1. How was life in a medieval city different from life in a Hellenistic city (Chapter 4), or life in Rome during the time of Augustus (Chapter 6)? In what ways was it similar? What problems did these cities confront that are still issues for cities today?
2. Historians have begun to turn their attention to the history of children and childhood. How were children's lives in the societies you have examined shaped by larger social structures and cultural forces? What commonalities do you see in children's lives across time?
3. Chapter 4 and this chapter both examine ways in which religion and philosophy shaped life for ordinary people and for the educated elite. How would you compare Hellenistic religious practices with those of medieval Europe? How would you compare the ideas of Hellenistic philosophers such as Epicurus or Zeno with those of Scholastic philosophers such as Thomas Aquinas?

FIGURE 7.3 Effective Use of Repetition

Information from A HISTORY OF WESTERN SOCIETY, Eleventh Edition (Boston: Bedford/St. Martin's, 2014). John P. McKay, Clare Haru Crowston, Merry E. Wiesner-Hanks, Joe Perry, p. 319.

- **To help readers understand the structure and hierarchy of the information.** As they navigate a document, readers should know where they are and how to get where they are headed. They should also be able to see the hierarchical relationship between one piece of information and another.
- **To help readers locate and access the information they need.** Usually, people don't read every word in a print document, and they don't study every screen of an online document. In print documents, design elements (such as tabs, icons, and color), page design, and typography help readers find the information they need quickly and easily. In online documents, design elements are critically important because readers can see only what is displayed on the screen.
- **To help readers understand the information.** Effective design can clarify information. For instance, designing a set of instructions so that the text describing each step is next to the accompanying graphic makes the instructions easier to understand. An online document with a navigation bar displaying the main sections is easier to understand than an online document without one.
- **To help readers remember the information.** An effective design helps readers create a visual image of the information, making it easier to remember. Text boxes, pull quotes, and similar design elements help readers remember important explanations and passages.

Planning the Design of Print and Online Documents

The first step in designing a print or online technical document is planning. Analyze your audience and purpose, and then determine your resources.

GUIDELINES Planning Your Design

Follow these four suggestions as you plan your design.

- **Analyze your audience.** Consider factors such as your readers' knowledge of the subject, their attitudes, their reasons for reading, the way they will be using the document, and the kinds of tasks they will perform. Think too about your audience's expectations. Readers expect to see certain kinds of information presented in certain ways. Plan to fulfill those expectations. For example, hyperlinks in websites are often emphasized in some fashion and presented in an alternative color.

For more about analyzing your audience and purpose, see Ch. 4.

- **Consider multicultural readers.** If you are writing for multicultural readers, keep in mind that many aspects of design vary from one culture to another. In memos, letters, reports, and manuals, you may see significant differences in design practice. The best advice, therefore, is to study documents from the culture you are addressing. Look for differences in paper size, text direction, typeface preferences, and color preferences.

(continued)

- **Consider your purpose.** For example, if you are creating a website for a new dental office, do you merely want to provide information on the hours and location, or do you also want to present dental information for patients? Let patients set up or change appointments? Ask a question? Each of these purposes affects the site design.
- **Determine your resources.** Think about your resources of time, money, and equipment. Short, informal documents and websites are usually produced in-house; more-ambitious projects are often subcontracted to specialists. If your organization has a technical-publications department, consult the people there about scheduling and budgeting. A sophisticated design might require professionals at service bureaus and print shops, and their services can require weeks or months and cost thousands of dollars.

Understanding Design Principles

Your biggest challenge in thinking about how to design a document is that, more than ever, readers control how the document appears. When they review documents online, through websites, apps, and other programs, readers can control many aspects of the design, including color and the size, shape, and location of objects on the screen. Perhaps the most significant variable that you have to consider is screen size. Some devices on which your readers will use your document will be as large as big-screen TVs, whereas others will be as small as wrist watches.

In this chapter, the term *print document* will be used to refer to documents that are designed to be printed on paper, such as letters, memos, and reports, regardless of whether readers hold pieces of paper in their hands or view the documents online. The term *online document* will be used to refer to documents that are designed to be used online, such as websites, apps, and other software programs.

Because there are so many different types of print and online documents used in so many different environments by so many different people for so many different purposes, it is impossible to provide detailed advice about "how to design" a technical document. Still, there are some powerful and durable principles that can help you design any kind of print or online document. The following discussion is based on Robin Williams's *The Non-designer's Design Book* (2015), which describes four principles of design: proximity, alignment, repetition, and contrast.

PROXIMITY

The principle of proximity is simple: if two items appear close to each other, the reader will interpret them as related to each other. If they are far apart, the reader will interpret them as unrelated. Text describing a graphic should be positioned close to the graphic, as shown in Figure 7.1.

Some software programs search for common nominalizations. With any word processor, however, you can identify most of them by searching for character strings such as *tion*, *ment*, *sis*, *ence*, *ing*, and *ance*, as well as the word *of*.

USE PARALLEL STRUCTURE

A sentence is parallel if its coordinate elements follow the same grammatical form: for example, all the clauses are either passive or active, all the verbs are either infinitives or participles, and so on. Parallel structure creates a recognizable pattern, making a sentence easier for the reader to follow. Nonparallel structure creates no such pattern, distracting and possibly confusing readers. In the following examples of nonparallel constructions, the verbs are not in the same form (verbs are italicized).

NONPARALLEL — Our present system *is costing* us profits and *reduces* our productivity.

PARALLEL — Our present system *is costing* us profits and *reducing* our productivity.

NONPARALLEL — The compositor *should follow* the printed directions; *do not change* the originator's work.

PARALLEL — The compositor *should follow* the printed directions and *should not change* the originator's work.

When using parallel constructions, make sure that parallel items in a series do not overlap, causing confusion or even changing the meaning of the sentence:

CONFUSING — The speakers will include partners of law firms, businesspeople, and civic leaders.

Partners of appears to apply to *businesspeople* and *civic leaders*, as well as to *law firms*. That is, *partners of* carries over to the other items in the series. The following revision solves the problem by rearranging the items so that *partners* can apply only to *law firms*.

CLEAR — The speakers will include businesspeople, civic leaders, and partners of law firms.

CONFUSING — We need to buy more lumber, hardware, tools, and hire the subcontractors.

The writer has linked two ideas inappropriately. The first idea is that we need to buy three things: lumber, hardware, and tools. The second is that we need to hire the subcontractors. Hiring is not in the same category as the items to be bought. In other words, the writer has structured and punctuated the sentence as if it contained a four-item series, when in fact it should contain a three-item series followed by a second verb phrase.

CLEAR — We need to buy more lumber, hardware, and tools, and we need to hire the subcontractors.

WEAK	It is hoped that testing the evaluation copies of the software will help us make this decision.
STRONG	We hope that testing the evaluation copies of the software will help us make this decision.

The second example uses the expletive *it* with the passive voice. The problem is that the sentence does not make clear who is doing the hoping.

For more about writing to a multicultural audience, see Ch. 4, p. 58.

Expletives are not errors. Rather, they are conversational expressions that can clarify meaning by emphasizing the information that follows them.

WITH THE EXPLETIVE	It is hard to say whether the downturn will last more than a few months.
WITHOUT THE EXPLETIVE	Whether the downturn will last more than a few months is hard to say.

The second version is harder to understand because the reader has to remember a long subject (*Whether the downturn will last more than a few months*) before getting to the verb (*is*). Fortunately, you can revise the sentence in other ways to make it easier to understand and to eliminate the expletive.

I don't know whether the downturn will last more than a few months.

Nobody knows whether the downturn will last more than a few months.

Use the search function of your software to locate both weak subjects (usually they precede the word *of*) and expletives (search for *it is*, *there is*, and *there are*).

FOCUS ON THE "REAL" VERB

A "real" verb, like a "real" subject, should stand out in every sentence. A common problem in technical communication is the inappropriate use of a *nominalized* verb—a verb that has been changed into a noun, then coupled with a weaker verb. *To install* becomes *to effect an installation*; *to analyze* becomes *to conduct an analysis*. Notice how nominalizing the verbs makes the following sentences both awkward and unnecessarily long (the nominalized verbs are italicized).

WEAK	Each *preparation* of the solution is done twice.
STRONG	Each solution is prepared twice.
WEAK	*Consideration* should be given to an acquisition of the properties.
STRONG	We should consider acquiring the properties.

Like expletives, nominalizations are not errors. In fact, many common nouns are nominalizations: *maintenance*, *requirement*, and *analysis*, for example. In addition, nominalizations often effectively summarize an idea from a previous sentence (in italics in the following extract).

The telephone-service provider decided not to replace the land lines that were damaged in the recent storm. This *decision* could prove a real problem for those residents who used land lines to connect to the Internet and for their medical-alert services.

Put references to time and space at the beginning of the sentence, where they can provide context for the main idea that the sentence expresses.

> *Since the last quarter of 2014*, we have experienced an 8 percent turnover rate in personnel assigned to the project.

> *On the north side of the building*, water from the leaking pipes has damaged the exterior siding and the sheetrock on some interior walls.

CHOOSE AN APPROPRIATE SENTENCE LENGTH

Sometimes sentence length affects the quality of the writing. In general, an average of 15 to 20 words per sentence is effective for most technical communication. A series of 10-word sentences would be choppy. A series of 35-word sentences would probably be too demanding. And a succession of sentences of approximately the same length would be monotonous.

In revising a draft, use your software to compute the average sentence length of a representative passage.

Avoid Overly Long Sentences How long is too long? There is no simple answer, because ease of reading depends on the vocabulary, sentence structure, and sentence length; the reader's motivation and knowledge of the topic; the purpose of the communication; and the conventions of the application you are using. For instance, you use shorter sentences in tweets and text messages than in reports.

Often a draft will include sentences such as the following:

> The construction of the new facility is scheduled to begin in March, but it might be delayed by one or even two months by winter weather conditions, which can make it impossible or nearly impossible to begin excavating the foundation.

To avoid creating such long sentences, say one thing clearly and simply before moving on to the next idea. For instance, to make this difficult 40-word sentence easier to read, divide it into two sentences:

> The construction of the new facility is scheduled to begin in March. However, construction might be delayed until April or even May by winter weather conditions, which can make it impossible or nearly impossible to begin excavating the foundation.

Sometimes an overly long sentence can be fixed by creating a list (see the Guidelines box on page 123).

Avoid Overly Short Sentences Just as sentences can be too long, they can also be too short and choppy, as in the following example:

> Customarily, environmental cleanups are conducted on a "time-and-materials" (T&M) basis. Using the T&M basis, the contractor performs the work. Then the contractor bills for the hours worked and the cost of equipment and materials used during the work. With the T&M approach, spending for environmental cleanups by private and government entities has been difficult to contain. Also, actual contamination reduction has been slow.

The problem here is that some of the sentences are choppy and contain too little information, calling readers' attention to how the sentences are constructed rather than to what the sentences say. In cases like this, the best way to revise is to combine sentences:

> Customarily, environmental cleanups are conducted on a "time-and-materials" (T&M) basis: the contractor performs the work, then bills for the hours worked and the cost of equipment and materials. With the T&M approach, spending for environmental cleanups by private and government entities has been difficult to contain, and contamination reduction has been slow.

Another problem with excessively short sentences is that they needlessly repeat key terms. Again, consider combining sentences:

SLUGGISH	I have experience working with various *microprocessor-based systems*. Some of these *microprocessor-based systems* include the T90, RCA 9600, and AIM 7600.
BETTER	I have experience working with various microprocessor-based systems, including the T90, RCA 9600, and AIM 7600.

FOCUS ON THE "REAL" SUBJECT

The conceptual, or "real," subject of the sentence should also be the grammatical subject. Don't disguise or bury the real subject in a prepositional phrase following a weak grammatical subject. In the following examples, the weak subjects obscure the real subjects. (The grammatical subjects are italicized.)

WEAK	The *use* of this method would eliminate the problem of motor damage.
STRONG	This *method* would eliminate the problem of motor damage.
WEAK	The *presence* of a six-membered lactone ring was detected.
STRONG	A six-membered lactone *ring* was detected.

In revising a draft, look for the real subject (the topic) and ask yourself whether the sentence would be more effective if the real subject was also the grammatical subject. Sometimes all that is necessary is to ask yourself this question: *What is the topic of this sentence?* The author of the first example above wasn't trying to say something about *using* a method; she was trying to say something about the method itself. Likewise, in the second example, it wasn't the *presence* of a lactone ring that was detected; rather, the lactone ring itself was detected.

Another way to make the subject of the sentence prominent is to reduce the number of grammatical expletives. *Expletives* are words that serve a grammatical function in a sentence but have no meaning. The most common expletives are *it* (generally followed by is) and *there* (generally followed by is or *are*).

WEAK	There is no alternative for us except to withdraw the product.
STRONG	We have no alternative except to withdraw the product.

Notice that you do not have to use a strikingly different color to show contrast. The human brain can easily tell the difference between the paler blue and the navy blue of the other boxes.

FIGURE 7.4 Effective Use of Contrast

This portion of a web page shows effective use of color contrast in the five navigation boxes. The pale blue screen behind the word Impact *helps visitors see which portion of the site they are viewing.*

Information from U.S. Department of State, 2013b: http://eca.state.gov/impact.

CONTRAST

The principle of contrast says that the human eye is drawn to—and the brain interprets—differences in appearance between two items. For example, the principle of contrast explains why black print is easier to read against a white background than against a dark gray background; why 16-point type stands out more clearly against 8-point type than against 12-point type; and why information printed in a color, such as red, grabs readers' attention when the information around it is printed in black. Figure 7.4 shows effective use of contrast.

Designing Print Documents

Before you design the individual pages of a printed document, design the overall document. Decide whether you are creating a document that looks like a book, with content on both sides of the page, or a document that looks like a report, with content on only one side of the page. Decide whether to use paper of standard size (8.5 × 11 inches) or another size, choose a grade of paper, and decide how you will bind the pages together. Decide on the accessing elements you will include, such as a table of contents, index, and tabs. You want the different elements to work together to accomplish your objectives, and you want to stay within your budget for producing and (perhaps) shipping. Then think about how to design the document pages.

NAVIGATION AIDS

In a well-designed document, readers can easily find the information they seek. Most navigation aids use the design principles of repetition and contrast to help readers use the document. The Choices and Strategies feature explains six common kinds of navigation aids.

PAGE LAYOUT

Every page has two kinds of space: white space and space devoted to text and graphics. The best way to design a page is to make a grid: a drawing of what the page will look like. In making a grid, you decide how to use white space and determine how many columns to have on the page.

CHOICES AND STRATEGIES Creating Navigation Aids

IF YOU WANT TO . . .	TRY USING THIS NAVIGATION AID	EXAMPLE
Symbolize actions or ideas	**Icons.** Icons are pictures that symbolize actions or ideas. Perhaps the most important icon is the stop sign, which alerts you to a warning. Icons depend on repetition: every time you see the warning icon, you know what kind of information the writer is presenting. Don't be too clever in thinking up icons. One computer manual uses a cocktail glass about to fall over to symbolize "tip." This is a bad idea, because the pun is not functional: when you think of a cocktail glass, you don't think of a tip for using computers. Don't use too many different icons, or your readers will forget what each one represents.	Media YouTube Watch, upload and share videos Books Search the full text of books News Search thousands of news stories Picasa Find, edit and share your photos Information from Google, 2013: www.google.com/intl/en/about/products.
Draw attention to important features or sections of the document	**Color.** Perhaps the strongest visual attribute is color (Keyes, 1993). Use color to draw attention to important features of the document, such as warnings, hints, major headings, and section tabs. But use it sparingly, or it will overpower everything else in the document. Color exploits the principles of repetition (every item in a particular color is logically linked) and contrast (items in one color contrast with items in another color). Use color logically. Third-level headings should not be in color, for example, if first- and second-level headings are printed in black. Using paper of a different color for each section of a document is another way to simplify access.	Here green is used to emphasize the titles of the sections, the box at the top left, and the bar along the edge of the page. Information from MEDIA & CULTURE: MASS COMMUNICATION IN A DIGITAL AGE, Ninth Edition (Boston: Bedford/St. Martin's, 2014). Richard Campbell, Christopher R. Martin, Bettina Fabos, p. 113.
Enable readers to identify and flip to sections	**Dividers and tabs.** You are already familiar with dividers and tabs from loose-leaf notebooks. A tab provides a place for a label, which enables readers to identify and flip to a particular section. Sometimes dividers and tabs are color-coded. Tabs work according to the design principle of contrast: the tabs literally stick out.	Image Credit: © 2014 Macmillan. Photo by Regina Tavani.

(continued)

IF YOU WANT TO . . .	TRY USING THIS NAVIGATION AID	EXAMPLE
Refer readers to related information within the document	**Cross-reference tables.** These tables, which exploit the principle of alignment, refer readers to related discussions.	*Read . . .* *To learn to . . .* Ch. 1 connect to the router Ch. 2 set up a firewall
Help readers see where they are in the document	**Headers and footers.** Headers and footers help readers see where they are in the document. In a book, for example, the headers on the left-hand pages might repeat the chapter number and title; those on the right-hand pages might contain the most recent first-level heading. Sometimes writers build other identifying information into the headers. For example, your instructor might ask you to identify your assignments with a header like the following: "Smith, Progress Report, English 302, page 6." Headers and footers work according to the principle of repetition: readers learn where to look on the page to see where they are in the document.	CARS Information from General Services Administration, 2013.
	Page numbering. For one-sided documents, use Arabic numerals in the upper right corner, although the first page of most documents does not have a number on it. For two-sided documents, put the page numbers near the outside margins. Complex documents often use two number sequences: lowercase Roman numerals (i, ii, and so on) for front matter and Arabic numerals for the body. There is no number on the title page, but the page following it is ii. Appendixes are often paginated with a letter and number combination: Appendix A begins with page A-1, followed by A-2, and so on; Appendix B starts with page B-1. Sometimes documents list the total number of pages in the document (so recipients can be sure they have all of them). The second page is "2 of 17," and the third page is "3 of 17." Documents that will be updated are sometimes numbered by section: Section 3 begins with page 3-1, followed by 3-2; Section 4 begins with 4-1. This way, a complete revision of one section does not affect the page numbering of subsequent sections.	Information from THE BEDFORD GUIDE FOR COLLEGE WRITERS, Tenth Edition (Boston: Bedford/St. Martin's, 2014). X. J. Kennedy, Dorothy M. Kennedy, and Marcia F. Muth, pp. xxxi, 703, A-21.

GUIDELINES Understanding Learning Theory and Page Design

In designing a page, create visual patterns that help readers find, understand, and remember information. Three principles of learning theory, the result of research into how people learn, can help you design effective pages: chunking, queuing, and filtering.

- **Chunking.** People understand information best if it is delivered to them in chunks—small units—rather than all at once. For single-spaced type, chunking involves double-spacing between paragraphs, as shown in Figure 7.5.

During the 18th century, there were many wars in Europe caused by the ambition of various kings to make their domains larger and to increase their own incomes. King Louis XIV of France had built up a very powerful kingdom. Brave soldiers and skillful generals spread his rule over a great part of what is Belgium and Luxemburg, and annexed to the French kingdom the part of Germany between the Rhine River and the Vosges (Vozh) Mountains.

Finally, the English joined with the troops of the Holy Roman Empire to curb the further growth of the French kingdom, and at the battle of Blenheim (1704), the English Duke of Marlborough, aided by the emperor's army, put an end to the further expansion of the French.

The 18th century also saw the rise of a new kingdom in Europe. You will recall that there was a county in Germany named Brandenburg, whose count was one of the seven electors who chose the emperor. The capital of this county was Berlin. It so happened that a number of Counts of Brandenburg, of the family of Hohenzollern, had been men of ambition and ability. The little county had grown by adding small territories around it. One of these counts, called "the Great Elector," had added to Brandenburg the greater part of the neighboring county of Pomerania. His son did not have the ability of his father, but was a very proud and vain man.

He happened to visit King William III of England, and was very much offended because during the interview, the king occupied a comfortable arm chair, while the elector, being simply a count, was given a chair to sit in which was straight-backed and had no arms. Brooding over this insult, as it seemed to him, he went home and decided that he too should be called a king. The question was, What should his title be? He could not call himself "King of Brandenburg," for Brandenburg was part of the Empire, and the emperor would not allow it. It had happened some one hundred years before, that, through his marriage with the daughter of the Duke of Prussia, a Count of Brandenburg had come into possession of the district known as East Prussia, at the extreme southeastern corner of the Baltic Sea.

The son of this elector who first called himself king had more energy and more character than his father. He ruled his country with a rod of iron, and built up a strong, well-drilled army. He was especially fond of tall soldiers, and had agents out all over Europe, kidnapping men who were over six feet tall to serve in his famous regiment of guards. He further increased the size of the Prussian kingdom.

His son was the famous Frederick the Great, one of the most remarkable fighters that the world has ever seen. This prince had been brought up under strict discipline by his father. The old king had been insistent that his son should be no weakling. It is told that one day, finding Frederick playing upon a flute, he seized the instrument and snapped it in twain over his son's shoulder.

France in the 18th Century

During the 18th century, there were many wars in Europe caused by the ambition of various kings to make their domains larger and to increase their own incomes. King Louis XIV of France had built up a very powerful kingdom. Brave soldiers and skillful generals spread his rule over a great part of what is Belgium and Luxemburg, and annexed to the French kingdom the part of Germany between the Rhine River and the Vosges (Vozh) Mountains.

Finally, the English joined with the troops of the Holy Roman Empire to curb the further growth of the French kingdom, and at the battle of Blenheim (1704), the English Duke of Marlborough, aided by the emperor's army, put an end to the further expansion of the French.

Prussia in the 18th Century

The 18th century also saw the rise of a new kingdom in Europe. You will recall that there was a county in Germany named Brandenburg, whose count was one of the seven electors who chose the emperor. The capital of this county was Berlin. It so happened that a number of Counts of Brandenburg, of the family of Hohenzollern, had been men of ambition and ability. The little county had grown by adding small territories around it. One of these counts, called "the Great Elector," had added to Brandenburg the greater part of the neighboring county of Pomerania. His son did not have the ability of his father, but was a very proud and vain man.

He happened to visit King William III of England, and was very much offended because during the interview, the king occupied a comfortable arm chair, while the elector, being simply a count, was given a chair to sit in which was straight-backed and had no arms. Brooding over this insult, as it seemed to him, he went home and decided that he too should be called a king. The question was, What should his title be? He could not call himself "King of Brandenburg," for Brandenburg was part of the Empire, and the emperor would not allow it. It had happened some one hundred years before, that, through his marriage with the daughter of the Duke of Prussia, a Count of Brandenburg had come into possession of the district known as East Prussia, at the extreme southeastern corner of the Baltic Sea.

The son of this elector who first called himself king had more energy and more character than his father. He ruled his country with a rod of iron, and built up a strong, well-drilled army. He was especially fond of tall soldiers, and had agents out all over Europe, kidnapping men who were over six feet tall to serve in his famous regiment of guards. He further increased the size of the Prussian kingdom.

a. Without chunking

b. With chunking

FIGURE 7.5 Chunking

Chunking emphasizes units of related information. Note how the use of headings creates clear chunks of information.

- **Queuing.** Queuing refers to creating visual distinctions to indicate levels of importance. More-emphatic elements—those with bigger type or boldface type—are more important than less-emphatic ones. Another visual element of queuing is alignment. Designers start more-important information closer to the left margin and indent less-important information. (An exception is titles, which are often centered in reports in the United States.) Figure 7.6 shows queuing.

(continued)

- **Filtering.** Filtering is the use of visual patterns to distinguish various types of information. Introductory material might be displayed in larger type, and notes might appear in italics, another typeface, and a smaller size. Figure 7.7 shows filtering.

STRATEGIC GOALS AND RESULTS

To assess FY 2012 results, program managers examined quantitative and qualitative indicators to determine whether indicators met previously established annual targets. Managers also considered how the results impact the achievement of the Department and USAID's strategic goals. A rating was then assigned to each indicator based on the analysis. In the Strategic Goals and Results sections that follows, 16 illustrative indicators are highlighted to provide the reader with timely data and analysis of key achievements reported during FY 2012.

STRATEGIC GOAL 1:
ACHIEVING PEACE AND SECURITY

Preserve international peace by preventing regional conflicts and transnational crime, combating terrorism and weapons of mass destruction, and supporting homeland security and security cooperation

PUBLIC BENEFIT

U.S. policy states that the security of U.S. citizens at home and abroad is best guaranteed when countries and societies are secure, free, prosperous, and at peace. The Department, USAID, and their partners seek to strengthen their diplomatic and development capabilities, as well as those of international partners and allies, to prevent or mitigate conflict, stabilize countries in crisis, promote regional stability, and protect civilians. In 2012, a profound and dramatic wave of change continued to sweep across the Middle East as people courageously stood up to their governments to express their legitimate aspirations for greater political participation and economic opportunity. Our close relationship with our interagency partners has enabled the United States to strengthen our national security and provide leadership in conflict areas, such as the Middle East, to promote democratic and political reforms and ensure a voice for all peoples, including women, in building stable and peaceful societies.

Secretary of State Clinton delivers remarks at the Global Counterterrorism Forum in Istanbul, Turkey, June 7, 2012.

SUMMARY OF PERFORMANCE AND RESOURCES

The Department and USAID allocated $16.928 billion toward this Strategic Goal in FY 2012, which is 32 percent of the total State-USAID budget supporting all strategic goals. The performance of the illustrative indicator is provided in the following section.

DEPARTMENT OF STATE USAID JOINT SUMMARY OF PERFORMANCE AND FINANCIAL INFORMATION • FISCAL YEAR 2012 | 11

The size of the type used for the various headings indicates their importance.

The largest type suggests that *Strategic Goals and Results* is a chapter heading.

The next largest type indicates that *Strategic Goal 1: Achieving Peace and Security* is an A head (the highest level within a chapter).

Public Benefit and *Summary of Performance and Resources* are B heads.

FIGURE 7.6 Queuing

Information from U.S. Department of State, 2013: www.state.gov/documents/organization/203937.pdf.

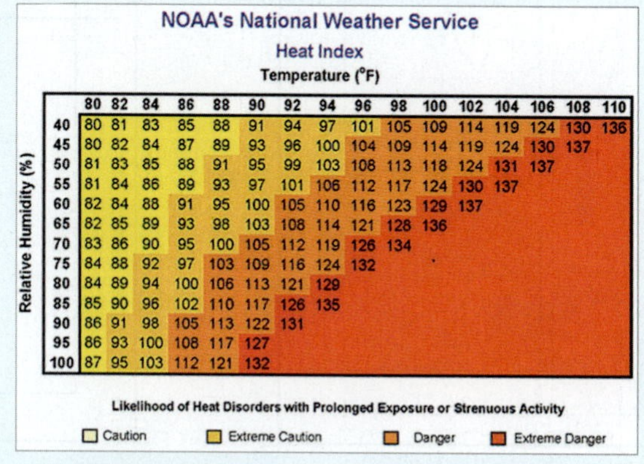

NOAA's National Weather Service

Heat Index

Temperature (°F)

Relative Humidity (%)	80	82	84	86	88	90	92	94	96	98	100	102	104	106	108	110
40	80	81	83	85	88	91	94	97	101	105	109	114	119	124	130	136
45	80	82	84	87	89	93	96	100	104	109	114	119	124	130	137	
50	81	83	85	88	91	95	99	103	108	113	118	124	131	137		
55	81	84	86	89	93	97	101	106	112	117	124	130	137			
60	82	84	88	91	95	100	105	110	116	123	129	137				
65	82	85	89	93	98	103	108	114	121	128	136					
70	83	86	90	95	100	105	112	119	126	134						
75	84	88	92	97	103	109	116	124	132							
80	84	89	94	100	106	113	121	129								
85	85	90	96	102	110	117	126	135								
90	86	91	98	105	113	122	131									
95	86	93	100	108	117	127										
100	87	95	103	112	121	132										

Likelihood of Heat Disorders with Prolonged Exposure or Strenuous Activity

Caution · Extreme Caution · Danger · Extreme Danger

FIGURE 7.7 Filtering

Effective technical communication presents data and explains what the data mean. In this table about the heat index, the writer uses color as a filtering device. In Western cultures, red signals danger.

Information from National Weather Service, 2014: http://nws.noaa.gov/os/heat/index.shtml.

Page Grids As the phrase suggests, a *page grid* is like a map on which you plan where the text, the graphics, and the white space will go. Many writers like to begin with a *thumbnail sketch*, a rough drawing that shows how the text and graphics will look on the page. Figure 7.8 shows thumbnail sketches of several options for a page from the body of a manual.

Experiment by sketching the different kinds of pages of your document: body pages, front matter, and so on. When you are satisfied, make page grids. You can use either a computer or a pencil and paper, or you can combine the two techniques.

Figure 7.9 shows two simple grids: one using *picas* (the unit that printing professionals use, which equals one-sixth of an inch) and one using inches. On the right is an example of a page laid out using the grid in the figure.

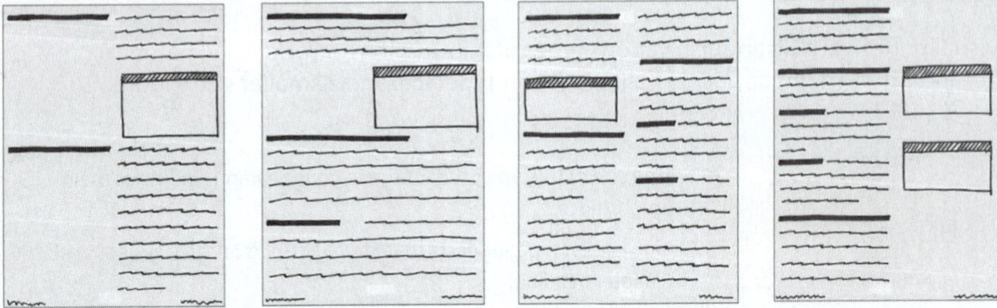

FIGURE 7.8 **Thumbnail Sketches**

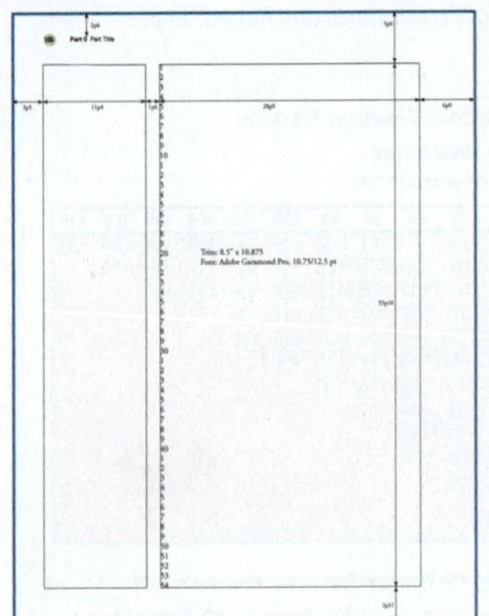

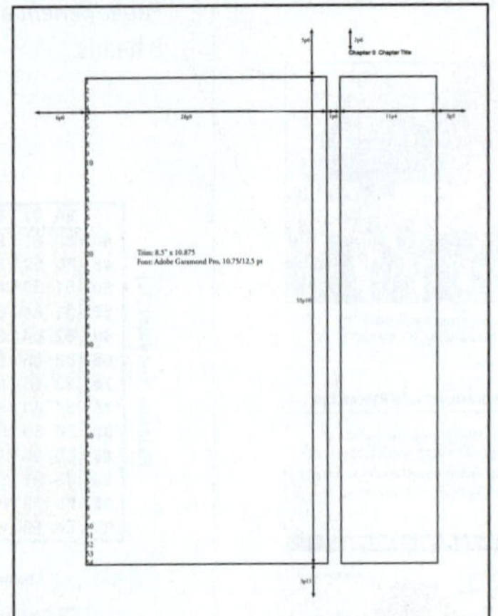

FIGURE 7.9 **Sample Grids Using Picas and Inches**

Information from O'Hair et al., *Real Communication: An Introduction*, Fourth Edition (Boston: Bedford/St. Martin's, 2018), p. 242.

TECH TIP

Why To Set Up Pages

An important key to meeting your audience's needs and expectations is effective page layout. Microsoft Word and other programs provide default page-layout settings, which you can easily adjust for any document you create in order to meet the needs of your audience and accomplish your purpose.

How To Set Up Pages

In the **Page Setup** group, use the **Page Setup** dialog box launcher to display the **Page Setup** dialog box.

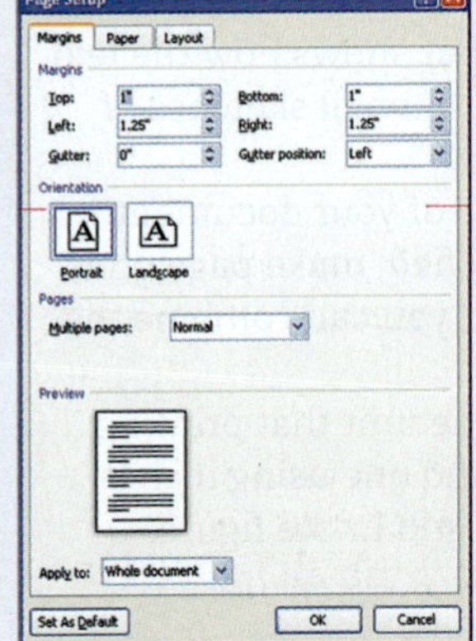

Use the **Margins**, **Paper**, and **Layout** tabs to specify such design elements as page margins, paper orientation, paper size, starting locations for new sections, and header and footer placement.

You can also use the drop-down menus on the **Page Setup** group to control many of the same design elements.

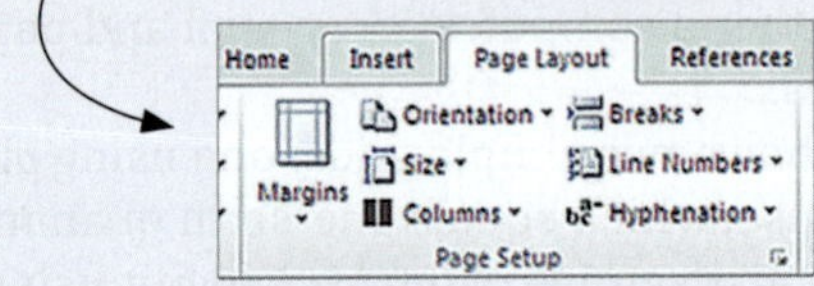

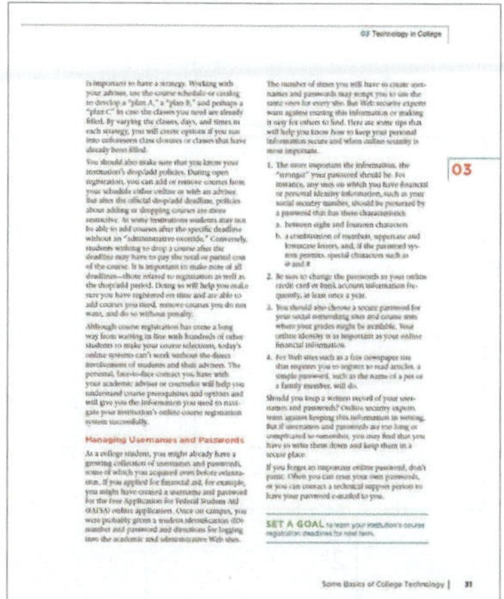

a. Double-column grid
Information from Gardner and Barefoot, YOUR COLLEGE EXPERIENCE: STUDY SKILLS EDITION, 10th ed., p. 31.

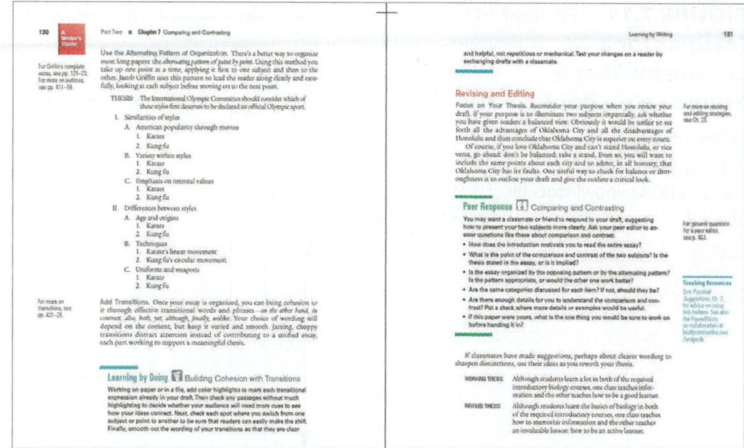

b. Two-page grid, with narrow outside columns for notes
Information from Kennedy, Kennedy, and Muth, THE BEDFORD GUIDE FOR COLLEGE WRITERS WITH READER, Tenth Edition, pp. 130–131.

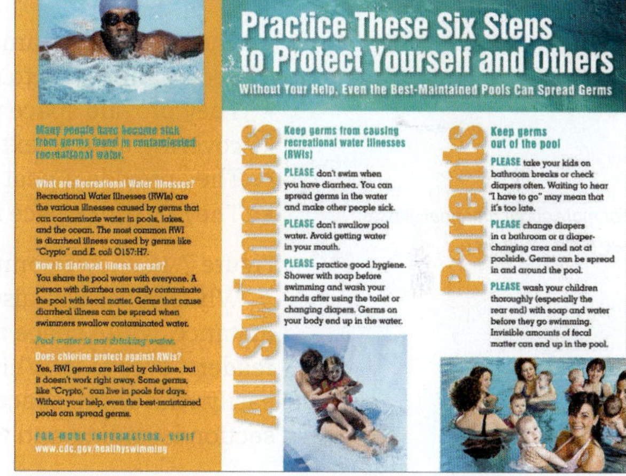

c. Three-panel brochure
Information from Centers for Disease Control and Prevention, 2014: www.cdc.gov/healthywater/pdf/swimming/resources/healthy-swimming-rwi-brochure.pdf.

FIGURE 7.10 Popular Grids

Create different grids until the design is attractive, meets the needs of your readers, and seems appropriate for the information you are conveying. Figure 7.10 shows some possibilities.

White Space Sometimes called *negative space*, *white space* is the area of the paper with no writing or graphics: the space between two columns of text, the space between text and graphics, and, most obviously, the margins.

FIGURE 7.11 Typical Margins for a Document That Is Bound Like a Book

Increase the size of the margins if the subject is difficult or if your readers are not knowledgeable about it.

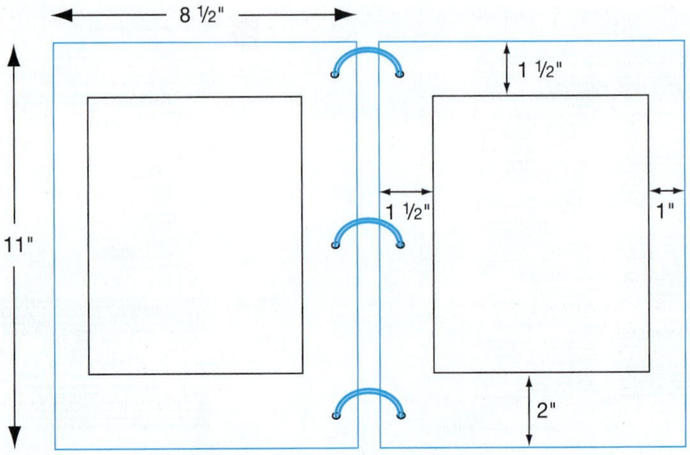

8 ½"
11"
1 ½"
1 ½"
1"
2"
Left-hand page
Right-hand page

Margins, which make up close to half the area on a typical page, serve four main purposes:

- They reduce the amount of information on the page, making the document easier to read and use.
- They provide space for binding and allow readers to hold the page without covering up the text.
- They provide a neat frame around the type.
- They provide space for marginal glosses.

For more about marginal glosses, see Table 7.1, p. 172.

Figure 7.11 shows common margin widths for an 8.5 × 11–inch document.

White space can also set off and emphasize an element on the page. For instance, white space around a graphic separates it from the text and draws readers' eyes to it. White space between columns helps readers read the text easily. And white space between sections of text helps readers see that one section is ending and another is beginning.

COLUMNS

Many workplace documents have multiple columns. A multicolumn design offers three major advantages:

- Text is easier to read because the lines are shorter.
- Columns enable you to fit more information on the page, because many graphics can fit in one column or extend across two or more columns. In addition, a multicolumn design enables you to put more words on a page than a single-column design.
- Columns enable you to use the principle of repetition to create a visual pattern, such as text in one column and accompanying graphics in an adjacent column.

TECH TIP

Why To Format Columns

A multicolumn format allows you to fit more text on the page, create easier-to-read and more visually interesting pages, and have more options when sizing and placing graphics.

How To Format Columns

To divide your document into multiple columns, select the **Page Layout** tab in Word to access the **Page Setup** group.

In the **Page Setup** group, select **Columns** and use the **Columns** drop-down menu to view **preset** layouts.

You can also select **More Columns** to launch the **Columns** dialog box. You can control the **number of columns** and specify the **width** and **spacing** yourself.

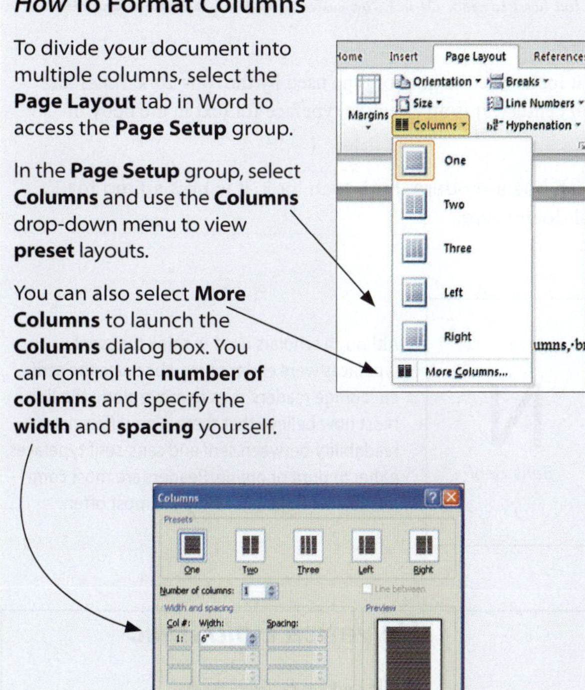

When you divide your document into columns, text flows from the bottom of one column to the top of the next column. Columns enable you to use the principle of repetition to create a visual pattern, such as text in one column and accompanying graphics in an adjacent column.

If you want to end a column of text in a specific location or create columns of equal length, use the **Breaks** drop-down menu to insert a **column break**. This action will move the text following the break to the next column.

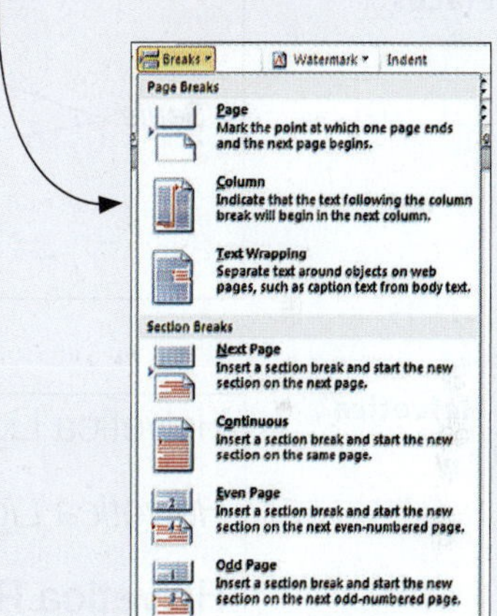

TYPOGRAPHY

Typography, the study of type and the way people read it, encompasses typefaces, type families, case, and type size, as well as factors that affect the white space of a document: line length, line spacing, and justification.

Typefaces A typeface is a set of letters, numbers, punctuation marks, and other symbols, all bearing a characteristic design. There are thousands of typefaces, and more are designed every year. Figure 7.12 shows three contrasting typefaces.

As Figure 7.13 illustrates, typefaces are generally classified into two categories: *serif* and *sans serif*.

Most of the time you will use a handful of standard typefaces such as Times New Roman, Cambria, Calibri, and Arial, which are included in your word-processing software and which your printer can reproduce.

Type Families Each typeface belongs to a family of typefaces, which consists of variations on the basic style, such as italic and boldface. Figure 7.14, for example, shows the Helvetica family.

FIGURE 7.12 Three Contrasting Typefaces

This paragraph is typed in French Script typeface. You are unlikely to see this style of font in a technical document because it is too ornate and too hard to read. It is better suited to wedding invitations and other formal announcements.

This paragraph is Times Roman. It looks like the kind of type used by the *New York Times* and other newspapers in the nineteenth century. It is an effective typeface for text in the body of technical documents.

This paragraph is Univers, which has a modern, high-tech look. It is best suited for headings and titles in technical documents.

FIGURE 7.13 Serif and Sans-Serif Typefaces

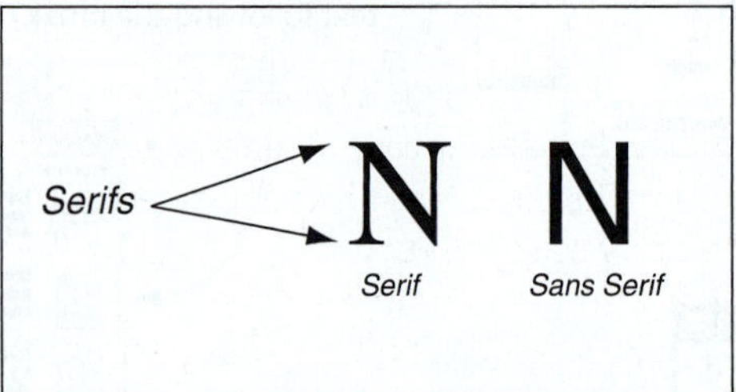

Although scholars used to think that serif typefaces were easier to read because the serifs encourage readers' eyes to move along the line, most now believe that there is no difference in readability between serif and sans-serif typefaces, either in print or online. Readers are most comfortable with the style they see most often.

FIGURE 7.14 Helvetica Family of Type

Helvetica Light	***Helvetica Bold Italic***
Helvetica Light Italic	**Helvetica Heavy**
Helvetica Regular	***Helvetica Heavy Italic***
Helvetica Regular Italic	Helvetica Regular Condensed
Helvetica Bold	*Helvetica Regular Condensed Italic*

FIGURE 7.15 Individual Variations in Lowercase and Uppercase Type

Lowercase letters are easier to read than uppercase because the individual variations from one letter to another are greater.

Individual variations are greater in lowercase words

THAN THEY ARE IN UPPERCASE WORDS.

Be careful not to overload your document with too many different members of the same family. Used sparingly and consistently, these variations can help you with filtering: calling attention to various kinds of text, such as warnings and notes. Use italics for book titles and other elements, and use bold type for emphasis and headings. Stay away from outlined and shadowed variations. You can live a full, rewarding life without ever using them.

Case To make your document easy to read, use uppercase and lowercase letters as you would in any other kind of writing (see Figure 7.15). Most people require 10 to 25 percent more time to read text using all uppercase letters than to read text using both uppercase and lowercase. In addition, uppercase letters take up as much as 35 percent more space than lowercase letters (Haley, 1991). If the text includes both cases, readers will find it easier to see where new sentences begin (Poulton, 1968).

Type Size Type size is measured with a unit called a *point*. There are 12 points in a pica and 72 points in an inch. In most technical documents, 10-, 11-, or 12-point type is used for the body of the text:

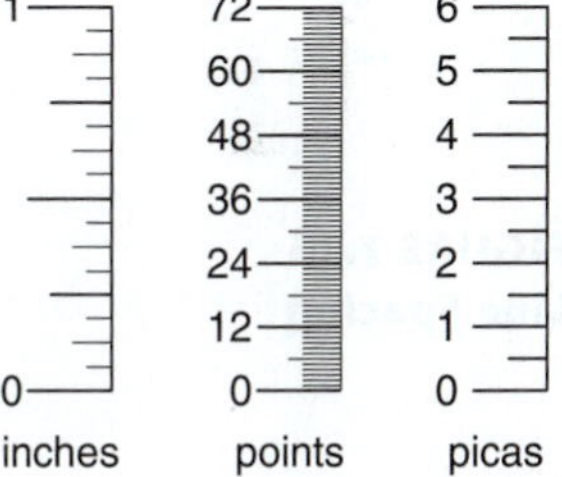

> This paragraph is printed in 10-point type. This size is easy to read, provided it is reproduced on a high-quality printer.
>
> This paragraph is printed in 12-point type. If the document will be read by people over age 40, 12-point type is a good size because it is more legible than a smaller size.
>
> This paragraph is printed in 14-point type. This size is appropriate for titles or headings.

Type sizes used for other parts of a document include the following:

footnotes	8- or 9-point type
indexes	2 points smaller than body text
slides	24- to 36-point type

In general, aim for at least a 2- to 4-point difference between the headings and the body. Too many size variations, however, suggest a sweepstakes advertisement rather than a serious text.

ETHICS NOTE

USING TYPE SIZES RESPONSIBLY

Text set in large type contrasts with text set in small type. It makes sense to use large type to emphasize headings and other important information. But be careful with small type. It is unethical (and, according to some court rulings, illegal) to use excessively small type (such as 6-point or smaller type) to present information that you *don't* want to stand out. When you read the fine print in an ad for cell-phone service, you get annoyed if you discover that the low rates are guaranteed for only three months or that you are committing to a long-term contract. You *should* get annoyed. Hiding information in tiny type is annoying. Don't do it.

Line Length The line length most often used on an 8.5 × 11-inch page—about 80 characters—is somewhat difficult to read. A shorter line of 50 to 60 characters is easier, especially in a long document (Biggs, 1980).

Line Spacing Sometimes called *leading* (pronounced "ledding"), *line spacing* refers to the amount of white space between lines or between a line of text and a graphic. If lines are too far apart, the page looks diffuse, the text loses coherence, and readers tire quickly. If lines are too close together, the page looks crowded and becomes difficult to read. Some research suggests that smaller type, longer lines, and sans-serif typefaces all benefit from extra line spacing. Figure 7.16 shows three variations in line spacing.

FIGURE 7.16
Line Spacing

a. Excessive line spacing

Aronomink Systems has been contracted by Cecil Electric Cooperative, Inc. (CECI) to design a solid waste management system for the Cecil County plant, Units 1 and 2, to be built in Cranston, Maryland. The system will consist of two 600 MW pulverized coal-burning units fitted with high-efficiency electrostatic precipitators and limestone reagent FGD systems.

b. Appropriate line spacing

Aronomink Systems has been contracted by Cecil Electric Cooperative, Inc. (CECI) to design a solid waste management system for the Cecil County plant, Units 1 and 2, to be built in Cranston, Maryland. The system will consist of two 600 MW pulverized coal-burning units fitted with high-efficiency electrostatic precipitators and limestone reagent FGD systems.

c. Inadequate line spacing

Aronomink Systems has been contracted by Cecil Electric Cooperative, Inc. (CECI) to design a solid waste management system for the Cecil County plant, Units 1 and 2, to be built in Cranston, Maryland. The system will consist of two 600 MW pulverized coal-burning units fitted with high-efficiency electrostatic precipitators and limestone reagent FGD systems.

Net (Cost)/Income *(Dollars in Millions)*	FY 2005	FY 2006	FY 2007	FY 2008	FY 2009
Earned Revenue	$ 1,372.8	$ 1,594.4	$ 1,735.7	$ 1,862.2	$ 1,927.1
Program Cost	(1,424.0)	(1,514.2)	(1,769.6)	(1,892.6)	(1,981.9)
Net (Cost)/Income	$ (51.2)	$ 80.2	$ (33.9)	$ (30.4)	$ (54.8)

STATEMENT OF NET COST

The Statement of Net Cost presents the USPTO's results of operations by the following responsibility segments – Patent, Trademark, and Intellectual Property Protection and Enforcement Domestically and Abroad. The above table presents the total USPTO's results of operations for the past five fiscal years. In FY 2005, the USPTO's operations resulted in a net cost. In FY 2006, the USPTO generated a net income due to the increased maintenance fees received and revenue recognition of previously deferred revenue collected subsequent to the fee increase on December 8, 2004. During FY 2007, FY 2008, and FY 2009 the USPTO's operations resulted in a net cost of $33.9 million, $30.4 million, and $54.8 million, respectively.

The Statement of Net Cost compares fees earned to costs incurred during a specific period of time. It is not necessarily an indicator of net income or net cost over the life of a patent or trademark. Net income or net cost for the fiscal year is dependent upon work that has been completed over the various phases of the production life cycle. The net income calculation is based on fees earned during the fiscal year being reported, regardless of when those fees were collected. Maintenance fees also play a large part in whether a total net income or net cost is recognized. Maintenance fees collected in FY 2009 are a reflection of patent issue levels 3.5, 7.5, and 11.5 years ago, rather than a reflection of patents issued in FY 2009. Therefore, maintenance fees can have a significant impact on matching costs and revenue.

During FY 2009, with the number of patent filings decreasing by 2.3 percent over the prior year, the backlog for patent applications likewise decreased, decreasing deferred revenue and increasing earned revenue. This was evidenced by the Patent organization disposing of 22.9 percent more applications than were disposed of during FY 2008.

During FY 2009, with the number of trademark applications decreasing by 12.3 percent over the prior year, the Trademark organization was able to continue to address the existing inventory and reduce pendency by 0.3 months from FY 2008. The Trademark organization was able to do this while recognizing a slight decrease in revenue earned.

EARNED REVENUE

The USPTO's earned revenue is derived from the fees collected for patent and trademark products and services. Fee collections are recognized as earned revenue when the activities to complete the work associated with the fee are completed. The table below presents the earned revenue for the past five years.

Earned revenue totaled $1,927.1 million for FY 2009, an increase of $64.9 million, or 3.5 percent, over FY 2008 earned revenue of $1,862.2 million. Of revenue earned during FY 2009, $454.3 million related to fee collections that were deferred for revenue recognition in prior fiscal years, $546.7 million related to maintenance fees collected during FY 2009, which were considered earned immediately, $920.7 million related to work performed for fees collected during FY 2009, and $5.4 million were not fee-related.

Earned Revenue *(Dollars in Millions)*	FY 2005	FY 2006	FY 2007	FY 2008	FY 2009
Patent	$ 1,197.8	$ 1,384.2	$ 1,507.0	$ 1,625.0	$ 1,697.4
Percentage Change in Patent Earned Revenue	*9.6%*	*15.6%*	*8.9%*	*7.8%*	*4.5%*
Trademark	175.0	210.2	228.7	237.2	229.7
Percentage Change in Trademark Earned Revenue	*19.5%*	*20.1%*	*8.8%*	*3.7%*	*(3.2)%*
Total Earned Revenue	$ 1,372.8	$ 1,594.4	$ 1,735.7	$ 1,862.2	$ 1,927.1
Percentage Change in Earned Revenue	*10.8%*	*16.1%*	*8.9%*	*7.3%*	*3.5%*

48 PERFORMANCE AND ACCOUNTABILITY REPORT: FISCAL YEAR 2009

The line spacing between two sections is greater than the line spacing within a section.

Line spacing is also used to separate the text from the graphics.

FIGURE 7.17 Line Spacing Used to Distinguish One Section from Another

Information from U.S. Patent and Trademark Office, 2010: www.uspto.gov/about/stratplan/ar/2009/2009annualreport.pdf.

Line spacing is usually determined by the kind of document you are writing. Memos and letters are single-spaced; reports, proposals, and similar documents are often double-spaced or one-and-a-half-spaced.

Figure 7.17 shows how line spacing can be used to distinguish one section of text from another and to separate text from graphics.

Justification Justification refers to the alignment of words along the left and right margins. In technical communication, text is often *left-justified* (also called *ragged right*). Except for the first line in each paragraph, which is sometimes indented, the lines begin along a uniform left margin but end on an irregular right margin. Ragged right is most common in word-processed text (even though word processors can justify the right margin).

In *justified* text, also called *full-justified text*, both the left and the right margin are justified. Justified text is seen most often in formal documents, such as books. The following passage (U.S. Department of Agriculture, 2002) is presented first in left-justified form and then in justified form:

Notice that the space between words is uniform in left-justified text.

We recruited participants to reflect the racial diversity of the area in which the focus groups were conducted. Participants had to meet the following eligibility criteria: have primary responsibility or share responsibility for cooking in their household; prepare food and cook in the home at least three times a week; eat meat and/or poultry; prepare meat and/or poultry in the home at least twice a week; and not regularly use a digital food thermometer when cooking at home.

In justified text, the spacing between words is irregular, slowing down the reader. Because a big space suggests a break between sentences, not a break between words, readers can become confused, frustrated, and fatigued.

Notice that the irregular spacing not only slows down reading but also can create "rivers" of white space. Readers are tempted to concentrate on the rivers running south rather than on the information itself.

We recruited participants to reflect the racial diversity of the area in which the focus groups were conducted. Participants had to meet the following eligibility criteria: have primary responsibility or share responsibility for cooking in their household; prepare food and cook in the home at least three times a week; eat meat and/or poultry; prepare meat and/or poultry in the home at least twice a week; and not regularly use a digital food thermometer when cooking at home.

Full justification can make the text harder to read in one more way. Some word processors and typesetting systems automatically hyphenate words that do not fit on the line. Hyphenation slows down and distracts the reader. Left-justified text does not require as much hyphenation as full-justified text.

TITLES AND HEADINGS

Titles and headings should stand out visually on the page because they introduce new ideas.

For more about titling your document, see Ch. 6, p. 105.

Titles Because the title is the most-important heading in a document, it should be displayed clearly and prominently. On a cover page or a title page, use boldface type in a large size, such as 18 or 24 points. If the title also appears at the top of the first page, make it slightly larger than the rest of the text—perhaps 16 or 18 points for a document printed in 12 point—but smaller than it is on the cover or title page. Many designers center titles on the page between the right and left margins.

Headings Readers should be able to tell when you are beginning a new topic. The most effective way to distinguish one level of heading from another is to use size variations (Williams & Spyridakis, 1992). Most readers will notice a 20-percent size difference between an A head (a first-level heading) and a B head (a second-level heading). Boldface also sets off headings effectively. The *least*-effective way to set off headings is underlining, because the underline obscures the *descenders*, the portions of letters that extend below the body of the letters, such as in *p* and *y*.

In general, the more important the heading, the closer it is to the left margin: A heads usually begin at the left margin, B heads are often indented a half inch, and C heads are often indented an inch. Indented C heads can also be run into the text.

For more about using headings, see Ch. 6, p. 106.

In designing headings, use line spacing carefully. A perceivable distance between a heading and the following text increases the impact of the heading. Consider these three examples:

Summary

In this example, the writer has skipped a line between the heading and the text that follows it. The heading stands out clearly.

Summary
In this example, the writer has not skipped a line between the heading and the text that follows it. The heading stands out, but not as emphatically.

Summary. In this example, the writer has begun the text on the same line as the heading. This run-in style makes the heading stand out the least.

OTHER DESIGN FEATURES

Table 7.1 shows five other design features that are used frequently in technical communication: rules, boxes, screens, marginal glosses, and pull quotes.

Analyzing Several Print-Document Designs

Figures 7.18 to 7.21 show typical designs used in print documents.

Designing Online Documents

Navigation aids are vitally important for online documents, because if your audience can't figure out how to find the information they want, they're out of luck. With a print document, they can at least flip through the pages.

The following discussion focuses on eight principles that can help you make it easy for readers to find and understand the information they seek:

- Use design to emphasize important information.
- Create informative headers and footers.
- Help readers navigate the document.
- Include extra features your readers might need.
- Help readers connect with others.
- Design for readers with disabilities.
- Design for multicultural readers.
- Aim for simplicity.

Although some of these principles do not apply to every type of online document, they provide a useful starting point as you think about designing your document.

TABLE 7.1 Additional Design Features for Technical Communication

Two types of rules are used here: vertical rules to separate the columns and horizontal rules to separate the items. Rules enable you to fit a lot of information on a page, but when overused they make the page look cluttered.	 Information from Institute of Scientific and Technical Communicators, "Industry News," in Communicator (Spring 2005).	**Rules.** *Rule* is a design term for a straight line. You can add rules to your document using the drawing tools in a word processor or as you define your styles. Horizontal rules can separate headers and footers from the body of the page or divide two sections of text. Vertical rules can separate columns on a multicolumn page or identify revised text in a manual. Rules exploit the principles of alignment and proximity.
	 OLDEST PIECES OF THE PLANET Information from J. W. Valley, "A cool early Earth?," Scientific American (October 2005): 58–65.	**Boxes.** Adding rules on all four sides of an item creates a box. Boxes can enclose graphics or special sections of text or can form a border for the whole page. Boxed text is often positioned to extend into the margin, giving it further emphasis. Boxes exploit the principles of contrast and repetition.

(*continued*)

TABLE 7.1 Additional Design Features for Technical Communication (*continued*)

The different-colored screens clearly distinguish the three sets of equations.	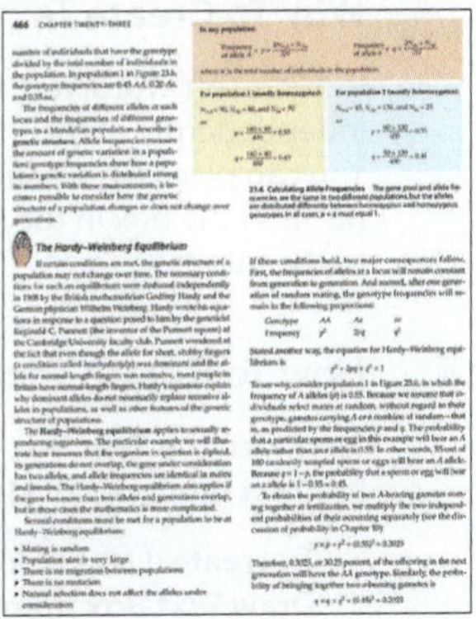 Information from W. K. Purves, D. Sadava, G. H. Orians, and H. C. Heller, LIFE: THE SCIENCE OF BIOLOGY, Seventh Edition, p. 466.	**Screens.** The background shading used behind text or graphics for emphasis is called a *screen*. The density of a screen can range from 1 percent to 100 percent; 5 to 10 percent is usually enough to provide emphasis without making the text illegible. You can use screens with or without boxes. Screens exploit the principles of contrast and repetition.
The marginal glosses present definitions of key words.	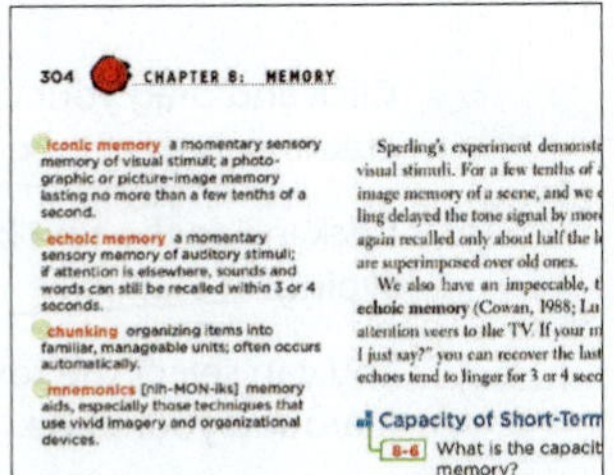 304 CHAPTER 8: MEMORY iconic memory a momentary sensory memory of visual stimuli; a photographic or picture-image memory lasting no more than a few tenths of a second. echoic memory a momentary sensory memory of auditory stimuli; if attention is elsewhere, sounds and words can still be recalled within 3 or 4 seconds. chunking organizing items into familiar, manageable units; often occurs automatically. mnemonics [nih-MON-iks] memory aids, especially those techniques that use vivid imagery and organizational devices. Information from PSYCHOLOGY, Tenth Edition (New York: Worth Publishers, 2013). David C. Myers, p. 304.	**Marginal glosses.** A marginal gloss is a brief comment on the main discussion. Marginal glosses are usually set in a different typeface—and sometimes in a different color—from the main discussion. Although marginal glosses can be helpful in providing a quick overview of the main discussion, they can also compete with the text for readers' attention. Marginal glosses exploit the principles of contrast and repetition.
This pull quote extends into the margin, but a pull quote can go anywhere on the page, even spanning two or more columns or the whole page.	Puritan leaders, however, interpreted an error caused either by a misguided r by the malevolent power of Satan. the cause, errors could lerated. As one Puritan proclaimed, "God doth in his word tolerate States, to give s to . . . adversaries of if they have power in s to suppress them. The saith . . . there is no one." Almost from the beginning, John Winthrop and other leaders had difficulty enforcing their views of Puritan orthodoxy. y after banishing Roger Winthrop confronted another dis- is time—as one New Englander ob- "Woman that Preaches better Gospell of your black-coates that have been at eversity." The woman was Anne n, a devout Puritan steeped in Scripture orbed by religious questions. n had received an excellent education Information from THE AMERICAN PROMISE: A HISTORY OF THE UNITED STATES, VOLUME I: TO 1877 (Boston: Bedford/St. Martin's, 2005). L. Roark, M. P. Johnson, P. C. Cohen, S. Stage, A. Lawson, and S. M. Hartman, p. 115.	**Pull quotes.** A pull quote is a brief quotation (usually just a sentence or two) that is pulled from the text, displayed in a larger type size and usually in a different typeface, and sometimes enclosed in a box. Newspapers and magazines use pull quotes to attract readers' attention. Pull quotes are inappropriate for reports and similar documents because they look too informal. They are increasingly popular, however, in newsletters. Pull quotes exploit the principles of contrast and repetition.

TECH TIP

Why To Create Borders and Screens

You can use background borders and screens (or shading) in Microsoft Word to emphasize page elements such as text and graphics. By emphasizing certain elements, you are letting your readers know what is important.

How To Create Borders and Screens

Start with the **Borders and Shading** dialog box.

To create a **border** around a page element or an entire page, select the area you want to format. Select the **Page Layout** tab, and then select **Page Borders** in the **Page Background** group.

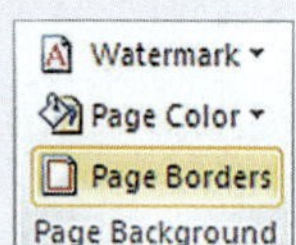

Select the **Borders** or **Page Border** tab.

You can specify the type of border, line style, color, and line width.

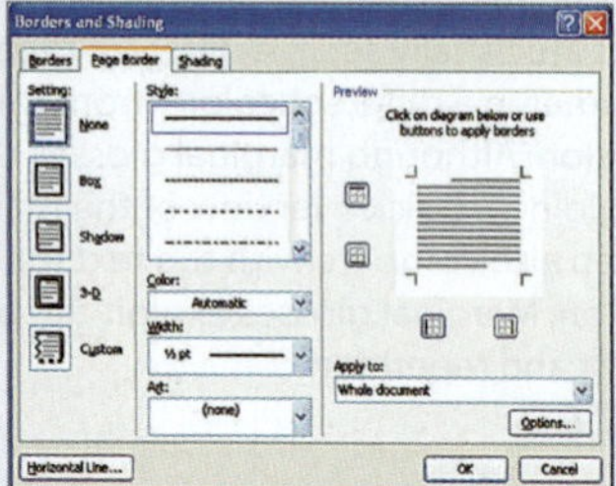

To create **shading**, also called a screen, select the area you want to format, and then select **Page Borders** on the **Page Background** group. Select the **Shading** tab.

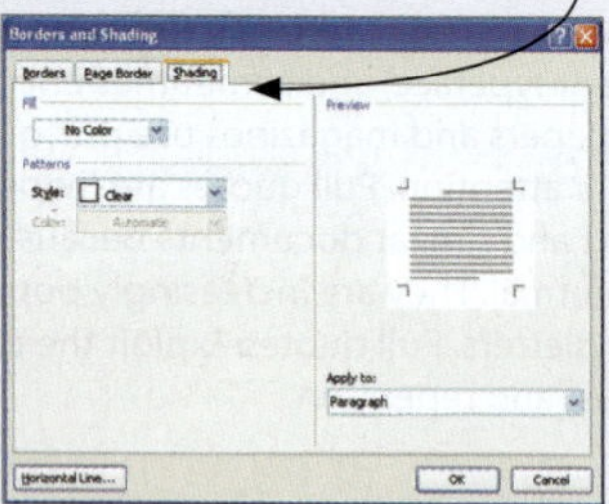

You can specify the color within the box as well as the style of the pattern.

TECH TIP

Why To Create Text Boxes

In Microsoft Word, text boxes allow you to position words independently of the margins and other settings on the page. You can place a quotation in a text box, for example, and then position the box in the middle of a column to draw attention to the quote.

How To Create Text Boxes

Start with the **Text Box** feature in the **Text** group on the **Insert** tab.

To **create** a text box, select **Draw Text Box** from the **Text Box** drop-down menu.

Click and drag your cursor to create your text box.

Click inside the text box and begin typing.

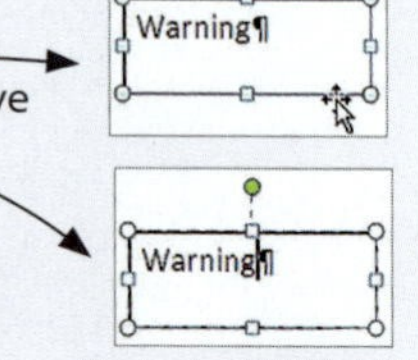

You can select the text box and move it around your page.

You can also insert a **built-in** text box from the **Text Box** drop-down menu.

To **format** your text box, select the box and then select the **Format Shape** dialog box launcher from the **Shape Styles** group on the **Format** tab.

The **Arrange** group enables you to specify design elements such as the text box's position in relation to other objects and the wrapping style of the surrounding text.

After selecting the box, you can also use buttons on the **Format** tab to specify such design elements as fill color, line color, font color, line style, and other effects.

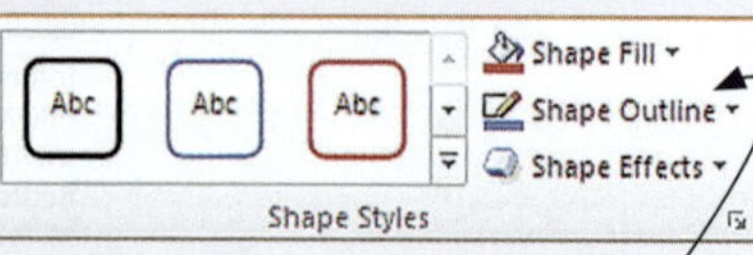

A multicolumn design enables you to present a lot of text and graphics of different sizes.

Financial Discussion and Analysis

Financial Highlights

The USPTO received an unqualified (clean) audit opinion from the independent public accounting firm of KPMG LLP on its FY 2012 financial statements, provided in the Financial Section of this report. This is the 20th consecutive year that the USPTO received a clean opinion. Our unqualified audit opinion provides independent assurance to the public that the information presented in the USPTO financial statements is fairly presented, in all material respects, in conformity with accounting principles generally accepted in the United States of America. In addition, KPMG LLP reported no material weaknesses in the USPTO's internal control, and no instances of non-compliance with laws and regulations affecting the financial statements. Refer to the Other Accompanying Information section for the Summary of Financial Statement Audit and Management Assurances.

The summary financial highlights presented in this section provide an analysis of the information that appears in the USPTO's FY 2012 financial statements. The USPTO financial management process ensures that management decision-making information is dependable, internal controls over financial reporting are effective, and that compliance with laws and regulations is maintained. The issuance of these financial statements is a component of the USPTO's objective to continually improve the accuracy and usefulness of its financial management information.

Balance Sheet and Statement of Changes in Net Position

At the end of FY 2012, the USPTO's consolidated Balance Sheet presents total assets of $1,982.1 million, total liabilities of $1,255.2 million, and a net position of $726.9 million.

Total assets increased 20.3 percent over the last four years, resulting largely from the increase in Fund Balance with Treasury. The decrease in Fund Balance with Treasury during FY 2009 is a result of the decrease in fee income. The following graph shows the changes in assets during this period.

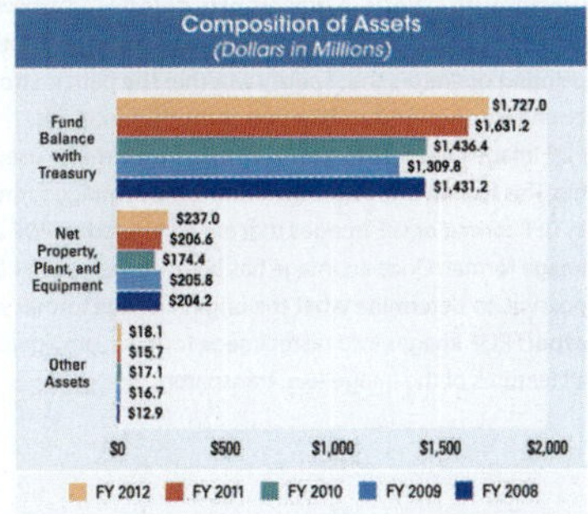

68

PERFORMANCE AND ACCOUNTABILITY REPORT: FISCAL YEAR 2012

Notice how the designer has used the whole width of the page for one graphic and a single column for a smaller graphic.

Note that the alley—the space between the two columns of text—need not be wide. Nor do you need to include a vertical rule to keep the columns separate. The human brain easily understands that each column is a separate space.

In this sample, the bar graph is exactly the width of the column in which it appears. But it doesn't have to be. It could break the shape of the column and extend into the other column or even into the margin. Or it could be narrower than its column, with the text wrapping around it. The design you see here looks neat and professional. If the graph were wider or narrower than the column, the design might appear somewhat more creative.

FIGURE 7.18 A Multicolumn Design

Information from U.S. Patent and Trademark Office, 2013: www.uspto.gov/about/stratplan/ar/USPTOFY2012PAR.pdf.

This page from a software company's white paper—a marketing document usually distributed on the web—shows one approach to a one-column design.

The main text column is relatively narrow, making the line easy to read.

The right margin is wide enough to accommodate text boxes, small graphics, or other items.

One goal of document design is to reduce the number of pages needed—but when you design a page, you want to make the text inviting and easy to read. Figuring out how to balance these two priorities is one of the major challenges of designing a page.

PDF Page Content

Page Description Language

PDF is a page description language, i.e. it describes how a page looks so that it can be reproduced for viewing and printing. The language resembles Postscript, but is much simpler to allow for more efficient processing. For example, it does not contain control structures like loops and „if" statements.

PDF Page Content Elements

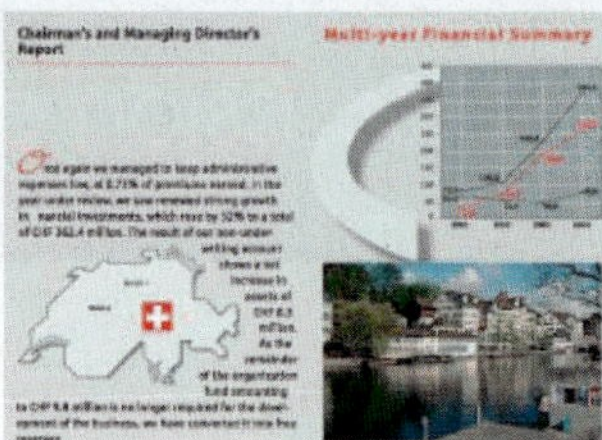

Basically PDF recognizes three types of page content elements:

- Text (fonts programs)
- Graphic paths (lines and curves)
- Images (raster samples)

The picture to the right shows examples of the three PDF content types.

Content Objects

PDF uses objects and object types to describe the content. Every string of text and all graphics and images are defined by one or several objects, created from one or more object types.

- Text Objects. Text objects are defined by a number of attributes including font family, style and size, a string of characters, and a position on a page. PDF does not recognize nor store objects for line breaks, headers, paragraphs, indentation etc. (i.e. paragraph formatting operators used in word processing applications like Microsoft Word). Text is broken down into fragments as small as single characters but not more than one line. The fragments can be randomly stored and are like pieces of a puzzle that all have to be placed in their correct location on the page to complete its appearance.
- Graphic Path Objects. A graphic path object is an arbitrary shape made up of straight lines, rectangles, and cubic Bézier curves. A graphic path object ends with one or more painting operators that specify whether the path is stroked, filled, used as a clipping boundary or some combination of these operations.
- PDF Image Objects. A PDF-specific image format is used for embedding images in a PDF file. This format is independent of the input image format. For example, scanned pages in TIFF format or GIF images that are converted to PDF are newly packaged into PDF image format. Once an image has been converted to PDF image format, it is usually not possible to determine what the original image format was. It is however possible to export PDF images into raster image formats, provided the raster image format supports all features of the image (e.g. transparency).

Unlike word processors, text is not continuous in a paragraph.

Text is defined in fragments with attributes such as font and font size, a string of characters and a location on a page.

Images that are imported into a PDF file are converted to PDF image format. They are not stored as TIFF, GIF etc. images.

FIGURE 7.19 A One-Column Design
Information from PDF Tools AG.

The Joint Inspection Team visits France and Italy's Concordia Station. *Photo by Evan Bloom.* **Below:** Then-Secretary of State Clinton speaks at the 2009 Joint Session of the Antarctic Treaty Consultative Meeting. *Department of State photo.*

inspections in East Antarctica required the team to travel more than 3,500 miles over six days by plane, truck, boat, helicopter, tracked vehicle and snowmobile.

The inspections, representing the first time either country had conducted a joint inspection in Antarctica, were called for in an agreement that Secretary of State Hillary Rodham Clinton signed with Russian Foreign Minister Sergey Lavrov in September 2012. The United States and Russia were architects of the Antarctic Treaty of 1959 and today conduct some of the most extensive and diverse scientific activities in Antarctica. Importantly, both countries reject territorial claims by other parties and are strong supporters of the Antarctic Treaty system. Working closely with our Russian counterparts provided an excellent opportunity to reinforce our shared objectives for the peace and science in Antarctica. The results of the inspection will be presented to all treaty parties at the May Antarctic Treaty Consultative Meeting.

Antarctica is an outstanding example of multilateral diplomatic success. Fifty years after the signing of the Antarctic Treaty, the continent is a global example of policy and scientific collaboration. The multinational science conducted in Antarctica informs global understanding of the Earth's history, processes and change, and policy and logistical cooperation there creates stronger ties among treaty parties. In the coming decades, the Antarctic Treaty system will continue to prove the resilience and value of multilateral cooperation. ◻

12 STATE MAGAZINE // MARCH 2013

This is a page from *State*, the magazine for employees of the U.S. State Department. Magazines for people who work together tend to feature a lot of photographs, including many showing people from that organization.

The large photograph extends to the top edge and the left edge of the page. Eliminating the margins in this way would be a mistake if the graphic were crammed with information that readers needed to study; a dense table of data, for instance, would be overwhelming. But in this case, the blue sky in the background acts as a decorative frame for the "information" in the bottom half of the photo.

Although this page uses a simple three-column design, note that one photo spans all three columns, another photo spans two columns, and the caption box spans one column. This creative use of the multicolumn design enables the designer to fill the page with content while keeping it visually interesting.

FIGURE 7.20 A Magazine Page Design

Information from U.S. Department of State, 2013: www.state.gov/documents/organization/205362.pdf.

The writer of this document hasn't designed the page. He or she has simply hit the Enter key repeatedly.

The full justification makes for a boxy appearance and irregular spacing between words.

The wide column results in long, difficult-to-read lines.

The two hierarchical levels—numbered and lettered—have the same design and therefore are difficult to distinguish from each other.

In the table, the second column is misaligned.

The footer, which includes the date and the page number, is a useful design feature, however.

(f) *Drug Purity - DEA Form 7*

(1) The presentence report will normally provide drug weight/purity information from DEA Form 7. This is a complicated form. "Total net weight" [normally in Item 31] refers to the amount of the pure drug. This is the weight used in calculation of the Commission's severity rating. For your information, "gross weight" is the weight of the drug plus adulterants plus the container (normally found in Item 24). Also normally found in Item 24 is "net weight" (the weight of the drug plus adulterants). "Strength" (the percent purity of the drug) is normally found in Item 28. Multiplying "net weight" x "strength" is how DEA arrives at the "total net weight". Remember, "total net weight" is the weight of the pure drug to be used in assessing the Commission's severity rating.

(2) If a presentence report does not specify "total net weight", the probation officer should be contacted for clarification (please be specific as to the clarification necessary; this will enhance feedback/training). Note: DEA lab reports (DEA Form 7), if necessary, also may be obtained directly from the DEA field office for the geographic area in which the offense occurred. If a request to the DEA field office is required, provide the subject's name, date of birth, place of offense, and dates of offense.

(g) If neither weight nor purity is available, but only a money value, DEA may be requested to provide an estimate of the amount of pure drug associated with that money value. In the absence of a specific estimate from DEA pertaining to the particular case, DEA publishes a report (Domestic Drug Prices) providing estimates of average drug prices by year and region from which an estimate may be obtained.

(h) *Determining Offense Severity Relative to Simple Possession of Drugs.* In certain cases, the Commission must determine whether the offense behavior should be considered as "simple possession" of a controlled substance or "possession with intent to distribute." In making this determination, the Commission shall examine a variety of factors (if available). These factors are shown below. The presence of any of the following factors may be considered as a presumption of possession with intent to distribute. However, this presumption may be rebutted if there are circumstances in the individual case which indicate that there was no intention to distribute.

(1) *Weight/amount/purity of the substance:* Possession of the following amounts of controlled substances are presumed to indicate possession with intent to distribute:

Heroin 1 gm. at 100% purity, or equivalent amount; or more
Cocaine 5 gms. at 100% purity, or equivalent amount; or more
Marijuana 10 lbs. or more
Hashish 3 lbs. or more
Hash Oil .3 lbs. or more
Drugs (other than above) 1,000 doses or more.

(2) *Other Factors*: The presence of any of the following factors may be considered indicative of intent to distribute: (A) the substance has been separated into multiple, individual packets; (B) the offender is a non-user of the substance in question; (C) the presence of instruments used in preparing

*Terms marked by an asterisk are defined in Chapter Thirteen.

11/15/07 Page 60 of 337

FIGURE 7.21 A Poorly Designed Page

Information from U.S. Department of Justice, 2010: www.justice.gov/uspc/rules_procedures/uspc-manual111507.pdf.

DOCUMENT ANALYSIS ACTIVITY

Analyzing a Page Design

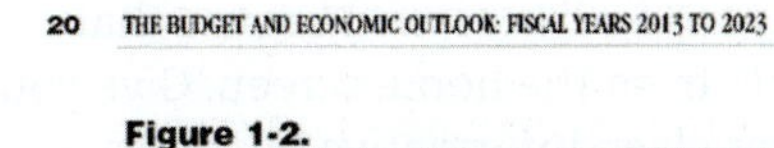

20 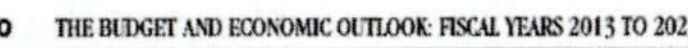
THE BUDGET AND ECONOMIC OUTLOOK: FISCAL YEARS 2013 TO 2023 FEBRUARY 2013

Figure 1-2.

Total Revenues and Outlays

(Percentage of gross domestic product)

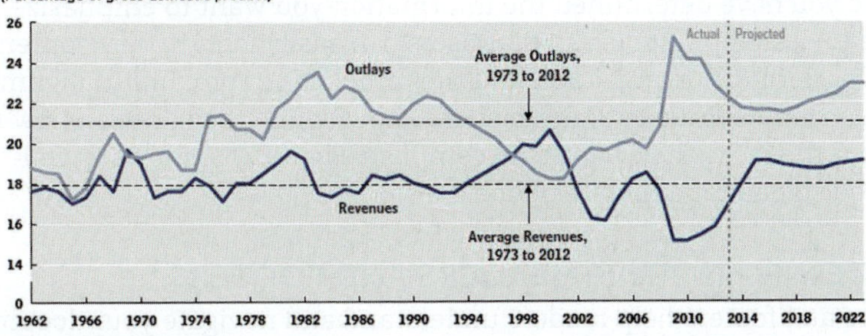

Source: Congressional Budget Office.

health care costs, and a significant expansion in eligibility for federal subsidies for health insurance, outlays for Social Security and the federal government's major health care programs are projected to rise substantially relative to the size of the economy over the next 10 years. In addition, growing debt and rising interest rates will boost net interest payments. Spending on all other programs—in the aggregate—is projected to decline relative to GDP between 2014 and 2023, primarily because of improving economic conditions and the spending limits in current law.

Revenues

CBO projects that, if current tax laws remain unchanged, revenues will rise relative to GDP over the next two years and then remain at about 19 percent of GDP through 2023. After 2015, increases in individual income tax receipts relative to GDP will roughly offset projected declines in corporate income tax receipts and declines in remittances from the Federal Reserve as a share of GDP.

Individual Income Taxes. CBO projects that, under current law, individual income tax receipts will rise from \$1.3 trillion this year to \$2.5 trillion in 2023—or from 7.9 percent to 9.8 percent of GDP. The projected increase in receipts relative to the economy in CBO's baseline reflects real (inflation-adjusted) bracket creep, the economic expansion, recent and scheduled changes in tax provisions, and other factors. In previous baselines, CBO had projected that those receipts would increase to a much higher percentage of GDP by the early part of the next decade, but the American Taxpayer Relief Act's permanent extension of most of the expiring income tax reductions has significantly reduced the amount of revenues anticipated under current law.

Real Bracket Creep. Increases in real income will push more income into higher tax brackets, which boosts revenues relative to GDP in CBO's projections by 0.9 percentage points over the next decade.[11]

Economic Recovery. CBO expects that the economic expansion and related factors will cause taxable incomes to rise faster than GDP, boosting individual income tax revenues as a share of GDP by about 0.4 percentage points over the next decade: most of that effect will occur by 2017. Certain components of taxable income—including wages and salaries, capital gains realizations, interest income, and proprietors' income—declined as a share of GDP over the past several years. CBO expects that, as the economy recovers, such income will rebound more quickly than the economy as a whole, increasing

11. Roughly three-quarters of that amount is a longer-term effect that results from increases in the potential output of the economy (that is, the maximum sustainable level of economic output), and the rest results from the return of output to its potential level over the next several years.

Information from U.S. Congressional Budget Office, 2013: www.cbo.gov/sites/default/files/cbofiles/attachments/43907-BudgetOutlook.pdf.

This page is from a government report. The questions below ask you to think about page design (as discussed on pp. 157–74).

1. How many levels of headings appear on this page? Are the different levels designed effectively so that they are easy to distinguish? If not, what changes would you make to the design?
2. How are rules used on this page? Are they effective? Would you change any of them?
3. Describe the design of the body text on this page, focusing on columns and alignment. Is the design of the body text effective? Would you change it in any way?

USE DESIGN TO EMPHASIZE IMPORTANT INFORMATION

The smaller the screen, the more cluttered it can become, making it difficult for readers to see what is truly important. In documents designed to be viewed on different-sized screens, you want readers to be able to find what they want quickly and easily. As you begin planning an online document, decide what types of information are most essential for your audience, and ensure that that content in particular is clearly accessible from the home screen. Give your buttons, tabs, and other navigational features clear, informative headings. For more guidance on emphasizing important information, see Chapter 6.

Once you have determined the information you want to emphasize, adhere to design principles so that users can easily identify key content. Use logical patterns of organization and the principles of proximity, alignment, repetition, and contrast so that readers know where they are and how to carry out the tasks they want to accomplish. Figure 7.22 shows a well-designed screen for a mobile phone.

CREATE INFORMATIVE HEADERS AND FOOTERS

Headers and footers help readers understand and navigate your document, and they help establish your credibility. You want readers to know that they are reading an official document from your organization and that it was created by professionals. Figure 7.23 shows a typical website header, and Figure 7.24 shows a typical footer.

HELP READERS NAVIGATE THE DOCUMENT

One important way to help readers navigate is to create and sustain a consistent visual design on every page or screen. Make the header, footer, background color or pattern, typography (typeface, type size, and color), and the placement of links the same on every page. That way, readers will know where to look for these items.

INCLUDE EXTRA FEATURES YOUR READERS MIGHT NEED

Because readers with a range of interests and needs will visit your site, consider adding some or all of the following features:

- **An FAQ page.** A list of frequently asked questions helps new readers by providing basic information, explaining how to use the site, and directing them to more-detailed discussions.
- **A search page or engine.** A search page or search engine enables readers to enter a keyword or phrase and find all the pages in the document that contain it.
- **Resource links.** If one of the purposes of your document is to educate readers, provide links to other sites.

FIGURE 7.22 Screen for a Mobile Application

Sepsis Clinical Guide mobile app, Escavo, Inc. (www.escavo.com). Used by permission.

This app helps physicians diagnose sepsis quickly and effectively. The information most crucial to evaluating the condition is easily accessible on the home screen. At the top of the screen, where the reader's eyes will initially fall, is an overview of the condition and, most importantly, the diagnostic tool. Less-essential items, such as resources and references, are located at the bottom of the screen. Supplementary information, such as a call for authors and a feedback form, is deeper in the app, behind the "More" tab.

This simple screen uses the principle of contrast effectively to highlight key content. Each of the eight main content areas has its own color and its own icon to distinguish it from the seven other areas. In addition, the four navigation items at the bottom of the screen use contrast in that the name and icon for the screen the reader is now viewing—in this case, the home page—are presented against a light blue screen, whereas those for the other three navigation buttons are presented against a dark blue screen.

FIGURE 7.23 Website Header

Notice that a header in a website provides much more accessing information than a header in a printed document. This header enables readers to search the site, as the header on almost every site does, but it also includes other elements that are particularly important to the Michael J. Fox Foundation. For instance, there is a link to drug trials that visitors might want to join, and there is a prominent link for donating to the foundation.

The Michael J. Fox Foundation for Parkinson's Research. Grand Central Station, P.O. Box 4777, New York, NY 10163-4777. Tel: 1-800-708-7644
Copyright © 2013 Michael J. Fox Foundation. Privacy Policy | Terms & Conditions | Sitemap

The top line of the footer contains the contact information for the site.

The bottom line contains the copyright notice, as well as links to the privacy policy, the terms and conditions for using the site, and a site map.

FIGURE 7.24 Website Footer

This simply designed footer presents all the links as text. Readers with impaired vision who use text-to-speech devices will be able to understand these textual links; they would not be able to understand graphical links without alternative text.

- **A printable version of your site.** Online documents are designed for a screen, not a page. A printable version of your document, with black text on a white background and all the text and graphics consolidated into one big file, will save readers paper and toner.
- **A text-only version of your document.** Many readers with impaired vision rely on text because their specialized software cannot interpret graphics. Consider creating a text-only version of your document for these readers, and include a link to it on your home page.

GUIDELINES Making Your Document Easy to Navigate

Follow these five suggestions to make it easy for readers to find what they want in your document.

- **Include a site map or index.** A site map, which lists the pages on the site, can be a graphic or a textual list of the pages, classified according to logical categories. An index is an alphabetized list of the pages. Figure 7.25 shows a portion of a site map.

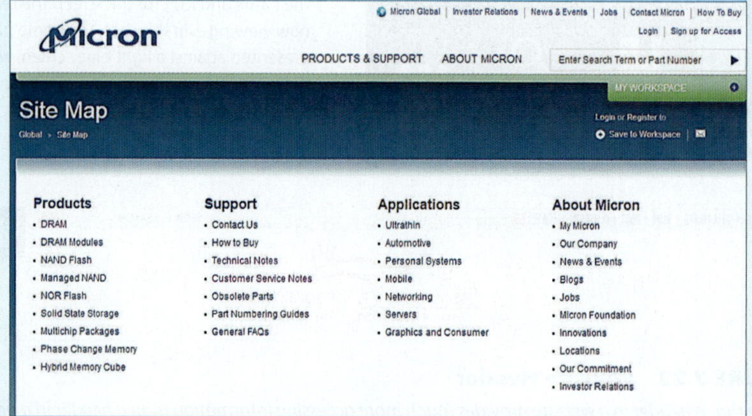

FIGURE 7.25 Site Map

For large websites, help your readers by organizing the site map rather than just presenting an alphabetical list of the pages. In this portion of a site map, Micron Technology classifies the pages in logical categories to help visitors find the pages they seek.

- **Use a table of contents at the top of long pages.** If your page extends for more than a couple of screens, include a table of contents—a set of links to the items on that page—so that your readers do not have to scroll down to find the topic they want. Tables of contents can link to information farther down on the same page or to information on separate pages. Figure 7.26 shows an excerpt from the table of contents at the top of a frequently asked questions (FAQ) page.

(continued)

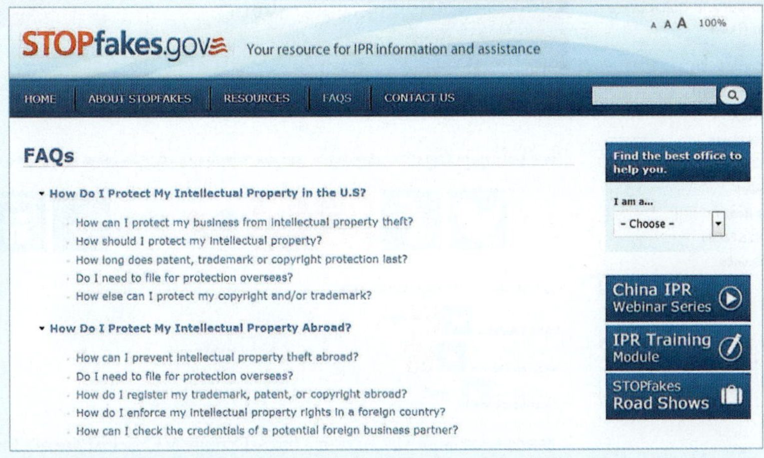

FIGURE 7.26 Table of Contents

The table of contents is classified by topic (first all the topics about protecting your intellectual property in the United States, then all the topics about protecting it outside the United States). For any online document, large or small, use the principles of organizing information presented in Chapter 6.

Information from U.S. Department of Commerce, 2013: www.stopfakes.gov/faqs.

- **Help readers get back to the top of long pages.** If a page is long enough to justify a table of contents, include a "Back to top" link (a textual link or a button or icon) before the start of each new chunk of information.
- **Include a link to the home page on every page.** This link can be a simple "Back to home page" textual link, a button, or an icon.
- **Include textual navigational links at the bottom of the page.** If you use buttons or icons for links, include textual versions of those links at the bottom of the page. Readers with impaired vision might use special software that reads the information on the screen. This software interprets text only, not graphics.

HELP READERS CONNECT WITH OTHERS

Organizations use their online documents, in particular their websites, to promote interaction with clients, customers, suppliers, journalists, government agencies, and the general public. For this reason, most organizations use their sites to encourage their various stakeholders to connect with them through social media such as discussion boards and blogs.

Use your online document to direct readers to interactive features of your own website, as well as to your pages on social-media sites such as Facebook or Twitter. Figure 7.27 shows a portion of NASA's community page.

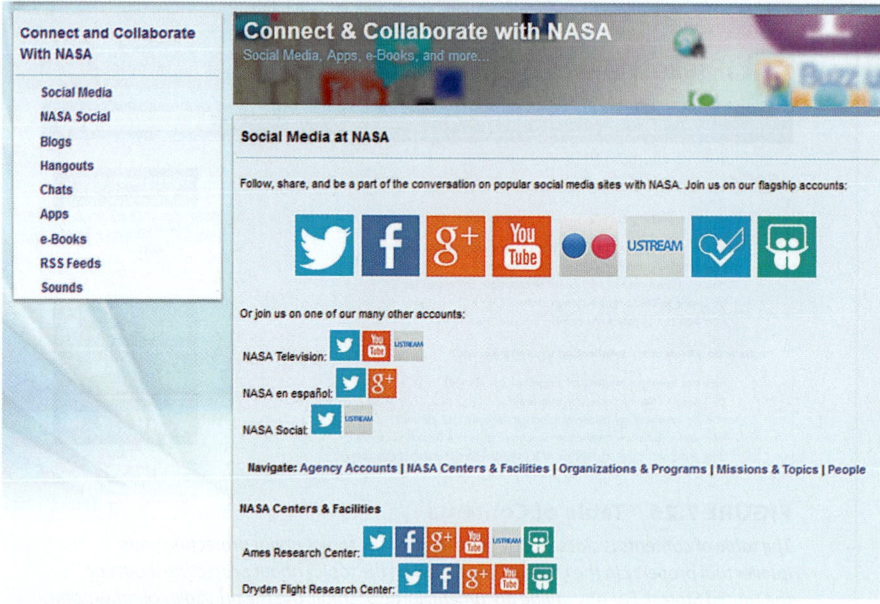

FIGURE 7.27 Maintaining Communities

As a federal agency with a mission that includes public education, NASA has a robust social-media presence. NASA has dozens of accounts on Facebook and the other popular social-media platforms, and it offers many opportunities for scientists and the general public alike to stay connected through blogs, podcasts, chat rooms, and educational programs and activities on such topics as how to view the International Space Station.

Information from National Aeronautics and Space Administration, 2013: www.nasa.gov/socialmedia/.

CONSIDER MATTERS OF ACCESSIBILITY

According to web Accessibility in Mind (WebAIM, 2016), an estimated 20 percent of the population has some form of visual, hearing, motor, or cognitive disability. Impairments vary wildly in type and severity, but a significant number of people will be denied access to online materials unless that material includes adaptations such as closed captioning of videos, links to audio for text files, and magnification of visuals. It's worth noting that such adaptations are also beneficial to people without disabilities who are in situations in which they need alternative accessibility—for example, when audio is unavailable or when they are using a very small screen.

Laws in the United States and throughout the world require that websites be accessible to people with disabilities. In addition to being a matter of ethics, accessibility also makes good business sense: why exclude any members of the population? Moreover, principles of accessible design generally coincide with the principles of good design in general. The Worldwide Web Consortium (W3C), the governing body of the web, has produced a list of international guidelines for good web design. WebAIM summarizes these guidelines as the "POUR" principles: documents should be perceivable, operable, understandable, and robust.

- **Perceivable:** Content is available to the senses (vision and hearing primarily) either through the browser or through assistive technologies (for example, screen readers, screen enlargers).
- **Operable:** Users can interact with all controls and interactive elements using either a mouse, a keyboard, or an assistive device.
- **Understandable:** Content is clear and limits confusion and ambiguity.
- **Robust:** A wide range of technologies (including old and new user agents and assistive technologies) can access the content.

When designing for web and mobile devices, it's helpful to consider matters of accessibility in the planning stages of your design. For example, keep in mind the space that may be needed for closed captioning to run at the bottom of the screen. Or think about the placement of labels on graphics that may be magnified up to 200 percent of the original size. A more detailed list of design principles is included in the Guidelines box "Designing Accessible websites."

GUIDELINES Designing Accessible Websites

The following key principles of accessible design have been adapted from WebAIM (2016). Most accessibility principles can be implemented very easily and will not affect the overall look or "feel" of your website. This list does not cover all accessibility issues, but by addressing these eleven basic principles, you can ensure that your web content is more accessible to everyone.

- **Provide appropriate alternative text.** Alternative text is a textual alternative to non-text content, such as the graphics in web pages. It is especially helpful for people who are blind and rely on a screen reader to read the content of the website to them.
- **Provide appropriate document structure.** Headings, lists, and other structural elements lend meaning and structure to web pages. They can also facilitate keyboard navigation within the page.
- **Provide headers for data tables.** Tables are used online for layout and to organize data. Tables that are used to organize data should have appropriate column headers. Data cells should be associated with their appropriate headers, making it easier for users of screen readers to navigate and understand the data table.
- **Ensure that users can complete and submit all forms.** Ensure that every element of a form (text field, checkbox, dropdown list, etc.) has a label. Also make sure the user can submit the form and recover from any errors, such as the failure to fill in all required fields.
- **Ensure that links make sense out of context.** Every link should make sense if the link text is read by itself. Users of screen readers may choose to read only the links on a web page. Certain phrases like "click here" and "more" should be avoided.
- **Caption and provide transcripts for media.** Videos and live audio must have captions and a transcript. With archived audio, a transcript may be sufficient.

(continued)

- **Ensure accessibility of non-HTML content.** PDF documents and other non-HTML content must be as accessible as possible. If you cannot make such content accessible, consider using HTML formatting instead or, at the very least, provide an accessible alternative. A PDF document should also include a series of tags to make it more accessible. A tagged PDF file looks the same as an untagged file, but it is almost always more accessible to a person using a screen reader.
- **Allow users to skip repetitive elements on the page.** You should provide a method that allows users to skip navigation or other elements that repeat on every page. This is usually accomplished by providing a "Skip to Main Content" or "Skip Navigation" link at the top of the page; such a link allows the user to jump to the main content of the page.
- **Do not rely on color alone to convey meaning.** The use of color can enhance comprehension, but do not use color alone to convey information. That information may not be available to a person who is colorblind and will be unavailable to users of screen readers.
- **Make sure content is clearly written and easy to read.** There are many ways to make your content easier to understand. Write clearly, use clear fonts, and use headings and lists appropriately.
- **Design to standards.** HTML-compliant and accessible pages are more robust and provide better search engine optimization. Cascading Style Sheets (CSS) allow you to separate content from presentation, providing more flexibility and making your content more accessible.

DESIGN FOR MULTICULTURAL AUDIENCES

Only about 9 percent of the people using the Internet are from North America (Internet World Stats, 2016). Therefore, it makes sense in planning your online documents to assume that many of your readers will not be proficient in English.

Planning for a multicultural website is similar to planning for a multicultural printed document:

- **Use common words and short sentences and paragraphs.**
- **Avoid idioms, both verbal and visual, that might be confusing.** For instance, don't use sports metaphors, such as *full-court press*, or a graphic of an American-style mailbox to suggest an email link.
- **If a large percentage of your readers speak a language other than English, consider creating a version of your site in that language.** The expense can be considerable, but so can the benefits.

ETHICS NOTE

DESIGNING LEGAL AND HONEST ONLINE DOCUMENTS

You know that the words and images that you see on the Internet are covered by copyright, even if you do not see a copyright symbol. The only exception is information that is in the public domain because it is not covered by copyright or because the copyright has expired, or because the creator of the information has explicitly stated that the information is in the public domain and you are free to copy it.

But what about the design of a site? Almost all web designers readily admit to spending a lot of time looking at other sites and pages for inspiration. And they admit to looking at the computer code to see how that design was achieved. This is perfectly ethical. So is copying the code for routine elements such as tables. But is it ethical to copy the code for a whole page, including the layout and the design, and then plug in your own data? No. Your responsibility is to create your own information, then display it with your own design.

For more about copyright law, see Ch. 2, "Obligations to Copyright Holders."

AIM FOR SIMPLICITY

Well-designed online documents are simple, with only a few colors and nothing extraneous. The text is easy to read and chunked effectively, and the links are written carefully so readers know where they are being directed.

Analyzing Several Online-Document Designs

The best way to learn about designing websites and their pages is to study them. Figures 7.28 to 7.30 offer examples of good web page design.

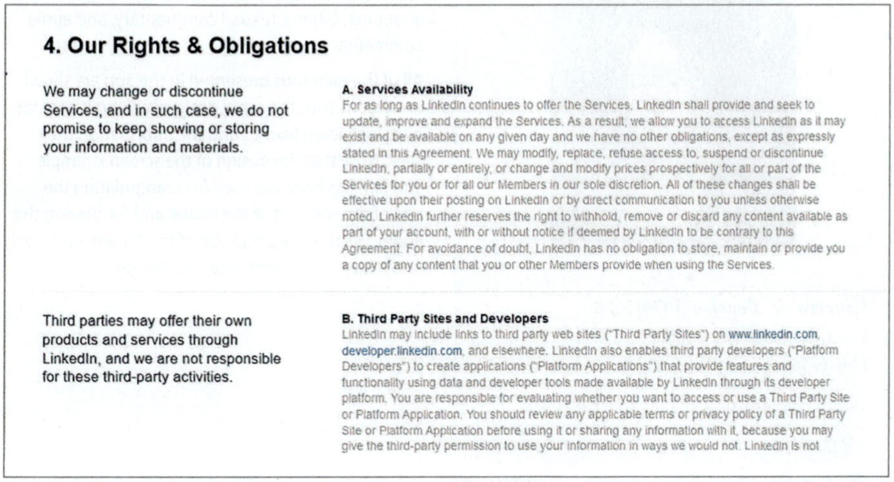

4. Our Rights & Obligations

We may change or discontinue Services, and in such case, we do not promise to keep showing or storing your information and materials.	**A. Services Availability** For as long as Linkedin continues to offer the Services, Linkedin shall provide and seek to update, improve and expand the Services. As a result, we allow you to access Linkedin as it may exist and be available on any given day and we have no other obligations, except as expressly stated in this Agreement. We may modify, replace, refuse access to, suspend or discontinue Linkedin, partially or entirely, or change and modify prices prospectively for all or part of the Services for you or for all our Members in our sole discretion. All of these changes shall be effective upon their posting on Linkedin or by direct communication to you unless otherwise noted. Linkedin further reserves the right to withhold, remove or discard any content available as part of your account, with or without notice if deemed by Linkedin to be contrary to this Agreement. For avoidance of doubt, Linkedin has no obligation to store, maintain or provide you a copy of any content that you or other Members provide when using the Services.
Third parties may offer their own products and services through LinkedIn, and we are not responsible for these third-party activities.	**B. Third Party Sites and Developers** Linkedin may include links to third party web sites ("Third Party Sites") on www.linkedin.com, developer.linkedin.com, and elsewhere. Linkedin also enables third party developers ("Platform Developers") to create applications ("Platform Applications") that provide features and functionality using data and developer tools made available by Linkedin through its developer platform. You are responsible for evaluating whether you want to access or use a Third Party Site or Platform Application. You should review any applicable terms or privacy policy of a Third Party Site or Platform Application before using it or sharing any information with it, because you may give the third-party permission to use your information in ways we would not. Linkedin is not

FIGURE 7.28 Making the Small Print a Little Larger

Information from LinkedIn, 2013: http://www.linkedin.com/legal/user-agreement.

Nobody likes user agreements, and few people read them carefully. LinkedIn, the online professional network, uses a simple table design to make its user agreement a little easier to read.

In this excerpt, the left column presents a simple overview of a portion of the agreement. The right column presents the "small print": the specific provision, including links to even more detailed information.

FIGURE 7.29 **An About Us Page**

The "About NIH" page on the National Institutes of Health website conveys its message simply but effectively.

The top row is reserved for the name of this government agency.

Below the agency's name is the main navigation pane, beginning with "Health Information."

Below the main navigation pane is the navigation pane for the section in which this page appears: "About NIH." The "About NIH" page has 18 sections, beginning with "Mission."

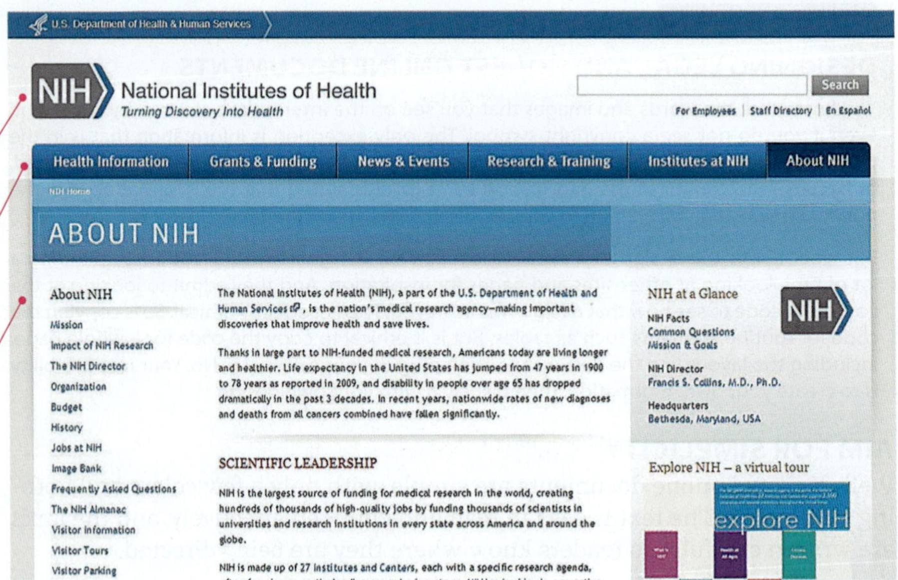

FIGURE 7.30 **An App Designed for a Small Screen**

Information from National Gallery of Art, 2013: http://apps.usa.gov/yourart.shtml.

Your Art, an app from the U.S. National Gallery of Art, enables museum visitors—and anyone with an Internet connection—to see and learn about many of the art treasures displayed in the museum. The app includes numerous features, including news about exhibitions, textual commentary, and audio commentary.

All of the paintings presented in the app are shown on screens with the same design, making it easy for people to learn how to find the information they seek. Note that the design of the screen is simple and familiar icons are used for manipulating the size of the text and of the image and for playing the audio. Despite the small size of the screen, users will find it easy to navigate and use the app.

GUIDELINES Designing Simple, Clear Web Pages

Follow these eight suggestions to make your design attractive and easy to use.

- **Use conservative color combinations to increase text legibility.** The greater the contrast between the text color and the background color, the more legible the text. The most legible color combination is black text against a white background. Bad idea: black on purple.
- **Avoid decorative graphics.** Don't waste space using graphics that convey no useful information. Think twice before you use clip art.
- **Use thumbnail graphics.** Instead of a large graphic, which takes up space and requires a long time to download, use a thumbnail that readers can click on if they wish to open a larger version.
- **Keep the text short.** Poor screen resolution makes reading long stretches of text difficult. In general, pages should contain no more than two or three screens of information.
- **Chunk information.** When you write for the screen, chunk information to make it easier to understand. Use frequent headings, brief paragraphs, and lists.
- **Make the text as simple as possible.** Use common words and short sentences to make the information as simple as the subject allows.
- **Structure your sentences as if there were no links in your text.**

AWKWARD	Click here to go to the Rehabilitation Center page, which links to research centers across the nation.
SMOOTH	The Rehabilitation Center page links to research centers across the nation.

- **Indicate what information a linked page contains.** Readers get frustrated if they wait for a web page to download and then discover that it doesn't contain the information they expected.

UNINFORMATIVE	See the Rehabilitation Center.
INFORMATIVE	See the Rehabilitation Center's hours of operation.

For more about chunking, see "Guidelines: Understanding Learning Theory and Page Design."

WRITER'S CHECKLIST

Did you

- ☐ analyze your audience: their knowledge of the subject, their attitudes, their reasons for reading, and the kinds of tasks they will be carrying out? *(p. 153)*
- ☐ consider the purpose or purposes you are trying to achieve? *(p. 154)*
- ☐ determine your resources in time, money, and equipment? *(p. 154)*

Designing Print Documents and Pages

Did you

- ☐ think about which accessing aids would be most appropriate, such as icons, color, dividers and tabs, and cross-reference tables? *(p. 158)*
- ☐ use color, if available, to highlight certain items, such as warnings? *(p. 158)*
- ☐ devise a style for headers and footers? *(p. 159)*
- ☐ devise a style for page numbers? *(p. 159)*
- ☐ draw thumbnail sketches and page grids that define columns and white space? *(p. 161)*
- ☐ choose typefaces that are appropriate for your subject? *(p. 165)*
- ☐ use appropriate styles from the type families? *(p. 165)*
- ☐ use type sizes that are appropriate for your subject and audience? *(p. 167)*
- ☐ choose a line length that is suitable for your subject and audience? *(p. 168)*
- ☐ choose line spacing that is suitable for your line length, subject, and audience? *(p. 168)*
- ☐ consider whether to use left-justified text or full-justified text? *(p. 169)*
- ☐ design your title for clarity and emphasis? *(p. 170)*
- ☐ devise a logical, consistent style for each heading level? *(p. 170)*
- ☐ use rules, boxes, screens, marginal glosses, and pull quotes where appropriate? *(p. 172)*

Designing Online Documents

Did you

- ☐ create informative headers and footers? *(p. 180)*
- ☐ include extra features your readers might need, such as an FAQ page, a search page or engine, resource links, a printable version of your site, or a text-only version? *(p. 180)*
- ☐ help readers navigate the site by including a site map, a table of contents, "Back to top" links, and textual navigation buttons? *(p. 182)*
- ☐ help readers connect with others through links to interactive portions of your site and to social-media sites? *(p. 183)*
- ☐ design for readers with vision, hearing, or mobility impairment? *(p. 184)*
- ☐ design for multicultural audiences? *(p. 186)*
- ☐ aim for simplicity in web page design by using simple backgrounds and conservative color combinations and by avoiding decorative graphics? *(p. 187)*
- ☐ make the text easy to read and understand by keeping it short, chunking information, and writing simply? *(p. 187)*

EXERCISES

For more about memos, see Ch. 9, p. 248.

1. Study the first and second pages of an article in a journal in your field. Describe ten design features on these two pages. Which design features are most effective for the audience and purpose? Which are least effective?
2. **TEAM EXERCISE** Form small groups for this collaborative exercise in analyzing design. Photocopy or scan a page from a book or a magazine. Choose a page that does not contain advertisements. Each person works independently for the first part of this project:
   - One person describes the design elements.
   - One person evaluates the design. Which aspects of the design are effective, and which could be improved?
   - One person creates a new design using thumbnail sketches.

   Then meet as a group and compare notes. Do all members of the group agree with the first member's description of the design? With the second member's evaluation of the design? Do all members like the third member's redesign? What have your discussions taught you about design? Write a memo to your instructor presenting your findings, and include the photocopy or scan of the page with your memo.

3. Study the excerpt from this Micron data flyer (2012, p. 1). Describe the designer's use of alignment as a design principle. How effective is it? How would you modify it? Present your analysis and recommendations in a brief memo to your instructor.

4. Find the websites of three manufacturers within a single industry, such as personal watercraft, cars, computers, or medical equipment. Study the three sites, focusing on one of these aspects of site design:

- use of color
- quality of the site map or index

CSN 33: Micron BGA Manufacturer's User Guide
Introduction

Customer Service Note

BGA Manufacturer's User Guide for Micron BGA Parts

Introduction

This customer service note provides information that will enable customers to easily integrate both leading-edge and legacy Micron® ball grid array (BGA) packages into their manufacturing processes. It is intended as a set of high-level guidelines and a reference manual describing typical package-related and manufacturing process-flow practices. The recommendations and suggestions provided in this customer service note serve as a guideline to help the end user to develop user-specific solutions. It is the responsibility of the end user to optimize the process to obtain the desired results.

Because the package landscape changes rapidly and information can become outdated very quickly, refer to the latest product specifications. Contact your sales representative for any additional questions not covered within this guide.

An overview of a typical BGA package and its components are shown in Figure 1.

Figure 1: Ball Grid Array Package (Dual Die, Wire Bonded)

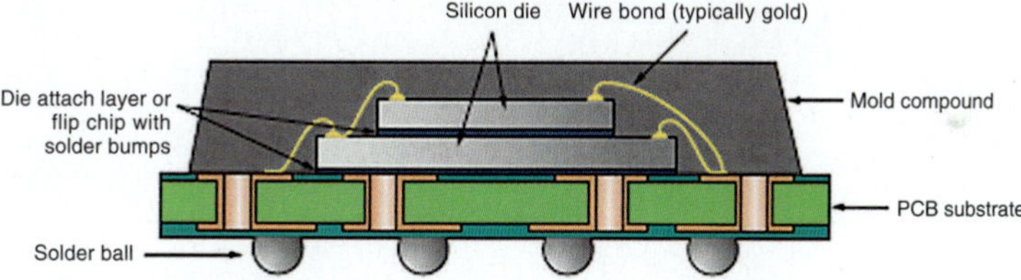

JEDEC Terminology

This document uses JEDEC terminology. JEDEC-based BGA devices in the semiconductor industry are identified by two key attributes:

- Maximum package height (profile)
- Ball pitch

For example: TFBGA - 1.2mm package height and less than 1.0mm ball pitch.

Package descriptors F1 through F6 have been added to provide more detailed ball pitch information for devices with a ball pitch of less than 0.8mm. Within the industry, many memory manufacturers continue to use only the "F" descriptor for any ball pitch of 1.0mm or less (see JEDEC JESD30E for additional information). Maximum package height profile and ball pitch codes based on the JEDEC standard are shown in Tables 1

PDF: 09005aef84793011/Source: 09005aef84792f7e
csn33_bga_user_guide.fm-Rev. A 12/12 EN1

1

Micron Technology, Inc., reserves the right to change products or specifications without notice.
©2011 Micron Technology, Inc. All rights reserved.

Products and specifications discussed herein are for evaluation and reference purposes only and are subject to change by Micron without notice. Products are only warranted by Micron to meet Micron's production data sheet specifications. All information discussed herein is provided on an "as is" basis, without warranties of any kind.

Information from Micron Technology, Inc.

- navigation, including the clarity and placement of links to other pages in the site
- accommodation of multicultural readers
- accommodation of people with disabilities
- phrasing of the links

Which of the three sites is most effective? Which is least effective? Why? Compare and contrast the three sites in terms of their effectiveness.

5. Find a website that serves the needs of people with a physical disability (for example, the Glaucoma Foundation, www.glaucomafoundation.org). Have the designers tried to accommodate the needs of visitors to the site? How effective do you think those attempts have been?

CASE 7: Designing a Flyer

As an employee in the educational information office in the U.S. Department of Education, you have been asked by your supervisor to design a flyer for international students hoping to complete graduate school in the United States. She has given you a text document with all of the relevant information; it's your job to turn that information into a visually appealing flyer that will catch students' attention. Your supervisor has asked you to write her a memo before you begin, describing and defending the design you have in mind. To get started designing your flyer, go to LaunchPad.

CHAPTER

8 Creating Graphics

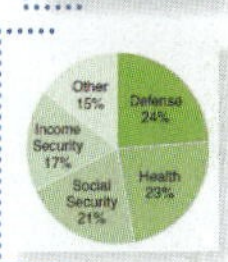

GRAPHICS ARE THE VISUALS in technical communication: drawings, maps, photographs, diagrams, charts, graphs, and tables. Graphics range from realistic, such as photographs, to highly abstract, such as organization charts. They range from decorative, such as clip art and stock photos that show people seated at a conference table, to highly informative, such as a schematic diagram of an electronic device.

Graphics are important in technical communication because they do the following:

- catch readers' attention and interest
- help writers communicate information that is difficult to communicate with words
- help writers clarify and emphasize information
- help nonnative speakers of English understand information
- help writers communicate information to multiple audiences with different interests, aptitudes, and reading habits

The Functions of Graphics

We have known for decades that graphics motivate people to study documents more closely. Some 83 percent of what we learn derives from what we see, whereas only 11 percent derives from what we hear (Gatlin, 1988).

Because we are good at acquiring information through sight, a document that includes a visual element in addition to the words is more effective than one that doesn't. People studying a document with graphics learn about one-third more than people studying a document without graphics (Levie & Lentz, 1982). And people remember 43 percent more when a document includes graphics (Morrison & Jimmerson, 1989). In addition, readers like graphics. According to one survey, readers of computer documentation consistently want more graphics and fewer words (Brockmann, 1990, p. 203).

Graphics offer five benefits that words alone cannot:

- **Graphics are indispensable in demonstrating logical and numerical relationships.** For example, an organization chart effectively represents the lines of authority in an organization. And if you want to communicate the number of power plants built in each of the last 10 years, a bar graph works better than a paragraph.
- **Graphics can communicate spatial information more effectively than words alone.** If you want to show the details of a bicycle derailleur, a diagram of the bicycle with a close-up of the derailleur is more effective than a verbal description.
- **Graphics can communicate steps in a process more effectively than words alone.** A troubleshooter's guide, a common kind of table, explains what might be causing a problem in a process and how you might fix it. And a diagram can show clearly how acid rain forms.
- **Graphics can save space.** Consider the following paragraph:

  In the Wilmington area, some 80 percent of the population aged 18 to 24 have watched streamed movies on their computers. They watch an average of 1.86 movies a week. Among 25- to 34-year-olds, the percentage is 72, and the average number of movies is 1.62. Among 35- to 49-year-olds, the percentage is 62, and the average number of movies is 1.19. Among the 50 to 64 age group, the percentage is 47, and the number of movies watched averages 0.50. Finally, among those people 65 years old or older, the percentage is 28, and the average number of movies watched weekly is 0.31.

  Presenting this information in a paragraph is uneconomical and makes the information hard to remember. Presented as a table, however, the information is more concise and more memorable.

AGE	PERCENTAGE WATCHING STREAMING MOVIES	NUMBER OF MOVIES WATCHED PER WEEK
18–24	80	1.86
25–34	72	1.62
35–49	62	1.19
50–64	47	0.50
65+	28	0.31

THINKING VISUALLY

Characteristics of an Effective Graphic

To be effective, graphics must be clear, understandable, and meaningfully related to the larger discussion. Follow these five principles when creating them.

A graphic should serve a PURPOSE.

Don't include a graphic unless it will help readers understand or remember information.

A graphic should be SIMPLE and UNCLUTTERED.

Three-dimensional bar graphs are easy to make, but they are harder to understand than two-dimensional ones.

A graphic should present a MANAGEABLE amount of information.

Presenting too much information can confuse readers. Consider what kinds of graphics your readers are familiar with, how much they already know about the subject, and what you want the document to do. Consider creating several simple graphics rather than a single complicated one.

A graphic should meet readers' FORMAT expectations.

Follow the conventions—for instance, use diamonds to represent decision points in a flowchart—unless you have a good reason not to.

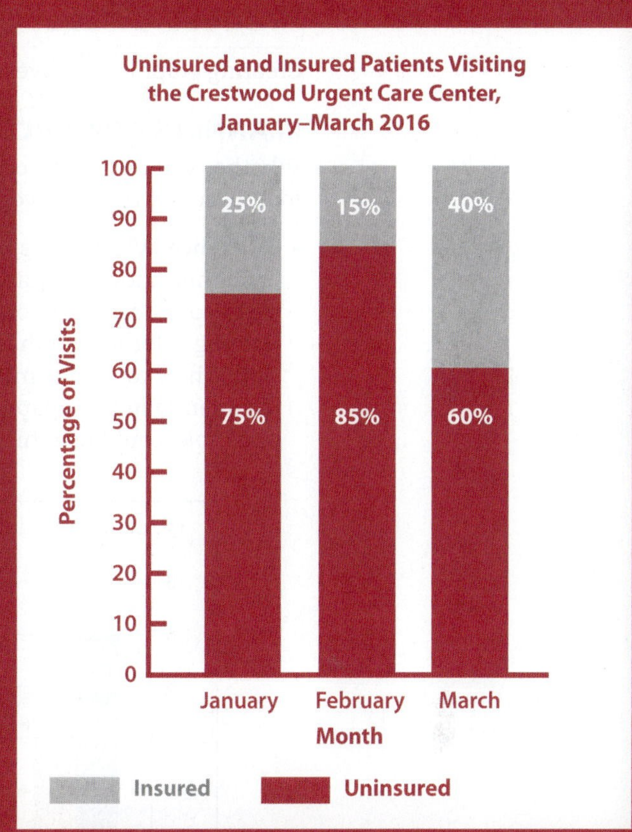

A graphic should be clearly LABELED.

Give every graphic (except a brief, informal one) a unique, clear, informative title. Fully label the columns of a table and the axes and lines of a graph. Don't make readers guess whether you are using meters or yards, or whether you are also including statistics from the previous year.

- **Graphics can reduce the cost of documents intended for international readers.** Translation costs more than 10 cents per word (ProZ.com, 2013). Used effectively, graphics can reduce the number of words you have to translate.

As you plan and draft your document, look for opportunities to use graphics to clarify, emphasize, summarize, and organize information.

Understanding the Process of Creating Graphics

Creating graphics involves planning, producing, revising, and citing.

PLANNING GRAPHICS

Whether you focus first on the text or the graphics, consider the following four issues as you plan your graphics.

- **Audience.** Will readers understand the kinds of graphics you want to use? Will they know the standard icons in your field? Are they motivated to read your document, or do you need to enliven the text—for example, by adding color for emphasis—to hold their attention? General audiences know how to read common types of graphics, such as those that appear frequently in newspapers or on popular websites. A general audience, for example, could use this bar graph to compare two bottles of wine:

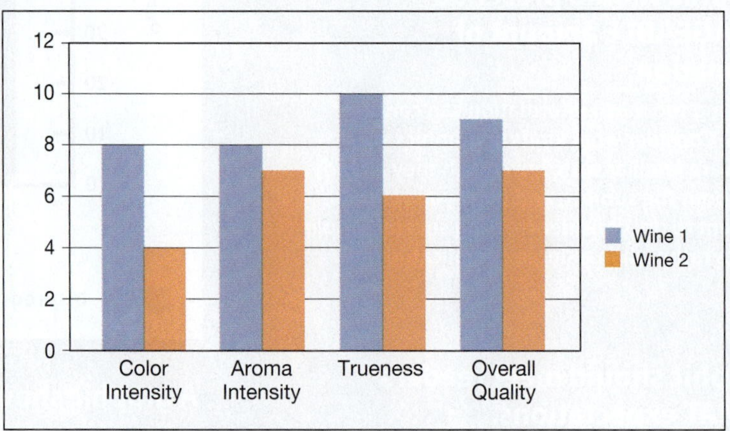

However, they would probably have trouble with the following radar graph:

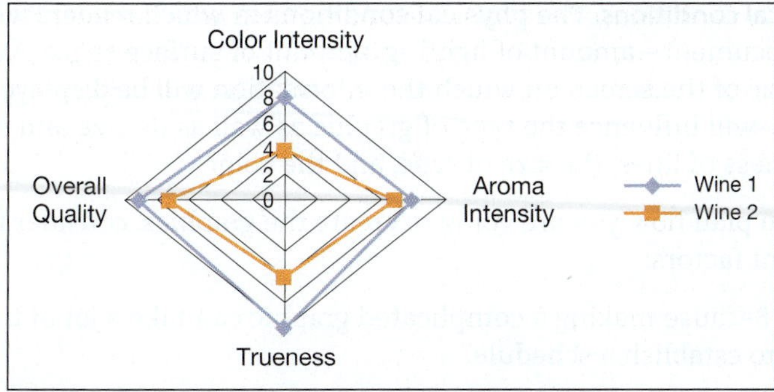

- **Purpose.** What point are you trying to make with the graphic? Imagine what you want your readers to know and do with the information. For example, if you want readers to know the exact dollar amounts spent on athletics by a college, use a table:

YEAR	MEN'S ATHLETICS ($)	WOMEN'S ATHLETICS ($)
2016	380,990	290,305
2017	420,400	300,080
2018	440,567	440,213

If you want readers to know how spending on athletics is changing over time, use a line graph:

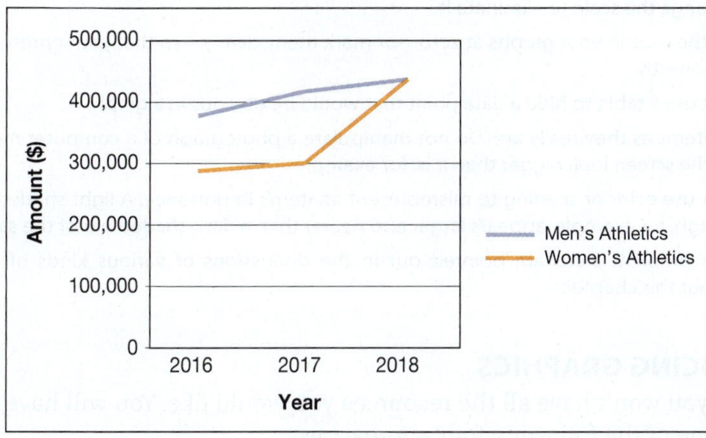

- **The kind of information you want to communicate.** Your subject will help you decide what type of graphic to include. For example, in writing about languages spoken by your state's citizens, you might use a table for the statistical data, a map for the patterns of language use, and a graph for statistical trends over time.

- **Physical conditions.** The physical conditions in which readers will use the document—amount of lighting, amount of surface space available, the size of the screen on which the information will be displayed, and so forth—will influence the type of graphic as well as its size and shape, the thickness of lines, the size of type, and the color.

As you plan how you are going to create the graphics, consider four important factors:

- **Time.** Because making a complicated graphic can take a lot of time, you need to establish a schedule.
- **Money.** Creating a high-quality graphic can be expensive. How big is the project budget? How can you use that money effectively?
- **Equipment.** Determine what tools and software you will require, such as spreadsheets for tables and graphs or graphics software for diagrams.
- **Expertise.** How much do you know about creating graphics? Do you have access to the expertise of others?

ETHICS NOTE

CREATING HONEST GRAPHICS

Follow these six suggestions to ensure that you represent data honestly in your graphics.

- If you did not create the graphic or generate the data, cite your source. If you want to publish a graphic that you did not create, obtain permission. For more on citing graphics, see page 200.
- Include all relevant data. For example, if you have a data point that you cannot explain, do not change the scale to eliminate it.
- Begin the axes in your graphs at zero—or mark them clearly—so that you represent quantities honestly.
- Do not use a table to hide a data point that would be obvious in a graph.
- Show items as they really are. Do not manipulate a photograph of a computer monitor to make the screen look bigger than it is, for example.
- Do not use color or shading to misrepresent an item's importance. A light-shaded bar in a bar graph, for example, appears larger and nearer than a dark-shaded bar of the same size.

Common problem areas are pointed out in the discussions of various kinds of graphics throughout this chapter.

PRODUCING GRAPHICS

Usually, you won't have all the resources you would like. You will have to choose one of the following four approaches:

- **Use existing graphics.** For a student paper that *will not be published*, some instructors allow the use of photocopies or scans of existing graphics; other instructors do not. For a document that *will be published*, whether written by a student or a professional, using an existing graphic is permissible if the graphic is in the public domain (that is, not under copyright), if it is the property of the writer's organization, or if the organization has obtained permission to use it. Be particularly careful about graphics you find on the web. Many people mistakenly think that anything on the web can be used without permission. The same copyright laws that apply to printed material apply to web-based material, whether words or graphics. For more on citing graphics, see page 200.

Aside from the issue of copyright, think carefully before you use existing graphics. The style of the graphic might not match that of the others you want to use; the graphic might lack some features you want or include some you don't. If you use an existing graphic, assign it your own number and title.

- **Modify existing graphics.** You can redraw an existing graphic or use a scanner to digitize the graphic and then modify it with graphics software.
- **Create graphics on a computer.** You can create many kinds of graphics using your spreadsheet software and the drawing tools on your word processor. Consult the Selected Bibliography, located in LaunchPad, for a list of books about computers and technical communication.
- **Have someone else create the graphics.** Professional-level graphics software can cost hundreds of dollars and require hundreds of hours of practice. Some companies have technical-publications departments with graphics experts, but others subcontract this work. Many print shops and service bureaus have graphics experts on staff or can direct you to them.

For more about work made for hire, see Ch. 2, p. 22.

TECH TIP

Why To Insert and Modify Graphics

You can insert graphics into your document in order to highlight, clarify, summarize, and/or organize information. You may have access to graphics, from online stock photo agencies or your company's graphics library, that could benefit from some modification.

How To Insert and Modify Graphics

To **insert a graphic**—such as a photograph, drawing, chart, or graph—place your cursor where you want to make the insertion.

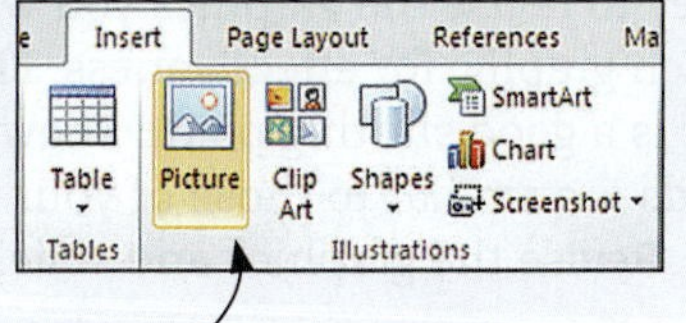

From the **Insert** tab in either Microsoft Word or Google Docs, choose the type of item you wish to insert.

To **modify a graphic**, select it and then work with the range of tools available in your program, which can usually be seen by right-clicking on the graphic. Typical commands relate to cropping, resizing, and making adjustments to color, brightness, and contrast.

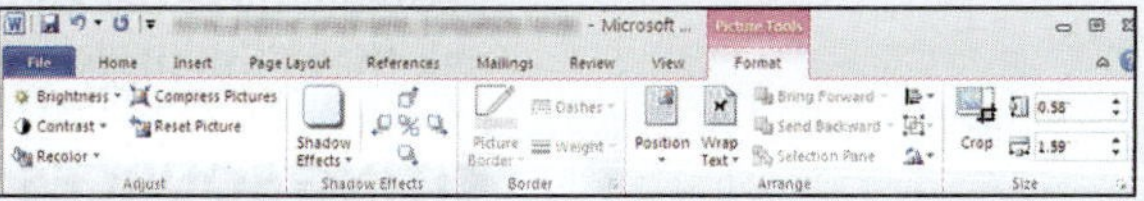

GUIDELINES Integrating Graphics and Text

It is not enough to add graphics to your text; you have to integrate the two.

- **Place the graphic in an appropriate location.** If readers need the graphic in order to understand the discussion, put it directly after the relevant point in the discussion or as soon after it as possible. If the graphic merely supports or elaborates a point, include it as an appendix.
- **Introduce the graphic in the text.** Whenever possible, refer to a graphic before it appears (ideally, on the same page). Refer to the graphic by number (such as "see Figure 7"). Do not refer to "the figure above" or "the figure below," because the graphic might move during the production process. If the graphic is in an appendix, cross-reference it: "For complete details of the operating characteristics, see Appendix B, page 19."
- **Explain the graphic in the text.** State what you want readers to learn from it. Sometimes a simple paraphrase of the title is enough: "Figure 2 compares the costs of the three major types of coal gasification plants." At other times, however, you might need to explain why the graphic is important or how to interpret it. If the graphic is intended to make a point, be explicit:

  > As Figure 2 shows, a high-sulfur bituminous coal gasification plant is more expensive than either a low-sulfur bituminous or an anthracite plant, but more than half of its cost is for cleanup equipment. If these expenses could be eliminated, high-sulfur bituminous would be the least expensive of the three types of plants.

  In addition to text explanations, graphics are often accompanied by captions, ranging from a sentence to several paragraphs.

For more about white space, rules, boxes, and screens, see Ch. 7, Figure 7.10 and Table 7.1. For more about lists of illustrations, see Ch. 13, "Table of Contents."

- **Make the graphic clearly visible.** Distinguish the graphic from the surrounding text by adding white space around it, placing rules (lines) above and below it, putting a screen behind it, or enclosing it in a box.
- **Make the graphic accessible.** If the document is more than a few pages long and contains more than four or five graphics, consider including a list of illustrations so that readers can find them easily.

REVISING GRAPHICS

As with any other aspect of technical communication, build in enough time and budget enough money to revise the graphics you want to use. Create a checklist and evaluate each graphic for effectiveness. The Writer's Checklist at the end of this chapter is a good starting point. Show your graphics to people whose backgrounds are similar to those of your intended readers and ask them for suggestions. Revise the graphics and solicit more reactions.

CITING SOURCES OF GRAPHICS

For more information about copyright, see Ch. 2.

If you wish to publish a graphic that is protected by copyright (even if you have revised it), you need to obtain written permission from the copyright holder. Related to the issue of permission is the issue of citation. Of course, you do not have to cite the source of a graphic if you created it yourself, if it is not protected by copyright, or if your organization owns the copyright.

In all other cases, however, you should include a source citation, even if your document is a course assignment and will not be published. Citing the sources of graphics, even those you have revised substantially, shows your instructor that you understand professional conventions and your ethical responsibilities.

If you are following a style manual, check to see whether it presents a format for citing sources of graphics. In addition to citing a graphic's source in the reference list, most style manuals call for a source statement in the caption:

For more about style manuals, see Appendix, Part A.

PRINT SOURCE

Source: Verduijn, 2015, p. 14. Copyright 2015 by Tedopres International B.V. Reprinted with permission.

ONLINE SOURCE

Source: Johnson Space Center Digital Image Collection. Copyright 2015 by NASA. Reprinted with permission.

If your graphic is based on an existing graphic, the source statement should state that the graphic is "based on" or "adapted from" your source:

Source: Adapted from Jonklaas et al., 2011, p. 771. Copyright 2008 by American Medical Association. Reprinted with permission.

Using Color Effectively

Color draws attention to information you want to emphasize, establishes visual patterns to promote understanding, and adds interest. But it is also easy to misuse. The following discussion is based on Jan V. White's excellent text *Color for the Electronic Age* (1990).

In using color in graphics and page design, keep these six principles in mind:

- **Don't overdo it.** Readers can interpret only two or three colors at a time. Use colors for small items, such as portions of graphics and important words. And don't use colors where black and white will work better.
- **Use color to emphasize particular items.** People interpret color before they interpret shape, size, or placement on the page. Color effectively draws readers' attention to a particular item or group of items on a page. In Figure 8.1, for example, color adds emphasis to different kinds of information.
- **Use color to create patterns.** The principle of repetition—readers learn to recognize patterns—applies in graphics as well as in document design. In creating patterns, also consider shape. For instance, use red for safety comments but place them in octagons resembling a stop sign. This way, you give your readers two visual cues to help them recognize the pattern.

For more about designing your document, see Ch. 7.

Color used to set off a title and the totals row in a table.

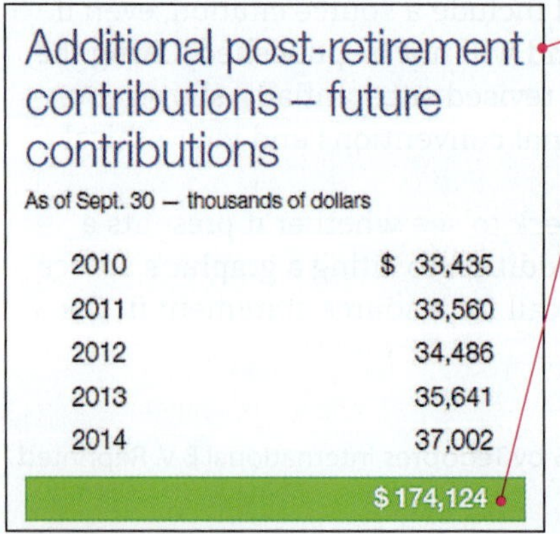

Additional post-retirement contributions – future contributions

As of Sept. 30 – thousands of dollars

2010	$ 33,435
2011	33,560
2012	34,486
2013	35,641
2014	37,002
	$ 174,124

Information from Bonneville, 2009: www.bpa.gov/Financial Information/Annual reports/Documents/AR2009.pdf.

Color used to emphasize one item among others.

Red Algae

Almost all **red algae** are multicellular (Figure 28.24). Some plant biologists now refer to the red algae as the "red plant kingdom." Their characteristic color is a result of the accessory photosynthetic pigment *phycoerythrin*, which is found in relatively large amounts in the chloroplasts of many species. In addition to phycoerythrin, red algae contain phycocyanin, carotenoids, and chlorophyll *a*.

LIFE: THE SCIENCE OF BIOLOGY 7E edited by W. K. Purves, D. Sadava, G. H. Orians, and H. C. Heller (2004): Figure: "Red Algae" (p. 560). By permission of Oxford University Press USA.

FIGURE 8.1 Color Used for Emphasis

Figure 8.2 shows the use of color to establish patterns. Color is also an effective way to emphasize design features such as text boxes, rules, screens, and headers and footers.

For more about presentation graphics, see Ch. 15, "Preparing Presentation Graphics."

- **Use contrast effectively.** The visibility of a color is a function of the background against which it appears (see Figure 8.3). The strongest contrasts are between black and white and between black and yellow. The need for effective contrast also applies to graphics used in presentations, as shown in Figure 8.4.
- **Take advantage of any symbolic meanings colors may already have.** In American culture, for example, red signals danger, heat, or electricity; yellow signals caution; and orange signals warning. Using these warm colors in ways that depart from these familiar meanings could be confusing. The cooler colors—blues and greens—are more conservative and subtle. (Figure 8.5 illustrates these principles.) Keep in mind, however, that people in different cultures interpret colors differently.
- Be aware that color can obscure or swallow up text.

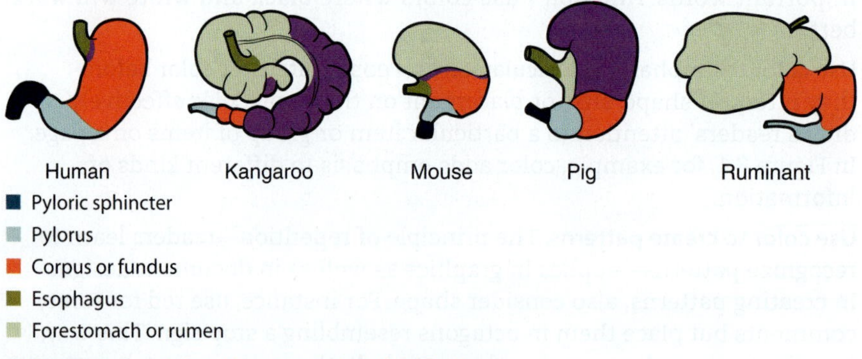

FIGURE 8.2 Color Used to Establish Patterns

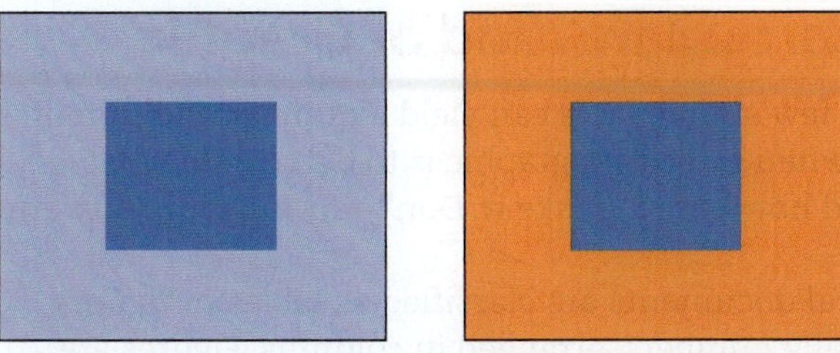

FIGURE 8.3 **The Effect of Background in Creating Contrast**

Notice that a color washes out if the background color is too similar.

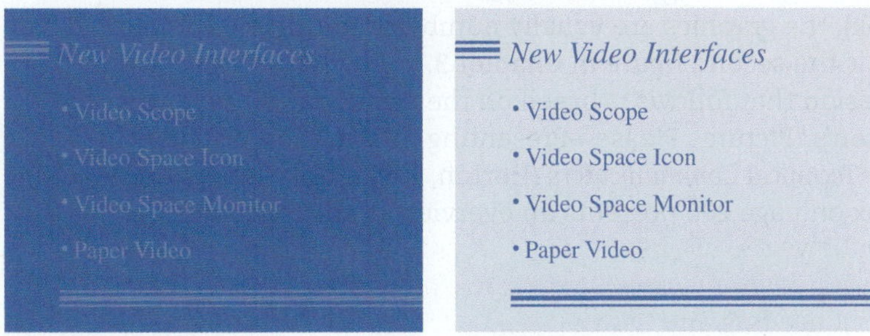

a. Insufficient contrast **b.** Effective contrast

FIGURE 8.4 **Effective Contrast Used in a Presentation Slide**

In graphic (a), the text is hard to read because of insufficient contrast. The greater contrast in graphic (b) makes the text easier to read.

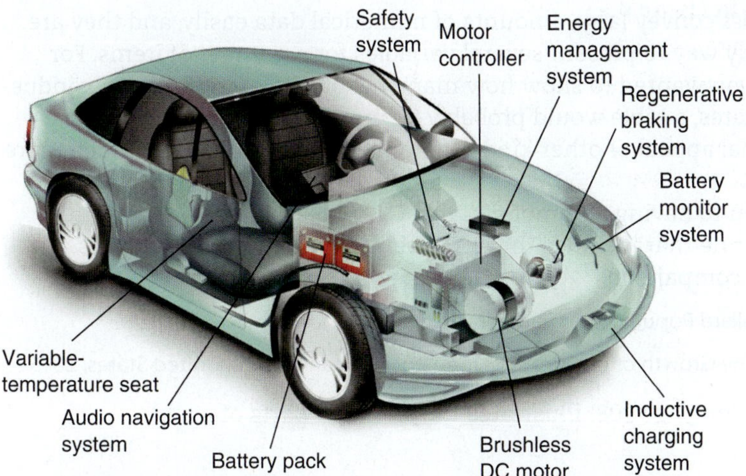

FIGURE 8.5 **Colors Have Clear Associations for Readers**

The batteries are red. The warm red contrasts effectively with the cool green of the car body.

Choosing the Appropriate Kind of Graphic

As Figure 8.6 shows, even a few simple facts can yield a number of different points. Your responsibility when creating a graphic is to determine what point you want to make and how best to make it. Don't rely on your software to do your thinking; it can't.

Graphics used in technical documents are classified as tables or figures. Tables are lists of data, usually numbers, arranged in columns. Figures are everything else: graphs, charts, diagrams, photographs, and the like. Typically, tables and figures are numbered separately: the first table in a document is Table 1; the first figure is Figure 1. In documents of more than one chapter (like this book), the graphics are usually numbered within each chapter. That is, Figure 3.2 is the second figure in Chapter 3.

The discussion that follows is based on the classification system in William Horton's "Pictures Please—Presenting Information Visually," in *Techniques for Technical Communicators* (Horton, 1992). The Choices and Strategies box on page 206 presents an overview of the following discussion.

ILLUSTRATING NUMERICAL INFORMATION

The kinds of graphics used most often to display numerical values are tables, bar graphs, infographics, line graphs, and pie charts.

Tables *Tables* convey large amounts of numerical data easily, and they are often the only way to present several variables for a number of items. For example, if you wanted to show how many people are employed in six industries in 10 states, a table would probably be most effective. Although tables lack the visual appeal of other kinds of graphics, they can handle much more information.

In addition to having a number ("Table 1"), tables are identified by an informative title that includes the items being compared and the basis (or bases) of comparison:

> Table 3. Mallard Population in Rangeley, 2012–2015
>
> Table 4.7. The Growth of the Robotics Industry in Japan and the United States, 2015

Figure 8.7 (on page 208) illustrates the standard parts of a table.

Rail Line	November		December		January	
	Disabled by electrical problems (%)	Total disabled	Disabled by electrical problems (%)	Total disabled	Disabled by electrical problems (%)	Total disabled
Bryn Mawr	19 (70)	27	17 (60)	28	20 (76)	26
Swarthmore	12 (75)	16	9 (52)	17	13 (81)	16
Manayunk	22 (64)	34	26 (83)	31	24 (72)	33

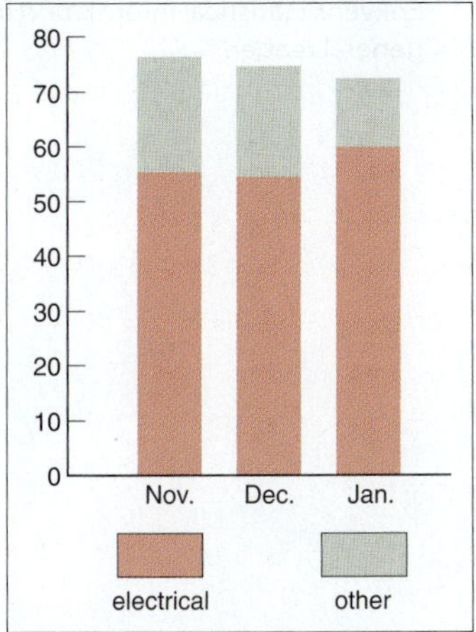

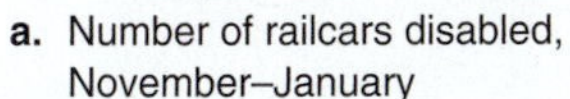

a. Number of railcars disabled, November–January

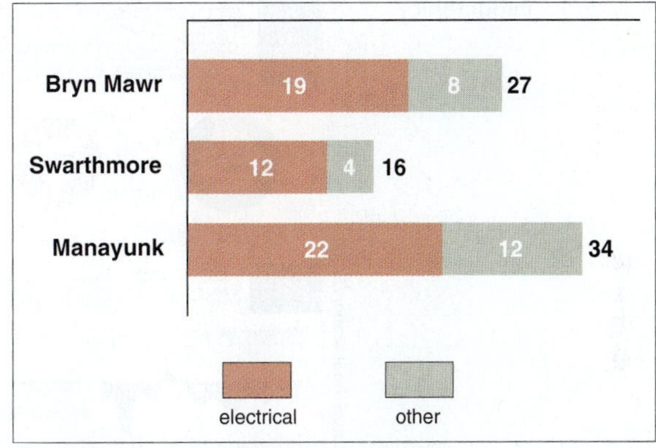

b. Number of railcars disabled in November

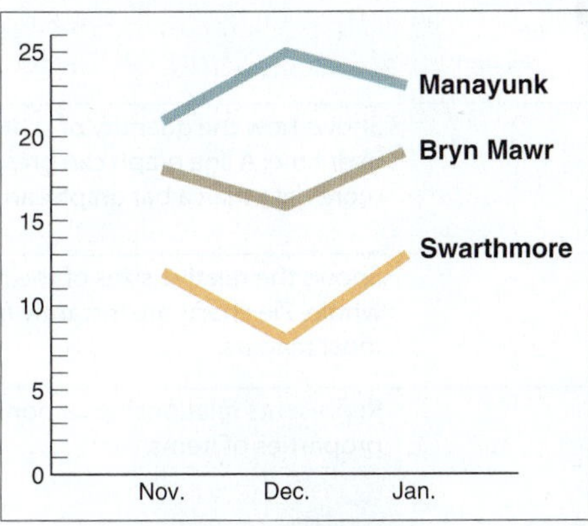

c. Number of railcars disabled by electrical problems, November–January

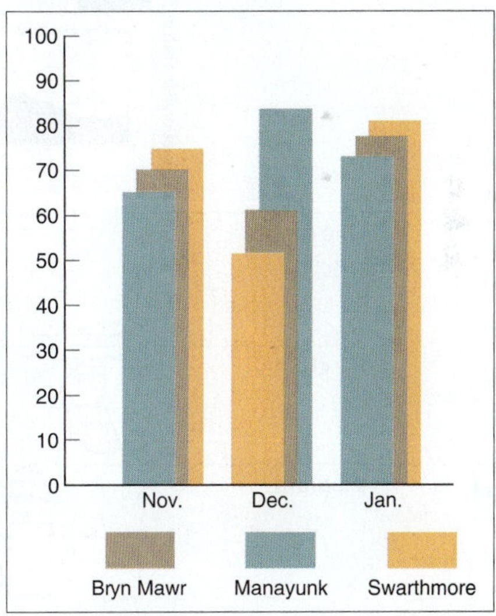

d. Range in percentage of railcars, by line, disabled by electrical problems, November–January

FIGURE 8.6 Different Graphics Emphasizing Different Points

Each of these four graphs emphasizes a different point derived from the data in the table. Graph (a) focuses on the total number of railcars disabled each month, classified by cause; graph (b) focuses on the three rail lines during one month; and so forth. For information on bar graphs, see pages 211–13; for information on line graphs, see pages 216–17.

CHOICES AND STRATEGIES Choosing the Appropriate Kind of Graphic

IF YOU WANT TO...	TRY THIS TYPE OF GRAPHIC	WHAT THIS GRAPHIC DOES BEST
Illustrate numerical information	Table	Shows large amounts of numerical data, especially useful when there are several variables for a number of items
	Bar graph	Shows the relative values of two or more items.
	Infographic (Information from https://www.cbo.gov/publication/52408.)	Enlivens statistical information for the general reader.
	Line graph	Shows how the quantity of an item changes over time. A line graph can present much more data than a bar graph can.
	Pie chart	Shows the relative sizes of the parts of a whole. Pie charts are instantly familiar to most readers.
Illustrate logical relationships	Diagram	Represents relationships among items or properties of items.
	Organization chart	Shows the lines of authority and responsibility in an organization or hierarchical relationships among items.

(*continued*)

<table>
<tr><th>IF YOU WANT TO...</th><th>TRY THIS TYPE OF GRAPHIC</th><th></th><th>WHAT THIS GRAPHIC DOES BEST</th></tr>
<tr><td>Illustrate process descriptions and instructions</td><td>Checklist</td><td>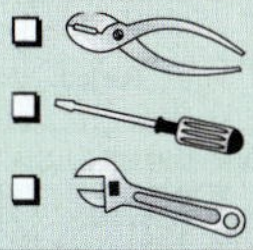</td><td>Lists or shows what equipment or materials to gather or describes an action.</td></tr>
<tr><td></td><td>Table</td><td>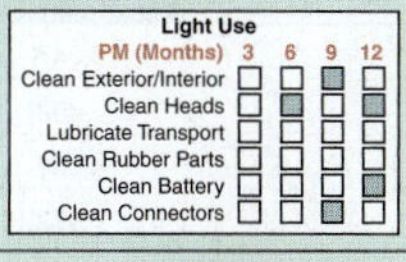
</td><td>Shows numbers of items or indicates the state (on/off) of an item.</td></tr>
<tr><td></td><td>Flowchart</td><td>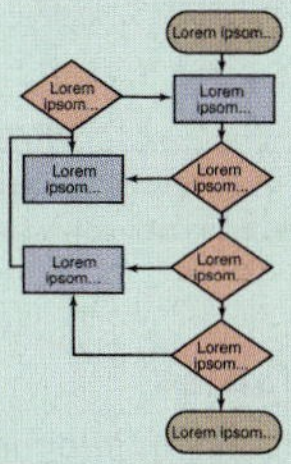
</td><td>Shows the stages of a procedure or a process.</td></tr>
<tr><td></td><td>Logic tree</td><td>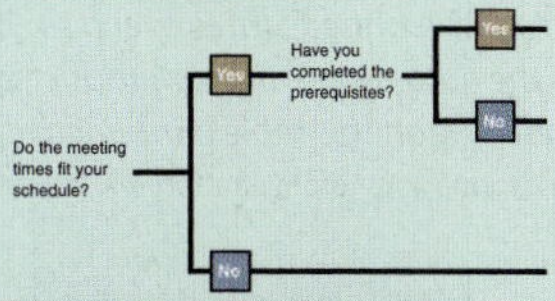
</td><td>Shows which of two or more paths to follow.</td></tr>
<tr><td>Illustrate visual and spatial characteristics</td><td>Photograph</td><td>
NASA</td><td>Shows precisely the external surface of objects.</td></tr>
<tr><td></td><td>Screen shot</td><td>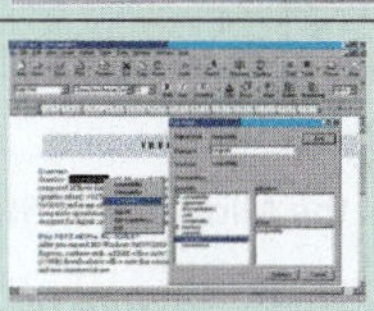</td><td>Shows what appears on a computer screen.</td></tr>
<tr><td></td><td>Line drawing</td><td>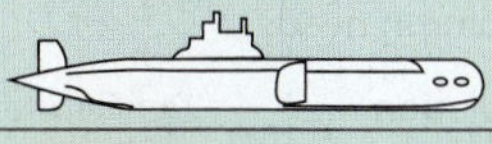</td><td>Shows simplified representations of objects.</td></tr>
<tr><td></td><td>Map</td><td></td><td>Shows geographic areas.</td></tr>
</table>

Information from W. Horton, "The Almost Universal Language: Graphics for International Documentation" from *Technical Communication* 40 (1993): 682–93.
Source: https://www.cbo.gov/publication/52408.

Tables are usually titled at the top because readers scan them from top to bottom.

Some tables include a stub head. The stub—the left-hand column—lists the items for which data are displayed. Note that indentation in the stub helps show relationships. The heading "Natural Gas Liquids" is left-aligned. This row functions as a Totals row. Indented beneath this heading are the two categories that make up the totals: pentanes plus and liquefied petroleum gases. Beneath the heading are rows for the four kinds of liquefied petroleum gases.

Note that the numbers are right-aligned.

Note that tables often contain one or more source statements and footnotes.

Natural Gas Plant Stocks
(Thousand Barrels)

Area: U.S. Period: Monthly

Download Series History | Definitions, Sources & Notes

Show Data By: ◉ Product ○ Area	Graph / Clear	Nov-12	Dec-12	Jan-13	Feb-13	Mar-13	Apr-13	View History
Natural Gas Liquids	☐	5,047	4,828	4,550	4,787	5,625	5,419	1993-2013
Pentanes Plus	☐	492	383	462	553	675	787	1993-2013
Liquefied Petroleum Gases	☐	4,555	4,445	4,088	4,234	4,950	4,632	1993-2013
Ethane	☐	672	837	871	812	1,212	822	1993-2013
Propane	☐	1,998	1,944	1,655	1,702	2,126	2,098	1993-2013
Normal Butane	☐	1,207	907	864	830	757	936	1993-2013
Isobutane	☐	678	757	698	890	855	776	1993-2013

- = No Data Reported; -- = Not Applicable; NA = Not Available; W = Withheld to avoid disclosure of individual company data.

Notes: See Definitions, Sources, and Notes link above for more information on this table.

Release Date: 6/27/2013
Next Release Date: Last Week of July 2013

FIGURE 8.7 Parts of a Table

This is an interactive table from the Energy Information Administration showing the amount of natural gas stocks in the United States over a six-month period. Note that you can use the radio buttons to show the data by location or product, and you can use the pull-down menu to specify a time frame. Even though this table is interactive, it functions much the way any table does.

Source: U.S. Energy Information Administration, 2013: www.eia.gov/dnav/pet/pet_stoc_gp_dcu_nus_m.htm.

GUIDELINES Creating Effective Tables

Follow these nine suggestions to make sure your tables are clear and professional.

- **Indicate the units of measure.** If all the data are expressed in the same unit, indicate that unit in the title:

  Farm Size in the Midwestern States (in Hectares)

  If the data in different columns are expressed in different units, indicate the units in the column heads:

Population (in Millions)	Per Capita Income (in Thousands of U.S. Dollars)

  If all the *data cells* in a column use the same unit, indicate that unit in the column head, not in each data cell:

  Speed (in Knots)
  15
  18
  14

(continued)

You can express data in both real numbers and percentages. A column head and the first data cell under it might read as follows:

Number of Students (Percentage)

53 (83)

- **In the stub—the left-hand column—list the items being compared.** Arrange the items in a logical order: big to small, more important to less important, alphabetical, chronological, geographical, and so forth. If the items fall into several categories, include the names of the categories in the stub:

Snowbelt States . . . .
    Connecticut . . . .
    New York . . . .
    Vermont . . . .
Sunbelt States . . . .
    Arizona . . . .
    California . . . .
    New Mexico . . . .

If you cannot group the items in the stub in logical categories, skip a line after every five rows to help the reader follow the rows across the table. Or use a screen (a colored background) for every other set of five rows. Also useful is linking the stub and the next column with a row of dots called *dot leaders*.

For more about screens, see Ch. 7, Table 7.1.

- **In the columns, arrange the data clearly and logically.** Use the decimal-tab feature to line up the decimal points:

3,147.4
365.7
46,803.5

In general, don't vary the units used in a column unless the quantities are so dissimilar that your readers would have a difficult time understanding them if expressed in the same units.

3.4 hr
12.7 min
4.3 sec

This list would probably be easier for most readers to understand than one in which all quantities were expressed in the same unit.

- **Do the math.** If your readers will need to know the totals for the columns or the rows, provide them. If your readers will need to know percentage changes from one column to the next, present them:

Number of Students (Percentage Change from Previous Year)

2013	2014	2015
619	644 (+4.0)	614 (–4.7)

(*continued*)

- **Use dot leaders if a column contains a "blank" spot—a place where there are no appropriate data:**

  3,147
  . . .
  46,803

  But don't substitute dot leaders for a quantity of zero.

- **Don't make the table wider than it needs to be.** The reader should be able to scan across a row easily. As White (1984) points out, there is no reason to make a table as wide as the text column in the document. If a column head is long—more than five or six words—stack the words:

  Computers Sold Without
  <u>a Memory-Card Reader</u>

- **Minimize the use of rules.** Grimstead (1987) recommends using rules only when necessary: to separate the title and the heads, the heads and the body, and the body and the notes. When you use rules, make them thin rather than thick.

- **Provide footnotes where necessary.** All the information your readers need in order to understand the table should accompany it.

- **If you did not generate the information yourself, indicate your source.** See the discussion of citing sources of graphics on pages 200–201.

Bar Graphs Like tables, *bar graphs* can communicate numerical values, but they are better at showing the relative values of two or more items. Figure 8.8 shows typical horizontal and vertical bar graphs that you can make easily using your spreadsheet software.

Horizontal bars are best for showing quantities such as speed and distance. Vertical bars are best for showing quantities such as height, size, and amount. However, these distinctions are not ironclad; as long as the axes are clearly labeled, readers should have no trouble understanding the graph.

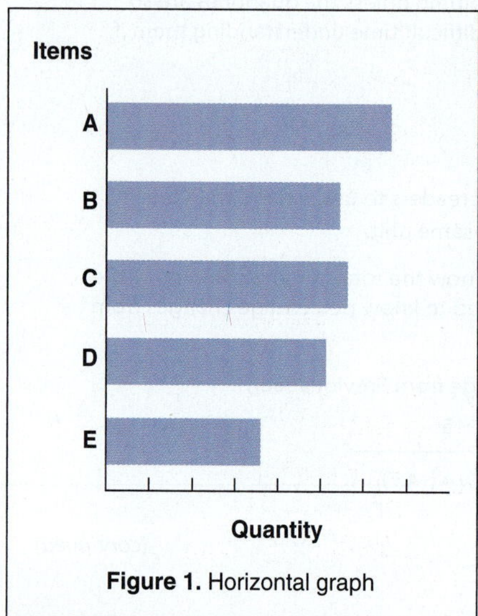

Figure 1. Horizontal graph

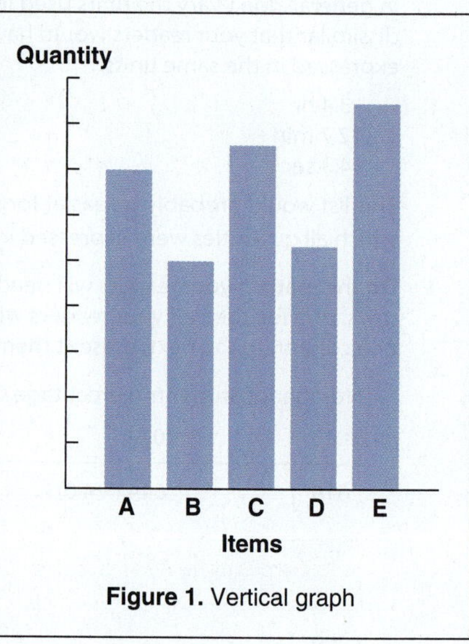

Figure 1. Vertical graph

FIGURE 8.8 Structures of Horizontal and Vertical Bar Graphs

Figure 8.9 shows an effective bar graph that uses grid lines.

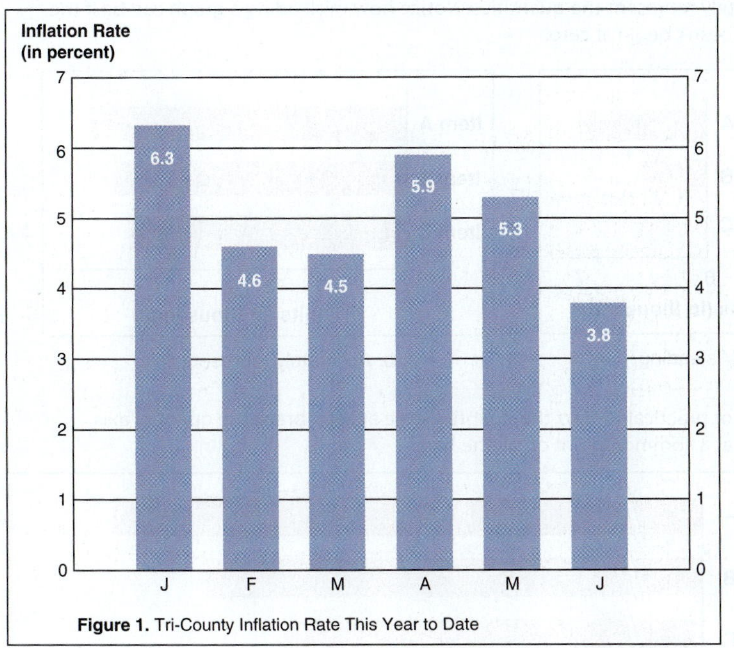

Figure 1. Tri-County Inflation Rate This Year to Date

FIGURE 8.9 Effective Bar Graph with Grid Lines

GUIDELINES Creating Effective Bar Graphs

- **Make the proportions fair.** Make your vertical axis about 25 percent shorter than your horizontal axis. An excessively long vertical axis exaggerates the differences in quantities; an excessively long horizontal axis minimizes the differences. Make all the bars the same width, and make the space between them about half as wide as a bar. Here are two poorly proportioned graphs:

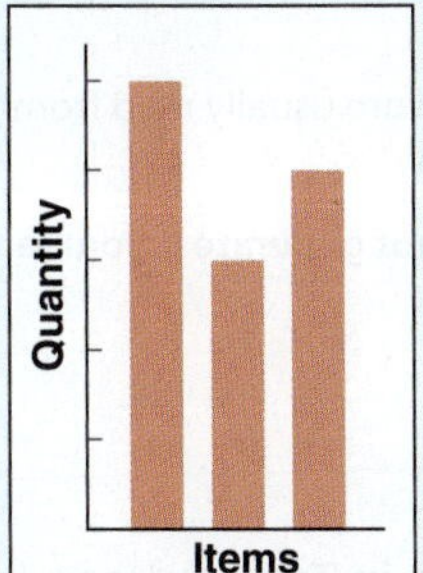

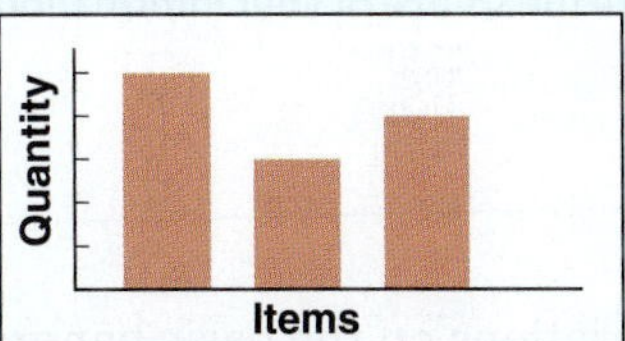

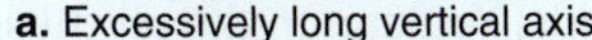

a. Excessively long vertical axis **b.** Excessively long horizontal axis

(continued)

- **If possible, begin the quantity scale at zero.** Doing so ensures that the bars accurately represent the quantities. Notice how misleading a graph can be if the scale doesn't begin at zero.

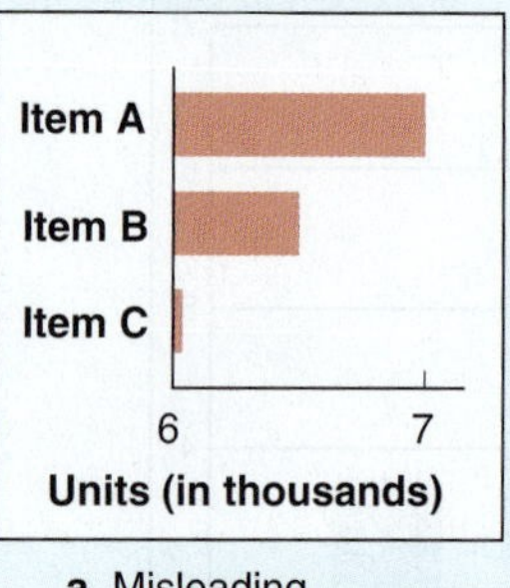

a. Misleading

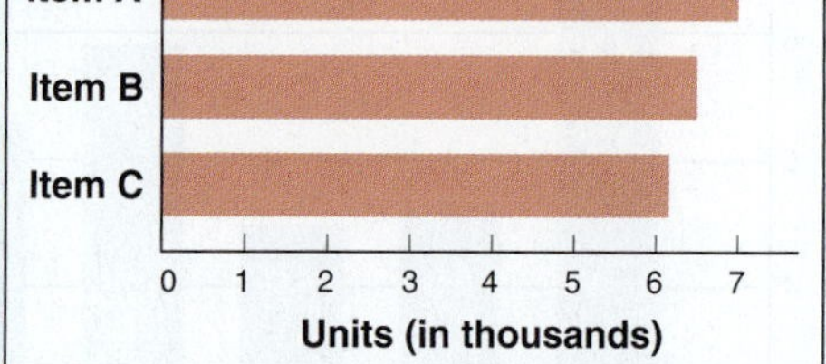

b. Accurately representative

If it is not practical to start the quantity scale at zero, break the quantity axis clearly at a common point on all the bars.

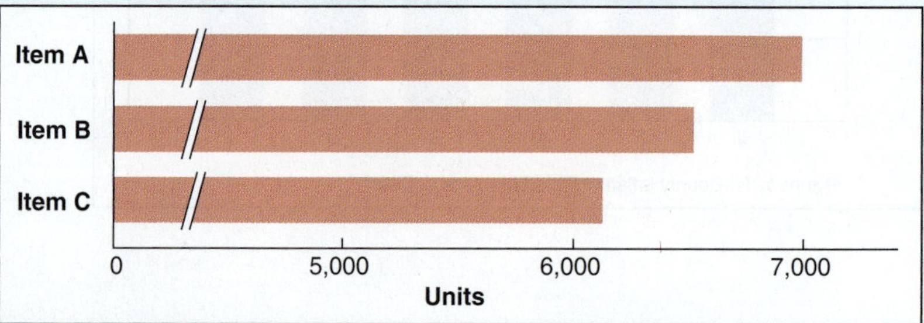

- **Use tick marks—marks along the axis—to signal the amounts.** Use grid lines—tick marks that extend through the bars—if the table has several bars, some of which are too far away from the tick marks to enable readers to gauge the quantities easily. (See Figure 8.9.)
- **Arrange the bars in a logical sequence.** For a vertical bar graph, use chronology if possible. For a horizontal bar graph, arrange the bars in order of descending size, beginning at the top of the graph, unless some other logical sequence seems more appropriate.
- **Place the title below the figure.** Unlike tables, which are usually read from top to bottom, figures are usually read from the bottom up.
- **Indicate the source of your information if you did not generate it yourself.**

The five variations on the basic bar graph shown in Table 8.1 can help you accommodate different communication needs. You can make all these types using your spreadsheet software.

TABLE 8.1 Variations on the Basic Bar Graph

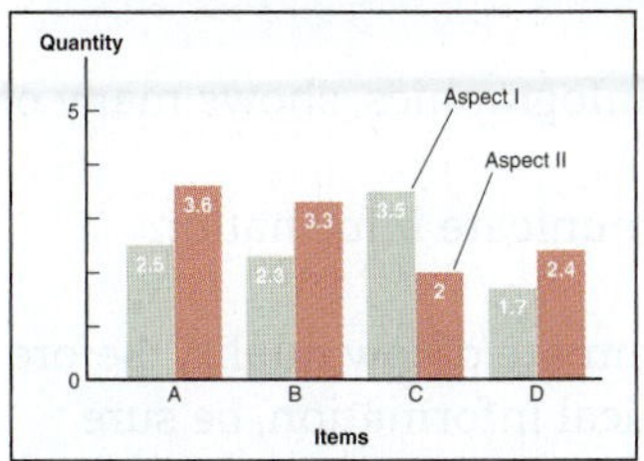

Grouped bar graph. The *grouped bar graph* lets you compare two or three aspects for each item. Grouped bar graphs would be useful, for example, for showing the numbers of full-time and part-time students at several universities. One bar could represent full-time students; the other, part-time students. To distinguish between the bars, use hatching (striping), shading, or color, and either label one set of bars or provide a key.

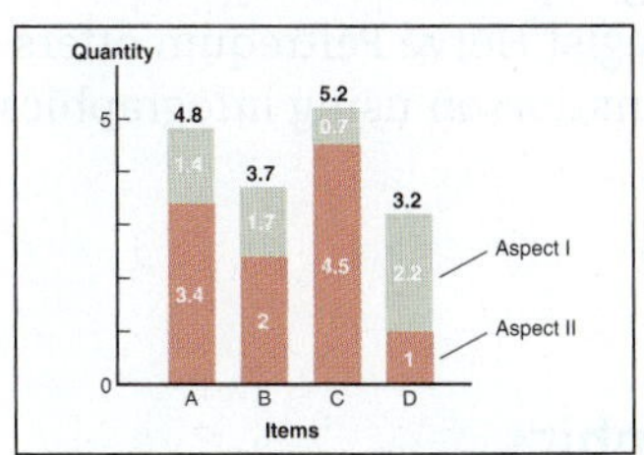

Subdivided bar graph. In the *subdivided bar graph*, Aspect I and Aspect II are stacked like wooden blocks placed on top of each other. Although totals are easy to compare in a subdivided bar graph, individual quantities are not.

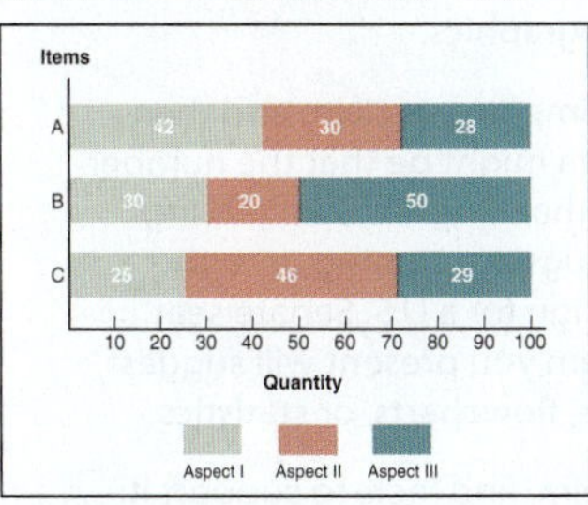

100-percent bar graph. The *100-percent bar graph*, which shows the relative proportions of the aspects that make up several items, is useful in portraying, for example, the proportion of full-scholarship, partial-scholarship, and no-scholarship students at a number of colleges.

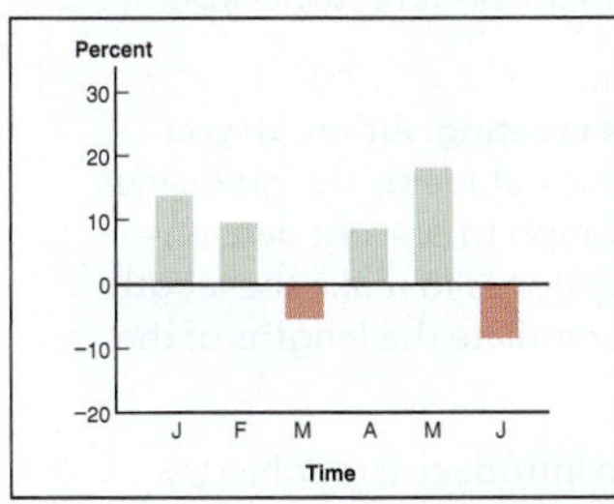

Deviation bar graph. The *deviation bar graph* shows how various quantities deviate from a norm. Deviation bar graphs are often used when the information contains both positive and negative values, such as profits and losses. Bars on the positive side of the norm line (above it) represent profits; bars on the negative side (below it), losses.

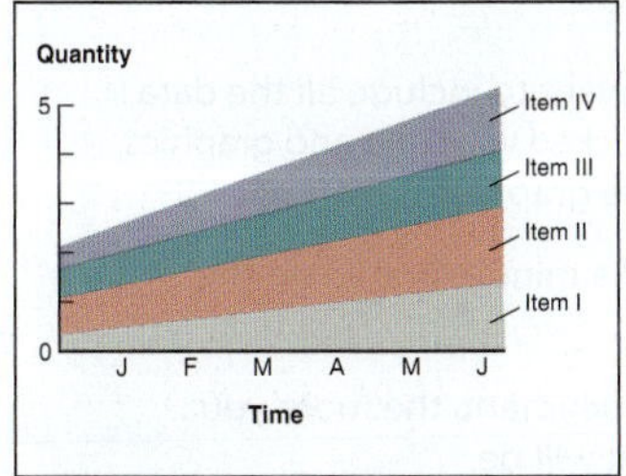

Stratum graph. The *stratum graph*, also called an *area graph*, shows the change in quantities of several items over time. Although stratum graphs are used frequently in business and scientific fields, general readers sometimes have trouble understanding how to read them.

Infographics An *infographic*—short for *information graphic*—is a combination of words and graphics used to present factual data about a subject in a visually interesting way.

Figure 8.10, a portion of an infographic about infographics, shows many of the techniques used in this type of display.

Infographics are also an effective way to communicate information through a visual/verbal argument.

Infographics are very popular, but many of them are of low quality. Before you create an infographic to communicate technical information, be sure you are not skewing your data or oversimplifying to promote an agenda. Doing so is unethical. In Figure 8.11, digital strategist Hervé Peitrequin offers a clever commentary on infographics. (For information on using infographics to create résumés, see Chapter 10.)

GUIDELINES Creating Effective Infographics

Follow these seven suggestions for making effective infographics.

- **Make a claim.** A good infographic states—or at least implies—a claim and then presents evidence to support it. For instance, the claim might be that the number of people accessing the Internet in a language other than English is increasing at an accelerating rate, that the pace at which new drugs are coming onto the market is slowing, or that the cost of waging a campaign for a U.S. Senate seat has increased tenfold in the last twenty years. The claim you present will suggest the theme of your graphics: you might consider maps, flowcharts, or statistics.

For more about research techniques, see Ch. 5.

- **Use accurate data.** Once you have settled on your claim, find facts to support it. Use reputable sources, and then check and re-check them. Be sure to cite your sources on the infographic itself.
- **Follow the guidelines for the type of graphic you are creating.** Although you want to express your creativity when you create graphics, abide by the guidelines for that type of graphic. For instance, if you use a bar graph to present data on the number of zebras born in captivity, your first obligation is to make the length of each bar reflect the quantity it represents; don't manipulate the lengths of the bars to make the graph look like a zebra.
- **Write concisely.** If you need more than a paragraph to introduce a graphic, try revising the text to get the word count down or see if you can break the idea into several smaller ones.
- **Don't present too much information.** It's natural to want to include all the data you have found, but if the infographic is too tightly packed with text and graphics, readers will be intimidated. Use white space to let the graphics breathe.
- **Don't go on forever.** Your readers will want to spend a minute or two on the infographic. They won't want to spend 15 minutes.
- **Test the infographic.** As with any kind of technical document, the more you revise, evaluate, and test the infographic, the better it will be.

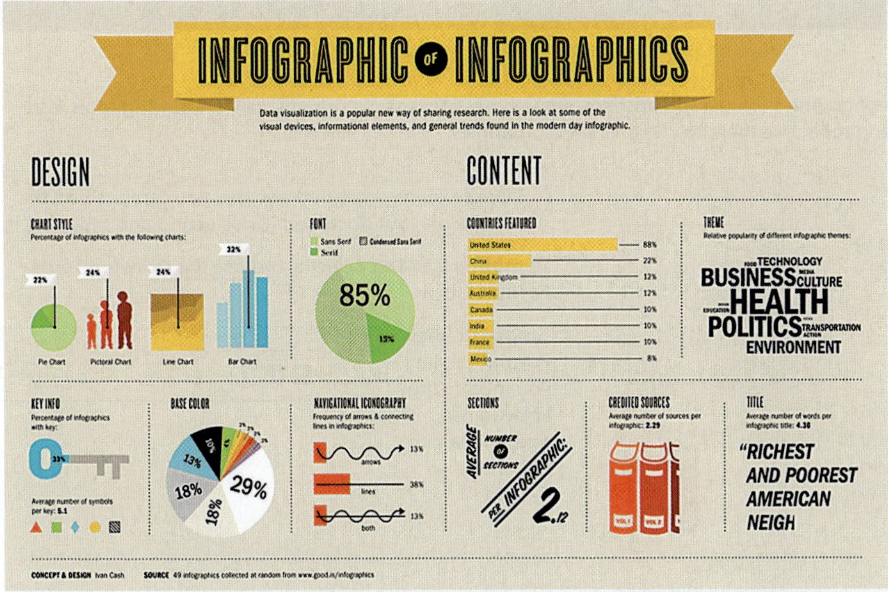

FIGURE 8.10 An Infographic About Infographics

Ivan Cash/Cash Studios.

Designer Ivan Cash created this infographic by collecting data about infographics and then creating graphics to make the data interesting and visually appealing.

Infographics are built around basic types of graphical display: pie charts, line graphs, bar graphs, and diagrams. In an effective infographic, each visual display adheres to the conventions of the graphic on which it is based. For instance, in the "Countries Featured" bar graph, the length of each bar accurately reflects the quantity of the item it represents.

The art makes the data visually interesting, but the most important characteristic of an infographic is accuracy: the data must be accurate and presented fairly.

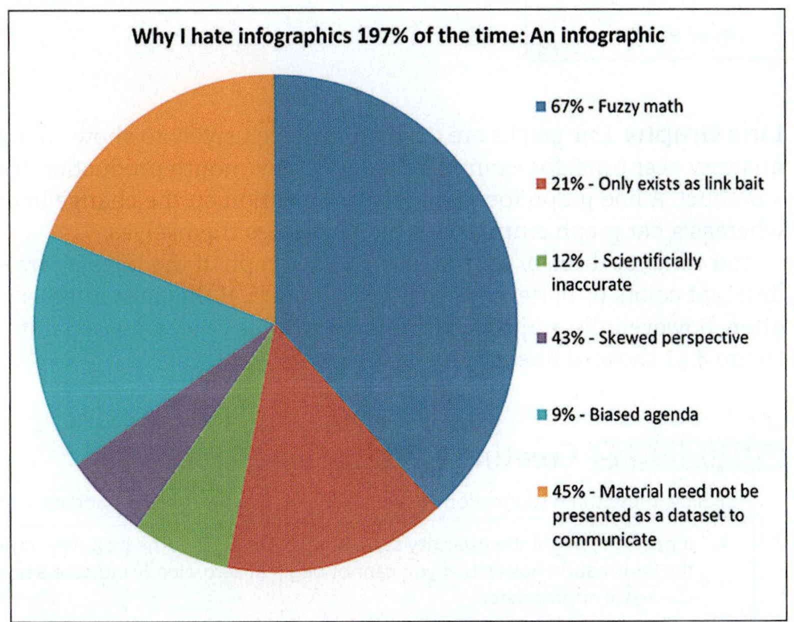

FIGURE 8.11 A Critique of Infographics

Information from Hervé Peitrequin.

TECH TIP

Why To Use Drawing Tools

Although you can make many types of graphics using a spreadsheet, some types, such as pictograms, call for drawing tools. Your word processor includes basic drawing tools. Use them to create and edit graphics.

How To Use Drawing Tools

To **create shapes** and **SmartArt**, use the **Illustrations** group on the **Insert** tab.

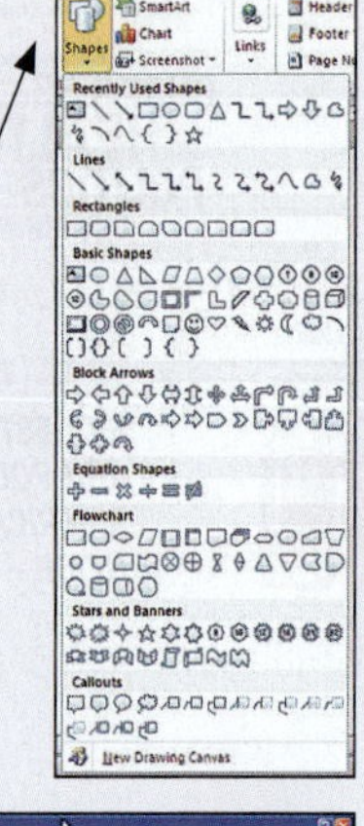

Use the **Shapes** drop-down menu to select a simple shape, such as a line, arrow, rectangle, or oval. Then drag your cursor to create the shape.

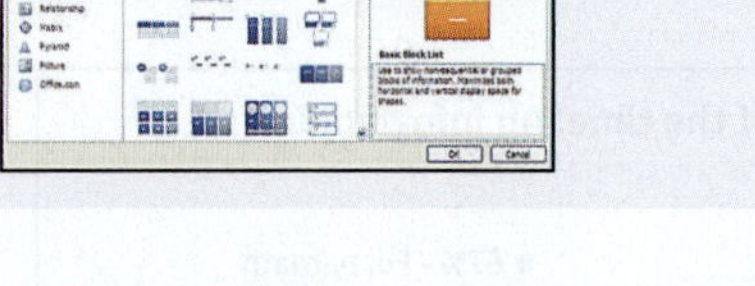

You can select complex shapes from the **SmartArt** drop-down menu in the **Illustrations** group.

Once you have created a shape, you can position the shape on your document by selecting and dragging it.

To **modify a shape**, select it and use the **Drawing Tools Format** tab.

Groups on the **Format** tab let you modify the appearance, size, and layout of a shape.

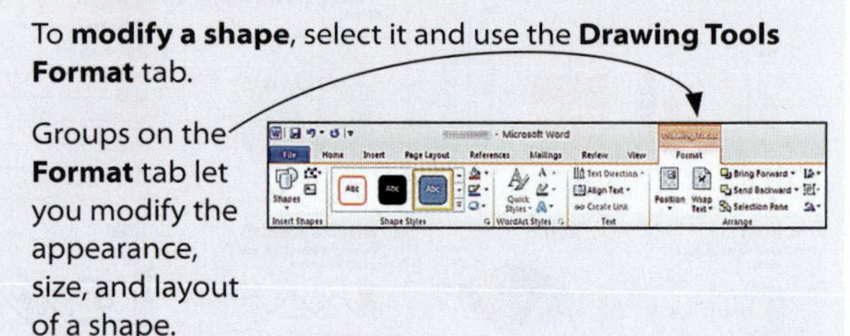

Line Graphs *Line graphs* are used almost exclusively to show changes in quantity over time, for example, the month-by-month production figures for a product. A line graph focuses readers' attention on the change in quantity, whereas a bar graph emphasizes the quantities themselves.

You can plot three or four lines on a line graph. If the lines intersect, use different colors or patterns to distinguish them. If the lines intersect too often, however, the graph will be unclear; in this case, draw separate graphs. Figure 8.12 shows a line graph.

GUIDELINES Creating Effective Line Graphs

Follow these three suggestions to create line graphs that are clear and easy to read.

- **If possible, begin the quantity scale at zero.** Doing so is the best way to portray the information honestly. If you cannot begin at zero, clearly indicate a break in the axis, if appropriate.
- **Use reasonable proportions for the vertical and horizontal axes.** As with bar graphs, make the vertical axis about 25 percent shorter than the horizontal axis.
- **Use grid lines—horizontal, vertical, or both—rather than tick marks when your readers need to read the quantities precisely.**

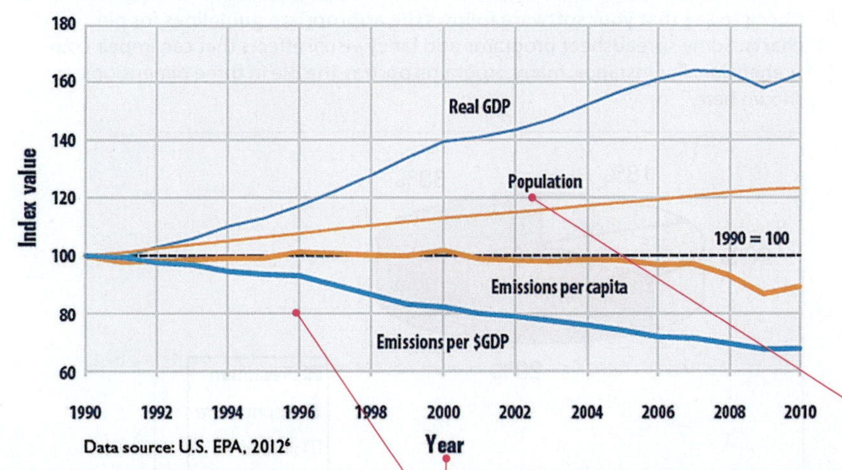

Figure 3. U.S. Greenhouse Gas Emissions per Capita and per Dollar of GDP, 1990–2010
This figure shows trends in greenhouse gas emissions from 1990 to 2010 per capita (heavy orange line), based on the total U.S. population (thin orange line). It also shows trends in emissions compared with the real GDP (heavy blue line). Real GDP is the value of all goods and services produced in the country during a given year, adjusted for inflation (thin blue line). All data are indexed to 1990 as the base year, which is assigned a value of 100. For instance, a real GDP value of 163 in the year 2010 would represent a 63 percent increase since 1990.
Index value
180
160
140
120
100
80
60
Real GDP
Population
1990 = 100
Emissions per capita
Emissions per $GDP
1990
1992
1994
1996
1998
2000
2002
2004
2006
2008
2010
Data source: U.S. EPA, 2012[6]
Year
Note that the title is lengthy because it specifically names the main variables presented in the graph. Name all the important data in the title; it is better for a title to be lengthy than to be imprecise or unclear.
The designer has included a caption that explains how to read the graph. Because this graph is illustrating several items that are measured in different units and therefore cannot be plotted on the same scale (including population and greenhouse gas emissions), the designer chose to have the y-axis express variations from a norm. In this case, the norm is represented by the quantity of each item in the year 1990. If this graph illustrated several items that were measured in the same units, such as the sales figures, in dollars, of several salespersons, the designer would start the y-axis at zero.
Because the four data lines are sufficiently far apart, the designer placed the appropriate data label next to each line. Alternatively, the designer could have used a separate color-coded legend.
Each axis is labeled clearly.
Using different colors and thicknesses for the lines helps readers distinguish them.
The grid lines—both vertical and horizontal—help readers see the specific value for any data point on the graph.

FIGURE 8.12 Line Graph
Information from U.S. Environmental Protection Agency, 2012.

Pie Charts The *pie chart* is a simple but limited design used for showing the relative sizes of the parts of a whole. You can make pie charts with your spreadsheet software. Figure 8.13 on page 219 shows typical examples.

GUIDELINES Creating Effective Pie Charts

Follow these eight suggestions to ensure that your pie charts are easy to understand and professional looking.

- **Restrict the number of slices to no more than seven.** As the slices get smaller, judging their relative sizes becomes more difficult.
- **Begin with the largest slice at the top and work clockwise in order of decreasing size, unless you have a good reason to arrange the slices otherwise.**

(continued)

- **If you have several very small quantities, put them together in one slice, to maintain clarity.** Explain its contents in a footnote. This slice, sometimes called "other," follows the other slices.
- **Place a label (horizontally, not radially) inside the slice, if space permits.** Include the percentage that each slice represents and, if appropriate, the raw number.
- **To emphasize one slice, use a bright, contrasting color or separate the slice from the pie.** Do this, for example, when you introduce a discussion of the item represented by that slice.
- **Check to see that your software follows the appropriate guidelines for pie charts.** Some spreadsheet programs add fancy visual effects that can impair comprehension. For instance, many programs portray the pie in three dimensions, as shown here.

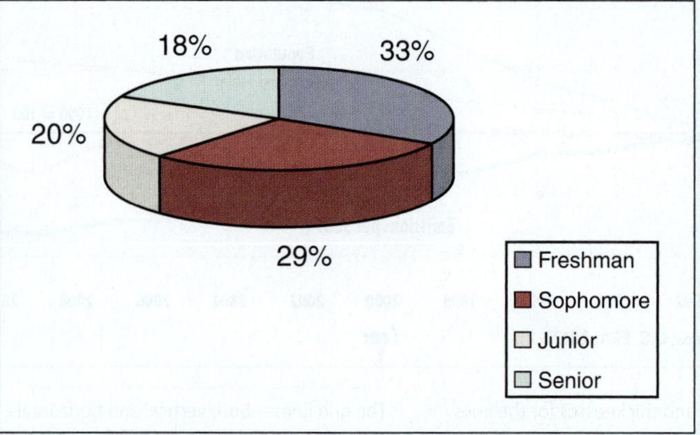

In this three-dimensional pie chart about the percentages of a college's student body, by year, the sophomore slice looks bigger than the freshman slice, even though it isn't, because it appears closer to the reader. To communicate clearly, make pie charts two-dimensional.

- **Don't overdo fill patterns.** Fill patterns are patterns, shades, or colors that distinguish one slice from another. In general, use simple, understated patterns or none at all.
- **Check that your percentages add up to 100.** If you are doing the calculations yourself, check your math.

ILLUSTRATING LOGICAL RELATIONSHIPS

Graphics can help you present logical relationships among items. For instance, in describing a piece of hardware, you might want to show its major components. The two kinds of graphics that best show logical relationships are diagrams and organization charts.

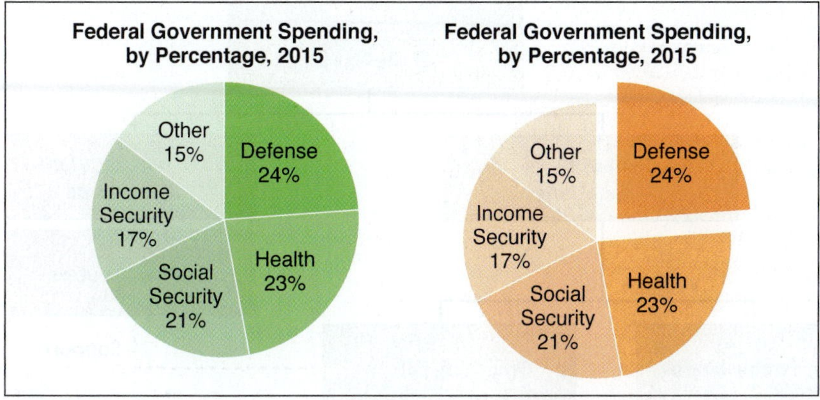

FIGURE 8.13 Pie Charts

You can set your software so that the slices use different saturations of the same color. This approach makes the slices easy to distinguish from each other—without any distractions or misrepresentations caused by a rainbow of colors.

You can set your software to emphasize one slice by separating it from the rest of the pie.

Diagrams A *diagram* is a visual metaphor that uses symbols to represent relationships among items or their properties. In technical communication, common kinds of diagrams are blueprints, wiring diagrams, and schematics. Figure 8.14 is a diagram.

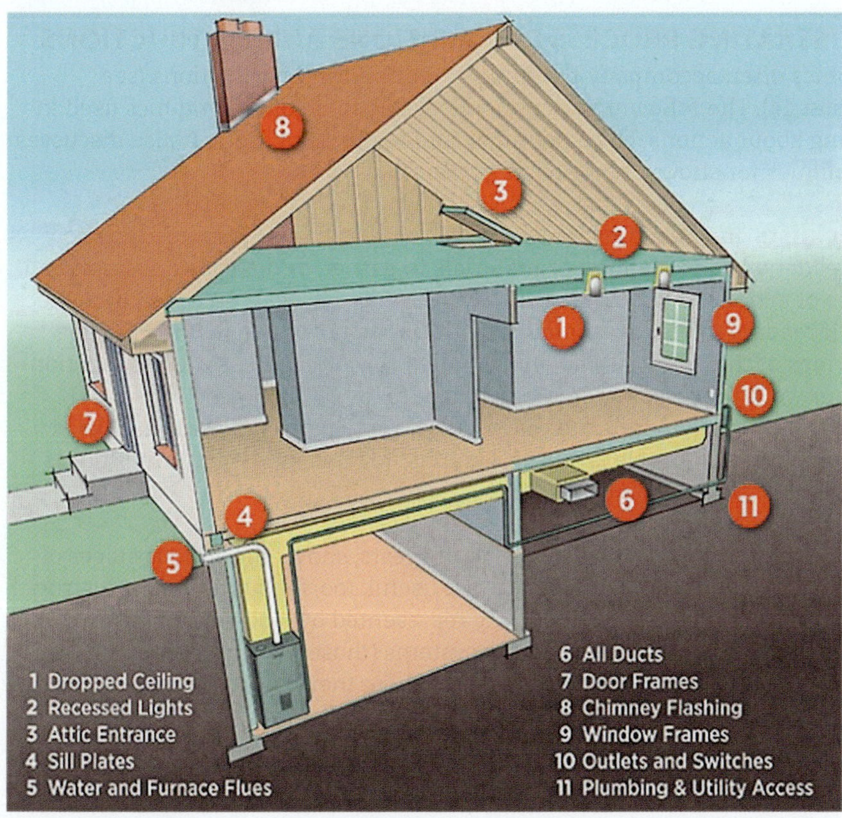

FIGURE 8.14 Diagram

Information from U.S. Department of Energy, 2014: http://energy.gov/energysaver/articles/tips-sealing-air-leaks.

The purpose of this diagram is to help people understand the different areas in their home that need to be insulated. In diagrams, items do not necessarily look realistic. Here the designer is trying to represent logical relationships, not the physical appearances of items.

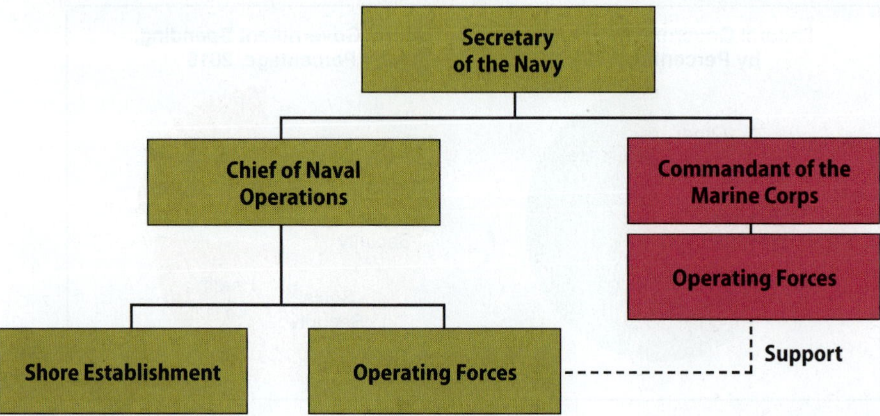

FIGURE 8.15
Organization Chart
Information adapted from "US Navy Organization—An Overview," United States Department of the Navy: http://www.navy.mil/navydata/organization/org-over.asp.

An organization chart is often used to show the hierarchy in an organization, with the most senior person in the organization in the box at the top.

Alternatively, an organization chart can show the functional divisions of a system, such as the human nervous system.

Organization Charts A popular form of diagram is the *organization chart*, in which simple geometric shapes, usually rectangles, suggest logical relationships, as shown in Figure 8.15. You can create organization charts with your word processor.

ILLUSTRATING PROCESS DESCRIPTIONS AND INSTRUCTIONS

Graphics often accompany process descriptions and instructions (see Chapter 14). The following discussion looks at some of the graphics used in writing about actions: checklists, flowcharts, and logic trees. It also discusses techniques for showing motion in graphics.

Checklists In explaining how to carry out a task, you often need to show the reader what equipment or materials to gather, or describe an action or a series of actions to take. A *checklist* is a list of items, each preceded by a check box. If readers might be unfamiliar with the items you are listing, include drawings of the items, as shown in Figure 8.16. You can use the list function in your word processor to create checklists.

Often you need to indicate that readers are to carry out certain tasks at certain intervals. A table is a useful graphic for this kind of information, as shown in Figure 8.17 on page 222.

Flowcharts A *flowchart*, as the name suggests, shows the various stages of a process or a procedure. Flowcharts are useful, too, for summarizing instructions. On a basic flowchart, stages are represented by labeled geometric shapes. Flowcharts can portray open systems (those that have a start and a finish) or closed systems (those that end where they began). Figure 8.18 on page 222 shows an open-system flowchart and a closed-system flowchart. Figure 8.19 on page 223 shows a deployment flowchart, which you can make using the drawing tools in your word processor.

DOCUMENT ANALYSIS ACTIVITY

Analyzing a Graphic

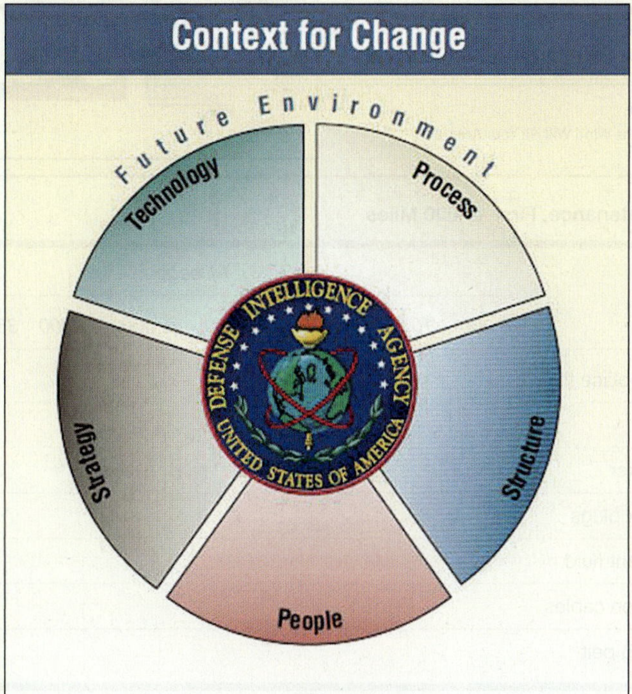

Information from Defense Intelligence Agency, 2003: www.dia.mil/thisisdia/DIA_Workforce_of_the_Future.pdf.

This diagram is from a government report. The questions below ask you to think about diagrams, as discussed on page 219.

1. This design resembles a pie chart, but it does not have the same function as a pie chart. What message does this design communicate? Is it effective?
2. Do the colors communicate any information, or are they merely decorative? If you think they are decorative, would you revise the design to change them in any way?
3. What does the phrase "Future Environment," above the graphic, mean? Is it meant to refer only to the "Technology" and "Process" shapes?
4. Is the explanation below the graphic clear? Would you change it in any way?

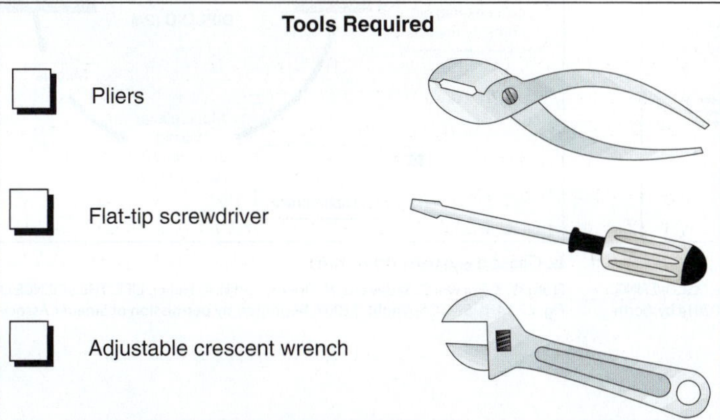

FIGURE 8.16 Checklist

DOCUMENT ANALYSIS ACTIVITY

To analyze an interactive graphic, go to LaunchPad.

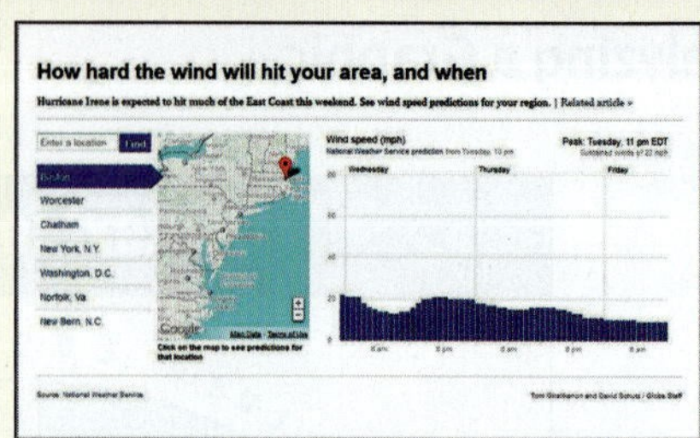

How Hard the Wind Will Hit Your Area, and When

Information from Tom Giratikanon and David Schutz, "How Hard the Wind Will Hit Your Area, and When," Boston Globe, August 26, 2011.

FIGURE 8.17 A Table Used to Illustrate a Maintenance Schedule

Regular Maintenance, First 40,000 Miles

	Mileage							
	5,000	*10,000*	*15,000*	*20,000*	*25,000*	*30,000*	*35,000*	*40,000*
Change oil, replace filter	✓	✓	✓	✓	✓	✓	✓	✓
Rotate tires	✓	✓	✓	✓	✓	✓	✓	✓
Replace air filter				✓				✓
Replace spark plugs				✓				✓
Replace coolant fluid								✓
Replace ignition cables								✓
Replace timing belt								✓

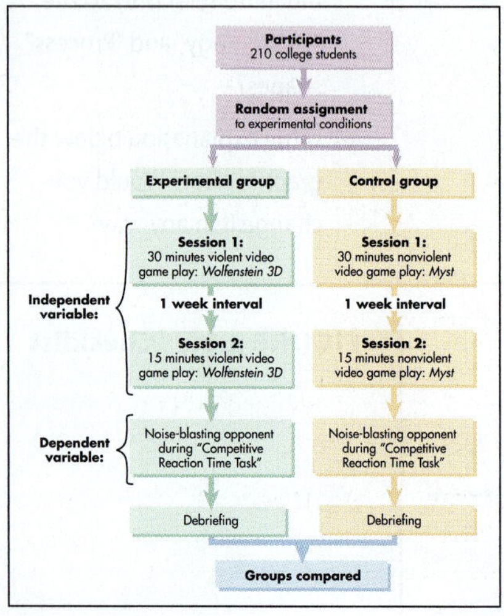

a. Open-system flowchart

From Don Hockenbury and Sandra E. Hockenbury, DISCOVERING PSYCHOLOGY, Sixth Edition, Figure 1.3. Copyright ©2014 by Worth Publishers. Used by permission of the publisher.

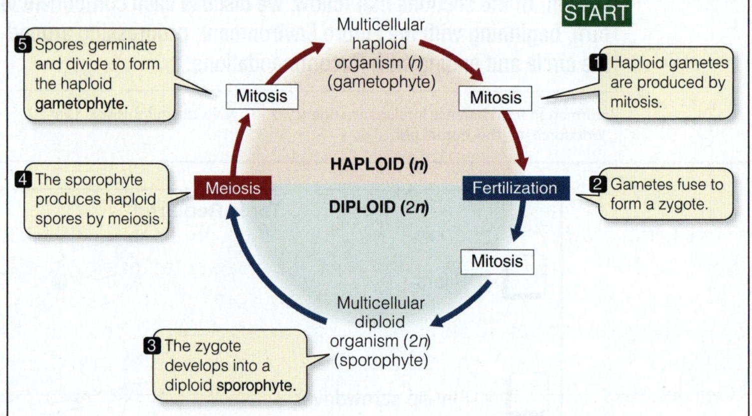

b. Closed-system flowchart

From W. K. Purves, D. Sadava, G. H. Orians, and H. C. Heller, LIFE: THE SCIENCE OF BIOLOGY, Eighth Edition, Fig. 27.14, p. 594. Copyright ©2007. Reprinted by permission of Sinauer Associates.

FIGURE 8.18 Flowcharts

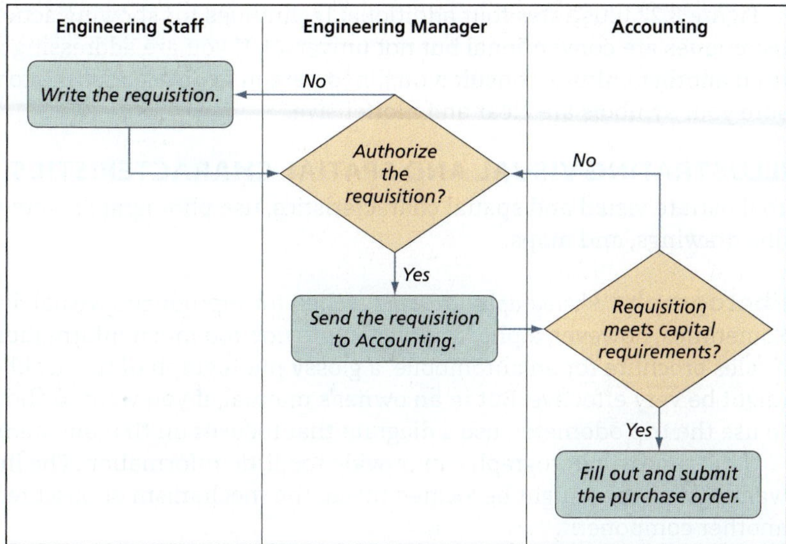

FIGURE 8.19 Deployment Flowchart

A deployment flowchart shows who is responsible for carrying out which tasks. Here the Engineering Staff writes the requisition, then send it to the Engineering Manager.

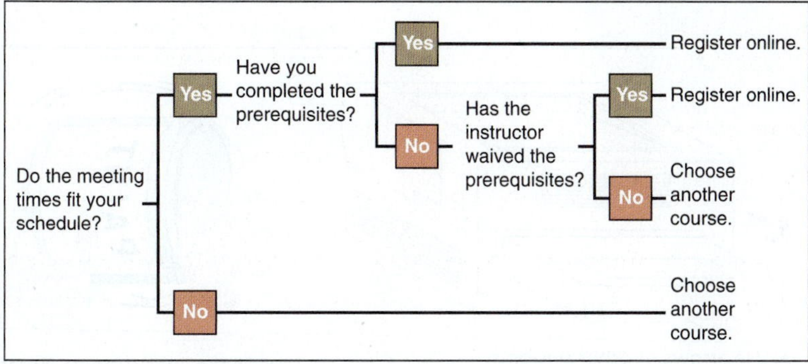

FIGURE 8.20 Logic Tree

Logic Trees *Logic trees* use a branching metaphor. The logic tree shown in Figure 8.20 helps students think through the process of registering for a course.

Techniques for Showing Action or Motion In some types of process descriptions and instructions, you will want to show action or motion. For instance, in an instruction manual for helicopter technicians, you might want to illustrate the process of removing an oil dipstick or tightening a bolt, or you might want to show a warning light flashing. Although animation and video are frequently used to illustrate action or motion in online documents, such processes still need to be communicated in static graphics for print documents.

If the reader is to perform the action, show the action from the reader's point of view, as in Figure 8.21.

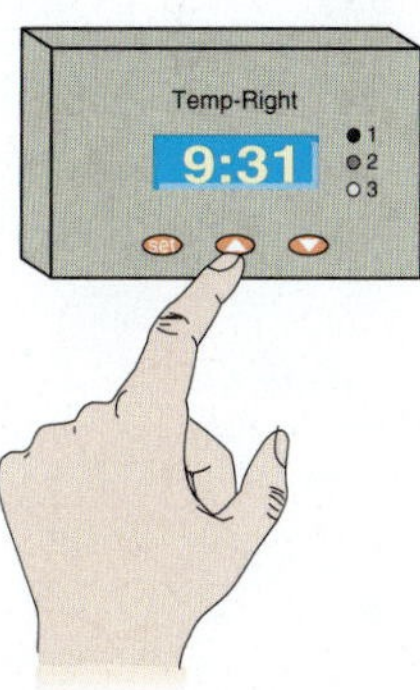

FIGURE 8.21 Showing Action from the Reader's Perspective

In many cases, you need to show only the person's hands, not the whole body.

Figure 8.22 illustrates four additional techniques for showing action. These techniques are conventional but not universal. If you are addressing readers from another culture, consult a qualified person from that culture to make sure your symbols are clear and inoffensive.

ILLUSTRATING VISUAL AND SPATIAL CHARACTERISTICS

To illustrate visual and spatial characteristics, use photographs, screen shots, line drawings, and maps.

For a tutorial on photo editing, go to LaunchPad.

Photographs *Photographs* are unmatched for reproducing visual detail. Sometimes, however, a photograph can provide too much information. In a sales brochure for an automobile, a glossy photograph of the dashboard might be very effective. But in an owner's manual, if you want to show how to use the trip odometer, use a diagram that focuses on that one item.

Sometimes a photograph can provide too little information. The item you want to highlight might be located inside the mechanism or obscured by another component.

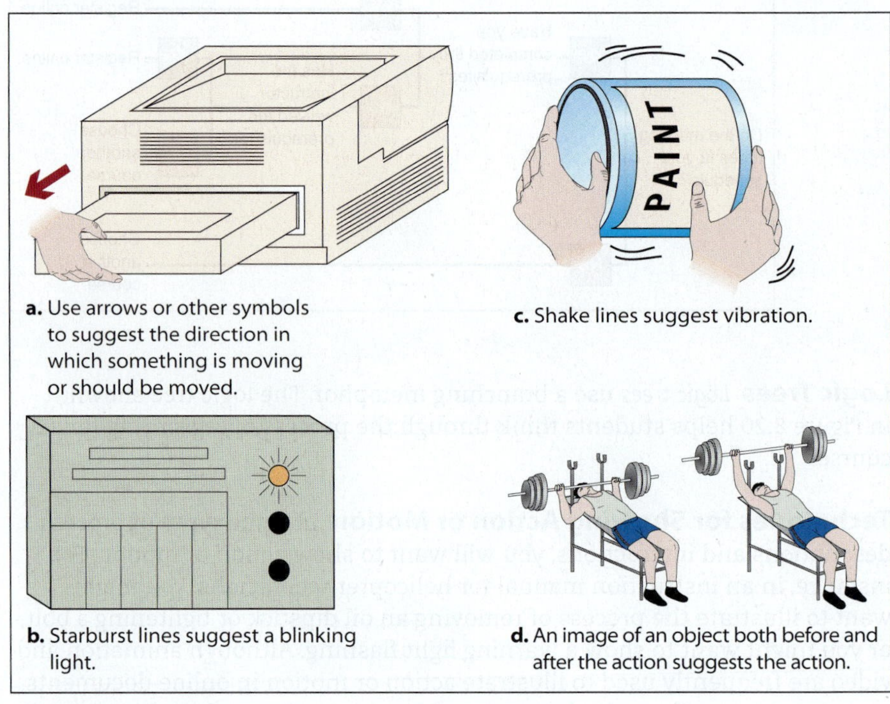

FIGURE 8.22 **Showing Action or Motion**

GUIDELINES Presenting Photographs Effectively

Follow these five suggestions to make sure your photographs are clear, honest, and easy to understand.

- **Eliminate extraneous background clutter that can distract readers.** Crop the photograph to delete unnecessary detail. Figure 8.23 shows examples of cropped and uncropped photographs.
- **Do not electronically manipulate the photograph.** There is nothing unethical about removing blemishes or cropping a digital photograph. However, manipulating a photograph—for example, enlarging the size of the monitor that comes with a computer system—*is* unethical.
- **Help readers understand the perspective.** Most objects in magazines and journals are photographed at an angle to show the object's depth as well as its height and width.
- **If appropriate, include some common object, such as a coin or a ruler, in the photograph to give readers a sense of scale.**
- **If appropriate, label components or important features.**

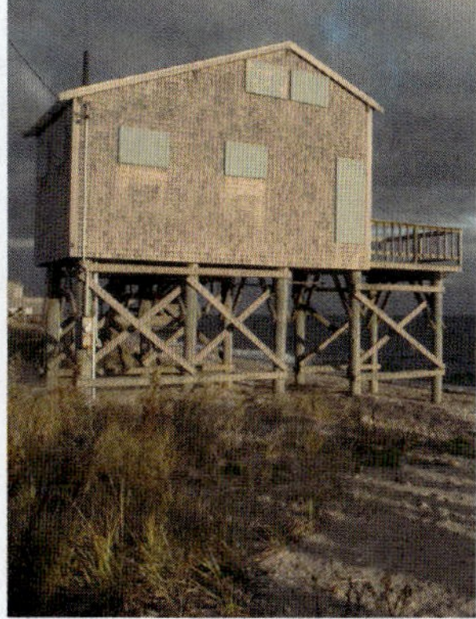

FIGURE 8.23 Cropping a Photograph

Sometimes, technical communicators crop photographs to save space. Ideally, you should crop a photo because it helps you make your point. If you want to show how vulnerable to natural forces the structure in the photograph is, the left-hand version is better because it emphasizes the vastness of the sea. But if you want to discuss the ways the structure has been designed and built to resist natural forces, the right-hand version is better.

KenWiedemann/Getty Images

Screen Shots *Screen shots*—images of what appears on a computer monitor or some other screen—are often used in manuals to show users what the screen will look like as they perform tasks with the device. Readers who see that the screen shot accurately portrays what appears on their own devices are reassured and therefore better able to concentrate on the task they are trying to perform. Figure 8.24 is an example of how screen shots are used.

Line Drawings *Line drawings* are simplified visual representations of objects. Line drawings offer three possible advantages over photographs:

- Line drawings can focus readers' attention on desired information better than a photograph can.
- Line drawings can highlight information that might be obscured by bad lighting or a bad angle in a photograph.
- Line drawings are sometimes easier for readers to understand than photographs are.

Figure 8.25 shows the effectiveness of line drawings.

FIGURE 8.24 **Screen Shot**
Information from NASA, 2013: https://www.nasa.gov/connect/sounds/iphone_install_directions.html.

This screen shot, from a user guide on NASA's website, shows users how to select a ringtone.

This drawing, which accompanies a manual about the Americans with Disabilities Act, illustrates the idea that "wheelchair seating locations must provide lines of sight comparable to those provided to other spectators." A photograph could not show this concept as clearly as this drawing does.

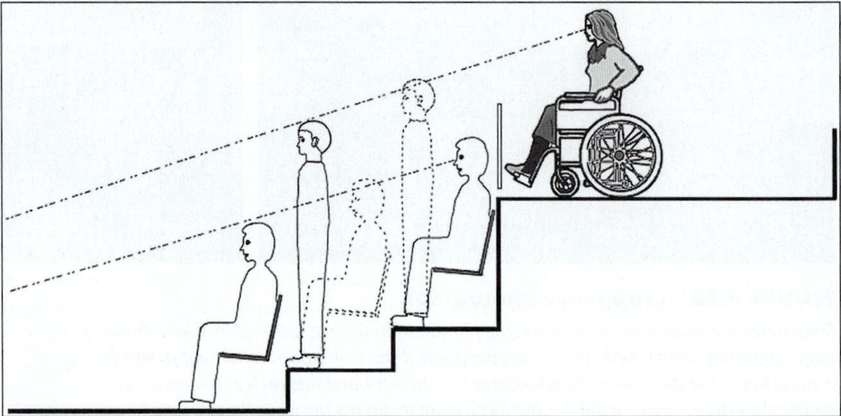

FIGURE 8.25 **Line Drawing**
Information from U.S. Department of Justice, 2010: www.ada.gov/stadium.pdf.

TECH TIP

Why To Create and Insert Screen Shots

You may find that you need to share an image of your computer with an instructor, a colleague, or a tech-support worker. To show your reader what appears in a window on your monitor, you can create a screen shot by using a program such as Microsoft Word or simply by using your Windows or Mac operating system. If you plan to create many screen shots or if you want more sophisticated functionality, including video capture, search the Internet for "screen capture apps" such as TechSmith's SnagIt.

How To Create and Insert Screen Shots

On Windows computers, press the **Print Screen** or **PrtScn** key on your keyboard to capture the entire screen, or **Alt + Print Screen** to capture only the active window. The screen shot will automatically be saved on your clipboard and can be inserted into a document using the **Paste** (or **Ctrl + V**) command. On Mac computers, press **Cmd + Shift + 3** to capture the entire screen, or **Cmd + Shift + 4** to define a screen selection to capture using your cursor. You can also use Microsoft Word and Adobe Acrobat to capture screen shots and then modify them by cropping or resizing.

Microsoft Word

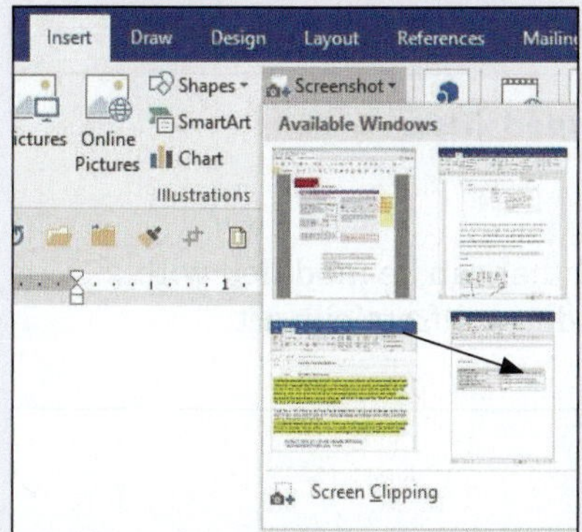

From Word's **Insert** tab, select **Screenshot** to see a small version of each window you have open on your desktop. Click on the screen you want to show your readers, and Word will insert the picture into your document. To define your own screen selection, select **Screen Clipping** and use your cursor to select the portion of your screen you want to capture. To modify a screen shot, select it and use the **Format** tab tools as you would for any image.

Adobe Acrobat Reader

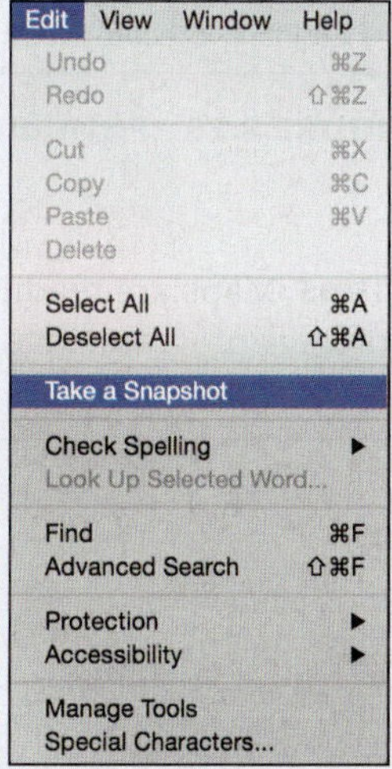

In Adobe Acrobat Reader, from the **Edit** menu, select **Take a Snapshot** and then use your cursor to select the portion of the screen you want to capture. The selection will automatically be saved to your clipboard. To select the entire page, simply click on the screen without dragging the mouse.

You have probably seen the three variations on the basic line drawing shown in Figure 8.26.

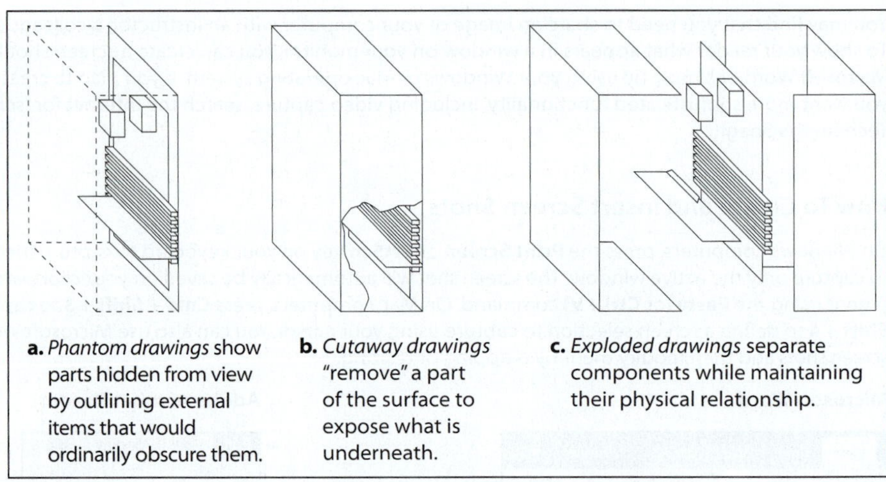

FIGURE 8.26 Phantom, Cutaway, and Exploded Views

Maps Maps are readily available as clip art that can be modified with a graphics program. Figure 8.27 shows a map derived from clip art.

Include a scale and a legend if the map is one that is not thoroughly familiar to your readers. Also, use conventional colors, such as blue for water.

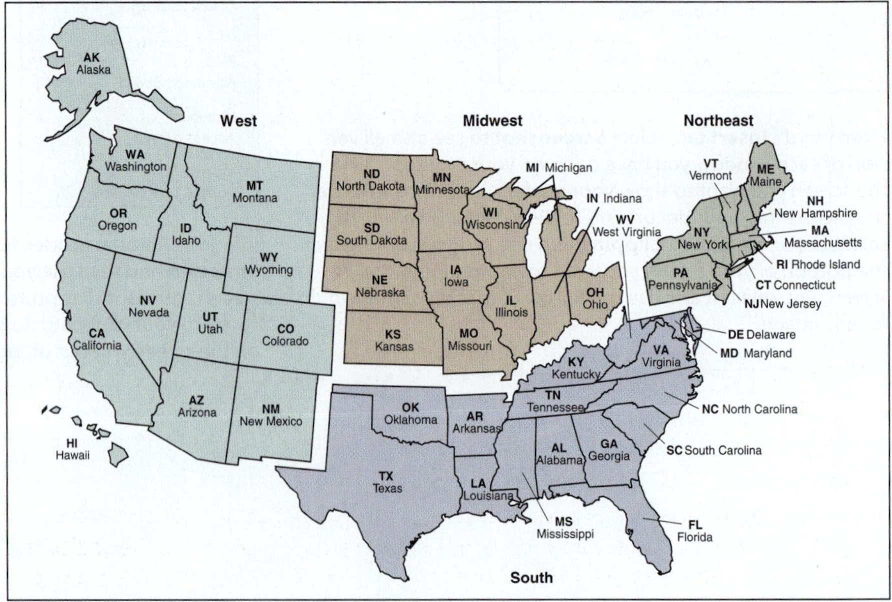

FIGURE 8.27 Map

Creating Effective Graphics for Multicultural Readers

Whether you are writing for people within your organization or outside it, consider the needs of readers whose first language is different from your own. Like words, graphics have cultural meanings. If you are unaware of these meanings, you could communicate something very different from what you intend. The following guidelines are based on William Horton's article "The Almost Universal Language: Graphics for International Documents" (1993).

- **Be aware that reading patterns differ.** In some countries, people read from right to left or from top to bottom. In some cultures, direction signifies value: the right-hand side is superior to the left, or the reverse. You need to think about how to sequence graphics that show action or where to put "before" and "after" graphics. If you want to show a direction, as in an informal flowchart, consider using arrows to indicate how to read the chart.
- **Be aware of varying cultural attitudes toward giving instruction.** Instructions for products made in Japan are highly polite and deferential: "Please attach the cable at this time." Some cultures favor spelling out general principles but leaving the reader to supply the details. To people in these cultures, instructions containing a detailed close-up of how to carry out a task might appear insulting.
- **Deemphasize trivial details.** Because common objects, such as plugs on the ends of power cords, come in different shapes around the world, draw them to look generic rather than specific to one country.
- **Avoid culture-specific language, symbols, and references.** Don't use a picture of a mouse (the furry rodent) to symbolize a computer mouse because the device is not known by that name everywhere. Avoid the casual use of national symbols (such as the maple leaf or national flags); any error in a detail might offend your readers. Use colors carefully: red means danger to most people from Western cultures, but it is a celebratory color to the Chinese.
- **Portray people very carefully.** Every aspect of a person's appearance, from clothing to hairstyle to physical features, is culture- or race-specific. A photograph of a woman in casual Western attire seated at a workstation would be ineffective in an Islamic culture where only a woman's hands and eyes may be shown. Horton (1993) recommends using stick figures or silhouettes that do not suggest any one culture, race, or sex.
- **Be particularly careful in portraying hand gestures.** Many Western hand gestures, such as the "okay" sign, are considered obscene in other cultures, and some people consider long red fingernails inappropriate. Use hands

in graphics only when necessary—for example, to illustrate carrying out a task—and obscure the person's sex and race.

Cultural differences are many and subtle. Learn as much as possible about your readers and about their culture and outlook, and have your graphics reviewed by a native of the culture.

WRITER'S CHECKLIST

- ☐ Does the graphic have a purpose? *(p. 195)*
- ☐ Is the graphic simple and uncluttered? *(p. 195)*
- ☐ Does the graphic present a manageable amount of information? *(p. 195)*
- ☐ Does the graphic meet readers' format expectations? *(p. 195)*
- ☐ Is the graphic clearly labeled? *(p. 195)*
- ☐ If you want to use an existing graphic, do you have the legal right to do so? *(p. 200)* If so, have you cited its source appropriately? *(p. 201)*
- ☐ Is the graphic honest? *(p. 198)*
- ☐ Does the graphic appear in a logical location in the document? *(p. 200)*
- ☐ Is the graphic introduced clearly in the text? *(p. 200)*
- ☐ Is the graphic explained in the text? *(p. 200)*
- ☐ Is the graphic clearly visible in the text? *(p. 200)*
- ☐ Is the graphic easily accessible to readers? *(p. 200)*
- ☐ Is the graphic inoffensive to your readers? *(p. 229)*

EXERCISES

For more about memos, see Ch. 9, "Writing Memos."

1. Find out from the admissions department at your college or university the number of students enrolled from the different states or from the different counties in your state. Present this information in four different kinds of graphics:

   a. map

   b. table

   c. bar graph

   d. pie chart

   In three or four paragraphs, explain why each graphic is appropriate for a particular audience and purpose and how each emphasizes different aspects of the information.

2. Design a flowchart for a process you are familiar with, such as applying for a summer job, studying for a test, preparing a paper, or performing some task at work. Your audience is someone who will be carrying out the process.

3. The following table provides statistics on federal research and development expenditures (U.S. Census Bureau, 2013, Table 815). Study the table, and then perform the following tasks:

   a. Create two different graphics, each of which compares federal R&D funding in 2011 and 2012.

   b. Create two different graphics, each of which compares defense and nondefense R&D funding in either 2011 or 2012.

Table 815. Federal Research and Development (R&D) by Federal Agency: Fiscal Year 2011 and 2012

[In millions of dollars (144,368 represents $144,368,000,000). For fiscal years ending September 30. R&D refers to actual research and development activities as well as R&D facilities. R&D facilities (also known as R&D plants) includes construction, repair, or alteration of physical plant used in the conduct of R&D. Based on Office of Management and Budget data]

Federal agency	2011	2012	Federal agency	2011	2012
Total research and development	**144,368**	**140,565**			
Defense R&D	83,193	78,745	Department of Veterans Affairs	1,160	1,164
Nondefense R&D	61,176	61,820	Department of Homeland Security	760	617
			Department of Transportation	954	945
Department of Defense	79,112	74,464	Department of Interior	757	796
Science and technology	12,751	13,530	U.S. Geological Survey	640	675
All other Department of Defense R&D	66,361	60,935	Environmental Protection Agency	582	568
Health and Human Services	31,183	31,143	Department of Education	362	392
National Institute of Health	29,831	30,046	Smithsonian	259	243
All other Health and Human Services R&D	1,352	1,097	International Assistance Programs	121	121
Department of Energy	10,673	11,019	Department of Housing and Urban Development	79	57
Atomic Energy Defense	4,081	4,281	Department of State	75	75
Office of Science	4,461	4,463	Nuclear Regulatory Commission	99	83
Energy R&D	2,131	2,275	Department of Justice	109	92
NASA	9,099	9,399	Social Security Administration	42	8
National Science Foundation	5,494	5,614	U.S. Postal Service	14	14
Department of Agriculture	2,135	2,331	Tennessee Valley Authority	18	15
Department of Commerce	1,217	1,263	Army Corps of Engineers	11	11
National Oceanic and Atmospheric Administration	629	581	Telecommunications Development Agency	7	4
National Institute of Standards and Technology	532	555	Department of Labor	4	4

Source: American Association for the Advancement of Science (AAAS), *AAAS Report XXXVI: Research and Development FY2013*, annual (copyright). See also <http://www.aaas.org/spp/rd/rdreport2013/>.

American Association for the Advancement of Science (AAAS) AAAS Report XXXVI. Research and Development FY 2013. Reprinted with permission from AAAS.

4. For each of the following four graphics, write a paragraph evaluating its effectiveness and describing how you would revise it.

a. Majors

	2012	**2013**	**2014**
Civil Engineering	236	231	253
Chemical Engineering	126	134	142
Comparative Literature	97	86	74
Electrical Engineering	317	326	401
English	714	623	592
Fine Arts	112	96	72
Foreign Languages	608	584	566
Materials Engineering	213	227	241
Mechanical Engineering	196	203	201
Other	46	42	51
Philosophy	211	142	151
Religion	86	91	72

b. Number of Members of the U.S. Armed Forces in 2012 (in Thousands)

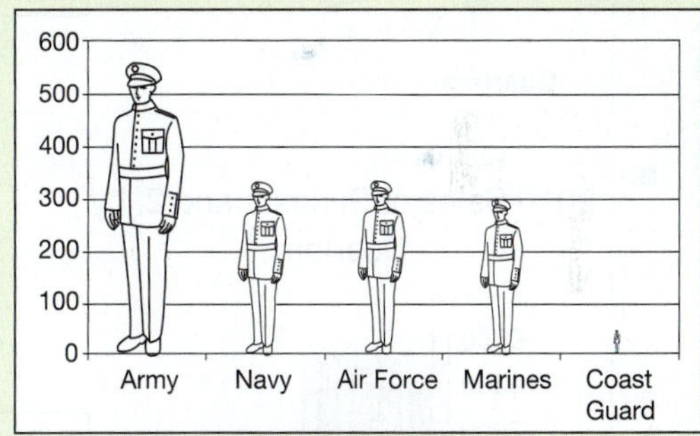

c. Expenses at Hillway Corporation

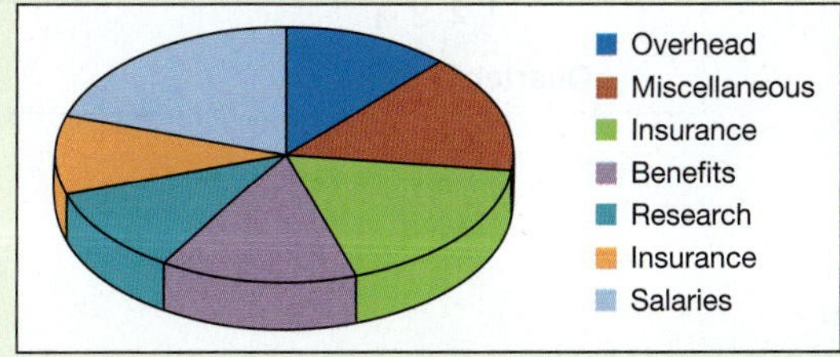

d. Costs of the Components of a PC

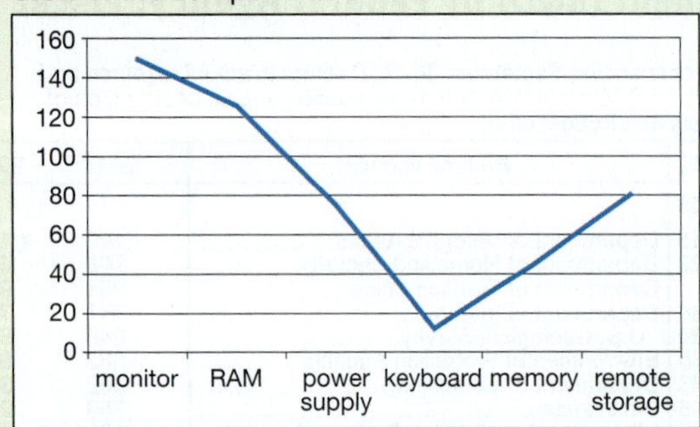

5. The following three graphs illustrate the sales of two products—Series 1 and Series 2—for each quarter of 2016. Which is the most effective in conveying the information? Which is the least effective? What additional information would make the most-effective graph better?

a.

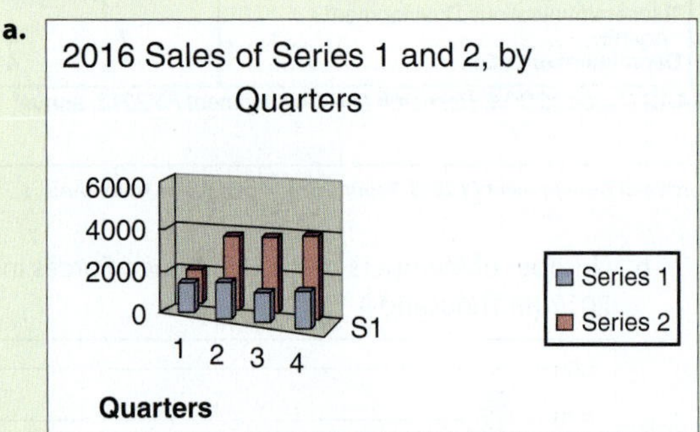

b.

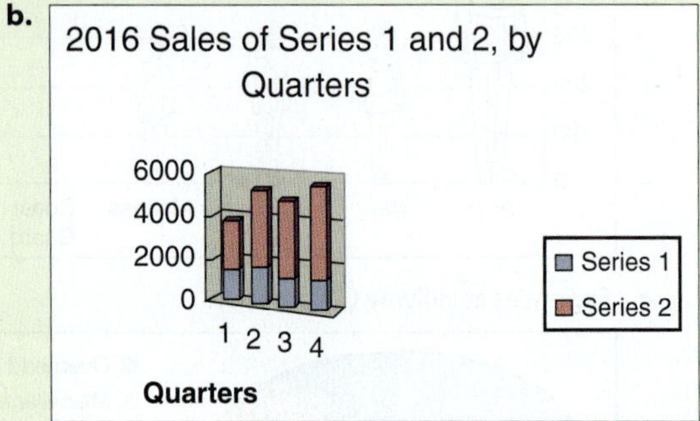

c.

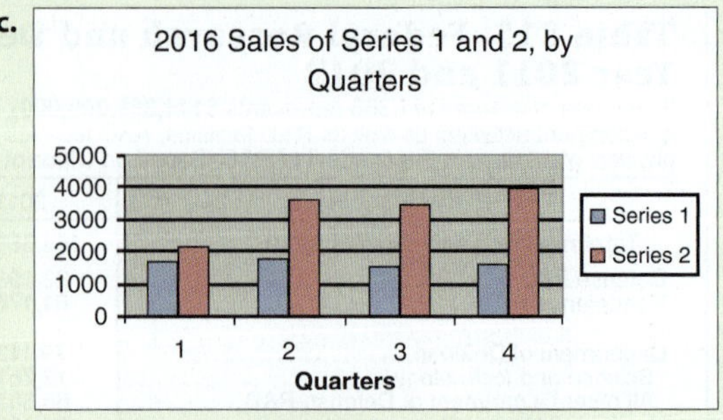

6. Search for "infographics college tuition." Find two infographics that present information on similar topics, such as tuition costs for public and private colleges and universities, average tuition costs in each of the fifty states, or the relationship between tuition costs and future earnings potential. Write a 1,000-word memo to your instructor in which you identify the audience and purpose of the infographics and compare and contrast them using such criteria as audience, purpose, clarity, visual appeal, use of different types of graphics, and citation of the data sources. Which infographic do you think is better? Why?

7. Locate a graphic on the web that you consider inappropriate for an international audience because it might be offensive or unclear to readers in some cultures. Imagine an intended audience for the graphic, such as people from the Middle East, and write a brief statement explaining the potential problem. Finally, revise the graphic so that it would be appropriate for its intended audience.

CASE 8: Creating Appropriate Graphics To Accompany a Report

Following a series of texting-related driving accidents, a representative in your state legislature has decided to introduce legislation restricting the use of cell phones while driving. As an assistant to the state's insurance commissioner, you have been asked to collect data on how cell-phone use affects driving. The representative hopes to use these data to make his case for the legislation. When you present your findings to your supervisor, she asks that you pare them down to the facts most relevant to the representative. She also asks that you consider presenting some of the data graphically. To get to work improving your research, go to LaunchPad.

Part 4

Yuri_Arcurs/Getty Images

Learning Important Applications

CHAPTER

9 Corresponding in Print and Online

Mail To: Doc Team
File Edit View Actions Tools
Send Cancel
Mail Send Options
From: Melissa Cartright
To: Doc Team
Subject: Input re Policy on

REGARDLESS OF THE APPLICATION you use, you will communicate in writing every day on the job. This chapter discusses the four major formats used for producing workplace correspondence: letters, memos, emails, and microblog posts. Throughout this chapter, the word *correspondence* refers to all these forms.

Understanding the Process of Writing Correspondence

The process of writing correspondence is essentially like that of writing any other kind of workplace document. The Focus on Process box below presents an overview of this process, focusing on letters, memos, and emails. The more formal the correspondence, the more time you are likely to spend on each of these steps.

When you need to correspond with others in the workplace, your first task is to decide on the appropriate application. You have four major choices: letters, memos, emails, and microblogs.

FOCUS ON PROCESS Writing Correspondence

When writing correspondence, pay special attention to these steps of the writing process.

PLANNING You will need to choose the appropriate type of correspondence for your writing situation.

DRAFTING For letters, memos, and email, clearly state your purpose, use headings to help your readers, summarize your message, provide adequate background, organize the discussion, and highlight action items. For microblogs, state your message or question clearly.

REVISING
EDITING
PROOFREADING

You might need to write correspondence quickly, but you still need to write carefully. Review the section of the Writer's Checklist at the end of this chapter that applies to your document. Edit and proofread everything before you publish or send it. See Appendix, Part B, p. 482, for help.

CHOICES AND STRATEGIES Choosing a Type of Correspondence

IF THE SITUATION IS . . .	AND YOU ARE WRITING TO . . .	TRY THIS TYPE OF CORRESPONDENCE
Formal	People outside or within your organization	**Letters.** Because letters use centuries-old conventions such as the salutation and complimentary close, they are the most formal of the four types of correspondence.
Moderately formal	People outside your organization	**Letters or email.** Letters are more formal than email, but consider how your readers will need to use the information. Recipients can store and forward an email easily, as well as capture the text and reuse it in other documents. In addition, you can attach other files to an email.
Moderately formal	People within your organization	**Memos or email.** Memos are moderately formal and can be sent in the body of an email or as an attachment.
Informal	People outside or within your organization	**Email or microblogs.** Email is good for quick, relatively informal communication with one or many recipients. Microblog posts such as Twitter tweets or Facebook status updates can be useful for quick questions addressed to a group. Microblogs are the most informal type of correspondence.

Presenting Yourself Effectively in Correspondence

When you write business correspondence, follow these five suggestions for presenting yourself as a professional:

- Use the appropriate level of formality.
- Communicate correctly.
- Project the "you attitude."
- Avoid correspondence clichés.
- Communicate honestly.

USE THE APPROPRIATE LEVEL OF FORMALITY

People are sometimes tempted to use informal writing in informal digital applications such as email and microblogs. Don't. Everything you write on the job is legally the property of the organization for which you work, and messages are almost always archived digitally, even after recipients have deleted them. Your documents might be read by the company president, or they might appear in a newspaper or in a court of law. Therefore, use a moderately formal tone to avoid potential embarrassment.

TOO INFORMAL	Our meeting with United went south right away when they threw a hissy fit, saying that we blew off the deadline for the progress report.
MODERATELY FORMAL	In our meeting, the United representative expressed concern that we had missed the deadline for the progress report.

However, you don't want to sound like a dictionary.

TOO FORMAL	It was indubitably the case that our team was successful in presenting a proposal that was characterized by quality of the highest order. My appreciation for your industriousness is herewith extended.
MODERATELY FORMAL	I think we put together an excellent proposal. Thank you very much for your hard work.

COMMUNICATE CORRECTLY

One issue closely related to formality is correctness. Correct writing is free of grammar, punctuation, style, usage, and spelling errors. Correctness problems occur most often in email and microblogs.

Some writers mistakenly think that they do not need to worry about correctness because these digital applications are meant for quick communication. They are wrong. You have to plan your digital correspondence just as you plan any other written communication, and you should revise, edit, and proofread it. Sending correspondence that contains language errors is unprofessional because it suggests a lack of respect for your reader—and for yourself. It also causes your reader to think that you are careless about your job.

PROJECT THE "YOU ATTITUDE"

Correspondence should convey a courteous, positive tone. The key to accomplishing this task is using the "you attitude"—looking at the situation from the reader's point of view and adjusting the content, structure, and tone to meet his or her needs. For example, if you are writing to a supplier who has failed to deliver some merchandise by the agreed-on date, the "you attitude" dictates that you not discuss problems you are having with other suppliers; those problems don't concern your reader. Instead, concentrate on explaining clearly and politely that the reader has violated your agreement and that not having the merchandise is costing you money. Then propose ways to expedite the shipment.

Following are two examples of thoughtless sentences, each followed by an improved version that shows the "you attitude."

ACCUSING You must have dropped the engine. The housing is badly cracked.

BETTER The badly cracked housing suggests that the engine must have fallen onto a hard surface from some height.

SARCASTIC You'll need two months to deliver these parts? Who do you think you are, the post office?

BETTER Surely you would find a two-month delay for the delivery of parts unacceptable in your business. That's how I feel, too.

A calm, respectful tone makes the best impression and increases the chances that you will achieve your goal.

AVOID CORRESPONDENCE CLICHÉS

Over the centuries, a group of words and phrases have come to be associated with business correspondence; one common example is *as per your request*. These phrases sound stilted and insincere. Don't use them. Figure 9.1 is a list of common clichés and their plain-language equivalents. Figure 9.2 shows two versions of the same letter: one written in clichés, the other in plain language.

For more about choosing the right words and phrases, see Ch. 6, "Choosing the Right Words and Phrases."

LETTER CLICHÉ	PLAIN-LANGUAGE EQUIVALENT
attached please find	attached is
enclosed please find	enclosed is
pursuant to our agreement	as we agreed
referring to your ("Referring to your letter of March 19, the shipment of pianos . . .")	"As you wrote in your letter of March 19, the . . ." (or subordinate the reference at the end of your sentence)
wish to advise ("We wish to advise that . . .")	(The phrase doesn't say anything. Just say what you want to say.)
the writer ("The writer believes that . . .")	"I believe . . ."

FIGURE 9.1 **Letter Clichés and Plain-Language Equivalents**

LETTER CONTAINING CLICHÉS	LETTER IN PLAIN LANGUAGE
Dear Mr. Smith: Referring to your complaint regarding the problem encountered with your new Trailrider Snowmobile, our Customer Service Department has just submitted its report. It is their conclusion that the malfunction is caused by water being present in the fuel line. It is our conclusion that you must have purchased some bad gasoline. We trust you are cognizant of the fact that while we guarantee our snowmobiles for a period of not less than one year against defects in workmanship and materials, responsibility cannot be assumed for inadequate care. We wish to advise, for the reason mentioned hereinabove, that we cannot grant your request to repair the snowmobile free of charge. Permit me to say, however, that the writer would be pleased to see that the fuel line is flushed at cost, $30. Your Trailrider would then give you many years of trouble-free service. Attached please find an authorization statement. Should we receive it, we shall perform the above-mentioned repair and deliver your snowmobile forthwith. Sincerely yours,	Dear Mr. Smith: Thank you for letting us know about the problem with your new Trailrider Snowmobile. Our Customer Service Department has found water in the fuel line. Apparently some of the gasoline was bad. While we guarantee our snowmobiles for one year against defects in workmanship and materials, we cannot assume responsibility for problems caused by bad gasoline. We cannot, therefore, grant your request to repair the snowmobile free of charge. However, no serious harm was done to the snowmobile. We would be happy to flush the fuel line at cost, $30. Your Trailrider would then give you many years of trouble-free service. If you will authorize us to do this work, we will have your snowmobile back to you within four working days. Just fill out the enclosed authorization statement and return it as an attachment in a reply to this email or fax it to the number shown on the form. Sincerely yours,

FIGURE 9.2 Sample Letters with and Without Clichés

The letter on the right avoids clichés and shows an understanding of the "you attitude." Instead of focusing on the violation of the warranty, it presents the conclusion as good news: the snowmobile is not ruined, and it can be repaired and returned in less than a week for a small charge.

COMMUNICATE HONESTLY

You should communicate honestly when you write any kind of document, and business correspondence is no exception. Communicating honestly shows respect for your reader and for yourself.

ETHICS NOTE

WRITING HONEST BUSINESS CORRESPONDENCE

Why is dishonesty a big problem in correspondence? Perhaps because the topics discussed in business correspondence often relate to the writer's professionalism and the quality of his or her work. For instance, when a salesperson working for a supplier writes to a customer explaining why a product did not arrive on time, he is tempted to make it seem as if his company—and he personally—is blameless. Similarly, when a manager has to announce a new policy that employees will dislike, she might be tempted to distance herself from the policy.

The professional approach is to tell the truth. If you mislead a reader in explaining why the shipment didn't arrive on time, the reader will likely double-check the facts, conclude that you are trying to avoid responsibility, and end your business relationship. If you try to convince readers that you had nothing to do with a new, unpopular policy, some of them will know you are being misleading, and you will lose your most important credential: your credibility.

Writing Letters

Letters are still a basic means of communication between organizations, with millions written each day. To write effective letters, you need to understand the elements of a letter, its format, and the common types of letters sent in the business world.

ELEMENTS OF A LETTER

Most letters include a heading, inside address, salutation, body, complimentary close, and signature. Some letters also include one or more of the following: attention line, subject line, enclosure line, and copy line. Figure 9.3 shows the elements of a letter.

DAVIS TREE CARE
1300 Lancaster Avenue
Berwyn, PA 19092
www.davisfortrees.com

May 11, 2019

Fairlawn Industrial Park
1910 Ridgeway Drive
Berwyn, PA 19092

Attention: Director of Maintenance

Subject: Fall pruning

Dear Director of Maintenance:

Do you know how much your trees are worth? That's right—your trees. As a maintenance director, you know how much of an investment your organization has in its physical plant. And the landscaping is a big part of your total investment.

Most people don't know that even the hardiest trees need periodic care. Like shrubs, trees should be fertilized and pruned. And they should be protected against the many kinds of diseases and pests that are common in this area.

At Davis Tree Care, we have the skills and experience to keep your trees healthy and beautiful. Our diagnostic staff is made up of graduates of major agricultural and forestry universities, and all of our crews attend special workshops to keep current with the latest information on tree maintenance. Add to this our proven record of 43 years of continuous service in the Berwyn area, and you have a company you can trust.

Image: Sapik/Shutterstock.

Heading. Most organizations use letterhead stationery with their heading printed at the top. This preprinted information and the date the letter is sent make up the heading. If you are using blank paper rather than letterhead, your address (without your name) and the date form the heading. Whether you use letterhead or blank paper for the first page, do not number it. Use blank paper for the second and all subsequent pages.

Inside Address. If you are writing to an individual who has a professional title—such as Professor, Dr., or, for public officials, Honorable—use it. If not, use Mr. or Ms. (unless you know the recipient prefers Mrs. or Miss). If the reader's position fits on the same line as the name, add it after a comma; otherwise, drop it to the line below. Spell the name of the organization the way the organization itself does: for example, International Business Machines calls itself IBM. Include the complete mailing address: street number and name, city, state, and zip code.

Attention Line. Sometimes you will be unable to address a letter to a particular person because you don't know (and cannot easily find out) the name of the individual who holds that position in the company.

Subject Line. The subject line is an optional element in a letter. Use either a project number (for example, "Subject: Project 31402") or a brief phrase defining the subject (for example, "Subject: Price quotation for the R13 submersible pump").

Salutation. If you decide not to use an attention line or a subject line, put the salutation, or greeting, two lines below the inside address. The traditional salutation is *Dear*, followed by the reader's courtesy title and last name and then a colon (not a comma):

Dear Ms. Hawkins:

FIGURE 9.3 Elements of a Letter

Header for second page.

Letter to Fairlawn Industrial Park
Page 2
May 11, 2019

Body. In most cases, the body contains at least three paragraphs: an introductory paragraph, a concluding paragraph, and one or more body paragraphs.

May we stop by to give you an analysis of your trees—absolutely without cost or obligation? Spending a few minutes with one of our diagnosticians could prove to be one of the wisest moves you've ever made. Just give us a call at 610-555-9187, and we'll be happy to arrange an appointment at your convenience.

Complimentary Close. The conventional phrases *Sincerely, Sincerely yours, Yours sincerely, Yours very truly,* and *Very truly yours* are interchangeable.

Sincerely yours,

Jasmine Brown

Jasmine Brown
President

Signature. Type your full name on the fourth line below the complimentary close. Sign the letter, in ink, above the printed name. Most organizations prefer that you include your position under your printed name.

Enclosure: Davis Tree Care brochure

c: Darrell Davis, Vice President

Enclosure Line. If the envelope contains documents other than the letter, include an enclosure line that indicates the number of enclosures. For more than one enclosure, add the number: "Enclosures (2)." In determining the number of enclosures, count only separate items, not pages. A three-page memo and a 10-page report constitute only two enclosures. Some writers like to identify the enclosures:
Enclosure: 2018 Placement Bulletin
Enclosures (2): "This Year at Ammex"
2018 Annual Report

Copy Line. If you want the primary recipient to know that other people are receiving a copy of the letter, include a copy line. Use the symbol c (for "copy") followed by a colon and the names of the other recipients (listed either alphabetically or according to organizational rank). If appropriate, use the symbol cc (for "courtesy copy") followed by the names of recipients who are less directly affected by the letter.

FIGURE 9.3 Elements of a Letter *(continued)*

Two other types of letters are discussed in this book: the job-application letter in Ch. 10, p. 281, and the transmittal letter in Ch. 13, p. 348.

COMMON TYPES OF LETTERS

Organizations send out many different kinds of letters. This section focuses on four types of letters written frequently in the workplace: inquiry, response to an inquiry, claim, and adjustment.

Inquiry Letter Figure 9.4 shows an inquiry letter, in which you ask questions.

14 Hawthorne Ave.
Belleview, TX 75234

November 2, 2019

Dr. Andrea Shakir
Director of Technical Services
Orion Corporation
721 West Douglas Avenue
Maryville, TN 31409

Dear Dr. Shakir:

I am writing to you because of Orion's reputation as a leader in the manufacture of adjustable x-ray tables. I am a graduate student in biomedical engineering at the University of Texas, and I am working on an analysis of diagnostic equipment for a seminar paper. Would you be able to answer a few questions about your Microspot 311?

1. Can the Microspot 311 be used with lead oxide cassettes, or does it accept only lead-free cassettes?
2. Are standard generators compatible with the Microspot 311?
3. What would you say is the greatest advantage, for the operator, of using the Microspot 311? For the patient?

Because my project is due on January 15, I would greatly appreciate your assistance in answering these questions by January 10. Of course, I would be happy to send you a copy of my report when it is completed.

Yours very truly,

Albert K. Stern

Albert K. Stern

You write an inquiry letter to acquire information. Explain who you are and why you are writing. Make your questions precise and clear, and therefore easy to answer. Explain what you plan to do with the information and how you can compensate the reader for answering your questions.

This writer's task is to motivate the reader to provide some information. That information is not likely to lead to a sale because the writer is a graduate student doing research, not a potential customer.

Notice the flattery in the first sentence.

The writer presents specific questions in a list format, making the questions easy to read and understand.

In the final paragraph, the writer politely indicates his schedule and requests the reader's response. Note that he offers to send the reader a copy of his report.

If the reader provides information, the writer should send a thank-you letter.

FIGURE 9.4 **Inquiry Letter**

Response to an Inquiry Figure 9.5 shows a response to the inquiry letter in Figure 9.4.

Claim Letter Figure 9.6 (on page 245) is an example of a claim letter that the writer scanned and attached to an email to the reader. The writer's decision to present his message in a letter rather than an email suggests that he wishes to convey the more-formal tone associated with letters—and yet he wants the letter to arrive quickly.

In responding to an inquiry letter, answer the questions if you can. If you cannot, either because you don't know the answers or because you cannot divulge proprietary information, explain the reasons and offer to assist with other requests.

ORION

721 WEST DOUGLAS AVE.
MARYVILLE, TN 31409

(615) 619-8132
www.orioninstruments.com

November 7, 2019

Mr. Albert K. Stern
14 Hawthorne Ave.
Belleview, TX 75234

Dear Mr. Stern:

The writer responds graciously.

I would be pleased to answer your questions about the Microspot 311. We think it is the best unit of its type on the market today.

The writer answers the three questions posed in the inquiry letter.

1. The 311 can handle lead oxide or lead-free cassettes.
2. At the moment, the 311 is fully compatible only with our Duramatic generator. However, special wiring kits are available to make the 311 compatible with our earlier generator models—the Olympus and the Saturn. We are currently working on other wiring kits.
3. For the operator, the 311 increases the effectiveness of the radiological procedure while at the same time cutting down the amount of film used. For the patient, it reduces the number of repeat exposures and therefore reduces the total dose.

The writer encloses other information to give the reader a fuller understanding of the product.

I am enclosing a copy of our brochure on the Microspot 311. If you would like additional information, please visit our website at www.orioninstruments.com/products/microspot311. I would be happy to receive a copy of your analysis when it is complete. Good luck!

Sincerely yours,

Andrea Shakir, M.D.

Andrea Shakir, M.D.
Director of Technical Services

The writer uses the enclosure notation to signal that she is attaching an item to the letter.

Enclosure

The writer indicates that she is forwarding a copy to her supervisor.

c: Robert Anderson, Executive Vice President

FIGURE 9.5 **Response to an Inquiry**

A claim letter is a polite, reasonable complaint. If you purchase a defective or falsely advertised product or receive inadequate service, you write a claim letter. If the letter is convincing, your chances of receiving a satisfactory settlement are good because most organizations realize that unhappy customers are bad for business. In addition, claim letters help companies identify weaknesses in their products or services.

255 Robbins Place, Centerville, MO 65101 | (417) 555-1850 | robbinsconstruction.com

August 17, 2018

Mr. David Larsyn
Larsyn Supply Company
311 Elmerine Avenue
Anderson, MO 63501

Dear Mr. Larsyn:

As steady customers of yours for over 15 years, we came to you first when we needed a quiet pile driver for a job near a residential area. On your recommendation, we bought your Vista 500 Quiet Driver, at $14,900. We have since found, much to our embarrassment, that it is not substantially quieter than a regular pile driver.

We received the contract to do the bridge repair here in Centerville after promising to keep the noise to under 90 dB during the day. The Vista 500 (see enclosed copy of bill of sale for particulars) is rated at 85 dB, maximum. We began our work and, although one of our workers said the driver didn't seem sufficiently quiet to him, assured the people living near the job site that we were well within the agreed sound limit. One of them, an acoustical engineer, marched out the next day and demonstrated that we were putting out 104 dB. Obviously, something is wrong with the pile driver.

I think you will agree that we have a problem. We were able to secure other equipment, at considerable inconvenience, to finish the job on schedule. When I telephoned your company that humiliating day, however, a Mr. Meredith informed me that I should have done an acoustical reading on the driver before I accepted delivery.

I would like you to send out a technician—as soon as possible—either to repair the driver so that it performs according to specifications or to take it back for a full refund.

Yours truly,

Jack Robbins

Jack Robbins, President
Enclosure

The writer indicates clearly in the first paragraph that he is writing about an unsatisfactory product. Note that he identifies the product by model name.

The writer presents the background, filling in specific details about the problem. Notice how he supports his earlier claim that the problem embarrassed him professionally.

The writer states that he thinks the reader will agree that there was a problem with the equipment.

Then the writer suggests that the reader's colleague did not respond satisfactorily.

The writer proposes a solution: that the reader take appropriate action. The writer's clear, specific account of the problem and his professional tone increase his chances of receiving the solution he proposes.

FIGURE 9.6 Claim Letter

Adjustment Letter Figures 9.7 and 9.8 show "good news" and "bad news" adjustment letters. The first is a reply to the claim letter shown in Figure 9.6.

An adjustment letter, a response to a claim letter, tells the customer how you plan to handle the situation. Your purpose is to show that your organization is fair and reasonable and that you value the customer's business.

If you can grant the request, the letter is easy to write. Express your regret, state the adjustment you are going to make, and end on a positive note by encouraging the customer to continue doing business with you.

Larsyn Supply Company

311 Elmerine Avenue
Anderson, MO 63501
(417) 555-2484
larsynsupply.com

August 22, 2019

Mr. Jack Robbins, President
Robbins Construction, Inc.
255 Robbins Place
Centerville, MO 65101

Dear Mr. Robbins:

I was very unhappy to read your letter of August 17 telling me about the failure of the Vista 500. I regretted most the treatment you received from one of my employees when you called us.

The writer wisely expresses regret about the two problems cited in the claim letter.

Harry Rivers, our best technician, has already been in touch with you to arrange a convenient time to come out to Centerville to talk with you about the driver. We will of course repair it, replace it, or refund the price. Just let us know your wish.

The writer describes the actions he has already taken and formally states that he will do whatever the reader wishes.

I realize that I cannot undo the damage that was done on the day that a piece of our equipment failed. To make up for some of the extra trouble and expense you incurred, let me offer you a 10 percent discount on your next purchase or service order with us, up to a $1,000 total discount.

The writer expresses empathy in making the offer of adjustment. Doing so helps to create a bond: you and I are both professionals who rely on our good reputations.

You have indeed been a good customer for many years, and I would hate to have this unfortunate incident spoil that relationship. Won't you give us another chance? Just bring in this letter when you visit us next, and we'll give you that 10 percent discount.

This polite conclusion appeals to the reader's sense of fairness and reflects good business practice.

Sincerely,

Dave Larsyn

Dave Larsyn, President

FIGURE 9.7 "Good News" Adjustment Letter

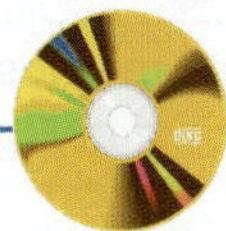

Quality Storage Media

2077 Highland, Burley, ID 84765
208 • 555 • 1613
qualstorage.com

February 3, 2019

Ms. Dale Devlin
1903 Highland Avenue
Glenn Mills, NE 69032

Dear Ms. Devlin:

Thank you for writing us about the external hard drive you purchased on January 11, 2019. I know from personal experience how frustrating it is when a drive fails.

According to your letter, you used the drive to store the business plan for your new consulting business. When you attempted to copy that file to your internal hard drive, the external drive failed, and the business plan was lost. You have no other copy of that file. You are asking us to reimburse you $1,500 for the cost of re-creating that business plan from notes and rough drafts.

As you know, our drives carry a lifetime guarantee covering parts and workmanship. We will gladly replace the defective external drive. However, the guarantee states that the manufacturer and the retailer will not assume any incidental liability. Thus we are responsible only for the retail value of the external drive, not for the cost of duplicating the work that went into making the files stored on the drive.

However, your file might still be recoverable. A reputable data-recovery firm might be able to restore the data from the file at a very reasonable cost. To prevent such problems in the future, we always recommend that you back up all valuable files periodically.

We have already sent out your new external drive by overnight delivery. It should arrive within the next two days.

Please contact us if we can be of any further assistance.

Sincerely yours,

Paul R. Blackwood

Paul R. Blackwood, Manager
Customer Relations

FIGURE 9.8 **"Bad News" Adjustment Letter**

If you are writing a "bad news" adjustment letter, salvage as much goodwill as you can by showing that you have acted reasonably. In denying a request, explain your side of the matter, thus educating the customer about how the problem occurred and how to prevent it in the future.

The writer does not begin by stating that he is denying the reader's request. Instead, he begins politely by trying to form a bond with the reader. In trying to meet the customer on neutral ground, be careful about admitting that the customer is right. If you say, "We are sorry that the engine you purchased from us is defective," it will bolster the customer's claim if the dispute ends up in court.

The writer summarizes the facts of the incident, as he sees them.

The writer explains that he is unable to fulfill the reader's request. Notice that the writer never explicitly denies the request. It is more effective to explain why granting the request is not appropriate. Also notice that the writer does not explicitly say that the reader failed to make a backup copy of the plan and therefore the problem is her fault.

The writer shifts from the bad news to the good news. The writer explains that he has already responded appropriately to the reader's request.

The writer ends with a polite conclusion. A common technique is to offer the reader a special discount on another, similar product.

Writing Memos

Like letters, memos have a characteristic format, which consists of the elements shown in Figure 9.9.

Write out the month instead of using the all-numeral format (6/12/19); multicultural readers might use a different notation for dates and could be confused.

List the names of persons receiving copies of the memo, either alphabetically or in descending order of organizational rank.

Most writers put their initials or signature next to the typed name (or at the end of the memo) to show that they have reviewed the memo and accept responsibility for it.

AMRO MEMO

To: B. Pabst
From: J. Alonso *J.A.*
Subject: MIXER RECOMMENDATION FOR PHILLIPS
Date: 12 June 2019

INTEROFFICE

To: C. Cleveland
From: H. Rainbow *H.R.*
Subject: Shipment Date of Blueprints to Collier
Date: 2 October 2019

c: B. Aaron
K. Lau
J. Manuputra
W. Williams

NORTHERN PETROLEUM COMPANY
INTERNAL CORRESPONDENCE

Date: January 3, 2019
To: William Weeks, Director of Operations
From: Helen Cho, Chemical Engineering Dept. *H. C.*
Subject: Trip Report—Conference on Improved Procedures for Chemical Analysis Laboratory

FIGURE 9.9 **Identifying Information in a Memo**
Some organizations prefer the full names of the writer and reader; others want only the first initials and last names. Some prefer job titles; others do not. If your organization does not object, include your job title and your reader's. The memo will then be informative for anyone who refers to it after either of you has moved on to a new position, as well as for others in the organization who do not know you.

As with letters, you can attach memos to emails and deliver them electronically. To preserve the memo format for the email recipient, save the memo as a PDF before sending it.

If you prefer to distribute hard copies, print the second and all subsequent pages of a memo on plain paper rather than on letterhead. Include three items in the upper right-hand or left-hand corner of each subsequent page: the name of the recipient, the date of the memo, and the page number. See the header in Figure 9.3 on page 242.

Dynacol Corporation

INTEROFFICE COMMUNICATION

To: G. Granby, R&D
From: P. Rabin, Technical Services *P.R.*
Subject: Trip Report—Computer Dynamics, Inc.
Date: September 21, 2019

The purpose of this memo is to present my impressions of the Computer Dynamics technical seminar of September 19. The goal of the seminar was to introduce their new PQ-500 line of high-capacity storage drives.

Summary
In general, I was impressed with the technical capabilities and interface of the drives. Of the two models in the 500 series, I think we ought to consider the external drives, not the internal ones. I'd like to talk to you about this issue when you have a chance.

Discussion
Computer Dynamics offers two models in its 500 series: an internal drive and an external drive. Both models have the same capacity (1T of storage), and they both work the same way: they extend the storage capacity of a server by integrating an optical disk library into the file system. The concept is that they move files between the server's faster, but limited-capacity storage devices (hard disks) and its slower, high-capacity storage devices (magneto-optical disks). This process, which they call data migration and demigration, is transparent to the user.

For the system administrator, integrating either of the models would require no more than one hour. The external model would be truly portable; the user would not need to install any drivers, as long as his or her device is docked on our network. The system administrator would push the necessary drivers onto all the networked devices without the user having to do anything.

Although the internal drive is convenient—it is already configured for the computer—I think we should consider only the external drive. Because so many of our employees do teleconferencing, the advantage of portability outweighs the disadvantage of inconvenience. The tech rep from Computer Dynamics walked me through the process of configuring both models. A second advantage of the external drive is that it can be salvaged easily when we take a computer out of service.

Recommendation
I'd like to talk to you, when you get a chance, about negotiating with Computer Dynamics for a quantity discount. I think we should ask McKinley and Rossiter to participate in the discussion. Give me a call (x3442) and we'll talk.

The subject line is specific: the reader can tell at a glance that the memo reports on a trip to Computer Dynamics, Inc. If the subject line read only "Computer Dynamics, Inc.," the reader would not know what the writer was going to discuss about that company.

The memo begins with a clear statement of purpose, as discussed in Ch. 4, p. 50.

Note that the writer has provided a summary, even though the memo is only one page long. The summary gives the writer an opportunity to convey his main request: he would like to meet with the reader.

The main section of the memo is the discussion, which conveys the detailed version of the writer's message. Often the discussion begins with the background: the facts that readers will need to know to understand the memo. In this case, the background consists of a two-paragraph discussion of the two models in the company's 500 series. Presumably, the reader already knows why the writer went on the trip.

Note that the writer ends this discussion with a conclusion, or statement of the meaning of the facts. In this case, the writer's conclusion is that the company should consider only the external drive.

A recommendation is the writer's statement of what he would like the reader to do next. In this case, the writer would like to sit down with the reader to discuss how to proceed.

FIGURE 9.10 Sample Memo
Image: Bumbim/Shutterstock.

Figure 9.10 (on page 249), a sample memo, is a trip report, a record of a business trip written after the employee returned to the office. Readers are less interested in an hour-by-hour narrative of what happened than in a carefully structured discussion of what was important. Although writer and reader appear to be relatively equal in rank, the writer goes to the trouble of organizing the memo to make it easy to read and refer to later.

GUIDELINES Organizing a Memo

When you write a memo, organize it so that it is easy to follow. Consider these five organizational elements.

- **A specific subject line.** "Breast Cancer Walk" is too general. "Breast Cancer Walk Rescheduled to May 14" is better.
- **A clear statement of purpose.** As discussed in Chapter 4 (p. 50), the purpose statement is built around a verb that clearly states what you want the readers to know, believe, or do.
- **A brief summary.** Even if a memo fits on one page, consider including a summary. For readers who want to read the whole memo, the summary is an advance organizer; for readers in a hurry, reading the summary substitutes for reading the whole memo.
- **Informative headings.** Headings make the memo easier to read by enabling readers to skip sections they don't need and by helping them understand what each section is about. In addition, headings make the memo easier to write because they prompt the writer to provide the kind of information readers need.
- **A prominent recommendation.** Many memos end with one or more recommendations. Sometimes these recommendations take the form of action steps: bulleted or numbered lists of what the writer will do or what the writer would like others to do. Here is an example:

  Action Items:
  I would appreciate it if you would work on the following tasks and have your results ready for the meeting on Monday, June 9.
  - Henderson: recalculate the flow rate.
  - Smith: set up meeting with the regional EPA representative for sometime during the week of May 13.
  - Falvey: ask Armitra in Houston for his advice.

Writing Emails

Before you write an email in the workplace, find out your organization's email policies. Most companies have written policies that discuss circumstances under which you may and may not use email, principles you should use in writing emails, and the monitoring of employee email. The Tech Tip box and Figure 9.11 show the basic elements of an email.

TECH TIP

Why To Use Email for Business Correspondence

Email allows for quick, direct correspondence with one or more recipients. It also provides a time- and date-stamped record of your communication, and users can create a "thread" or "string" to keep track of questions, comments, and responses. Although email itself is somewhat informal, it is a convenient time- and cost-saving option, especially for correspondence being sent internationally, and formal documents can easily be attached.

How To Use Email for Business Correspondence

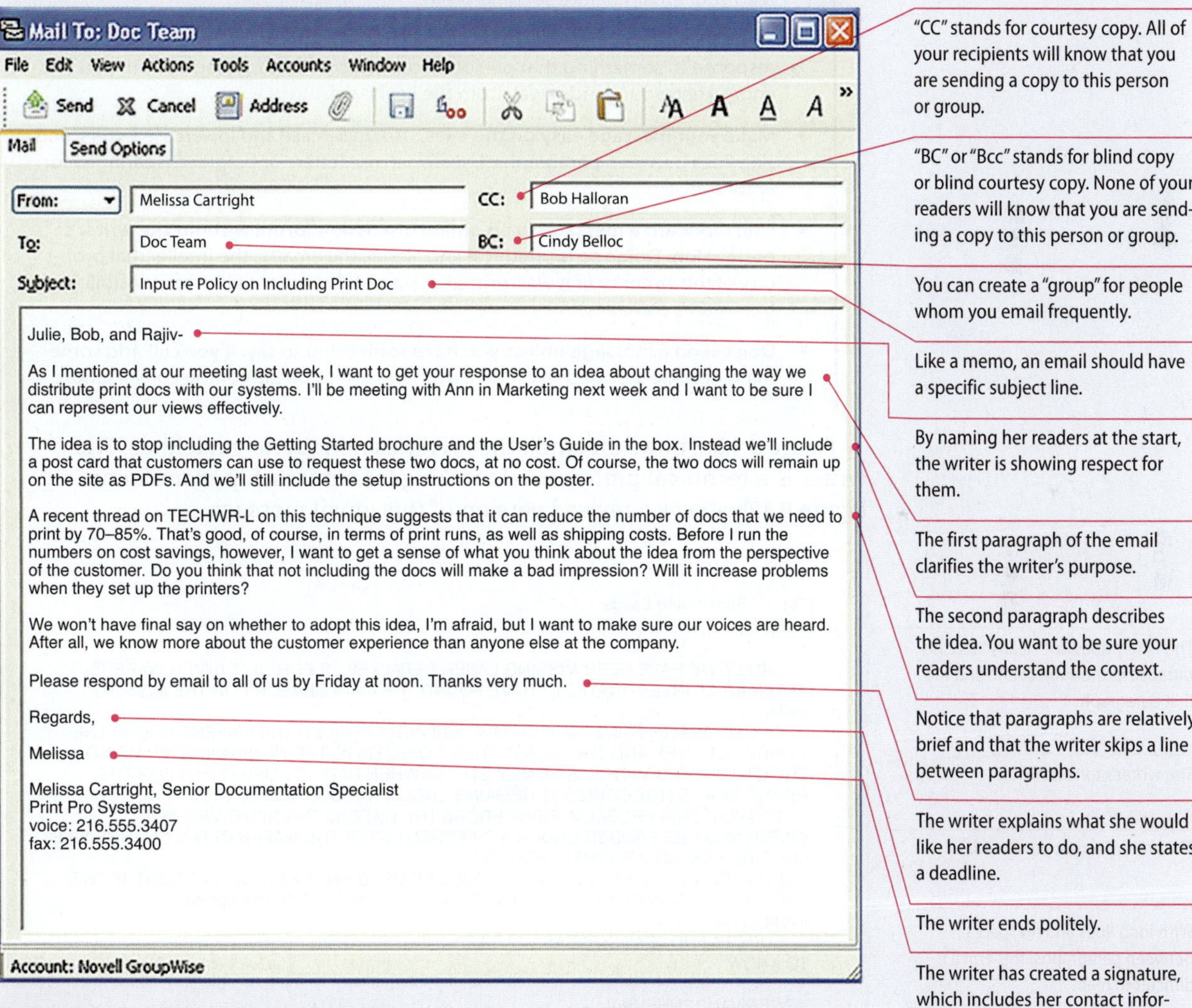

FIGURE 9.11 Elements of an Email

GUIDELINES Following Netiquette

When you write email in the workplace, adhere to the following netiquette guidelines. *Netiquette* refers to etiquette on a network.

- **Stick to business.** Don't send jokes or other nonbusiness messages.
- **Use the appropriate level of formality.** As discussed on page 238, avoid informal writing.
- **Write correctly.** As discussed on page 238, remember to revise, edit, and proofread your emails before sending them.
- **Don't flame.** To *flame* is to scorch a reader with scathing criticism, usually in response to something that person wrote in a previous message. When you are angry, keep your hands away from the keyboard.
- **Make your message easy on the eyes.** Use uppercase and lowercase letters, and skip lines between paragraphs. Use uppercase letters or boldface (sparingly) for emphasis.
- **Don't forward a message to an online discussion forum without the writer's permission.** Doing so is unethical and illegal; the email is the intellectual property of the writer or (if it was written as part of the writer's work responsibilities) the writer's company.
- **Don't send a message unless you have something to say.** If you can add something new, do so, but don't send a message just to be part of the conversation.

Figure 9.12a shows an email that violates netiquette guidelines. The writer is a technical professional working for a microchip manufacturer. Figure 9.12b shows a revised version of this email message.

The writer does not clearly state his purpose in the subject line and the first paragraph.

The writer's tone is hostile.

The writer has not proofread.

With long lines and no spaces between paragraphs, this email is difficult to read.

To: Supers and Leads
Subject:

LATELY, WE HAVE BEEN MISSING LASER REPAIR FILES FOR OUR 16MEG WAFERS. AFTER BRIEF INVESTIGATION, I HAVE FOUND THE MAIN REASON FOR THE MISSING DATA.
OCCASIONALLY, SOME OF YOU HAVE WRONGLY PROBED THE WAFERS UNDER THE CORRELATE STEP AND THE DATA IS THEN COPIED INTO THE NONPROD STEP USING THE QTR PROGRAM. THIS IS REALLY STUPID. WHEN DATE IS COPIED THIS WAY THE REPAIR DATA IS NOT COPIED. IT REMAINS UNDER THE CORRELATE STEP.
TO AVOID THIS PROBLEM, FIRST PROBE THE WAFERS THE RIGHT WAY. IF A WAFER MUST BE PROBED UNDER A DIFFERENT STEP, THE WAFER IN THE CHANGE FILE MUST BE RENAMED TO THE ** FORMAT.
EDITING THE WAFER DATA FILE SHOULD BE USED ONLY AS A LAST RESORT, IF THIS BECOMES A COMMON PROBLEM, WE COULD HAVE MORE PROBLEMS WITH INVALID DATA THAT THERE ARE NOW.
SUPERS AND LEADS: PLEASE PASS THIS INFORMATION ALONG TO THOSE WHO NEED TO KNOW.

ROGER VANDENHEUVAL

a. Email that violates netiquette guidelines

FIGURE 9.12 Netiquette

To: Supers and Leads
Subject: Fix for Missing Laser Repair Files for 16MB Wafers

Supers and Leads:

Lately, we have been missing laser repair files for our 16MB wafers. In this email I want to briefly describe the problem and recommend a method for solving it.

Here is what I think is happening. Some of the wafers have been probed under the correlate step; this method copies the data into the nonprod step and leaves the repair data uncopied. It remains under the correlate step.

To prevent this problem, please use the probing method outlined in Spec 344-012. If a wafer must be probed using a different method, rename the wafer in the CHANGE file to the *.* format. Edit the wafer data file only as a last resort.

I'm sending along copies of Spec 344-012. Would you please pass along this email and the spec to all of your operators?

Thanks. Please get in touch with me if you have any questions.

Roger Vandenheuval

The writer has edited and proofread the email.

The subject line and first paragraph clearly state the writer's purpose.

Double-spacing between paragraphs and using short lines make the email easier to read.

The writer concludes politely.

b. Email that adheres to netiquette guidelines

FIGURE 9.12 **Netiquette** *(continued)*

Writing Microblogs

As discussed earlier in this chapter, microblog posts are different from letters, memos, and email in that they are often extremely brief and quite informal in tone. However, the fact that microblog posts are fast and informal does not mean that anything goes. When you write microblog posts, you are creating communication that will be archived and that will reflect on you and your organization. In addition, anything you write is subject to the same laws and regulations that pertain to all other kinds of documents. Many of the guidelines for following netiquette (see p. 252) apply to microblog posts as well as email. Take care, especially, not to flame. Become familiar with your microblog's privacy settings, and be aware of which groups of readers may view and share your posts. The best way to understand your responsibilities when you write a microblog post at work is to study your organization's guidelines.

DOCUMENT ANALYSIS ACTIVITY

Following Netiquette in an Email Message

This message was written in response to a question emailed to several colleagues by a technical communicator seeking advice on how to write meeting minutes effectively. A response to an email message should adhere to the principles of effective emails and proper netiquette. The questions below ask you to think about these principles (explained on pp. 250–52).

1. How effectively has the writer stated her purpose?
2. How effectively has the writer projected a "you attitude" (explained on p. 239)?
3. How effectively has the writer made her message easy to read?

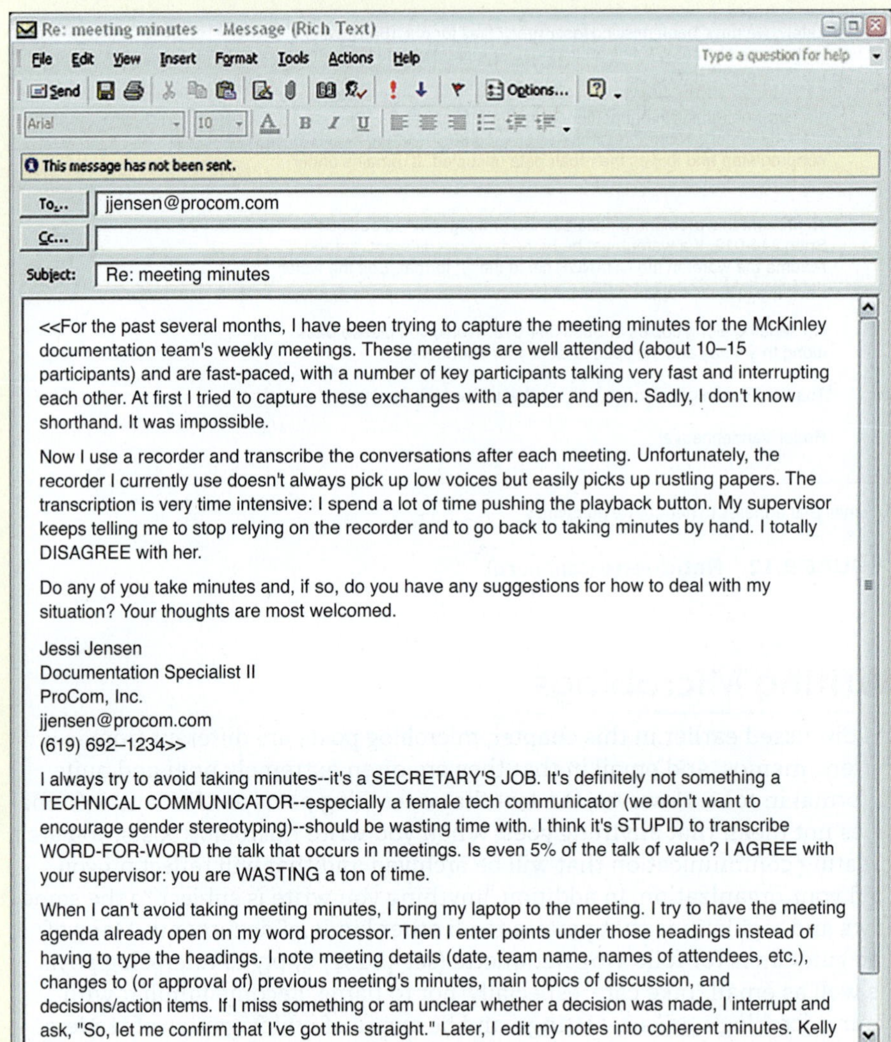

Re: meeting minutes - Message (Rich Text)

File Edit View Insert Format Tools Actions Help

This message has not been sent.

To... jjensen@procom.com

Cc...

Subject: Re: meeting minutes

<<For the past several months, I have been trying to capture the meeting minutes for the McKinley documentation team's weekly meetings. These meetings are well attended (about 10–15 participants) and are fast-paced, with a number of key participants talking very fast and interrupting each other. At first I tried to capture these exchanges with a paper and pen. Sadly, I don't know shorthand. It was impossible.

Now I use a recorder and transcribe the conversations after each meeting. Unfortunately, the recorder I currently use doesn't always pick up low voices but easily picks up rustling papers. The transcription is very time intensive: I spend a lot of time pushing the playback button. My supervisor keeps telling me to stop relying on the recorder and to go back to taking minutes by hand. I totally DISAGREE with her.

Do any of you take minutes and, if so, do you have any suggestions for how to deal with my situation? Your thoughts are most welcomed.

Jessi Jensen
Documentation Specialist II
ProCom, Inc.
jjensen@procom.com
(619) 692–1234>>

I always try to avoid taking minutes--it's a SECRETARY'S JOB. It's definitely not something a TECHNICAL COMMUNICATOR--especially a female tech communicator (we don't want to encourage gender stereotyping)--should be wasting time with. I think it's STUPID to transcribe WORD-FOR-WORD the talk that occurs in meetings. Is even 5% of the talk of value? I AGREE with your supervisor: you are WASTING a ton of time.

When I can't avoid taking meeting minutes, I bring my laptop to the meeting. I try to have the meeting agenda already open on my word processor. Then I enter points under those headings instead of having to type the headings. I note meeting details (date, team name, names of attendees, etc.), changes to (or approval of) previous meeting's minutes, main topics of discussion, and decisions/action items. If I miss something or am unclear whether a decision was made, I interrupt and ask, "So, let me confirm that I've got this straight." Later, I edit my notes into coherent minutes. Kelly

GUIDELINES Representing Your Organization on a Microblog

If you use a microblog at work to communicate with people outside your own organization, such as vendors and customers, you want to use it in such a way that people are encouraged to like, respect, and trust you. These ten suggestions can help.

- **Decide on your audience and your purpose.** Are you connecting with clients, providing customer service, helping people understand your company's goals and vision? You might want to have different accounts if you have several different audiences and purposes.
- **Learn the technology.** Know how to use hashtags, how to mention other users in your tweets, how to reply publicly and privately, how to integrate images and videos, and how to cross-post to your other social-media accounts should you need to.
- **Learn the culture of the community.** Listen and learn before you post. Most communities have a distinct culture, which influences how and when people post, link, and reply. For instance, in some communities, people stick close to the technical topic; in others, they roam more freely and include personal comments.
- **Share, don't sell.** Post about incidents and developments that reinforce your organization's core principles, such as environmental awareness or making technology available around the world. Talk about leadership, teamwork, and cooperation. Don't try to sell products.
- **Help educate readers and solve their problems.** Regardless of whether you're responding to individual questions and complaints or helping people understand your company's culture or goals, focus on helping people learn and solve problems.
- **Sound like a person.** Use an informal tone. Readers are especially pleased when high-ranking employees show their human side, such as when the Zappos CEO posted, "Dropped my laptop on floor this morning. I usually drop my phone, so good to know I'm moving on to bigger and better things" (Hall, 2009).*
- **Apologize when you make a mistake.** At the start of a basketball game against their rivals, the Dallas Mavericks, the Houston Rockets sent out a tweet with a gun emoji pointed at a horse. Within two hours, after receiving heavy criticism, the Rockets apologized and removed the tweet (Meyer, 2016).
- **Link generously.** When you want to talk about something you've learned online, don't paraphrase. Rather, link back to the original source. If you're using a platform with limited space, use a URL shortener such as Bitly or TinyURL so that the link won't take up too many of your characters.
- **Get your facts right.** Like anything online, your post is permanent. Double-check your facts before you post. Otherwise, you could embarrass yourself and erode people's trust in your professionalism.
- **Edit and proofread before you post.** You should be informal, but you shouldn't be sloppy. It sends the wrong message.

*Hall, Susan. "Twitter fan companies say it helps them get closer to customers," *ITBusinessEdge*. Retrieved November 11, 2013. Reproduced with permission.

For more about cultural variables, see Ch. 4, "Communicating Across Cultures."

Writing Correspondence to Multicultural Readers

The four types of business correspondence discussed in this chapter are used in countries around the world. The ways they are used, however, can differ significantly from the ways they are used in the United States. These differences fall into three categories:

- **Cultural practices.** As discussed in Chapter 4, cultures differ in a number of ways, such as whether they focus on individuals or groups, the distance between power ranks, and attitudes toward uncertainty. Typically, a culture's attitudes are reflected in its business communication. For example, in Japan, which has a high power distance—that is, people in top positions are treated with great respect by their subordinates—a reader might be addressed as "Most Esteemed Mr. Director." Some cultural practices, however, are not intuitively obvious even if you understand the culture. For example, in Japanese business culture, it is considered rude to reply to an email by using the reply function in the email software; it is polite to begin a new email (Sasaki, 2010).
- **Language use and tone.** In the United States, writers tend to use contractions, the first names of their readers, and other instances of informal language. In many other countries, this informality is potentially offensive. Also potentially offensive is U.S. directness. A writer from the United States might write, for example, that "14 percent of the products we received from you failed to meet the specifications." A Korean would more likely write, "We were pleased to note that 86 percent of the products we received met the specifications." The writer either would not refer to the other 14 percent (assuming that the reader would get the point and replace the defective products quickly) or would write, "We would appreciate replacement of the remaining products." Many other aspects of business correspondence differ from culture to culture, such as preferred length, specificity, and the use of seasonal references in the correspondence.
- **Application choice and use.** In cultures in which documents tend to be formal, letters might be preferred to memos, or face-to-face meetings to phone calls or email. In Asia, for instance, a person is more likely to walk down the hall to deliver a brief message in person because doing so shows more respect. In addition, the formal characteristics of letters, memos, and emails are different in different cultures. The French, for instance, use indented paragraphs in their letters, whereas in the United States, paragraphs are typically left-justified. The ordering of the information in the inside address and complimentary close of letters varies widely. In many countries, emails are structured like memos, with the "to," "from," "subject," and "date" information added at the top, even though this information is already present in the routing information.

Try to study business correspondence written by people from the culture you will be addressing. When possible, have important documents reviewed by a person from that culture before you send them.

WRITER'S CHECKLIST

Letter Format

- [ ] Is the first page printed on letterhead stationery? *(p. 241)*
- [ ] Is the date included? *(p. 241)*
- [ ] Is the inside address complete and correct? *(p. 241)*
- [ ] Is the appropriate courtesy title used? *(p. 241)*
- [ ] If appropriate, is an attention line included? *(p. 241)*
- [ ] If appropriate, is a subject line included? *(p. 241)*
- [ ] Is the salutation appropriate? *(p. 241)*
- [ ] Is the complimentary close typed with only the first word capitalized? *(p. 242)*
- [ ] Is the signature legible, and is the writer's name typed beneath the signature? *(p. 242)*
- [ ] If appropriate, is an enclosure line included? *(p. 242)*
- [ ] If appropriate, is a copy and/or courtesy-copy line included? *(p. 242)*

Types of Letters

Does the inquiry letter

- [ ] explain why you chose the reader to receive the inquiry? *(p. 243)*
- [ ] explain why you are requesting the information and how you will use it? *(p. 243)*
- [ ] specify the date when you need the information? *(p. 243)*
- [ ] list the questions clearly and, if appropriate, provide room for the reader's responses? *(p. 243)*
- [ ] offer, if appropriate, the product of your research? *(p. 243)*

Does the response to an inquiry letter

- [ ] answer the reader's questions? *(p. 244)*
- [ ] explain why, if any of the reader's questions cannot be answered? *(p. 244)*

Does the claim letter

- [ ] identify specifically the unsatisfactory product or service? *(p. 245)*
- [ ] explain the problem(s) clearly? *(p. 245)*
- [ ] propose an adjustment? *(p. 245)*
- [ ] conclude courteously? *(p. 245)*

Does the "good news" adjustment letter

- [ ] express your regret about the problem? *(p. 246)*
- [ ] explain the adjustment you will make? *(p. 246)*
- [ ] conclude on a positive note? *(p. 246)*

Does the "bad news" adjustment letter

- [ ] meet the reader on neutral ground, expressing regret for the problem but not apologizing? *(p. 247)*
- [ ] explain why the company is not at fault? *(p. 247)*
- [ ] clearly imply that the reader's request is denied? *(p. 247)*
- [ ] attempt to create goodwill? *(p. 247)*

Memos

- [ ] Does the identifying information adhere to your organization's standards? *(p. 248)*
- [ ] Did you include a specific subject line? *(p. 249)*
- [ ] Did you clearly state your purpose at the start of the memo? *(p. 249)*
- [ ] If appropriate, did you summarize your message? *(p. 249)*
- [ ] Did you provide appropriate background for the discussion? *(p. 249)*
- [ ] Did you organize the discussion clearly? *(p. 249)*
- [ ] Did you include informative headings to help your readers? *(p. 249)*
- [ ] Did you highlight items requiring action? *(p. 249)*

Email

- [ ] Did you refrain from sending jokes or other nonbusiness messages? *(p. 252)*
- [ ] Did you keep the email as brief as possible and send it only to appropriate people? *(p. 252)*
- [ ] Did you use the appropriate level of formality? *(p. 252)*
- [ ] Did you write correctly? *(p. 252)*

- [ ] Did you avoid flaming? (*p. 252*)
- [ ] Did you write a specific, accurate subject line? (*p. 250*)
- [ ] Did you use uppercase and lowercase letters? (*p. 252*)
- [ ] Did you skip lines between paragraphs? (*p. 252*)
- [ ] Did you check with the writer before forwarding his or her message? (*p. 252*)

Microblogs

- [ ] Did you study your organization's policy on which microblog sites you may use and how you should use them? (*p. 255*)

EXERCISES

1. You are the head of research for a biological research organization. Six months ago, you purchased a $2,000 commercial refrigerator for storing research samples. Recently, you suffered a loss of more than $600 in samples when the thermostat failed and the temperature in the refrigerator rose to more than 48 degrees over the weekend. Inventing any reasonable details, write a claim letter to the manufacturer of the refrigerator.
2. As the recipient of the claim letter described in Exercise 1, write an adjustment letter granting the customer's request.
3. You are the manager of a private swimming club. A member has written saying that she lost a contact lens (value $75) in your pool and she wants you to pay for a replacement. The contract that all members sign explicitly states that the management is not responsible for the loss of personal possessions. Write an adjustment letter denying the request. Invent any reasonable details.
4. As the manager of a retail electronics store, you guarantee that the store will not be undersold. If a customer finds another retailer selling the same equipment at a lower price within one month of his or her purchase, you will refund the difference. A customer has written to you and enclosed an ad from another store showing that it is selling a router for $26.50 less than he paid at your store. The advertised price at the other store was a one-week sale that began five weeks after the date of his purchase. He wants a $26.50 refund. Inventing any reasonable details, write an adjustment letter denying his request. You are willing, however, to offer him an 8-GB USB drive worth $9.95 if he would like to come pick it up.
5. **TEAM EXERCISE** Form small groups for this exercise on claim and adjustment letters. Have each member of your group study the following two letters. Then meet and discuss your reactions to the two letters. How effectively does the writer of the claim letter present his case? How effective is the adjustment letter? Does its writer succeed in showing that the company's procedures for ensuring hygiene are effective? Does its writer succeed in projecting a professional tone? Write a memo to your instructor discussing the two letters. Attach a revision of the adjustment letter to the memo.

Seth Reeves
19 Lowry's Lane
Morgan, TN 30610

April 13, 2016

Sea-Tasty Tuna
Route 113
Lynchburg, TN 30563

Gentlemen:

I've been buying your tuna fish for years, and up to now it's been OK.

But this morning I opened a can to make myself a sandwich. What do you think was staring me in the face? A fly. That's right, a housefly. That's him you see taped to the bottom of this letter.

What are you going to do about this?

Yours very truly,

Sea-Tasty Tuna
Route 113
Lynchburg, TN 30563
www.seatastytuna.com

April 20, 2019

Mr. Seth Reeves
19 Lowry's Lane
Morgan, TN 30610

Dear Mr. Reeves:

We were very sorry to learn that you found a fly in your tuna fish.

Here at Sea-Tasty we are very careful about hygiene at our plant. The tuna are scrubbed thoroughly as soon as we receive them. After they are processed, they are inspected visually at three different points. Before we pack them, we rinse and sterilize the cans to ensure that no foreign material is sealed in them.

Because of these stringent controls, we really don't see how you could have found a fly in the can. Nevertheless, we are enclosing coupons good for two cans of Sea-Tasty tuna.

We hope this letter restores your confidence in us.

Truly yours,

6. Louise and Paul work for the same manufacturing company. Louise, a senior engineer, is chairing a committee to investigate ways to improve the hiring process at the company. Paul, a technical editor, also serves on the committee. The excerpts quoted in Louise's email are from Paul's email to all members of the committee in response to Louise's request that members describe their approach to evaluating job-application materials. How would you revise Louise's email to make it more effective?

To: Paul
From: Louise

Sometimes I just have to wonder what you're thinking, Paul.

>Of course, it's not possible to expect perfect resumes. But I
>have to screen them, and last year I had to read over 200. I'm
>not looking for perfection, but as soon as I spot an error I
>make a mental note of it and, when I hit a second and
>then a third error I can't really concentrate on the writer's
>credentials.
Listen, Paul, you might be a sharp editor, but the rest of us have a different responsibility: to make the products and move them out as soon as possible. We don't have the luxury of studying documents to see if we can find errors. I suggest you concentrate on what you were hired to do, without imposing your "standards" on the rest of us.

>From my point of view, an error can include a
>misused tradmark.

Misusing a "tradmark," Paul? Is that Error Number 1?

7. Because students use email to communicate with other group members when they write collaboratively, your college or university would like to create a one-page handout on how to use email responsibly. Using a search engine, find three or four netiquette guides on the Internet that focus on email. Study these guides and write a one-page student guide to using email to communicate with other students. Somewhere in the guide, be sure to list the sites you studied, so that students can visit them for further information about netiquette.

CASE 9: Setting Up and Maintaining a Professional Microblog Account

As the editor-in-chief of your college newspaper, you have recently been granted permission to create a Twitter account. The newspaper's faculty advisor has requested that, before you set up the account, you develop a statement of audience and purpose based on your school's own social-media policy statement and statements from other schools, newspapers, and organizations. To begin putting together a bibliography to guide your research and craft your statement, go to LaunchPad.

CHAPTER 10 Applying for a Job

GETTING HIRED has always involved writing. Whether you apply online through a company's website, reply to a post on LinkedIn, or send a formal letter and résumé through the mail, you will use words to make the case that the organization should offer you a position.

You will probably make that case quite a few times. According to the U.S. Department of Labor (2015), the typical American worker holds more than 11 different jobs while he or she is between the ages of 18 and 48. Obviously, most of those jobs don't last long. Even when American workers begin a new job between the ages of 40 and 48, a third of those workers will no longer be with that company at the end of one year, and two-thirds will no longer be there in five years.

For most of you, looking for professional work is the first nonacademic test of your technical-communication skills. And it's an important test. Kyle Wiens, CEO of two tech companies, iFixit and Dozuki, requires all new employees to pass a writing test. His reason? "If it takes someone more than 20 years to notice how to properly use 'it's,' then that's not a learning curve I'm comfortable with" (Bowers, 2013).

Establishing Your Professional Brand

Being a successful job seeker requires a particular frame of mind. Think of yourself not as a student at this college or an employee of that company but rather as a professional with a brand to establish and maintain. For instance, say your name is Amber Cunningham, and you work as a human-resources officer for Apple. Don't think of yourself as an Apple human-resources officer. Instead, think of yourself as Amber Cunningham, a human-resources specialist who has worked for several companies (including Apple) and who has a number of marketable skills and a substantial record of accomplishments. Your *professional brand* (sometimes referred to as a "personal brand") is Amber Cunningham.

To present your professional brand successfully, you need to understand what employers are looking for, and then you need to craft the materials that will present that brand to the world.

There is really no mystery about what employers want in an employee. Across all fields, employers want a person who is honest, hard-working, technically competent, skilled at solving problems, able to work effectively alone and in teams, willing to share information with others, and eager to keep learning.

You need to find the evidence that you can use to display these qualities. Begin by thinking about everything you have done throughout your college career (courses, internships, projects, service-learning experiences, organizations, leadership roles) and your professional career (job responsibilities, supervision of others, accomplishments, awards). And don't forget your volunteer activities; through these activities, many people acquire what are called *transferable skills*—skills that are useful or even necessary in seemingly unrelated jobs. For instance, volunteering for Habitat for Humanity says something important not only about your character but also about your ability to work effectively in a team and to solve problems. Even if you will never swing a hammer on the job, you will want to refer to this experience. Make a list—a long list—of your experiences, characteristics, skills, and accomplishments that will furnish the kinds of evidence that you can use in establishing your professional brand.

With your long list of characteristics, experiences, skills, and accomplishments in hand, it's time to start creating the materials that will display your professional brand. Many of those materials will be found online.

In making job offers, employers often rely on information they learn about potential new employees on the Internet. According to a 2015 study commissioned by CareerBuilder, 52 percent of companies research job applicants on social media (Tarpey, 2015). The good news: 32 percent of those companies were motivated to seek out an applicant because of the positive information they found online, including background information that supported qualifications, a personality that fit with company culture, a professional-looking website, great communication skills, and creativity.

The bad news: 48 percent found information that made them reject an applicant. The employers who rejected applicants most often cited the following five problems:

- provocative or inappropriate photos
- suggestion of drug or alcohol use
- negative comments about a former employer or co-worker
- poor communication skills
- discriminatory comments about race, gender, or religion

Search online for your own name. Look at what potential employers will see and ask yourself whether your online personal brand is what you want to display. If it isn't, start to change it.

GUIDELINES Building the Foundation of Your Professional Brand

Follow these six guidelines in developing your professional brand.

- **Research what others have done.** What kinds of information do they present about themselves online? On which social-media sites are they active? What kinds of comments and questions do they post? How do they reply to what others have posted?
- **Tell the truth.** Statistics about how many people lie and exaggerate in describing themselves in the job search vary, but it is probably between a third and a half. Companies search online themselves or hire investigators to verify the information you provide about yourself, to see if you are honest.
- **Communicate professionally.** Show that you can write clearly and correctly, and remember that it is inappropriate (and in some cases illegal) to divulge trade secrets or personal information about colleagues.
- **Highlight your job skills.** Employers want to see that you have the technical skills that the job requires. They look for degrees, certifications, speeches and publications, and descriptions of what you do in your present position and have done in previous positions.
- **Focus on problem-solving and accomplishments.** The most compelling evidence that you would be a good hire is a solid record of identifying problems and devising solutions that met customers' needs, reduced costs, increased revenues, improved safety, and reduced environmental impact. Numbers tell the story: try to present your accomplishments as quantifiable data.
- **Participate actively online.** One way to show you are a professional who would generously and appropriately share information and work well in a team is to display those characteristics online. Participate professionally through sites such as LinkedIn, Facebook, and Twitter.

To watch a tutorial on crafting your professional brand, go to LaunchPad.

GUIDELINES Presenting Your Professional Brand

The following six guidelines can help you display your professional brand.

- **Create a strong online presence.** The best online presence is your own website, which functions as your online headquarters. All your other online activities will link back to this one site, the only site on the Internet that is all about you. Register a site and try to name it *yourname.com* (you will be required to pay a small fee to secure the domain name). If you aren't experienced designing and creating sites, try a drag-and-drop site builder like Weebly or Squarespace, or use a template from a free blogging site such as WordPress. Upload to your site everything you want potential employers to see: contact information, a professional history, work samples, documents, and links to your accounts on social-media sites.

- **Participate on LinkedIn.** LinkedIn is the major social-media site used by employers to find employees. Set up a LinkedIn account and create a profile that includes the keywords that will attract potential employers. Rather than calling yourself a "programmer at ADP," which describes your current situation, call yourself "an experienced programmer in various programming languages (Java, C, C^{++}, and PHP) and scripting languages (JavaScript, Perl, WSH, and UNIX shells) who understands interactive web pages and web-based applications, including JavaServer Pages (JSP), Java servlets, Active Server Pages (ASP), and ActiveX controls." Including keywords makes it easier for potential employers to find you when they search for employees. In addition, remember to list specific skills in the "Skills and Abilities" section of your profile. Potential employers searching for specific skills can then locate you more easily, and colleagues who know your work can endorse you for various skills. Participate actively on LinkedIn by linking to articles or videos you find useful and contributing to discussion forums. Make connections with people and endorse them honestly.

- **Participate on Facebook.** You probably already have a Facebook account. Within your account, you have the option to create separate pages for specific interests. Create a public Facebook page and use it only for professional activities. Share information that will be interesting and useful to other professionals.

- **Participate on Twitter.** Follow influential people in your industry on Twitter to see the kinds of activities, conferences, and publications that interest them. Comment on and retweet useful tweets, link to the best items you see in the media, and reply when others send you messages.

- **Create a business card.** Having a business card if you're a student might seem odd, but a card is the best way to direct people to your website when you meet them in person. Your card should have your contact information, a few phrases highlighting your skills, and the URL of your website. Some people add a QR code (a Quick Response code, the square barcode that smartphones can read) to allow others to link to their websites instantly. (Search for "QR code generator" to find free sites that will help you generate a QR code.)

- **Practice an "elevator pitch."** An elevator pitch is a brief oral summary of your credentials. At less than 20 seconds long, it's brief enough that you can say it if you find yourself in an elevator with a potential employer. After the pitch, you hand the person your business card, which contains all the information he or she needs to get to your website.

ETHICS NOTE

WRITING HONEST JOB-APPLICATION MATERIALS

Many résumés contain lies or exaggerations. Job applicants say they attended colleges they didn't and were awarded degrees they weren't, give themselves inflated job titles, say they were laid off when they were really fired for poor performance, and inflate their accomplishments. A CareerBuilder survey found that 38 percent of employees have embellished their job responsibilities at some point, and 18 percent have lied about their skills (Lorenz, 2012). Economist Steven D. Levitt, co-author of *Freakonomics*, concludes that more than 50 percent of job applicants lie on their résumés (Isaacs, 2012).

Companies take this problem seriously. They hire agencies that verify an applicant's education and employment history and check for a criminal record. They do their own research online. They phone people whose names the candidate has provided. If they find any discrepancies, they do not offer the candidate a position. If the person is already working for the company when discrepancies arise, they fire the employee.

Understanding Four Major Ways to Look for a Position

Once you have established your personal brand, you can start to look for a position. There are four major ways to find a job.

- **Through an organization's website.** Most organizations list their job offerings in a careers section on their websites and explain how to apply. If you are interested in a particular organization, start with its own site.
- **Through a job site on the Internet.** Job boards are sites sponsored by federal agencies, Internet service providers, and private organizations. Some sites merely list positions; you respond to such listings by email. Other sites let you upload your résumé electronically, so that employers can get in touch with you. Some job boards offer resources on how to prepare job-application materials; others do not. Among the biggest job boards are the following:

  — AfterCollege

  — CareerBuilder

  — CareerMag

  — CareerOneStop (sponsored by the U.S. Department of Labor)

  — Glassdoor

  — Indeed.com (a metasearch engine for job seekers)

  — Monster

  One caution about using job boards: once you upload your résumé to an Internet site, you probably have lost control of it. Here are four questions to consider before you post to a job board:

  — Who has access to your résumé? You might want to remove your home address and phone number from it if anyone can view it.

— How will you know if an employer requests your résumé? Will you be notified by the job board?

— Can your current employer see your résumé? If your employer discovers that you are looking for a new job, your current position could be in jeopardy.

— Can you update your résumé at no cost? Some job boards charge you each time you update your résumé.

- **Through your network.** A relative or an acquaintance can exert influence to help you get a job, or at least point out a new position. Other good contacts include past employers and professors. Also consider becoming active in the student chapter of a professional organization in your field, through which you can meet professionals in your local area. Many people use Twitter, Facebook, and—in particular—LinkedIn to connect with their contacts, as well as to try to identify hiring officers and other professionals who can help them apply. Figure 10.1 shows an excerpt from one professional's LinkedIn profile.
- **Through a college or university placement office or professional placement bureau.** College and university placement offices bring companies and students together. Student résumés are made available to representatives of business, government, and industry, who arrange on-campus interviews. Students who do well in the campus interviews are then invited by the representatives to visit the organization for a tour and another interview. A professional placement bureau offers essentially the same service but charges a fee (payable by either the employer or the person who is hired for a job). Placement bureaus cater primarily to more-advanced professionals who are changing jobs.

Writing Résumés

Although you will present your credentials on LinkedIn and other sites, you will also need to create a résumé, which you will upload to a job board or a company's website, email to the company, or paste into a web-based form.

Many students wonder whether to write their résumés themselves or use a résumé-preparation agency. It is best to write your own résumé, for three reasons:

- **You know yourself better than anyone else does.** No matter how professional the work of a résumé-preparation agency is, you can do a better job communicating important information about yourself.
- **Employment officers know the style of the local agencies.** Readers who recognize that you did not write your own résumé might wonder whether you are hiding any deficiencies.
- **If you write your own résumé, you will be more likely to adapt it to different situations.** You are unlikely to return to a résumé-preparation agency and pay an additional fee to make a minor revision.

Everything in this excerpt from Jonathan Bednar's LinkedIn profile makes the argument that he is talented, hard-working, and ambitious.

Jonathan's profile uses a clear, professional-looking photo that makes him look eager and ready to work.

The summary statement gives a brief description of his qualifications and experience. Keywords such as "data analysis" and "usage monitoring" will help potential employers locate his profile more easily.

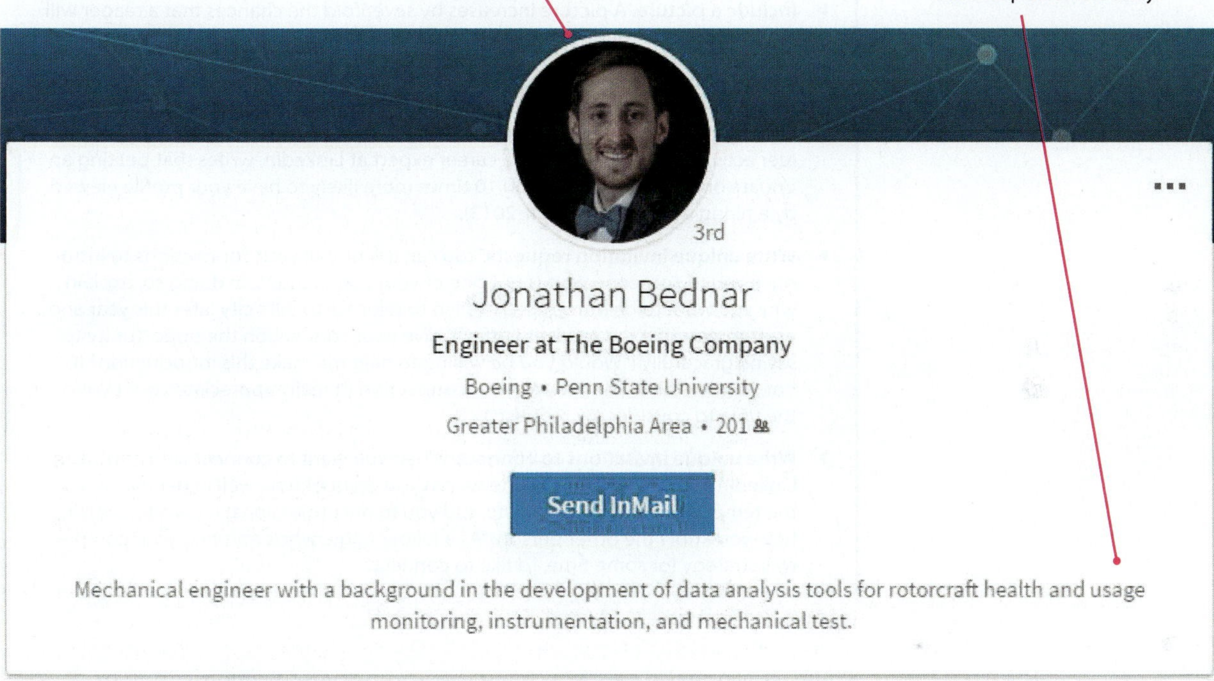

FIGURE 10.1 Excerpts from a Professional's LinkedIn Profile
Jonathan Bednar.

Because most companies use résumé-application software to scan résumés into databases and search for keywords, a good résumé includes the right keywords. Only after a résumé has made it through that initial electronic pass will it be read by a person. Résumé consultant Ramsey Penegar puts it this way: "If your résumé doesn't have the keywords that match their job requirements, your résumé may hit the 'no' pile early in the process" (Auerbach, 2012).

GUIDELINES Using LinkedIn's Employment Features

Some 93 percent of hiring officers use LinkedIn to search for candidates (Fottrell, 2014). The following five guidelines can help you take advantage of the employment features on the world's most influential networking site for professionals.

- **Use the profile section fully.** The profile section includes information from your résumé, but unlike a résumé, which needs to be concise and contains only words, the profile section can include any kind of digital file, such as presentation slides

(continued)

or videos. Describe your education and important jobs in detail; remember that the keywords in your descriptions will enable potential employers to find you as they search for employees. If you add "skills" to your profile, others have an opportunity to "endorse" those skills, adding credibility to your profile.

- **Include a picture.** A picture increases by sevenfold the chances that a reader will read your profile (Halzack, 2013).
- **Post updates.** Post information about interesting articles you have read, conferences you are attending, and other professional activities. Be generous in praising co-workers and others you follow on the Internet. Mention your volunteer activities. Nicole Williams, a career expert at LinkedIn, writes that posting an update once a week makes you 10 times more likely to have your profile viewed by a hiring manager (Halzack, 2013).
- **Write unique invitation requests.** You can ask one of your connections to introduce you to someone who is not one of your connections. In doing so, explain why you want to be introduced ("I plan to relocate to Bill's city later this year and want to describe the services I offer"), give your connection the opportunity to say no gracefully ("Would you be willing to help me make this introduction? If not, I understand"), and thank your connection ("I really appreciate your taking the time to consider my request").
- **Write unique invitations to connect.** When you want to connect with another LinkedIn member, especially one whom you do not know well in person, avoid the template invitation, "I'd like to add you to my professional network." Explain how you know the other person: "As a fellow Aggie who's admired your company's strategy for some time, I'd like to connect."

The best way to be sure you have the appropriate keywords in your résumé is to study the job description in the actual job posting you want to respond to. Then find ten other ads for similar positions and identify the terms that come up frequently. Think in terms of job titles, names of products, companies, technologies, and professional organizations. For instance, if the job is to develop web pages, you will likely see many references to "web page," "Internet," "XHTML," "HTML5," "Java," "W3C," and "CSS." Also include keywords that refer to your communication skills, such as "public speaking," "oral communication," and "communication skills."

But don't just list the keywords. Instead, integrate them into sentences about your skills and accomplishments. For instance, a computer-science student might write, "Wrote applications for migrating data between systems/databases using C#, XML, and Excel Macros." A chemical engineer might write, "Worked with polymers, mixing and de-gassing polydimethylsiloxane."

How long should a résumé be? It should be long enough to include all pertinent information but not so long that it bores or irritates the reader. Although some hiring consultants have guidelines (such as that a student's résumé should be no longer than one page, or that applicants who are vice presidents at companies can write two-page résumés), the consensus is that length is unimportant. If an applicant has more experience, the résumé will be longer; if an applicant has less experience, it will be shorter. If all the information in the résumé helps make the case that the applicant is an excellent fit for the position, it's the right length.

The information that goes into a résumé is commonly ordered either chronologically or by skills. In a *chronological résumé*, you use time as the organizing pattern for each section, including education and experience, and discuss your responsibilities for each job you have held. In a *skills résumé* (sometimes called a *functional résumé*), you merely list your previous jobs but include a skills section in which you describe your talents, skills, and achievements.

A chronological résumé focuses on the record of employment, giving an applicant the opportunity to describe the duties and accomplishments related to each job. The skills résumé highlights the skills (such as supervising others, managing a large department, reducing production costs) that the candidate demonstrated at several different companies. The skills résumé is a popular choice for applicants who have a gap in their employment history, who are re-entering the workforce, or who have changed jobs frequently.

In both types of résumé, you use reverse chronology; that is, you present the most recent jobs and degrees first, to emphasize them.

ELEMENTS OF THE CHRONOLOGICAL RÉSUMÉ

Most chronological résumés have five basic elements: identifying information, summary of qualifications, education, employment history, and interests and activities. Sometimes writers include a sixth section: references. In filling in these basic sections, remember that you want to include the keywords that will attract employers.

Identifying Information If you are submitting your résumé directly to a company, include your full name, address, phone number, and email address. Use your complete address, including the zip code. If your address during the academic year differs from your home address, list both and identify them clearly. An employer might call during an academic holiday to arrange an interview.

However, if you are posting your résumé to an Internet job board, where it can be seen by anyone, you will be more vulnerable to scammers, spammers, and identity thieves. Don't include a mailing address or phone number, and use an email address that does not identify you.

Summary Statement After the identifying information, add a summary statement, a brief paragraph that highlights three or four important skills or accomplishments. For example:

Summary

Six years' experience creating testing documentation to qualify production programs that run on Automated Test and Handling Equipment. Four years' experience running QA tests on software, hardware, and semiconductor products. Bilingual English and Italian. Secret security clearance.

Education If you are a student or a recent graduate, place the education section next. If you have substantial professional experience, place the employment-history section before the education section.

Include at least the following information in the education section:

- **Your degree.** After the degree abbreviation (such as BS, BA, AA, or MS), list your academic major (and, if you have one, your minor)—for example, "BS in Materials Engineering, minor in General Business."
- **The institution.** Identify the institution by its full name: "Louisiana State University," not "LSU."
- **The location of the institution.** Include the city and state.
- **The date of graduation.** If your degree has not yet been granted, add "Anticipated date of graduation" or a similar phrase.
- **Information about other schools you attended.** List any other institutions you attended beyond high school, even those from which you did not earn a degree. The description for other institutions should include complete information, as in the main listing. Arrange entries in reverse chronological order: that is, list first the school you attended most recently.

GUIDELINES Elaborating on Your Education

The following four guidelines can help you develop the education section of your résumé.

- **List your grade-point average.** If your average is significantly above the median for the graduating class, list it. Or list your average in your major courses, or all your courses in the last two years. Calculate it however you wish, but be honest and clear.
- **Compile a list of courses.** Include courses that will interest an employer, such as advanced courses in your major or courses in technical communication, public speaking, or organizational communication. For example, a list of business courses on an engineer's résumé might show special knowledge and skills. But don't bother listing required courses; everyone else in your major took the same courses. Include the substantive titles of listed courses. Employers won't know what "Chemistry 450" is; call it by its official title: "Chemistry 450. Organic Chemistry."
- **Describe a special accomplishment.** If you completed a special senior design or research project, present the title and objective of the project, any special or advanced techniques or equipment you used, and, if you know them, the major results: "A Study of Shape Memory Alloys in Fabricating Actuators for Underwater Biomimetic Applications—a senior design project to simulate the swimming styles and anatomy of fish." A project description makes you seem more like a professional: someone who designs and carries out projects.
- **List honors and awards you received.** Scholarships, internships, and academic awards suggest exceptional ability. If you have received a number of such honors, or some that were not exclusively academic, you might list them separately (in a section called "Honors" or "Awards") rather than in the education section. Decide where this information will make the best impression.

The education section is the easiest part of the résumé to adapt in applying for different positions. For example, a student majoring in electrical engineering who is applying for a position requiring strong communication skills can emphasize communication courses in one version of the résumé and advanced electrical engineering courses in another version. As you compose the education section, emphasize those aspects of your background that meet the requirements for the particular job.

Employment History Present at least the basic information about each job you have held: the dates of employment, the organization's name and location, and your position or title. Then add carefully selected details. Readers want to know what you did and accomplished. Provide at least a two- to three-line description for each position. For particularly important or relevant jobs, write more, focusing on one or more of the following factors:

- **Skills.** What technical skills did you use on the job?
- **Equipment.** What equipment did you operate or oversee? In particular, mention computer equipment or software with which you are familiar.
- **Money.** How much money were you responsible for? Even if you considered your data-entry position fairly easy, the fact that the organization grossed, say, $2 million a year shows that the position involved real responsibility.
- **Documents.** What important documents did you write or assist in writing, such as brochures, reports, manuals, proposals, or websites?
- **Personnel.** How many people did you supervise?
- **Clients.** What kinds of clients, and how many, did you do business with in representing your organization?

Whenever possible, emphasize *accomplishments*. If you reorganized the shifts of the weekend employees you supervised, state the results:

> Reorganized the weekend shift, resulting in a cost savings of more than $3,000 per year.
>
> Wrote and produced (with Adobe InDesign) a 56-page parts catalog that is still used by the company and that increased our phone inquiries by more than 25 percent.

When you describe positions, functions, or responsibilities, use the active voice ("supervised three workers") rather than the passive voice ("three workers were supervised by me"). The active voice highlights action. Note that writers often omit the *I* at the start of sentences: "Prepared bids," rather than "I prepared bids." Whichever style you use, be consistent. Figure 10.2 lists some strong verbs to use in describing your experience.

For more about using strong verbs, see Ch. 6, p. 134.

administered	coordinated	evaluated	maintained	provided
advised	corresponded	examined	managed	purchased
analyzed	created	expanded	monitored	recorded
assembled	delivered	hired	obtained	reported
built	developed	identified	operated	researched
collected	devised	implemented	organized	solved
completed	directed	improved	performed	supervised
conducted	discovered	increased	prepared	trained
constructed	edited	instituted	produced	wrote

FIGURE 10.2 **Strong Action Verbs Used in Résumés**

Here is a sample description from an employment history:

> June–September 2016: Student Dietitian
>
> Millersville General Hospital, Millersville, TX
>
> Gathered dietary histories and assisted in preparing menus for a 300-bed hospital.
>
> Received "excellent" on all seven items in evaluation by head dietitian.

In just a few lines, you can show that you sought and accepted responsibility and that you acted professionally. Do not write, "I accepted responsibility"; instead, present facts that lead the reader to that conclusion.

Naturally, not all jobs entail professional skills and responsibilities. Many students find summer work as laborers, sales clerks, and so forth. If you have not held a professional position, list the jobs you have held, even if they were unrelated to your career plans. If the job title is self-explanatory, such as restaurant server or service-station attendant, don't elaborate. If you can write that you contributed to your tuition or expenses, such as by earning 50 percent of your annual expenses through a job, employers will be impressed by your self-reliance.

If you have held a number of nonprofessional as well as several professional positions, group the nonprofessional ones:

> Other Employment: clerk (summer 2016); salesperson (part-time, 2015); cashier (summer 2014)

This strategy prevents the nonprofessional positions from drawing the reader's attention away from the more important positions.

If you have gaps in your employment history—because you were raising children, attending school, or recovering from an accident, or for other reasons—consider using a skills résumé, which focuses more on your skills and less on your job history. You can explain the gaps in the job-application letter (if you write one) or in an interview. For instance, you could say, "I spent 2014 and part of 2015 caring for my elderly parent, but during that time I was able to do some substitute teaching and study at home to prepare for my A+ and Network+ certification, which I earned in late 2015." Do not lie or mislead about your dates of employment.

If you have had several positions with the same employer, you can present one description that encompasses all the positions or present a separate description for each position.

PRESENTING ONE DESCRIPTION

Blue Cross of Iowa, Ames, Iowa (January 2011–present)

- *Internal Auditor II (2015–present)*
- *Member Service Representative/Claims Examiner II (2013–2015)*
- *Claims Examiner II (2011–2013)*

As Claims Examiner II, processed national account inquiries and claims in accordance with . . . After promotion to Member Service Representative/Claims Examiner II position, planned policies and procedures . . . As Internal Auditor II, audit claims, enrollment, and inquiries; run dataset population and sample reports . . .

This format enables you to mention your promotions and to create a clear narrative that emphasizes your progress within the company.

PRESENTING SEPARATE DESCRIPTIONS

Blue Cross of Iowa, Ames, Iowa (January 2011–present)

- *Internal Auditor II (2015–present)*
  Audit claims, enrollment, and inquiries . . .
- *Member Service Representative/Claims Examiner II (2013–2015)*
  Planned policies and procedures . . .
- *Claims Examiner II (2011–2013)*
  Processed national account inquiries and claims in accordance with . . .

This format, which enables you to create a fuller description of each position, is effective if you are trying to show that each position is distinct and you wish to describe the more-recent positions more fully.

Interests and Activities The interests-and-activities section of the résumé is the appropriate place for several kinds of information about you:

- participation in community-service organizations, such as Big Brothers/Big Sisters or volunteer work in a hospital
- hobbies related to your career (for example, electronics for an engineer)
- sports, especially those that might be socially useful in your professional career, such as tennis, racquetball, and golf
- university-sanctioned activities, such as membership on a team, work on the college newspaper, or election to a responsible position in an academic organization or a residence hall

Do not include activities that might create a negative impression, such as gambling. And always omit such activities as meeting people and reading. Everybody does these things.

References Potential employers will want to learn more about you from your professors and previous employers. These people who are willing to speak or write on your behalf are called *references*.

Some applicants list their references on their résumé. The advantage of this strategy is that the potential employer can contact the references without having to contact the applicant. Other applicants prefer to wait until the potential employer has asked for the list. The advantage of this strategy is that the applicant can assemble a different set of references for each position without having to create different résumés. Although applicants in the past added a note stating "References available upon request" at the end of their résumés, many applicants today do not do so because they think the comment is unnecessary: employers assume that applicants can provide a list of references—and that they would love to do so.

Regardless of whether you list your references on your résumé, choose your references carefully. Solicit references only from those who know your work best and for whom you have done your best work—for instance, a previous employer with whom you worked closely or a professor from whom you received A's. Don't ask prominent professors who do not know your work well; they will be unable to write informative letters.

Do not simply assume that someone is willing to serve as a reference for you. Give the potential reference writer an opportunity to decline gracefully. The person might not have been as impressed with your work as you think. If you simply ask the person to serve as a reference, he or she might accept and then write a lukewarm letter. It is better to ask, "Would you be able to write an enthusiastic letter for me?" or "Do you feel you know me well enough to write a strong recommendation?" If the person shows any signs of hesitation or reluctance, withdraw the request. It may be a little embarrassing, but it is better than receiving a weak recommendation.

In listing their references, some applicants add, for each reference, a sentence or two describing their relationship with the person, as shown in this sample listing for a reference.

Dr. Dale Cletis
Professor of English
Boise State University
Boise, ID 83725
208.555.2637
dcletis@boisestate.edu

Dr. Cletis was my instructor in three literature courses, as well as my adviser.

Other Elements The sections discussed so far appear on almost everyone's résumé. Other sections are either optional or appropriate for only some job seekers.

- **Computer skills.** Classify your skills in categories such as hardware, software, languages, and operating systems. List any professional certifications you have earned.
- **Military experience.** If you are a veteran, describe your military service as if it were a job, citing dates, locations, positions, ranks, and tasks. List positive job-performance evaluations.
- **Language ability.** A working knowledge of another language can be very valuable, particularly if the potential employer has international interests and you could be useful in translation or foreign service. List your proficiency, using terms such as *beginner*, *intermediate*, and *advanced*. Some applicants distinguish among reading, writing, and speaking abilities. Don't overstate your abilities; you could be embarrassed—and without a job—when the potential employer hands you a business letter written in the language you say you know, or invites a native speaker of that language to sit in on the interview.
- **Willingness to relocate.** If you are willing to relocate, say so. Many organizations will find this flexibility attractive.

ELEMENTS OF THE SKILLS RÉSUMÉ

A skills résumé differs from a chronological résumé in that it includes a separate section, usually called "Skills" or "Skills and Abilities," that emphasizes job skills and knowledge. In a skills résumé, the employment section becomes a brief list of information about your employment history: companies, dates of employment, and positions. Here is an example of a skills section.

Skills and Abilities

Management
Served as weekend manager of six employees in a retail clothing business. Also trained three summer interns at a health-maintenance organization.

Writing and Editing
Wrote status reports, edited performance appraisals, participated in assembling and producing an environmental impact statement using desktop publishing.

Teaching and Tutoring
Tutored in the university writing center. Taught a two-week course in electronics for teenagers. Coach youth basketball.

In a skills section, you choose the headings, the arrangement, and the level of detail. Your goal, of course, is to highlight the skills an employer is seeking.

PREPARING A PLAIN-TEXT RÉSUMÉ

Most companies use computerized *applicant-tracking systems*, such as RESUMate, Bullhorn, or HRsmart, to evaluate the dozens, hundreds, or even thousands of job applications they receive every day. The information from these applications is stored in databases, which can be searched

GUIDELINES Formatting a Plain-Text Résumé

Start with the résumé that you prepared in Word or with some other word-processing software. Save it as "Plain text" and then paste it into Notepad or another text editor. Revise the Notepad version so that it has these five characteristics:

- **It has no special characters.** It uses only the letters, numbers, and basic punctuation marks visible on your keyboard. That is, it *does not* use boldface, italics, bullets, or tabs.
- **It has a line length of 65 or fewer characters.** Use the space bar to break longer lines or, in Notepad, set the left and right margins (in **File/Page Setup**) to 1.5 inches.
- **It uses a non-proportional typeface such as Courier.** (A non-proportional typeface is one in which each letter takes up the same amount of space on the line; narrow letters are surrounded by a lot of space, whereas wider letters are surrounded by a smaller space.) Using a non-proportional typeface makes it easier to keep the line length to 65 characters.
- **Most of the information is left-justified.** If you want, you can use the space bar (not the Tab key) to move text to the right. For instance, you might want to center the main headings.
- **It uses ALL UPPERCASE or repeated characters for emphasis.** For example, a series of equal signs or hyphens might signal a new heading.

electronically for keywords to generate a pool of applicants for specific positions. Once a pool of candidates has been generated, someone at the company reads their résumés. Prepare a plain-text résumé so that you can survive this two-stage process.

A *plain-text résumé*, also called a *text résumé*, *ASCII résumé*, or *electronic résumé*, is a résumé that uses a very limited character set and has little formatting so that it can be stored in any database and read by any software. It will not be as attractive as a fully formatted document created with a word processor, but if you prepare it carefully it will say what you want it to say and be easy to read.

You might want to create two versions of your plain-text résumé: a version using Word Wrap (in Notepad's **Format** tab) to be attached to an email, and a version *not* using Word Wrap to be pasted into the body of an email.

Check each new version to be sure the information has converted properly. Copy and paste the version not using Word Wrap into an email and send it to yourself, and then review it. Attach the new file using Word Wrap to an email, open it in your text editor, and review it.

Figures 10.3 and 10.4 show a plain-text chronological résumé and a plain-text skills résumé.

Many of the job sites listed on p. 265 include samples of résumés.

```
CARL OPPENHEIMER
3109 Vista Street
Philadelphia, PA 19136
(215) 555-3880
coppen@dragon.du.edu
++++++++++++++++++++++++++++++++++++++++++++++++++++++++++
SUMMARY
Recent BSEE graduate with experience as an electrical
engineer intern for RCA Advanced Technology Laboratory.
Analytical, technical, and communication skills for lab-
oratory and customer-facing applications. Strong under-
standing of large-scale integrated systems and CMOS appli-
cations.
++++++++++++++++++++++++++++++++++++++++++++++++++++++++++
EDUCATION
BS in Electrical Engineering 6/2016
Drexel University, Philadelphia, PA
Grade-Point Average: 3.67 (on a scale of 4.0)
Senior Design Project: "Enhanced Path-Planning Software
for Robotics"
+++Advanced Engineering Courses
Digital Signal Processing
Computer Hardware
Introduction to Operating Systems I, II
Systems Design
Digital Filters
Computer Logic Circuits I, II
++++++++++++++++++++++++++++++++++++++++++++++++++++++++++
EMPLOYMENT
6/2013-1/2014 Electrical Engineering Intern II
RCA Advanced Technology Laboratory, Moorestown, NJ
Designed ultra-large-scale integrated circuits using
VERILOG and VHDL hardware description languages. Assisted
senior engineer in CMOS IC layout, modeling, parasitic
capacitance extraction, and PSPICE simulation operations.
+++6/2012-1/2013 Electrical Engineering Intern I
RCA Advanced Technology Laboratory, Moorestown, NJ
Verified and documented several integrated circuit
designs. Used CAD software and hardware to simulate,
check, and evaluate these designs. Gained experience with
Mathcad.
++++++++++++++++++++++++++++++++++++++++++++++++++++++++++
HONORS AND ORGANIZATIONS
Eta Kappa Nu (Electrical Engineering Honor Society)
Tau Beta Pi (General Engineering Honor Society)
IEEE
```

The writer provides his contact information, including his email address.

This plain-text résumé uses only plus signs to signal new headings. Notice that all information is left-justified.

The writer presents a summary statement. Some applicants find it awkward to praise themselves, describing their skills, but it is important to have keywords such as "analytical skills" in the résumé, particularly if the job ad mentioned them.

The writer chooses to emphasize his advanced engineering courses. For another job, he might emphasize other courses.

The writer wisely creates a category that calls attention to his academic awards and his membership in his field's major professional organization.

The writer does not include his references or write "References available upon request." If the reader invites him to proceed to the next step in the process, Carl will send a list of references, with their contact information.

FIGURE 10.3 Chronological Résumé of a Traditional Student

In a skills résumé, you present the skills section at the start. This organization lets you emphasize your professional attributes. Notice that the writer uses specific details — names of software, number of credits, types of documents, kinds of activities — to make her case.

The employment section contains a list of positions rather than descriptions of what the writer did in each position.

The volunteer position says something about the writer's character.

The writer believes that the skills required in raising children are relevant in the workplace. Other applicants might think that because a résumé describes job credentials, this information should be omitted.

```
Alice P. Linder
1781 Weber Road
Rawlings, MT 59211
(406) 555-3999
linderap423@gmail.com
''''''''''''''''''''''''''''''''''''''''''''''''''''''''''''
SUMMARY
Biotechnology major with broad laboratory experience at
GlaxoSmithKline, analyzing molecular data and writing C# pro-
grams. Working toward Certification in laboratory specialty
through ASCP. Strong written and oral communication skills.
Extensive volunteer experience in physical therapy for children.
Experience managing business office.
''''''''''''''''''''''''''''''''''''''''''''''''''''''''''''
SKILLS AND ABILITIES
Laboratory Skills
-Analyzed molecular data on E&S PS300, Macintosh, and IBM PCs.
Wrote programs in C#.
-Have taken 12 credits in biology and chemistry labs.
Communication Skills
-Wrote a user's guide for an instructional computing package.
-Trained and consulted with scientists and delivered in-house
briefings.
Management Skills
-Managed 12-person office in $1.2 million company.
''''''''''''''''''''''''''''''''''''''''''''''''''''''''''''
EDUCATION
Central Montana State University, Rawlings, MT
BS in Bioscience and Biotechnology
Expected Graduation Date: 6/2016
Related Course Work
General Chemistry I, II, III
Biology I, II, III
Organic Chemistry I, II
Calculus I, II
Statistical Methods for Research
Physics I, II
Technical Communication
''''''''''''''''''''''''''''''''''''''''''''''''''''''''''''
EMPLOYMENT EXPERIENCE
6/2013-present (20 hours per week): Laboratory Assistant Grade 3
GlaxoSmithKline, Rawlings, MT
8/2011-present: Volunteer, Physical Therapy Unit
Rawlings Regional Medical Center, Rawlings, MT
6/2012-1/2013: Office Manager
Anchor Products, Inc., Rawlings, MT
''''''''''''''''''''''''''''''''''''''''''''''''''''''''''''
HONORS
Awarded three $5,000 tuition scholarships (2012-2014) from the
Gould Foundation.
''''''''''''''''''''''''''''''''''''''''''''''''''''''''''''
ADDITIONAL INFORMATION
Member, CMSU Biology Club, Yearbook Staff
Raising three school-age children
Tuition 100% self-financed
```

FIGURE 10.4 **Skills Résumé of a Nontraditional Student**

Although fewer and fewer employers request a formatted résumé, some still do. Some applicants send formatted résumés in addition to their plain-text résumés. Figure 10.5 shows a formatted version of the résumé presented in Figure 10.4.

Alice P. Linder	1781 Weber Road (406) 555-3999 Rawlings, MT 59211 linderap423@gmail.com
Summary	Biotechnology major with broad laboratory experience at GlaxoSmithKline, analyzing molecular data and writing C# programs. Working toward Certification in laboratory specialty through ASCP. Strong written and oral communication skills. Extensive volunteer experience in physical therapy for children. Experience managing business office.
Skills and Abilities	*Laboratory Skills* • Analyzed molecular data on E&S PS300, Macintosh, and IBM PCs. Wrote programs in C#. • Have taken 12 credits in biology and chemistry labs. *Communication Skills* • Wrote a user's guide for an instructional computing package. • Trained and consulted with scientists and delivered in-house briefings. *Management Skills* • Managed 12-person office in a $1.2 million company.
Education	Central Montana State University, Rawlings, MT BS in Bioscience and Biotechnology Expected Graduation Date: June 2016 *Related Course Work* General Chemistry I, II, III — Biology I, II, III Organic Chemistry I, II — Statistical Methods for Research Physics I, II — Technical Communication Calculus I, II
Employment Experience	6/2013–present (20 hours/week) *GlaxoSmithKline, Rawlings, MT* Laboratory Assistant Grade 3 8/2011–present *Rawlings Regional Medical Center, Rawlings, MT* Volunteer, Physical Therapy Unit 6/2012–1/2013 *Anchor Products, Inc., Rawlings, MT* Office Manager
Honors	Awarded three $5,000 tuition scholarships (2012–2014) from the Gould Foundation.
Additional Information	Member, CMSU Biology Club, Yearbook Staff Raising three school-age children Tuition 100 percent self-financed

FIGURE 10.5
Formatted Version of a Skills Résumé

You have many more formatting options when you present your résumé on paper.

This writer has used a two-column table to organize the information. The left column presents the headings; the right column presents the data. The advantage of using a table rather than moving text using tabs is that you can use different text attributes (for instance, the headings can be boldfaced, set in a different typeface, or set in a different size) without having to worry about whether the different attributes will alter the line spacing. In addition, if you use a table, you can easily revise and edit; with tabs, your editing will create awkward line breaks and alignment.

DOCUMENT ANALYSIS ACTIVITY

This résumé was written by a graduating college senior who wanted to work for a wildland firefighting agency such as the U.S. Bureau of Land Management or U.S. Forest Service. The writer plans to save the résumé as a .txt file and enter it directly into these agencies' employment databases. The questions below ask you to think about electronic résumés (as discussed on pp. 275–76).

1. How effectively has the writer formatted this résumé?
2. What elements are likely to be problematic when the writer saves this résumé as a .txt file?
3. What is the function of the industry-specific jargon in this résumé?
4. Why does the writer place the education section below the sections on career history and fire and aviation qualifications?

BURTON L. KREBS

34456 West Jewell St. — 208-555-9627

Boise, ID 83704 — burtonkrebs@mail.com

Objective

Lead crew position on rappel crew.

Career History

- Senior Firefighter, Moyer Rappel Crew, 05/16–present
- Senior Firefighter, Boise Helitack, 05/15–10/15
- Hotshot Crew Member, Boise Interagency Hotshot Crew, 07/14–09/14
- Helirappel Crew Member, Moyer Rappel Crew, 06/10–09/13

Fire and Aviation Qualifications

Crew Boss (T)
Helicopter Manager
Helicopter Rappeller
Helirappel Spotter
Helispot Manager
Type 2 Helibase Manager (T)
Incident Commander Type 4 (T)

Education

Bachelor of Arts in Communication Training and Development, Boise State University, Boise, Idaho, GPA 3.57, May 2016

Skills

- Excellent oral and written communication skills
- Proficient in Word, Excel, and PowerPoint
- Knowledgeable of helicopter contract administration
- Perform daily and cumulative flight invoice cost summaries

Awards

"Outstanding Performance" Recognition, U.S. Bureau of Land Management, 2015

"Outstanding Performance" Recognition, U.S. Forest Service, 2012, 2013, 2014

DOCUMENT ANALYSIS ACTIVITY

To analyze an online professional portfolio, go to LaunchPad.

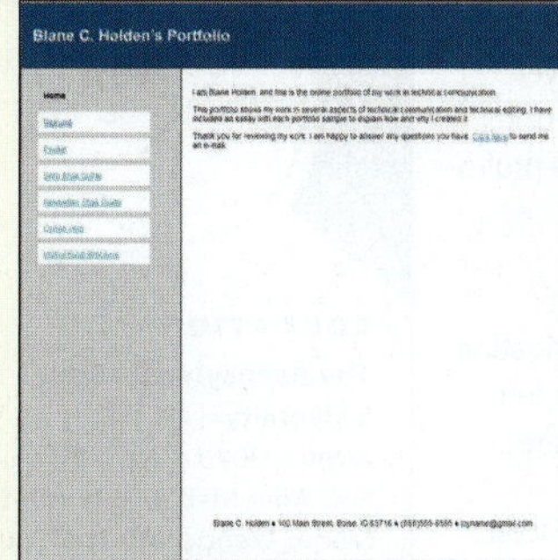

Blane C. Holden's Online Portfolio
Used by permission of Blane Holden.

CONSIDERING NONTRADITIONAL RÉSUMÉS

In applying for a job, you have the option of submitting a nontraditional résumé. Nontraditional résumés include elements of chronological and skills résumés, but draw on a broader range of characteristics from print and digital media. Some approaches expand the use of visual design techniques to organize and emphasize résumé content. Others use video to present résumé content or a professional introduction. You will need to think carefully about your job-search contexts and if nontraditional résumés are an appropriate option. Figures 10.6, 10.7 (on page 283), and 10.8 (on page 284) show a visually enhanced résumé, an infographic résumé, and a video résumé.

Writing Job-Application Letters

Some experts argue that the job-application letter (sometimes called a *cover letter*) is obsolete, but opinions vary widely. The Society for Human Resources Management (SHRM) surveyed over 400 human-resources professionals and found that more than 40 percent consider a cover letter "important" or "very important." Organizations with fewer than 500 employees in particular continue to value a cover letter (SHRM, 2014).* Applicants can explain more clearly in a letter than in a résumé how their qualifications match the employer's requirements. They can explain their professional relationship with someone in the employer's organization or gaps in their employment history. Perhaps most important, applicants can show that they can write well. Figure 10.9 (on page 285) shows a job-application letter.

Many of the job sites listed on p. 265 include samples of job-application letters.

*SHRM Survey Findings: Résumés, Cover Letters and Interviews, 2014. Reprinted by permission of the Society for Human Resource Management.

The writer uses a geometric design to add visual interest.

The logo on the résumé also appears on the writer's cover letter and professional website, helping to communicate a professional brand.

In the PDF version, the left column functions as a navigation column, linking to contact information and social media sites.

The left column uses both icons and text to represent links.

The main content area emphasizes professional experience and includes elements typical to chronological and skills résumé, such as headings, bulleted lists, and action verbs.

EDUCATION

The Pennsylvania State University
August 2013 - May 2017
B.A. Advertising, minor in Digital Media Trends and Analytics
GPA 3.56

CONTACT

(202) 555-0172
astridclark@imail.com
www.astridclark.web

SOCIAL MEDIA

fb.me/astridxc23
@astridclark23
@astridclark23
linkedin.com/in/astridxc

HONORS

Advertising Award of Excellence | Awarded to graduates who achieve academic distinction | May 2017

Alpha Delta Sigma Honor Society, Member | National Honor Society sponsored by the American Advertising Federation, recognizing scholastic achievement in advertising studies | Fall 2017 - Present

Dean's List | GPA of 3.5 or higher each semester | Spring 2013 - Spring 2017

ASTRID CLARK

PROFESSIONAL EXPERIENCE

SOCIAL MEDIA INTERN — August 2016 - June 2017
The Undergraduate Admissions Office | University Park, PA
- Produced and created original videos, editorial content, and photos for the social media pages, specifically Facebook, Twitter and Instagram.
- Managed and scheduled publication of content on the student blog section of the Admissions website.
- Coordinated and performed interviews with various students and faculty about their Penn State experience.
- Co-lead meetings with the department to discuss new content.

DIRECTOR OF COMMUNICATIONS — August 2016 - May 2017
State College Communications | University Park, PA
- Created and coordinated posts for State College Communications's social media accounts, including Facebook, Instagram, Twitter and LinkedIn.
- Copyedited and proofread all final content, including blog posts, social media posts, and client assets.
- Assisted in hiring account associates and account executives. Lead monthly firm-wide meetings.

ADVERTISING INTERN — June 2016 - August 2016
D.C. Relations Inc. | Washington, D.C.
- Designed flyers, newsletters, email campaigns, and online promotions.
- Helped the Marketing and Advertising departments target new media audiences.
- Researched and analyzed competitor marketing and communication strategies.
- Collaborated on team presentations and pitches for management and clients.

DIGITAL WEB INTERN — May 2015 - August 2015
Delicious Dining Magazine | Los Angeles, CA
- Wrote two to three 500-1000 word articles per week.
- Fact-checked and transcribed publication-wide interviews, recipes, and stories.
- Assisted in designing and producing how-to recipe videos.

WEB WRITER — December 2014 - May 2016
The Collegiate Lion Magazine | University Park, PA
- Pitched, researched, and wrote one 400-600 word article per week.
- Conducted interviews with students, faculty, visitors, and locals in the Penn State community.
- Covered timely events both on and off campus.

PROFESSIONAL SKILLS
- Technical Writing
- Technical Editing
- Graphic Design
- Video Production
- Photography
- Media Management

SOFTWARE SKILLS
- Microsoft Office
- Adobe Creative Cloud
- WordPress
- Google Analytics
- Hootsuite
- MediaWiki

FIGURE 10.6 Visually Enhanced Résumé

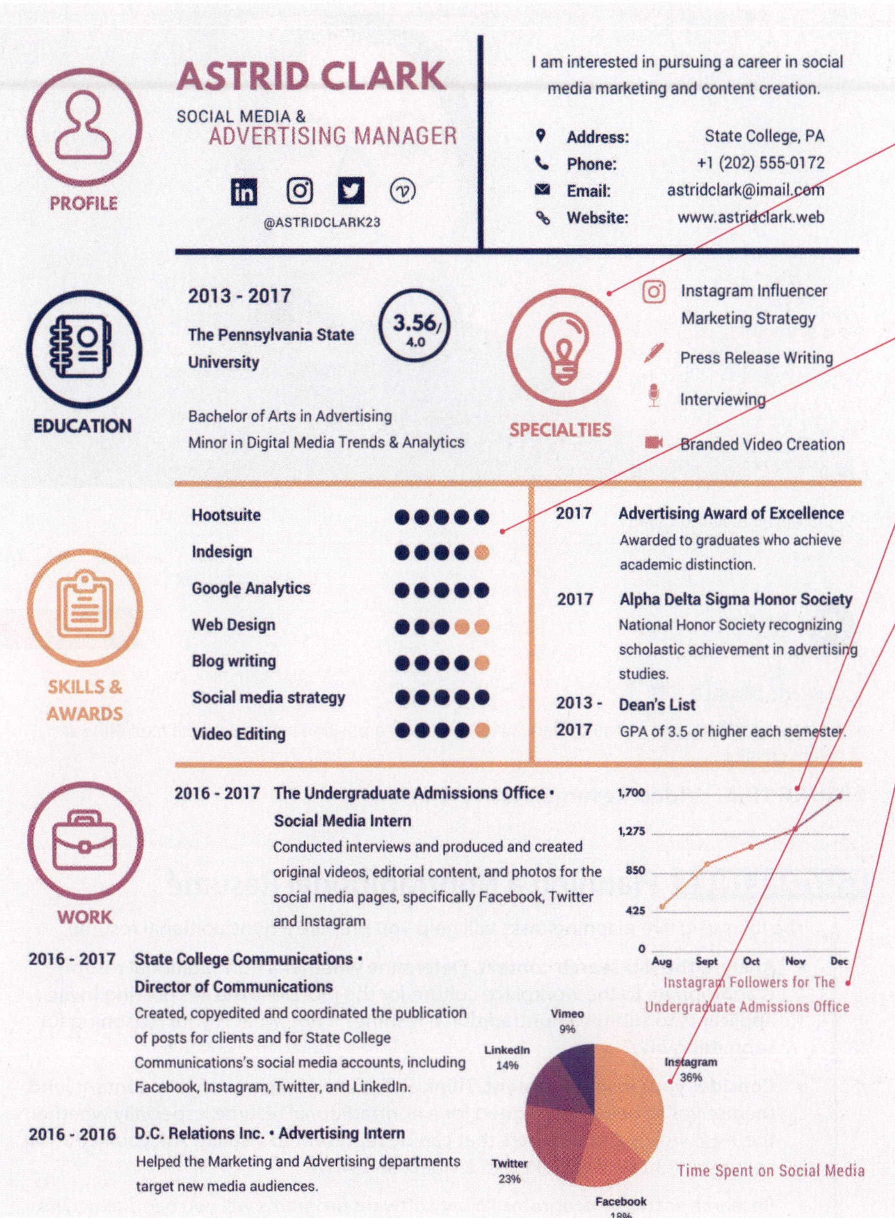

FIGURE 10.7 **Infographic Résumé**

The video is brief, just a few minutes long. It includes a beginning that serves as an introduction, a middle that highlights key experiences, and an end that serves as a conclusion.

The video uses text overlays to provide context or additional information.

The title for the video is meaningful to audiences.

The video includes a closed captioning option, to accommodate audiences with hearing impairments.

The video does not allow audiences to leave public comments.

The description of the video provides additional context.

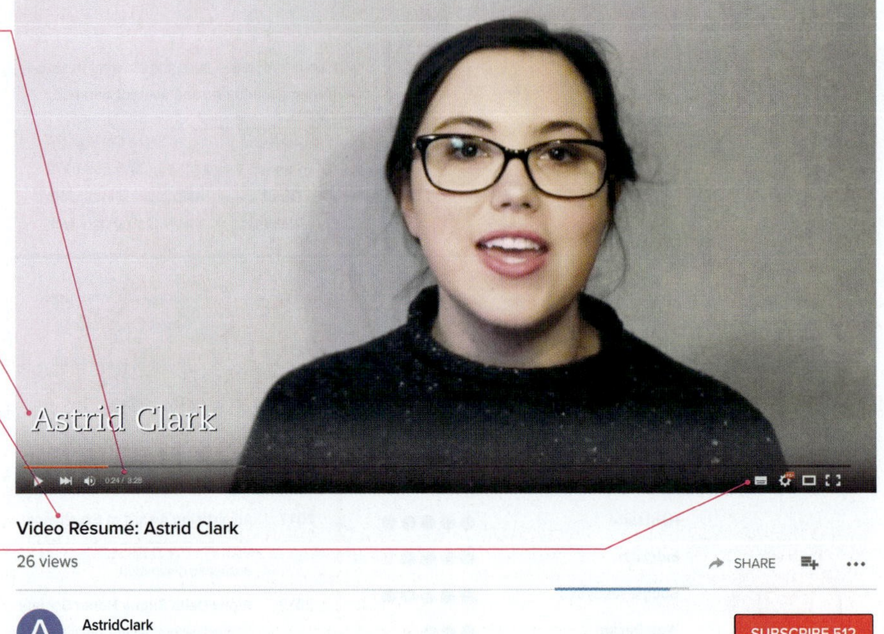

COMMENTS DISABLED SORT BY

This is the video résumé of Astrid Clark, who is seeking a position in social media marketing or content creation.

FIGURE 10.8 **Video Résumé: Astrid Clark**

GUIDELINES Planning a Nontraditional Résumé

The following five planning tasks will help you prepare a nontraditional résumé.

- **Analyze the job-search context.** Determine whether a nontraditional résumé is appropriate to the workplace culture for the job. Does the job posting invite applicants to submit a nontraditional résumé? If not, what is your reasoning for submitting one?
- **Consider your résumé content.** Think about which aspects of your content lend themselves to being redesigned for a nontraditional résumé, especially whether there are words and numbers that can be represented visually. Ask yourself what content might be well-suited to a video, and why.
- **Research software programs.** What software programs will you need to access in order to produce a nontraditional résumé? Will your current programs work, or will you need to learn new programs? Keep in mind that if you are using free programs, there may be limits to what you can do with them.
- **Research delivery options.** Consider whether you will host a video résumé on your own website, through a streaming service such as YouTube or Vimeo, or by using a job search portal that allows candidates to upload videos. What are the advantages and disadvantages of the different options?
- **Draft an element.** Test your ideas by drafting an element of a nontraditional résumé. What's a good starting point? Why? You may need to change your plans depending on the results of your draft.

3109 Vista Street
Philadelphia, PA 19136

January 20, 2019

Mr. Stephen Spencer, Director of Personnel
Department 411
Boeing Naval Systems
103 Industrial Drive
Wilmington, DE 20093

Dear Mr. Spencer:

I am writing in response to your advertisement in the January 16 *Philadelphia Inquirer*. Would you please consider me for the position in Signal Processing? I believe that my academic training in electrical engineering at Drexel University, along with my experience with the RCA Advanced Technology Laboratory, qualifies me for the position.

My education at Drexel has given me a strong background in computer hardware and system design. I have concentrated on digital and computer applications, developing and designing computer and signal-processing hardware in two graduate-level engineering courses. For my senior design project, I am working with four other undergraduates in using OO programming techniques to enhance the path-planning software for an infrared night-vision robotics application.

While working at the RCA Advanced Technology Laboratory, I was able to apply my computer experience to the field of DSP. I designed ultra-large-scale integrated circuits using VERILOG and VHDL hardware description languages. In addition, I assisted a senior engineer in CMOS IC layout, modeling, parasitic capacitance extraction, and PSPICE simulation operations.

The enclosed résumé provides an overview of my education and experience. Could I meet with you at your convenience to discuss my qualifications for this position? Please leave a message any time at (215) 555-3880 or email me at coppen@dragon.du.edu.

Yours truly,

Carl Oppenheimer

Carl Oppenheimer

Enclosure (1)

FIGURE 10.9 **Job-Application Letter**

For information about letter formatting, see Figure 9.3.

Notice that the writer's own name does not appear at the top of his letter.

In the inside address, he uses the reader's courtesy title, "Mr."

The introductory paragraph identifies the writer's source of information about the job, identifies the position he is applying for, states that he wishes to be considered, and forecasts the rest of the letter.

In a letter, you can't discuss everything in the résumé. Rather, you select a few key points from the résumé to emphasize.

Note that both the education paragraph and the employment paragraph begin with a clear topic sentence.

The writer points out that he has taken two graduate courses, and he discusses his senior design project, which makes him look more like an engineer solving a problem than a recent graduate.

Notice how the writer makes a smooth transition from the discussion of his college education to the discussion of his internship experience.

A concluding paragraph usually includes a reference to the résumé, a polite but confident request for an interview, and the writer's contact information.

The enclosure notation refers to the writer's résumé. Do not use an enclosure notation unless you are literally enclosing something along with the letter in the envelope.

Many of the job sites listed on p. 265 include samples of follow-up letters for different situations that occur during a job search.

Writing Follow-up Letters or Emails After an Interview

After an interview, you should write a letter or email of appreciation. If you are offered the job, you also might have to write a letter accepting or rejecting the position.

- **Letter of appreciation after an interview.** Thank the organization's representative for taking the time to see you, and emphasize your particular qualifications. You can also restate your interest in the position and mention a specific topic of conversation you found particularly interesting or a fact about the position you found exciting. A follow-up letter can do more good with less effort than any other step in the job-application procedure because so few candidates take the time to write one.

  Dear Mr. Weaver:

  Thank you for taking the time yesterday to show me your facilities and to introduce me to your colleagues.

  Your company's advances in piping design were particularly impressive. As a person with hands-on experience in piping design, I can appreciate the advantages your design will have.

  The vitality of your projects and the good fellowship among your employees further confirm my initial belief that Cynergo would be a fine place to work. I would look forward to joining your staff.

  Sincerely yours,

  Harriet Bommarito

  Harriet Bommarito

- **Letter accepting a job offer.** This one is easy: express appreciation, show enthusiasm, and repeat the major terms of your employment.

  Dear Mr. Weaver:

  Thank you very much for the offer to join your staff. I accept.

  I look forward to joining your design team on Monday, July 19. The salary, as you indicate in your letter, is $48,250.

  As you have recommended, I will get in touch with Mr. Matthews in Personnel to get a start on the paperwork.

  I appreciate the trust you have placed in me, and I assure you that I will do what I can to be a productive team member at Cynergo.

  Sincerely yours,

  Mark Greenberg

  Mark Greenberg

- **Letter rejecting a job offer.** If you decide not to accept a job offer, express your appreciation for the offer and, if appropriate, explain why you are declining it. Remember, you might want to work for this company at some time in the future.

  Dear Mr. Weaver:

  I appreciate very much the offer to join your staff.

  Although I am certain that I would benefit greatly from working at Cynergo, I have decided to take a job with a firm in Baltimore, where I have been accepted at Johns Hopkins to pursue my Master's degree at night.

  Again, thank you for your generous offer.

  Sincerely yours,

  Cynthia O'Malley

  Cynthia O'Malley

- **Letter acknowledging a rejection.** Why write back after you have been rejected for a job? To maintain good relations. You might get a phone call the next week explaining that the person who accepted the job has had a change of plans and offering you the position.

  Dear Mr. Weaver:

  I was disappointed to learn that I will not have a chance to join your staff, because I feel that I could make a substantial contribution. However, I realize that job decisions are complex, involving many candidates and many factors.

  Thank you very much for the courtesy you have shown me.

  Sincerely yours,

  Paul Goicochea

  Paul Goicochea

WRITER'S CHECKLIST

Establishing Your Professional Brand

Did you

- ☐ research how others in your field developed and maintain their online presence? *(p. 263)*
- ☐ ensure that all information you present about yourself online is accurate and honest? *(p. 263)*
- ☐ communicate clearly and professionally, maintaining confidentiality when necessary? *(p. 263)*
- ☐ clearly present your skills, degrees, certifications, and other evidence of your qualifications? *(p. 263)*
- ☐ actively participate on LinkedIn and other social networks? *(p. 263)*

Have you

- ☐ created a website? *(p. 264)*
- ☐ created a business card? *(p. 264)*
- ☐ developed and practiced an elevator pitch? *(p. 264)*

Résumé

- ☐ Does the résumé include appropriate keywords? *(p. 267)*
- ☐ Does the identifying information contain your name, address(es), phone number(s), and email address(es)? *(p. 269)*
- ☐ Does the résumé include a clear summary of your qualifications? *(p. 269)*
- ☐ Does the education section include your degree, your institution and its location, and your (anticipated) date of graduation, as well as any other information that will help a reader appreciate your qualifications? *(p. 270)*
- ☐ Does the employment section include, for each job, the dates of employment, the organization's name and location, and (if you are writing a chronological résumé) your position or title, as well as a description of your duties and accomplishments? *(p. 271)*
- ☐ Does the interests-and-activities section include relevant hobbies or activities, including extracurricular interests? *(p. 273)*
- ☐ Does the résumé include any other appropriate sections, such as skills and abilities, military service, language abilities, or willingness to relocate? *(p. 275)*
- ☐ Have you omitted any personal information that might reflect poorly on you? *(p. 274)*

Job-Application Letter

- ☐ Does the letter look professional? *(p. 285)*
- ☐ Does the introductory paragraph identify your source of information and the position you are applying for, state that you wish to be considered, and forecast the rest of the letter? *(p. 285)*
- ☐ Does the education paragraph respond to your reader's needs with a unified idea introduced by a topic sentence? *(p. 285)*
- ☐ Does the employment paragraph respond to your reader's needs with a unified idea introduced by a topic sentence? *(p. 285)*
- ☐ Does the concluding paragraph include a reference to your résumé, a request for an interview, your phone number, and your email address? *(p. 285)*
- ☐ Does the letter include an enclosure notation? *(p. 285)*

Follow-up Letters

- ☐ Does your letter of appreciation for a job interview thank the interviewer and briefly restate your qualifications? *(p. 286)*
- ☐ Does your letter accepting a job offer show enthusiasm and repeat the major terms of your employment? *(p. 286)*
- ☐ Does your letter rejecting a job offer express your appreciation for the offer and, if appropriate, explain why you are declining it? *(p. 287)*
- ☐ Does your letter acknowledging a rejection have a positive tone that will help you maintain good relations? *(p. 287)*

EXERCISES

For more about memos, see Ch. 9, p. 248.

1. Browse a job-search website such as Indeed.com. Then, list and briefly describe five positions being offered in a field that interests you. What skills, experience, and background does each position require? What is the salary range for each position?
2. Locate three job-search websites that provide interactive forms for creating a résumé automatically. In a brief memo to your instructor, note the three URLs and describe the strengths and weaknesses of each site. Which job board appears to be the easiest to use? Why?
3. The following résumé was submitted in response to this ad: "CAM Technician to work with other technicians and manage some GIS and mapping projects. Also perform updating of the GIS database. Experience required." In a brief memo to your instructor, analyze the effectiveness of the résumé. What are some of its problems?

Kenneth Bradley

530 Maplegrove	Bozeman, Mont. 59715 (406)-484-2916
Objective	Entry level position as a CAM Technician. I am also interested in staying with the company until after graduation, possibly moving into a position as a Mechanical Engineer.
Education	Enrolled at Montana State University August 2016 Present

Employment Fred Meyer
65520 Chinden
Garden City, MT
(208)-323-7030
Janitor- 7/15-6/16
Responsible for cleaning entire store, as well as equipment maintenance and floor maintenance and repair.
Assistant Janitorial Manager- 6/16-9/16
Responsible for cleaning entire store, equipment maintenance, floor maintenance and repair, scheduling, and managing personnel
Head of Freight- 9/16-Present
In charge of braking down all new freight, stocking shelves, cleaning the stock room, and managing personnel

Montana State University
Bozeman, MT
Teachers Aide ME 120- 1/15-5/15
Teachers Aide ME 120
In charge of keeping students in line and answering any questions related to drafting.

References Timothy Rayburn
Janitorial Manager
(406)-555-8571
Eduardo Perez
Coworker
(406)-555-2032

4. The following job-application letter responds to this ad: "CAM Technician to work with other technicians and manage some GIS and mapping projects. Also perform updating of the GIS database. Experience required." In a brief memo to your instructor, analyze the effectiveness of the letter and suggest how it could be improved.

530 Maplegrove
Bozeman, Mont. 59715
November 11, 2016

Mr. Bruce Hedley
Adecco Technical
Bozeman, Mont. 59715

Dear Mr. Hedley,

I am writing you in response to your ad on Monsterjobs.com. Would you please consider me for the position of CAM technician? I believe that my academic schooling at Montana State University, along with my work experience would make me an excellent candidate for the position.

While at Montana State University, I took one class in particular that applies well to this job. It was a CAD drafting class, which I received a 97% in. The next semester I was a Teachers Aid for that same class, where I was responsible for answering questions about drafting from my peers. This gave me a much stronger grasp on all aspects of CAD work than I could have ever gotten from simply taking the class.

My employment at Fred Meyer is also a notable experience. While there is no technical aspects of either positions I have held, I believe that my experience there will shed light on my work ethic and interpersonal skills. I started out as a graveyard shift janitor, with no previous experience. All of my coworkers were at least thirty years older than me, and had a minimum of five years of janitorial experience. However after working there for only one year I was promoted to assistant manager. Three months after I received this position, I was informed that Fred Meyer was going to contract out the janitorial work and that all of us would be losing our jobs. I decided that I wanted to stay within the company, and I was able to receive a position as head of freight.

The enclosed resumé provides an overview of my education and work experience. I would appreciate an opportunity to meet with you at your convience to disscuss my qualifications for this position. Please write me at the above address or leave a message any time. If you would like to contact me by email, my email address is kbradley@montanastate.edu.

Yours truly,
Ken Bradley

5. How effective is the following letter of appreciation? How could it be improved? Present your findings in a brief memo to your instructor.

914 Imperial Boulevard
Durham, NC 27708
November 13, 2016

Mr. Ronald O'Shea
Division Engineering
Safeway Electronics, Inc.
Holland, MI 49423

Dear Mr. O'Shea:

Thanks very much for showing me around your plant. I hope I was able to convince you that I'm the best person for the job.

Sincerely yours,
Robert Harad

6. In a newspaper or journal or on the Internet, find an ad for a position in your field for which you might be qualified. Write a résumé and a job-application letter in response to the ad; include a copy of the job ad. You will be evaluated not only on the content and appearance of the materials, but also on how well you have targeted them to the job ad.

CASE 10: Identifying the Best-of-the-Best Job-Search Sites

After the director of the Career Center at your school visits one of your classes, you decide to visit the Career Center's website. You study the available resources, but you find that they don't offer as much support as you and your classmates need navigating the online job-search environment for the first time. You decide to write the director to express your concerns. She agrees with your critique and asks you to research online information about job searching and begin putting together a new job-search resource for seniors at your college. To get started with your project, go to LaunchPad.

CHAPTER 11 Writing Proposals

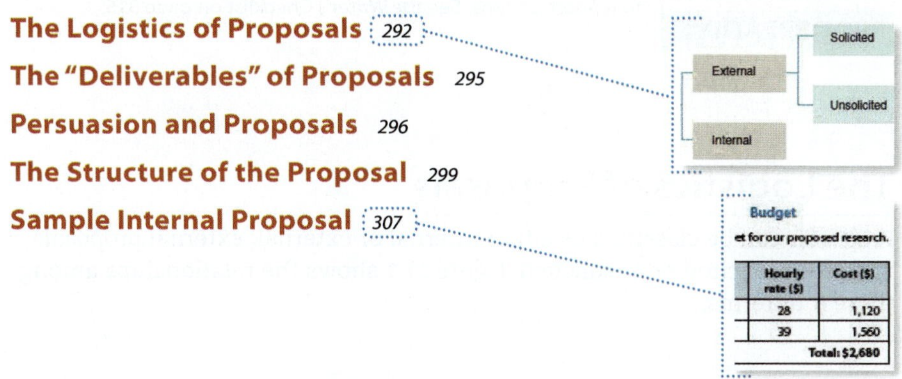

A PROPOSAL IS an offer to carry out research or to provide a product or service. For instance, a physical therapist might write a proposal to her supervisor for funding to attend a convention to learn about current rehabilitation practices. A defense contractor might submit a proposal to design and build a fleet of drones for the Air Force. A homeless shelter might submit a proposal to a philanthropic organization for funding to provide more services to the homeless community. Whether a project is small or big, within your own company or outside it, it is likely to call for a proposal.

FOCUS ON PROCESS Writing Proposals

When writing a proposal, pay special attention to these steps in the writing process.

PLANNING Consider your readers' knowledge about and attitudes toward what you are proposing. Use the techniques discussed in Chapters 4 and 5 to learn as much as you can about your readers' needs and about the subject. Also consider whether you have the personnel, facilities, and equipment to do what you propose to do.

(continued)

FOCUS ON PROCESS Writing Proposals

DRAFTING Collaboration is critical in large proposals because no one person has the time and expertise to do all the work. See Chapter 3 for more about collaboration. In writing the proposal, follow the instructions in any request for proposal (RFP) or information for bid (IFB) from the prospective customer. If there are no instructions, follow the structure for proposals outlined in this chapter.

REVISING
EDITING
PROOFREADING
External proposals usually have a firm deadline. Build in time to revise, edit, and proofread the proposal thoroughly and still get it to readers on time. See the Writer's Checklist on page 315.

The Logistics of Proposals

Proposals can be classified as either internal or external; external proposals are either solicited or unsolicited. Figure 11.1 shows the relationships among these four terms.

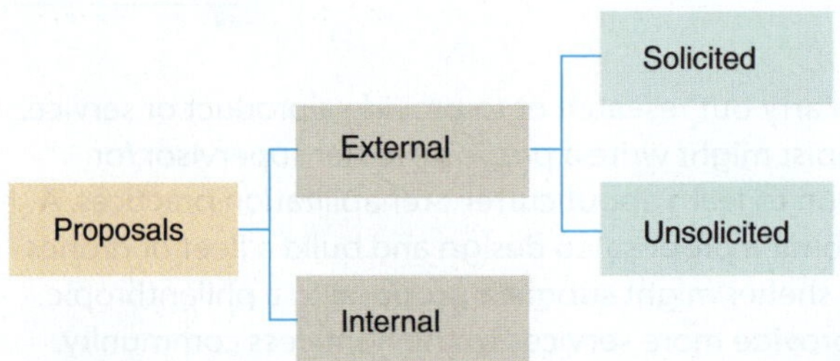

FIGURE 11.1 The Logistics of Proposals

INTERNAL AND EXTERNAL PROPOSALS

Internal proposals are submitted to the writer's own organization; external proposals are submitted to another organization.

Internal Proposals An internal proposal is an argument, submitted within an organization, for carrying out an activity that will benefit the organization. An internal proposal might recommend that the organization conduct research, purchase a product, or change some aspect of its policies or procedures.

For example, while working on a project in the laboratory, you realize that if you had a fiber-curl measurement system, you could do your job better and faster. The increased productivity would save your company the cost of the system in a few months. Your supervisor asks you to write a memo describing what you want, why you want it, how you plan to use it, and what it costs; if your request seems reasonable and the money is available, you'll likely get the new system.

Often, the scope of a proposal determines its format. A request for a small amount of money might be conveyed orally or by email or a brief memo. A request for a large amount, however, is likely to be presented in a formal report.

External Proposals No organization produces all the products or provides all the services it needs. websites need to be designed, written, and maintained; inventory databases need to be created; facilities need to be constructed. Sometimes projects require unusual expertise, such as sophisticated market analyses. Because many companies supply these products and services, most organizations require a prospective supplier to compete for the business by submitting a proposal, a document arguing that it deserves the business.

SOLICITED AND UNSOLICITED PROPOSALS

External proposals are either solicited or unsolicited. A *solicited proposal* is submitted in response to a request from the prospective customer. An *unsolicited proposal* is submitted by a supplier who believes that the prospective customer has a need for goods or services.

Solicited Proposals When an organization wants to purchase a product or service, it publishes one of two basic kinds of statements:

- An *information for bid* (IFB) is used for standard products. When a state agency needs desktop computers, for instance, it informs computer manufacturers of the configuration it needs. All other things being equal, the supplier that offers the lowest bid for a product with that configuration wins the contract. When an agency solicits bids for a specific brand and model, the solicitation is sometimes called a *request for quotation* (RFQ).
- A *request for proposal* (RFP) is used for more-customized products or services. For example, if the Air Force needs an "identification, friend or foe" system, the RFP it publishes might be a long and detailed set of technical specifications. The supplier that can design, produce, and deliver the device most closely resembling the specifications—at a reasonable price—will probably win the contract.

Most organizations issue IFBs and RFPs in print and online. Government solicitations are published on the FedBizOpps website. Figure 11.2 shows a portion of an RFQ.

FIGURE 11.2
Excerpt from an RFQ
Information from FedBizOpps.gov, 2013.

Solicitation Number: NIHCCOPC13014476

The Clinical Center at the National Institutes of Health (NIH) in Bethesda, Maryland, is one of the 27 institutes and centers that comprise NIH. The NIH Clinical Center is the nation's largest hospital devoted entirely to clinical research. With patients from all over the United States and some from abroad, the Clinical Center provides medical care and research services to support patients participating in over 1500 active protocols sponsored by the NIH Institutes. The Office of Patient Recruitment (OPR) reports to the Clinical Center Office of the Director. OPR is responsible for supporting the NIH intramural program by providing patient recruitment services for Institutes, staffing a call center to receive public inquiries, providing services to support the clinical research volunteer program, and serving as the NIH Intramural Liaison for Research Match (RM), an on-line recruitment tool.

The NIH has a requirement for call center services. Please see the attached Statement of Work. This is anticipated for up to four (4) full time positions onsite at the NIH Clinical Center. Offerors are to submit resumes with documentation for medical terminology training. Offerors may submit multiple resumes for positions.

Evaluation Criteria

Approach 30%: Proposal meets requirements and indicates an exceptional approach and understanding of the requirements. Risk of unsuccessful performance is very low. Adequacy of ability to meet the volume of contacts. Ability to provide consistent staffing and standards of staff performance.

Staffing 30%: Evidence of a broad pool of applicants that have been placed in the metropolitan DC area with the skill set desired within the last 2 years. Proposed staff meets the requirements (experience and credentialing) as stated in the Statement of Work. Staff has relevant experience with volume of contacts. Staff are recruited and retained. Ability to meet the 48 hours replacement time.

Corporate Experience 25%: Adequacy of the contractor's prior experience in providing support. The information shall include sufficient information to demonstrate previous effectiveness of staffing and oversight on performance.

Past Performance 15%: The quotation will be evaluated on 3 recent references (within the last 5 years).

Price Best Value

The government will evaluate the total price contained in the contractor's proposal. With the understanding and ability to project costs which are reasonable and indicates that the contractor understands the nature and extent of the work to be performed.

Period of Performance is one year September, 2013 to September, 2014 with up to 4 successive options.

This is an excerpt from an RFQ from the National Institutes of Health (NIH). The NIH is seeking to hire up to four call-center personnel. The specific details of the skills and experience required for the four persons are provided in a separate file.

This RFQ spells out the evaluation criteria by which the "offerors" will be judged.

The RFQ states that the solicitation is for "price best value." In other words, the contract will not necessarily be awarded to the lowest bidder. It will be awarded to the bidder that offers what the evaluators consider to be the best value, regardless of price.

Unsolicited Proposals An unsolicited proposal is like a solicited proposal except that it does not refer to an RFP. In most cases, even though the potential customer did not formally request the proposal, the supplier was invited to submit the proposal after people from the two organizations met and discussed the project. Because proposals are expensive to write, suppliers are reluctant to submit them without assurances that they will be considered carefully. Thus, the word *unsolicited* is only partially accurate.

The "Deliverables" of Proposals

A *deliverable* is what a supplier will deliver at the end of a project. Deliverables can be classified into two major categories: research or goods and services.

RESEARCH PROPOSALS

In a research proposal, you are promising to perform research and then provide a report about it. For example, a biologist for a state bureau of land management writes a proposal to the National Science Foundation requesting resources to build a window-lined tunnel in the forest to study tree and plant roots and the growth of fungi. The biologist also wishes to investigate the relationship between plant growth and the activity of insects and worms. The deliverable will be a report submitted to the National Science Foundation and, perhaps, an article published in a professional journal.

Research proposals often lead to two other applications: progress reports and recommendation reports.

After a proposal has been approved and the researchers have begun work, they often submit one or more *progress reports*, which tell the sponsor of the project how the work is proceeding. Is it following the plan of work outlined in the proposal? Is it going according to schedule? Is it staying within budget?

For more about progress reports and recommendation reports, see Ch. 12, and Ch. 13.

At the end of the project, researchers prepare a *recommendation report*, often called a *final report*, a *project report*, a *completion report*, or simply a *report*. A recommendation report tells the whole story of a research project, beginning with the problem or opportunity that motivated it and continuing with the methods used in carrying it out, the results, and the researchers' conclusions and recommendations.

People carry out research projects to satisfy their curiosity and to advance professionally. Organizations often require that their professional employees carry out research and publish in appropriate journals or books. Government researchers and university professors, for instance, are expected to remain active in their fields. Writing proposals is one way to get the resources—time and money for travel, equipment, and assistants—to carry out research.

GOODS AND SERVICES PROPOSALS

A *goods and services proposal* is an offer to supply a tangible product (a fleet of automobiles), a service (building maintenance), or some combination of the two (the construction of a building).

A vast network of goods and services contracts spans the working world. The U.S. government, the world's biggest customer, spent over $270 billion in 2015 buying military equipment from organizations that submitted proposals (USAspending.gov, 2016). But goods and services contracts are by no means limited to government contractors. An auto manufacturer might buy its engines from another manufacturer; a company that makes spark plugs might buy its steel and other raw materials from another company.

Another kind of goods and services proposal requests funding to support a local organization. For example, a women's shelter might receive some of its funding from a city or county but might rely on grants from private philanthropies. Typically, an organization such as a shelter would apply for a grant to fund increased demand for its services due to a natural disaster or an economic slowdown in the community. Or it might apply for a grant to fund a pilot program to offer job training at the shelter. Most large corporations have philanthropic programs offering grants to help local colleges and universities, arts organizations, and social-service agencies.

Persuasion and Proposals

A proposal is an argument. You must convince readers that the future benefits will outweigh the immediate and projected costs. Basically, you must persuade your readers of three things:

- that you understand the context and what your readers need
- that you have already determined what you plan to do and that you are able to do it
- that you are a professional and are committed to fulfilling your promises

UNDERSTANDING CONTEXTS

For more about analyzing your audience, see Ch. 4.

The most crucial element of the proposal is the definition of the problem or opportunity to which the proposed project responds. Although this point seems obvious, people who evaluate proposals agree that the most common weakness they see is an inadequate or inaccurate understanding of the problem or opportunity.

Internal Contexts Writing an internal proposal is both simpler and more complicated than writing an external one. It is simpler because you have greater access to internal readers than you do to external readers and you can get information more easily. However, it is more complicated because you might find it hard to understand the situation in your organization. Some colleagues will not tell you that your proposal is a long shot or that your ideas might threaten someone in the organization. Before you write an internal proposal, discuss your ideas with as many potential readers as you can to learn what those in the organization really think of them.

External Contexts When you receive an RFP, study it thoroughly. If you don't understand something in it, contact the organization. They will be happy to clarify it: a proposal based on misunderstood needs wastes everyone's time.

When you write an unsolicited proposal, analyze your audience carefully. How can you define the problem or opportunity so that readers will understand it? Keep in mind readers' needs and, if possible, their backgrounds. Concentrate on how the problem has decreased productivity or quality or how your ideas would create new opportunities. When you submit an unsolicited proposal, your task in many cases is to convince readers that a need exists. Even if you have reached an understanding with some of your potential customer's representatives, your proposal will still have to persuade other officials in the company. Most readers will reject a proposal as soon as they realize that it doesn't address their needs.

When you are preparing a proposal to be submitted to an organization in another culture, keep in mind the following six suggestions (Newman, 2011):

- **Understand that what makes an argument persuasive can differ from one culture to another.** Paying attention to the welfare of the company or the community might be more persuasive than offering a low bottom-line price. Representatives of an American company were surprised to learn that the Venezuelan readers of their proposal had selected a French company whose staff "had been making personal visits for years, bringing their families, and engaging in social activities long before there was any question of a contract" (Thrush, 2000).
- **Budget enough time for translating.** If your proposal has to be translated into another language, build in plenty of time. Translating long technical documents is a lengthy process because, even though some of the work can be done by computer software, the machine translation needs to be reviewed by native speakers of the target language.
- **Use simple graphics, with captions.** To reduce the chances of misunderstanding, use a lot of simple graphics, such as pie charts and bar graphs. Include captions so that readers can understand the graphics easily, without having to look through the text to see what each graphic means.

For more about graphics, see Ch. 8.

- **Write short sentences, using common vocabulary.** Short sentences are easier to understand than long sentences. Choose words that have few meanings. For example, use the word *right* as the opposite of *left*; use *correct* as the opposite of *incorrect*.

- **Use local conventions regarding punctuation, spelling, and mechanics.** Be aware that these conventions differ from place to place, even in the English-speaking world.
- **Ask if the prospective customer will do a read-through.** A *read-through* is the process of reading a draft of a proposal to look for any misunderstandings due to language or cultural differences. Why do prospective customers do this? Because it's in everyone's interest for the proposal to respond clearly to the customer's needs.

DESCRIBING WHAT YOU PLAN TO DO

Once you have shown that you understand what needs to be done and why, describe what you plan to do. Convince your readers that you can respond effectively to the situation you have just described. Discuss procedures and equipment you would use. If appropriate, justify your choices. For example, if you say you want to do ultrasonic testing on a structure, explain why, unless the reason is obvious.

Present a complete picture of what you would do from the first day of the project to the last. You need more than enthusiasm and good faith; you need a detailed plan showing that you have already started to do the work. Although no proposal can anticipate every question about what you plan to do, the more planning you have done before you submit the proposal, the greater the chances you will be able to do the work successfully if it is approved.

DEMONSTRATING YOUR PROFESSIONALISM

Once you have shown that you understand readers' needs and can offer a well-conceived plan, demonstrate that you are the kind of person (or that yours is the kind of organization) that is committed to delivering what you promise. Convince readers that you have the pride, ingenuity, and perseverance to solve the problems that are likely to occur. In short, show that you are a professional.

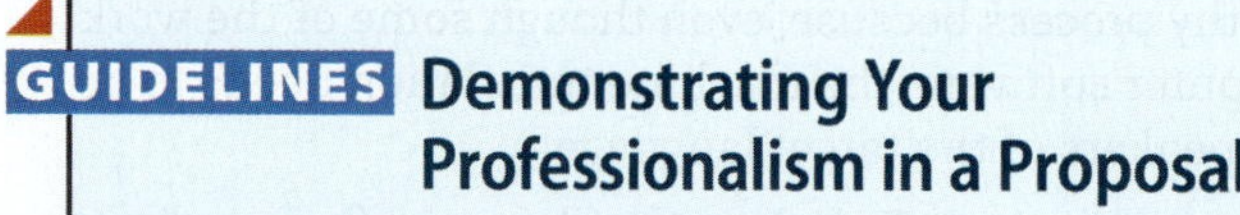

GUIDELINES Demonstrating Your Professionalism in a Proposal

In your proposal, demonstrate your ability to carry out the project by providing four kinds of information:

- **Credentials and work history.** Show that you know how to do this project because you have done similar ones. Who are the people in your organization with the qualifications to carry out the project? What equipment and facilities do you have that will enable you to do the work? What management structure will you use to coordinate the activities and keep the project running smoothly?

(continued)

- **Work schedule.** Sometimes called a *task schedule*, a work schedule is a graph or chart that shows when the various phases of the project will be carried out. The work schedule reveals more about your attitudes toward your work than about what you will be doing on any given day. A detailed work schedule shows that you have tried to foresee problems that might threaten the project.
- **Quality-control measures.** Describe how you will evaluate the effectiveness and efficiency of your work. Quality-control procedures might consist of technical evaluations carried out periodically by the project staff, on-site evaluations by recognized authorities or by the prospective customer, or progress reports.
- **Budget.** Most proposals conclude with a detailed budget, a statement of how much the project will cost. Including a budget is another way of showing that you have done your homework on a project.

ETHICS NOTE

WRITING HONEST PROPOSALS

When an organization approves a proposal, it needs to trust that the people who will carry out the project will do it professionally. Over the centuries, however, dishonest proposal writers have perfected a number of ways to trick prospective customers into thinking the project will go smoothly:

- saying that certain qualified people will participate in the project, even though they will not
- saying that the project will be finished by a certain date, even though it will not
- saying that the deliverable will have certain characteristics, even though it will not
- saying that the project will be completed under budget, even though it will not

Copying from another company's proposal is another common dishonest tactic. Proposals are protected by copyright law. An employee may not copy from a proposal he or she wrote while working for a different company.

There are three reasons to be honest in writing a proposal:

- to avoid serious legal trouble stemming from breach-of-contract suits
- to avoid acquiring a bad reputation, thus ruining your business
- to do the right thing

The Structure of the Proposal

Proposal structures vary greatly from one organization to another. A long, complex proposal might have 10 or more sections, including introduction, problem, objectives, solution, methods and resources, and management. If the authorizing agency provides an IFB, an RFP, an RFQ, or a set of guidelines, follow it closely. If you have no guidelines, or if you are writing an unsolicited proposal, use the structure shown here as a starting point. Then modify it according to your subject, your purpose, and the needs of your audience. An example of a proposal is presented on pages 308–14.

SUMMARY

For more about summaries, see Ch. 13, p. 351.

For a proposal of more than a few pages, provide a summary. Many organizations impose a length limit—such as 250 words—and ask the writer to present the summary, single-spaced, on the title page. The summary is crucial, because it might be the only item that readers study in their initial review of the proposal.

The summary covers the major elements of the proposal but devotes only a few sentences to each. Define the problem in a sentence or two. Next, describe the proposed program and provide a brief statement of your qualifications and experience. Some organizations wish to see the completion date and the final budget figure in the summary; others prefer that this information be presented separately on the title page along with other identifying information about the supplier and the proposed project.

INTRODUCTION

The purpose of the introduction is to help readers understand the context, scope, and organization of the proposal.

PROPOSED PROGRAM

In the section on the proposed program, sometimes called the *plan of work*, explain what you want to do. Be specific. You won't persuade anyone by saying that you plan to "gather the data and analyze it." *How* will you gather and analyze the data? Justify your claims. Every word you say—or don't say—will give your readers evidence on which to base their decision.

If your project concerns a subject written about in the professional literature, show your familiarity with the scholarship by referring to the pertinent studies. However, don't just string together a bunch of citations. For example, don't write, "Carruthers (2016), Harding (2017), and Vega (2016) have all researched the relationship between global warming and groundwater contamination." Rather, use the recent literature to sketch the necessary background and provide the justification for your proposed program. For instance:

> Carruthers (2016), Harding (2017), and Vega (2016) have demonstrated the relationship between global warming and groundwater contamination. None of these studies, however, included an analysis of the long-term contamination of the aquifer. The current study will consist of. . . .

GUIDELINES Introducing a Proposal

The introduction to a proposal should answer the following seven questions:

- **What is the problem or opportunity?** Describe the problem or opportunity in specific monetary terms, because the proposal itself will include a budget, and you want to convince your readers that spending money on what you propose is smart. Don't say that a design problem is slowing down production; say that it is costing $4,500 a day in lost productivity.
- **What is the purpose of the proposal?** The purpose of the proposal is to describe a solution to a problem or an approach to an opportunity and propose activities that will culminate in a deliverable. Be specific in explaining what you want to do.
- **What is the background of the problem or opportunity?** Although you probably will not be telling your readers anything they don't already know, show them that you understand the problem or opportunity: the circumstances that led to its discovery, the relationships or events that will affect the problem and its solution, and so on.
- **What are your sources of information?** Review the relevant literature, ranging from internal reports and memos to published articles or even books, so that readers will understand the context of your work.
- **What is the scope of the proposal?** If appropriate, indicate not only what you are proposing to do but also what you are not proposing to do.
- **What is the organization of the proposal?** Explain the organizational pattern you will use.
- **What are the key terms that you will use in the proposal?** If you will use any specialized or unusual terms, define them in the introduction.

You might include only a few references to recent research. However, if your topic is complex, you might devote several paragraphs or even several pages to recent scholarship.

For more about researching a subject, see Ch. 5.

Whether your project calls for primary research, secondary research, or both, the proposal will be unpersuasive if you haven't already done a substantial amount of research. For instance, say you are writing a proposal to do research on purchasing new industrial-grade lawn mowers for your company. Simply stating that you will visit Walmart, Lowe's, and Home Depot to see what kinds of lawn mowers they carry would be unpersuasive for two reasons:

- You need to justify why you are going to visit those three retailers rather than others. Anticipate your readers' questions: Why did you choose these three retailers? Why didn't you choose specialized dealers?

DOCUMENT ANALYSIS ACTIVITY

Writing the Proposed Program

The following project description is excerpted from a sample grant proposal seeking funding to begin a project to help police officers stay healthy (Ohio Office of Criminal Justice Services, 2003). The questions in the margin ask you to think about how to describe the project in a proposal.

1. The writer has used a lettering system to describe the four main tasks that will be undertaken if the project receives funding. What are the advantages of a lettering system?
2. How effective is the description of Task A? What factors contribute to the description's effectiveness or lack of effectiveness?
3. The descriptions of the tasks do not include cost estimates. Where would those estimates be presented in the proposal? Why would they be presented there?
4. How effective is the description of Task D? What additional information would improve its effectiveness?

PROJECT DESCRIPTION

The proposed project is comprised of several different, but related activities:

A. Physical Evaluation of the Officers

The first component of this project is the physical examination of all Summerville P.D. sworn employees. Of special interest for purposes of the project are resting pulse rate, target pulse rate, blood pressure, and percentage of body fat of the program participants. Dr. Feinberg will perform the physical examinations of all participating officers. The measurement of body fat will be conducted at the University of Summerville's Health Center under the direction of Dr. Farron Updike.

B. Renovation of Basement

Another phase of this project involves the renovation of the basement of police headquarters. The space is currently being used for storing Christmas decorations for City Hall.

The main storage room will be converted into a gym. This room will accommodate the Universal weight machine, the stationary bike, the treadmill and the rowing machine. Renovation will consist of first transferring all the Christmas decorations to the basement of the new City Hall. Once that is accomplished, it will be necessary to paint the walls, install indoor/outdoor carpeting and set up the equipment.

A second, smaller room will be converted into a locker room. Renovation will include painting the floors and the installation of lockers and benches.

To complete the fitness center, a third basement room will be equipped as a shower room. A local plumber will tap into existing plumbing to install several showerheads.

C. Purchase of Fitness Equipment

The Department of Public Safety has identified five vendors of exercise equipment in the greater Summerville area. Each of these vendors submitted bids for the following equipment:

- Universal Weight Machine
- Atlas Stationary Bike
- Yale Rowing Machine
- Speedster Treadmill

D. Training of Officers

Participating officers must be trained in the safe, responsible use of the exercise equipment. Dr. Updike of the University of Summerville will hold periodic training sessions at the Department's facility.

Information from Ohio Office of Criminal Justice Services, 2003.

DOCUMENT ANALYSIS ACTIVITY

To analyze a proposal delivered as a Prezi presentation, go to LaunchPad.

Marketing Proposal Presentation
Used by permission of Andrew Washuta.

- You should already have determined what stores carry what kinds of lawn mowers and completed any other preliminary research. If you haven't done the homework, readers have no assurance that you will in fact do it or that it will pay off. If your supervisor authorizes the project and then you learn that none of the lawn mowers in these stores meets your organization's needs, you will have to go back and submit a different proposal—an embarrassing move.

Unless you can show in your proposed program that you have done the research—and that the research indicates that the project is likely to succeed—the reader has no reason to authorize the project.

QUALIFICATIONS AND EXPERIENCE

After you have described how you would carry out the project, show that you can do it. The more elaborate the proposal, the more substantial the discussion of your qualifications and experience has to be. For a small project, include a few paragraphs describing your technical credentials and those of your co-workers. For larger projects, include the résumés of the project leader, often called the *principal investigator*, and the other primary participants.

External proposals should also discuss the qualifications of the supplier's organization, describing similar projects the supplier has completed successfully. For example, a company bidding on a contract to build a large suspension bridge should describe other suspension bridges it has built. It should also focus on the equipment and facilities the company already has and on the management structure that will ensure the project will go smoothly.

BUDGET

Good ideas aren't good unless they're affordable. The budget section of a proposal specifies how much the proposed program will cost.

Budgets vary greatly in scope and format. For simple internal proposals, add the budget request to the statement of the proposed program: "This study will take me two days, at a cost of about $400" or "The variable-speed recorder currently costs $225, with a 10 percent discount on orders of five or more." For more-complicated internal proposals and for all external proposals, include a more-explicit and complete budget.

Many budgets are divided into two parts: direct costs and indirect costs. *Direct costs* include such expenses as salaries and fringe benefits of program personnel, travel costs, and costs of necessary equipment, materials, and supplies. *Indirect costs* cover expenses that are sometimes called *overhead*: general secretarial and clerical expenses not devoted exclusively to any one project, as well as operating expenses such as costs of utilities and maintenance. Indirect costs are usually expressed as a percentage—ranging from less than 20 percent to more than 100 percent—of the direct expenses.

APPENDIXES

Many types of appendixes might accompany a proposal. Most organizations have boilerplate descriptions of the organization and of the projects it has completed. Another item commonly included in an appendix is a supporting letter: a testimonial to the supplier's skill and integrity, written by a reputable and well-known person in the field. Two other kinds of appendixes deserve special mention: the task schedule and the description of evaluation techniques.

Task Schedule A *task schedule* is almost always presented in one of three graphical formats: as a table, a bar chart, or a network diagram.

Tables The simplest but least informative way to present a schedule is in a table, as shown in Figure 11.3. As with all graphics, provide a textual reference that introduces and, if necessary, explains the table.

Although displaying information in a table is better than writing it out in sentences, readers still cannot "see" the information. They have to read the table to figure out how long each activity will last, and they cannot tell whether any of the activities are interdependent. They have no way of

Task Schedule

ACTIVITY	START DATE	FINISH DATE
Design the security system	4 Oct. 19	21 Oct. 19
Research available systems etc.	4 Oct. 19	3 Jan. 19

FIGURE 11.3 **Task Schedule Presented as a Table**

Schedule for Parking Analysis Project

Number	Task	1/14	1/21	1/28	2/4	2/11
1	Perform research					
2	Identify options					
3	Analyze options					
4	Test options					
5	Collect and analyze data					
6	Formulate recommendations					
7	Prepare report					

FIGURE 11.4 **Task Schedule Presented as a Bar Chart**

determining what would happen to the overall project schedule if one of the activities faced delays.

Bar Charts Bar charts, also called *Gantt charts* after the early twentieth-century civil engineer who first used them, are more informative than tables. The basic bar chart shown in Figure 11.4 allows readers to see how long each task will take and whether different tasks will occur simultaneously. Like tables, however, bar charts do not indicate the interdependence of tasks.

Network Diagrams Network diagrams show interdependence among various activities, clearly indicating which must be completed before others can begin. However, even a relatively simple network diagram, such as the one shown in Figure 11.5, can be difficult to read. You would probably not use this type of diagram in a document intended for general readers.

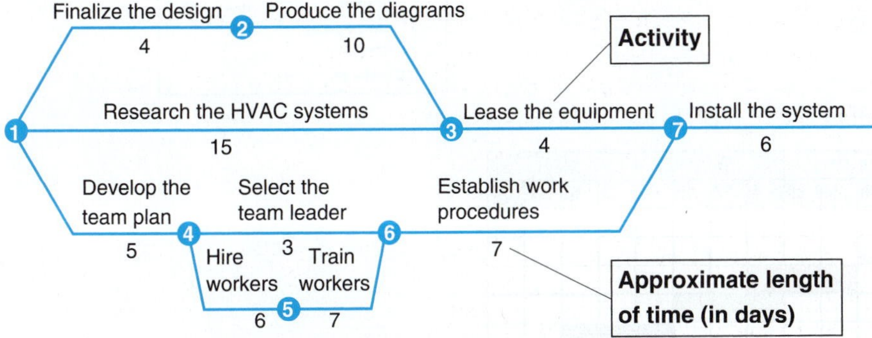

FIGURE 11.5 **Task Schedule Presented as a Network Diagram**

A network diagram provides more useful information than either a table or a bar chart.

TECH TIP

Why To Create a Gantt Chart

A Gantt chart is useful for showing how activities occur over time. Although they do not indicate the interdependence of tasks, Gantt charts can show overlaps in a schedule. In addition, color coding can visually indicate task priorities, task completion, task responsibilities, and so on.

How To Create a Gantt Chart

Using a program such as Word, Excel, or Google Spreadsheets, create a table or worksheet with enough cells to include all your tasks and dates.

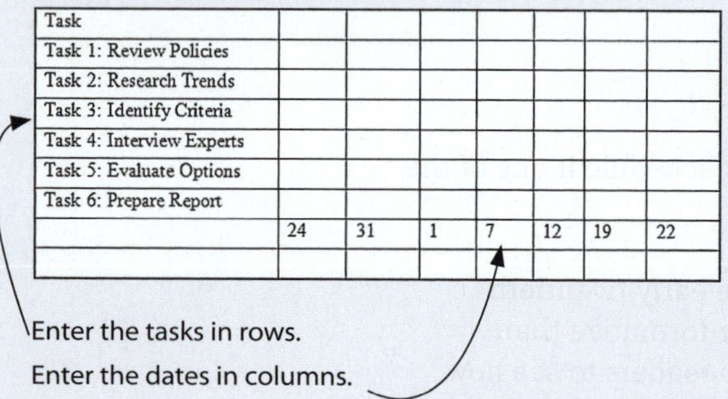

Task							
Task 1: Review Policies							
Task 2: Research Trends							
Task 3: Identify Criteria							
Task 4: Interview Experts							
Task 5: Evaluate Options							
Task 6: Prepare Report							
	24	31	1	7	12	19	22

Use the **Fill** command to color-code cells.

Word and Excel

Google Spreadsheets

To create cells that span several columns, use the **Merge Cells** function. In Word and Excel, select the cells you wish to merge and right-click, then choose **Merge Cells** from the pop-up menu. In Google Spreadsheets, choose from the **Merge Cells** options in the **Format** tab.

Word

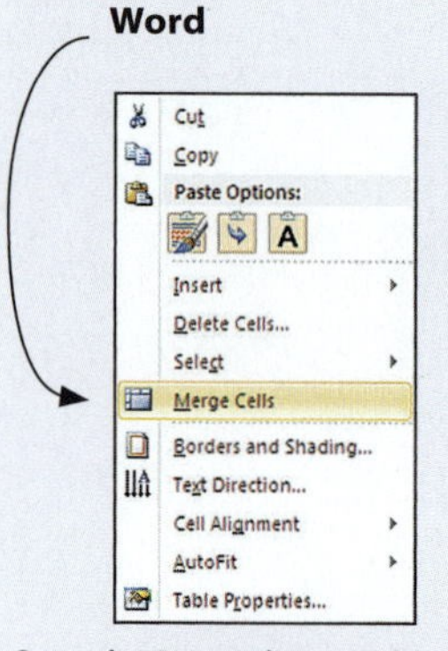

Excel

Google Spreadsheets

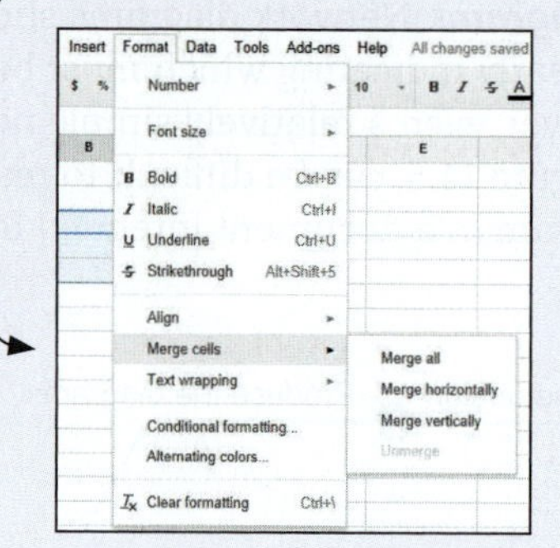

Sample Gantt chart with merged cells and color coding:

#	Task	March				April				May			
		7	15	22	29	5	12	19	26	3	10	17	24
1	Conceptualization												
2	Engineering Phase												
3	Pre-Production												
4	Product Testing												
5	Production												
6	Release												

Description of Evaluation Techniques Although *evaluation* can mean different things to different people, an *evaluation technique* typically refers to any procedure used to determine whether the proposed program is both effective and efficient. Evaluation techniques can range from writing simple progress reports to conducting sophisticated statistical analyses. Some proposals call for evaluation by an outside agent, such as a consultant, a testing laboratory, or a university. Other proposals describe evaluation techniques that the supplier will perform, such as cost-benefit analyses.

The issue of evaluation is complicated by the fact that some people think in terms of *quantitative evaluations*—tests of measurable quantities, such as production increases—whereas others think in terms of *qualitative evaluations*—tests of whether a proposed program is improving, say, the workmanship on a product. And some people include both qualitative and quantitative testing when they refer to evaluation. An additional complication is that projects can be tested while they are being carried out (*formative evaluations*) as well as after they have been completed (*summative evaluations*).

When an RFP calls for "evaluation," experienced proposal writers contact the prospective customer to determine precisely what the word means.

Sample Internal Proposal

The following example of an internal proposal has been formatted as a memo rather than as a formal proposal. (See Chapter 12, pp. 325–32, for the progress report written after this project was under way and Chapter 13, pp. 357–80, for the recommendation report.)

In most professional settings, writers use letterhead stationery for memos.

Proposals can be presented as memos or as reports. Memos are more popular for brief documents (fewer than five pages), whereas reports are more popular for longer documents.

The writers include their titles and that of their primary reader. This way, future readers will be able to readily identify the reader and writers.

The subject heading indicates the subject of the memo (the tablet study at Rawlings Regional Medical Center) and the purpose of the memo (proposal).

As discussed in Ch. 9, memos of more than one page should begin with a clear statement of purpose. Here, the writers communicate the primary purpose of the document in one sentence.

Memos of more than one page should contain a summary to serve as an advance organizer or to help readers who want only an overview of the document.

Although the writers are writing to Dr. Bremerton, they refer to her in the third person to suggest the formality of their relationship.

The background of the problem. Don't assume that your reader knows what you are discussing, even if it was the reader who suggested the project in the first place.

The problem at the heart of the project.

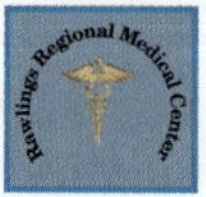

Rawlings Regional Medical Center
7500 Bannock Avenue
Rawlings, MT 59211

Date: October 6, 2018
To: Jill Bremerton, M.D.
Chief Executive Officer
Rawlings Regional Medical Center
From: Jeremy Elkins, Director of Information Technology
Eloise Carruthers, Director of Nursing
Rawlings Regional Medical Center
Subject: Proposal for the Tablet Study at RRMC

Purpose

The purpose of our proposal is to request authorization to conduct a study to determine the best course of action for integrating tablet computers into the RRMC clinical setting.

Summary

On September 16, 2018, Dr. Jill Bremerton, RRMC Chief Executive Officer, asked us to develop a plan to study national trends on tablet use, determine clinical-staff knowledge of and attitudes toward tablets, examine administrative models for tablet use, devise criteria for assessing tablets, and present our findings, including a recommendation.

Currently, RRMC has no formal policy on tablet usage by clinical staff. By default, we are following a bring-your-own-device (BYOD) approach. More than half of our clinical staff use their personal tablets in their work. This situation is not ideal because not all clinical staff are taking advantage of the enormous potential for improving patient care and reducing costs by using tablets, and IT is struggling to keep up with the work needed to ensure that all the different tablets are working properly and that information-security protocols required by HIPAA and current health care laws are not being violated.

Memo to Jill Bremerton, M.D. October 6, 2018 Page 2

Therefore, Dr. Bremerton wanted us to determine the best approach to making tablets available to all our clinical staff. Specifically, Dr. Bremerton asked that we develop a plan to determine how tablets are being used by clinical staff across the nation, determine the RRMC clinical staff's current knowledge of and attitudes toward tablet use, determine how hospitals administer the use of tablets in a clinical setting, establish criteria by which we might evaluate tablets for RRMC, and assess available tablets based on our criteria.

We propose to research tablet use in clinical settings and present our findings to Dr. Bremerton. To perform these tasks, we would carry out secondary and primary research. We would study the literature on tablet use, distribute a questionnaire to RRMC clinical staff, and interview Dr. Bremerton. Then, we would collect and analyze our data and write the report.

To perform this research and present a recommendation report, we estimate that we would each require approximately 40 hours over the next two months, at a cost of $2,680. Jeremy Elkins, Director of Information Technology, has been with RRMC for 9 years and has overseen numerous IT feasibility studies. Eloise Carruthers has been with RRMC for 13 years, the last 8 of which have been as Director of Nursing.

If this proposal is authorized, we would begin our research immediately, submitting to Dr. Bremerton a progress report on November 14, 2018, and a recommendation report on December 14, 2018. The recommendation report would include the details of our research and recommendations regarding how to proceed with the feasibility study.

Introduction

On September 16, 2018, Dr. Jill Bremerton, RRMC Chief Executive Officer, asked us to develop a plan to determine the best course of action for integrating tablet computers into the RRMC clinical setting.

Currently, RRMC has no formal policy on tablet usage by clinical staff. By default, we are following a bring-your-own-device (BYOD) approach. More than half of our clinical staff use their personal tablets in their work. This situation is not ideal because not all clinical staff are taking advantage of the enormous potential for improving patient care and reducing costs

The proposal. The writers have already begun to plan what they will do if the proposal is accepted, but they use the conditional tense ("would") because they cannot assume that their proposal will be approved.

A summary of the schedule and the credentials of the writers. Because the reader will likely want to read this entire proposal, the summary functions as an advance organizer.

A brief statement of the context for the proposal.

An explanation of the problem: the current situation is inadequate because the medical center is not taking full advantage of tablets and because IT is spending a lot of time ensuring that the tablets in use are in compliance with federal requirements.

This same paragraph appeared in the "Summary" section of the proposal. In technical communication, writers often present the same information several times because some readers will read only selected portions of the document.

The writers paraphrase text from a memo Dr. Bremerton had written to them. Often in technical communication, you will quote or paraphrase your reader's words to remind him or her of the context and to show that you are carrying out your tasks professionally — and to give the reader the opportunity to change the direction of the study before it officially begins.

The introduction concludes with an advance organizer for the rest of the proposal.

By presenting the project as a set of tasks, the writers show that they are well organized. This organization by tasks will be used in the progress report (see Ch. 12, pp. 325–32) and the recommendation report (see Ch. 13, pp. 357–80).

Memo to Jill Bremerton, M.D. October 6, 2018 Page 3

by using tablets, and IT is struggling to keep up with the work needed to ensure that all the different tablets are working properly and that information-security protocols required by HIPAA and current health care laws are not being violated.

Therefore, Dr. Bremerton wanted us to determine the best approach to making tablets available to all our clinical staff. Specifically, Dr. Bremerton asked that we develop a plan to perform five tasks:

- Determine how tablets are being used by clinical staff across the nation.
- Determine the RRMC clinical staff's current knowledge of and attitudes toward tablet use.
- Determine how hospitals administer the use of tablets in a clinical setting.
- Establish criteria by which we might evaluate tablets for RRMC.
- Assess available tablets based on our criteria.

In the following sections, we provide additional details about the proposed tasks, schedule, and budget, as well as our credentials and references.

Proposed Tasks

With Dr. Bremerton's approval, we would perform the following six tasks to help determine the best course of action for integrating tablet computers into the RRMC clinical setting.

Task 1. Acquire a basic understanding of tablet use by clinical staff across the nation

We have already begun our research by interviewing Dr. Bremerton, who emphasized that we need to maintain our focus on our priorities — patient care and service to the community — and not let technical questions about the tablets distract us from the needs of our clinical staff. "We're not going to do anything without the approval of the doctors and nurses," she said.

We also discovered an article that corroborated what Dr. Bremerton had told us (Narisi, 2013). Two keys to doing the research were to focus on security features — data-privacy issues mandated in HIPAA and in current health care laws — and to get the clinical staff's input.

Dr. Bremerton pointed us to a number of resources on tablet use in clinical settings. In addition, we have begun to conduct our own literature review. Most of the research we studied falls into one of four categories:

- general introductions to tablet use in trade magazines and general-interest periodicals
- more-focused articles about tablets used in health care
- technical specifications of tablets provided in trade magazines and on manufacturers' websites
- trade-magazine articles about best practices for managing the use of tablets in clinical settings

As we expected, the information we acquired is a mix of user opinions, benchmark-test results, and marketing. We would rely most heavily on case studies from hospital administrators and technical specialists in health IT. Because of the unreliability of information on manufacturers' websites, we are hesitant to rely on claims about product performance.

Task 2. Determine the RRMC clinical staff's knowledge of and attitudes toward tablet use

Next week, we propose to send all clinical staff members an email linking to a four-question Qualtrix survey. The email would explain that we were seeking opinions about tablet use by clinical-staff members who already own tablets and would make clear that the survey would take less than two minutes to complete.

Task 3. Assess the BYOD and hospital-owned tablet models

Our research has already revealed that hospitals currently use one of two models for giving clinical staff access to tablets: the bring-your-own-device (BYOD) approach and purchasing tablets to distribute to staff. We have found literature that assesses the advantages and disadvantages of each of these models (Jackson, 2011a, 2011b).

For statistics on the popularity of each of these administrative models, we would rely on a survey (Terry, 2011).

The proposal sounds credible because the writers have already begun their secondary research. Readers are reluctant to approve proposals unless they are sure that the writers have at least begun their research.

Following the recommendation from Dr. Bremerton, the writers start by outlining the secondary research they plan to do. The logic is obvious: if they are to present sensible recommendations, they need to understand their subject.

By stating that they know that their sources are a mixture of different kinds of information, not all of which are equally useful for every kind of question that needs to be answered, the writers suggest that they are careful analysts.

The writers show that they have applied the insights they gathered from their secondary research. Now they propose doing primary research to determine whether the RRMC clinical staff share the attitudes of clinical staff across the country. The logic is clear: if they do, the hospital administrators will know that they can rely on the national data.

The writers cite their sources throughout the proposal.

Memo to Jill Bremerton, M.D. October 6, 2018 Page 5

Task 4. Establish criteria for evaluating tablets

We (Jeremy Elkins and Eloise Carruthers) have both begun to study the voluminous literature on tablets. Jeremy Elkins has met informally with his five IT colleagues to discuss the data, and Eloise Carruthers has met informally with her nursing staff and with selected physicians, including several who already own tablets and use them in the clinic.

Often you will begin your project with a cost criterion: your recommended solution must not cost more than a certain amount.

We would begin with the first criterion: cost. Dr. Bremerton told us in our interview that the budget for the project (assuming that RRMC would supply a tablet to each member of the clinical staff) would be $800 per device, fully configured with any commercial software needed to operate it. For this reason, we would not thoroughly examine any tablets that do not meet this criterion.

In addition, we would pay particular attention to the complexities of the current tablet market, focusing on whether the various devices would work seamlessly with our health-records system and other security features ("Top Five," 2013), on the need to be able to disinfect the tablets (Carr, 2011), and on durability (Narisi, 2013). Furthermore, we know from experience with all kinds of portable information technology that the question of battery life would be problematic because it can vary so much depending on load and other factors.

We anticipate that two factors might be critically important: operating system and availability of relevant apps.

Task 5. Assess available tablets based on our criteria

We would begin our study of the tablets by examining trade magazines. We realize, however, that several of our likely criteria — namely, the ability to disinfect the tablet, as well as durability and battery life — might not be addressed adequately in the literature because each hospital has its own specific needs. If the literature cannot help us complete our assessments, we would need to carry out on-site evaluations at RRMC.

Preparing the recommendation report is part of the project because the report is the deliverable.

Task 6. Analyze our data and prepare a recommendation report

We would draft our recommendation report and upload it to a wiki to make it convenient for the other IT staff members and interested clinical staff members to help us revise it. We would then incorporate our

Memo to Jill Bremerton, M.D. October 6, 2018 Page 6

colleagues' suggestions and present a final draft of the report on the wiki to gather any final editing suggestions.

Schedule

Figure 1 is a schedule of the tasks we would complete for this project.

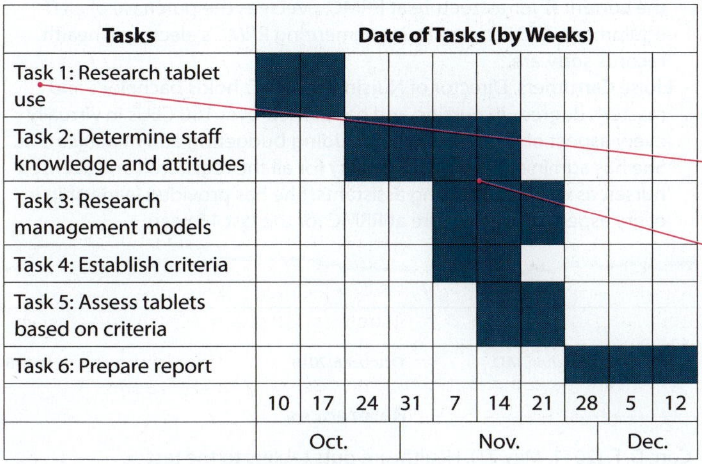

Tasks	Date of Tasks (by Weeks)									
Task 1: Research tablet use										
Task 2: Determine staff knowledge and attitudes										
Task 3: Research management models										
Task 4: Establish criteria										
Task 5: Assess tablets based on criteria										
Task 6: Prepare report										
	10	17	24	31	7	14	21	28	5	12
	Oct.				Nov.				Dec.	

Figure 1 Schedule of Project Tasks

Budget

Following is an itemized budget for our proposed research.

Name	Hours	Hourly rate ($)	Cost ($)
Jeremy Elkins	40	28	1,120
Eloise Carruthers	40	39	1,560
			Total: $2,680

Organizing the project by tasks makes it easy for the writers to present a Gantt chart. In addition, the task organization will help the writers stay on track if the proposal is approved and they continue their research.

Each task is presented with parallel grammar, which shows that the writers are careful and professional.

Some tasks overlap in time: researchers often work on several tasks simultaneously.

The Tech Tip on p. 306 explains how to create a Gantt chart.

The writers summarize their credentials. Strong credentials help reinforce the writers' professionalism.

Memo to Jill Bremerton, M.D. October 6, 2018 Page 7

Experience

We are experienced professionals who have participated in numerous studies both here at RRMC and elsewhere.

- Jeremy Elkins, Director of Information Technology, has chaired the Technology Infrastructure Committee and served as *ad hoc* member of the Steering Committee at RRMC for 9 years. He has designed the current IT infrastructure at RRMC, oversees the purchase of all IT equipment, and is currently implementing RRMC's electronic health records software.
- Eloise Carruthers, Director of Nursing at RRMC, holds bachelor's and master's degrees in nursing and has earned over 100 CEUs in virtually every aspect of the profession, including budgeting and management. She has administrative responsibility for all the registered and licensed nurses, as well as all nursing assistants. She has provided leadership in every aspect of patient care at RRMC for the last 13 years.

This list of references follows the APA documentation style, which is discussed in Appendix, Part A, p. 452. The APA documentation system calls for References to begin on a new page. Check with your instructor.

Memo to Jill Bremerton, M.D. October 6, 2018 Page 8

References

Carr, D. F. (2011, May 21). Healthcare puts tablets to the test. *InformationWeek Healthcare.* Retrieved October 1, 2018, from www.informationweek.com/healthcare/mobile-wireless/healthcare-puts-tablets-to-the-test/229503387.

Jackson, S. (2011a, August 15). Five reasons hospitals should buy tablets for physicians. *FierceMobileHealthcare.* Retrieved October 2, 2018, from www.fiercemobilehealthcare.com/story/5-reasons-why-hospitals-should-buy-tablets-physicians/2011-08-15.

Jackson, S. (2011b, August 3). Why physicians should buy their own mobile devices. *FierceMobileHealthcare.* Retrieved October 4, 2013, from www.fiercemobilehealthcare.com/story/why-physicians-should-buy-their-own-mobile-devices/2011-08-03.

Narisi, S. (2013, April 24). Choosing the best tablets for doctors: 3 keys. *Healthcare Business & Technology.* Retrieved October 3, 2018, from www.healthcarebusinesstech.com/best-tablets-for-doctors.

Terry, K. (2011, Dec. 9). Apple capitalizes on doctors' iPad romance. *InformationWeek Healthcare.* Retrieved October 2, 2018, from www.informationweek.com/healthcare/mobile-wireless/apple-capitalizes-on-doctors-ipad-roman/232300218.

Top Five Tablet Security Features. (2013). *Healthcare Data Solutions.* Retrieved October 2, 2018, from www.healthcaredatasolutions.com/top-five-tablet-security-features.html.

WRITER'S CHECKLIST

The following checklist covers the basic elements of a proposal. Guidelines established by the recipient of the proposal should take precedence over these general suggestions.

Does the summary provide an overview of

- ☐ the problem or the opportunity? *(p. 300)*
- ☐ the proposed program? *(p. 300)*
- ☐ your qualifications and experience? *(p. 303)*

Does the introduction indicate

- ☐ the problem or opportunity? *(p. 300)*
- ☐ the purpose of the proposal? *(p. 301)*
- ☐ the background of the problem or opportunity? *(p. 301)*
- ☐ your sources of information? *(p. 301)*
- ☐ the scope of the proposal? *(p. 301)*
- ☐ the organization of the proposal? *(p. 301)*
- ☐ the key terms that you will use in the proposal? *(p. 301)*

☐ Does the description of the proposed program provide a clear, specific plan of action and justify the tasks you propose performing? *(p. 300)*

Does the description of qualifications and experience clearly outline

- ☐ your relevant skills and past work? *(p. 303)*
- ☐ the skills and background of the other participants? *(p. 303)*
- ☐ your department's (or organization's) relevant equipment, facilities, and experience? *(p. 303)*

Is the budget

- ☐ complete? *(p. 303)*
- ☐ correct? *(p. 304)*
- ☐ accompanied by an in-text reference? *(p. 304)*

☐ Do the appendixes include the relevant supporting materials, such as a task schedule, a description of evaluation techniques, and evidence of other successful projects? *(p. 304)*

EXERCISES

For more about memos, see Ch. 9, p. 248.

1. Study the National Science Foundation's (NSF) Grant Proposal Guide (www.nsf.gov/pubs/policydocs/pappguide/nsf15001/gpg_print.pdf). In what important ways does the NSF's guide differ from the advice provided in this chapter? What accounts for these differences? Present your findings in a 500-word memo to your instructor.
2. **TEAM EXERCISE** Form groups according to major. Using the FedBizOpps website (www.fbo.gov), find and study an RFP for a project related to your academic field. What can you learn about the needs of the organization that issued the RFP? How effectively does it describe what the issuing organization expects to see in the proposal? Is it relatively general or specific? What sorts of evaluation techniques does it call for? In your response, include a list of questions that you would ask the issuing organization if you were considering responding to the RFP. Present your results in a memo to your instructor.
3. Write a proposal for a research project that will constitute a major assignment in this course. Your instructor will tell you whether the proposal is to be written individually or collaboratively. Start by defining a technical subject that interests you. (This subject could be one that you are involved with at work or in another course.) Using abstract services and other bibliographic tools, compile a bibliography of articles and books on the subject. (See Chapter 5 for a discussion of finding information.) Create a reasonable real-world context. Here are three common scenarios from the business world:
   - Our company uses Technology X to perform Task A. Should we instead be using Technology Y to perform Task A? For instance, our company uses traditional surveying tools in its contracting business. Should we be using surveying tools that incorporate GPS instead?

- Our company has decided to purchase a tool to perform Task A. Which make and model of the tool should we purchase, and from which supplier should we buy it? For instance, our company has decided to purchase 10 laptop computers. Which brand and model should we buy, and from whom should we buy them? Is leasing the tool a better option than purchasing?
- Our company does not currently perform Function X. Is it feasible to perform Function X? For instance, we do not currently offer day care for our employees. Should we? What are the advantages and disadvantages of doing so? What forms can day care take? How is it paid for?

Following are some additional ideas for topics:

- the value of using social media to form ties with students in a technical-communication class on another campus
- the need for expanded opportunities for internships or service-learning in your major
- the need to create an advisory board of industry professionals to provide expertise about your major
- the need to raise money to keep the college's computer labs up to date
- the need to evaluate the course of study offered by your university in your major to ensure that it is responsive to students' needs
- the advisability of starting a campus branch of a professional organization in your field
- the need to improve parking facilities on campus
- the need to create or improve organizations for minorities or women on campus

CASE 11: Revising a Brief Proposal

You work for an educational software company. Because most school districts are facing tough economic times, they need financial assistance in order to buy your company's products. The company has prepared a proposal guide for schools to use in applying for grants that will enable them to purchase the company's products. Your supervisor notes that this information hasn't been updated in a few years. He asks you to take a look at the effectiveness of the current sample proposal letter and suggest ways it could be improved. He'd also like you to annotate the letter to explain its various sections. To get started revising the letter, go to LaunchPad.

CHAPTER 12 Writing Informational Reports

Kamil Abbasov/ Shutterstock.

COMPLEX, EXPENSIVE PROJECTS call for a lot of documents. Before a project begins, a vendor might write a *proposal* to interest prospective clients in its work. After a project is completed, an organization might write a *completion report* to document the project or a *recommendation report* to argue for a future course of action. In between, many people will write various *informational reports.*

Whether they are presented as memos, emails, reports, or web pages, informational reports share one goal: to describe something that has happened or is happening now. Their main purpose is to provide clear, accurate, specific information to an audience. Sometimes, informational reports also analyze the situation. An *analysis* is an explanation of why something happened or how it happened. For instance, in an incident report about an accident on the job, the writer might speculate about how and why the accident occurred.

FOCUS ON PROCESS Writing Informational Reports

In writing informational reports, pay special attention to these steps in the writing process.

PLANNING In some cases, determining your audience and to whom to address the report is difficult. Choosing the appropriate format for your report can also be difficult. Consider whether your organization has a preferred format for reports and whether your

(continued)

FOCUS ON PROCESS Writing Informational Reports

For more about analyzing an audience from another culture, see Ch. 4, "Communicating Across Cultures."

report will be read by readers from other cultures who might expect a formal style and application. See Chapter 4 for more about analyzing your audience.

DRAFTING Some informational reports are drafted on site. For instance, an engineer might use a tablet computer to "draft" a report as she walks around a site. For routine reports, you can sometimes use sections of previous reports or boilerplate. In a status report, for instance, you can copy the description of your current project from the previous report and then update it as necessary. See Chapter 2, page 16, for more about boilerplate.

REVISING
EDITING
PROOFREADING Informal does not mean careless. Revise, edit, and proofread. Even informal reports should be free of errors.

This chapter discusses five kinds of informational reports:

- A supervisor writes a *directive* explaining a company's new policy on recycling and describing informational sessions that the company will offer to help employees understand how to implement the policy.
- An insurance adjuster writes a *field report* presenting the results of his inspection of a building after a storm caused extensive damage.
- A research team writes a *progress report* explaining what the team has accomplished in the first half of the project, speculating on whether it will finish on time and within budget, and describing how it has responded to unexpected problems.
- A worker at a manufacturing company writes an *incident report* after a toxic-chemical spill.
- A recording secretary writes a set of *meeting minutes* that will become the official record of what occurred at a meeting of the management team of a government agency.

Another type of informational report is the *recommendation report* (see Chapter 13).

Writing Directives

In a *directive*, you explain a policy or a procedure you want your readers to follow. Even though you have the authority to require your readers to follow the policy, you want to explain why the policy is desirable or at least necessary. You will be most persuasive if you present clear, compelling evidence (in the form of commonsense arguments, numerical data, and examples); consider opposing arguments effectively; and present yourself as cooperative, moderate, fair-minded, and modest. Figure 12.1 is an example of a directive.

NOTICE TO EMPLOYEES

Research has shown that minors find it easy to buy tobacco products even though state law prohibits sales to anyone under 18. To stop the sale of tobacco to minors and to comply with state law, we are implementing the following policy immediately:

> **THIS COMPANY WILL NOT SELL CIGARETTES, CHEWING TOBACCO, SMOKELESS TOBACCO, OR SMOKING PARAPHERNALIA TO ANYONE UNDER THE AGE OF 18.**
>
> **YOU CAN BE FINED $100 PLUS COURT COSTS AND FEES FOR SELLING ANY OF THESE PRODUCTS TO ANYONE UNDER THE AGE OF 18.**

Under this new policy, you are required to request valid photo identification for anyone attempting to purchase tobacco products *who appears to be under the age of 27.*

If a customer questions this policy, please explain that state law prohibits the sale of tobacco products to those under the age of 18, and therefore we refuse to sell to minors.

A copy of the law is posted near the cash register. Please read the law carefully and, if you have questions, confer with your supervisor.

Any employee who does not follow this policy will be subject to disciplinary action. Thank you for your cooperation.

The writer, the owner of a convenience store, begins with a clear explanation of the problem the directive addresses. Presenting the reasons for the new policy shows respect for the readers and therefore makes the directive more persuasive.

The writer uses a polite but official tone because the new policy is a policy, not a request. Notice that the directive specifies a penalty for not adhering to the policy and directs readers to their supervisors if they have questions.

FIGURE 12.1 A Directive

DOCUMENT ANALYSIS ACTIVITY

To analyze a report presented as a website and a report presented through an interactive graphic, go to LaunchPad.

High Plains Water-Level Monitoring Study

Information from United States Geological Survey, 2014: http://ne.water.usgs.gov/ogw/hpwlms/.

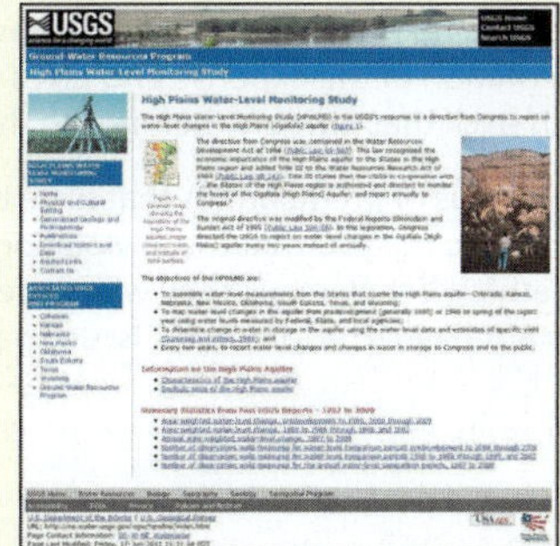

"Global Forest Change" Interactive Map

Information from Hansen/UMD/Google/USGS/NASA. Hansen, M. C., P. V. Potapov, R. Moore, M. Hancher, S. A. Turubanova, A. Tyukavina, D. Thau, S. V. Stehman, S. J. Goetz, T. R. Loveland, A. Kommareddy, A. Egorov, L. Chini, C. O. Justice, and J. R. G. Townshend. 2013. "High-Resolution Global Maps of 21st-Century Forest Cover Change." *Science* 342 (15 November): 850–53. Data available online from: http://earthenginepartners.appspot.com/science-2013-global-forest.

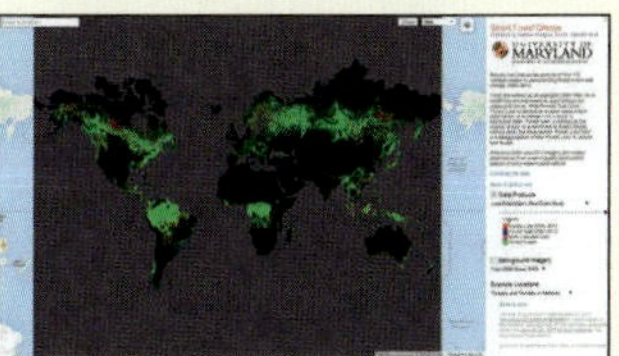

DOCUMENT ANALYSIS ACTIVITY

Writing a Persuasive Directive

This directive was sent to the members of a Montana government department. The questions below ask you to think about the process of writing persuasive directives (as discussed on p. 318).

1. How would you describe the tone used by the writer? Provide an example to support your claim.
2. The writer presents examples of what he calls violations of the state travel policy. Do these examples provide solid evidence that violations of the policy have in fact occurred?
3. How effectively has the writer encouraged his staff to abide by the travel policy? How might he improve the persuasiveness of the directive?

To: Members, Budget Allocation Office
From: Harold J. Jefferson, Director
Subject: Travel Policy
Date: January 23, 2016

It has come to my attention that certain members of this office are not abiding by the travel policies approved by the state. Let me offer a few examples.

The Montana Revised Code includes this statement in its introduction:

> Persons who travel on State business are encouraged to incur the lowest practical and reasonable expense while still traveling in an efficient and timely manner. Those traveling on State business are expected to avoid impropriety, or the appearance of impropriety, in any travel expense. They must conduct State business with integrity, in compliance with applicable laws, and in a manner that excludes consideration of personal advantage.

Yet I have learned from official sources that on four occasions in the last fiscal year, employees of this office have acted in flagrant violation of this policy. Two occasions involved choosing a flight that left at a more convenient time in the morning but that cost almost $160 more. One involved selecting a cab, rather than a shuttle bus, for a trip from the airport to downtown (a $24 difference), and one involved using room service when the motel has a café (a $14 difference).

Another provision of the travel policy that has been violated on more than one occasion is the following:

> Travel expenses are not paid in advance except for airfare charged to the State air travel card, for online (Internet) air or train ticket purchases, and for conference registrations.

Two employees have on more than three occasions each received reimbursements for air and/or train reservations made using their personal credit cards. As you know, using personal credit cards leaves the State without official documentation of the expense and gives the traveler bonus miles and/or cash back that properly belongs to the State.

These are just two of the kinds of irregularities that have been brought to my attention. I do not need to tell you that these violations constitute serious breaches of public ethics. If they recur, they will be dealt with harshly. I sincerely hope that I do not have to address this issue again.

Writing Field Reports

A common kind of informational report describes inspections, maintenance, and site studies. Such reports, often known as *field reports*, explain problems, methods, results, and conclusions, but they deemphasize methods and can include recommendations. The report in Figure 12.2 illustrates a possible variation on this standard report structure.

LOBATE CONSTRUCTION
3311 Industrial Parkway
Speonk, NY 13508
Quality Construction Since 1957

April 11, 2019

Ms. Christine Amalli, Head
Civil Engineering
New York Power
Smithtown, NY 13507

Dear Ms. Amalli:

We are pleased to report the results of our visual inspection of the Chemopump after Run #9, a 30-day trial on Kentucky #10 coal.

The inspection was designed to determine if the new Chemopump is compatible with Kentucky #10, the lowest-grade coal that you anticipate using. In preparation for the 30-day test run, the following three modifications were made by your technicians:

- New front-bearing housing buffer plates of tungsten carbide were installed.
- The pump-casting volute liner was coated with tungsten carbide.
- New bearings were installed.

Our summary is as follows. A number of small problems with the pump were observed, but nothing serious and nothing surprising. Normal break-in accounts for the wear. The pump accepted the Kentucky #10 well.

The following four minor problems were observed:

- The outer lip of the front-end bell was chipped along two-thirds of its circumference.
- Opposite the pump discharge, the volute liner received a slight wear groove along one-third of its circumference.
- The impeller was not free-rotating.
- The holes in the front-end bell were filled with insulating mud.

The following three components showed no wear:

- 5½″ impeller
- suction neck liner
- discharge neck liner

Our conclusion is that the problems can be attributed to normal break-in for a new Chemopump. The Kentucky #10 coal does not appear to have caused any extraordinary problems. In general, the new Chemopump seems to be operating well.

(*continued*)

FIGURE 12.2
A Field Report
Image: Kamil Abbasov/Shutterstock.

Because the writer and the reader work for different companies, a letter is the appropriate format for this brief informational report.

The word *visual* describes the methods.

The writer states the purpose of the inspection.

The writer has chosen to incorporate the words *summary* and *conclusion* in the body of the letter rather than use headings as a method of organization.

Informational reports sometimes include recommendations.

The writer concludes politely.

page 2

We would recommend, however, that the pump be modified as follows:

1. Replace the front-end bell with a tungsten carbide-coated front-end bell.
2. Replace the bearings on the impeller.
3. Install insulation plugs in the holes in the front-end bell.

Further, we recommend that the pump be reinspected after another 30-day run on Kentucky #10.

If you have any questions or would like to authorize these modifications, please call me at 555-1241. As always, we appreciate the trust you have placed in us.

Sincerely,

Marvin Littridge

Marvin Littridge
Director of Testing and Evaluation

FIGURE 12.2 **A Field Report** *(continued)*

GUIDELINES Responding to Readers' Questions in a Field Report

When you write a field report, be sure to answer the following six questions:

- What is the purpose of the report?
- What are the main points covered in the report?
- What were the problems leading to the decision to perform the procedure?
- What methods were used?
- What were the results?
- What do the results mean?

If appropriate, also discuss what you think should be done next.

Writing Progress and Status Reports

A *progress report* describes an ongoing project. A *status report*, sometimes called an *activity report*, describes the entire range of operations of a department or division. For example, the director of marketing for a manufacturing company might submit a monthly status report.

A progress report is an intermediate communication between a proposal (the argument that a project be undertaken) and a completion report (the comprehensive record of a completed project) or a recommendation report (an argument to take further action). Progress reports let you check in with your audience.

Regardless of how well the project is proceeding, explain clearly and fully what has happened and how those activities or events will affect the overall project. Your tone should be objective, neither defensive nor casual. Unless your own ineptitude or negligence caused a problem, you're not to blame. Regardless of the news you are delivering—good, bad, or mixed—your job is the same: to provide a clear and complete account of your activities and to forecast the next stage of the project.

When things go wrong, you might be tempted to cover up problems and hope that you can solve them before the next progress report. This course of action is unwise and unethical. Chances are that problems will multiply, and you will have a harder time explaining why you didn't alert your readers earlier.

ETHICS NOTE

REPORTING YOUR PROGRESS HONESTLY

Withholding bad news is unethical because it can mislead readers. As sponsors or supervisors of the project, readers have a right to know how it is going. If you find yourself faced with any of the following three common problems, consider responding in these ways:

- **The deliverable—the document or product you will submit at the end of the project—won't be what you thought it would be.** Without being defensive, describe the events that led to the situation and explain how the deliverable will differ from what you described in the proposal.
- **You won't meet your schedule.** Explain why you are going to be late, and state when the project will be complete.
- **You won't meet the budget.** Explain why you need more money, and state how much more you will need.

ORGANIZING PROGRESS AND STATUS REPORTS

The time pattern and the task pattern, two organizational patterns frequently used in progress and status reports, are illustrated in Figure 12.3.

In the time pattern, you describe all the work that you have completed in the present reporting period and then sketch in the work that remains. Some writers include a section on present work, which enables them to focus on a long or complex task still in progress.

THE TIME PATTERN	THE TASK PATTERN
Discussion	Discussion
A. Past Work	A. Task 1
B. Future Work	1. Past work
	2. Future work
	B. Task 2
	1. Past work
	2. Future work

The task pattern enables you to describe, in order, what has been accomplished on each task. Often a task-oriented structure incorporates the chronological structure.

FIGURE 12.3 Organizational Patterns in Reports

A status report is usually organized according to task; by its nature, this type of report covers a specified time period.

CONCLUDING PROGRESS AND STATUS REPORTS

In the conclusion of a progress or status report, evaluate how the project is proceeding. In the broadest sense, there are two possible messages: things are going well, or things are not going as well as anticipated.

If appropriate, use appendixes for supporting materials, such as computations, printouts, schematics, diagrams, tables, or a revised task schedule. Be sure to cross-reference these appendixes in the body of the report, so that readers can find them easily.

GUIDELINES Projecting an Appropriate Tone in a Progress or Status Report

Whether the news is positive or negative, these two suggestions will help you sound like a professional.

- **If the news is good, convey your optimism but avoid overstatement.**

  OVERSTATED We are sure the device will do all that we ask of it, and more.

  REALISTIC We expect that the device will perform well and that, in addition, it might offer some unanticipated advantages.

- **Beware of promising early completion.** Such optimistic forecasts rarely prove accurate, and it is embarrassing to have to report a failure to meet an optimistic deadline.

- **Don't panic if the preliminary results are not as promising as you had planned or if the project is behind schedule.** Even the best-prepared proposal writers cannot anticipate all problems. As long as the original proposal was well planned and contained no wildly inaccurate computations, don't feel responsible. Just do your best to explain unanticipated problems and the status of the project. If your news is bad, at least give the reader as much time as possible to deal with it effectively.

Sample Progress Report

The following progress report was written for the project proposed on pages 308-14 in Chapter 11. (The recommendation report for this study is on pages 357-80 in Chapter 13.)

Rawlings Regional Medical Center
7500 Bannock Avenue
Rawlings, MT 59211

Date: November 14, 2018
To: Jill Bremerton, M.D.
Chief Executive Officer
Rawlings Regional Medical Center
From: Jeremy Elkins, Director of Information Technology
Eloise Carruthers, Director of Nursing
Rawlings Regional Medical Center
Subject: Progress Report for the Tablet Study at RRMC

Purpose

This is a progress report on our study to recommend the best course of action for integrating tablet computers into the RRMC clinical setting.

Summary

On October 8, 2018, Dr. Jill Bremerton, RRMC Chief Executive Officer, approved our proposal to study national trends in tablet use, determine clinical-staff knowledge of and attitudes toward tablets, examine administrative models for tablet use, devise criteria for assessing tablets, and present our findings, including a recommendation.

We have completed Tasks 1 and 2 (understanding tablet use in a clinical setting and determining the clinical staff's knowledge of and attitudes toward tablet use), as well as part of Task 3 (assessing the bring-your-own-device and hospital-owned tablet models).

Our study is currently on schedule, and we expect to submit a recommendation report on December 14, 2018, as indicated in our proposal dated October 6, 2018.

In most professional settings, writers use letterhead stationery for memos.

Progress reports can be presented as memos or as reports.

The writers include their titles and that of their primary reader. This way, future readers will be able to readily identify the reader and writers.

The subject heading indicates the subject of the memo (the tablet study at Rawlings Regional Medical Center) and the purpose of the memo (progress report).

Memos of more than one page should begin with a clear statement of purpose. Here, the writers communicate the primary purpose of the document in one sentence.

Memos of more than one page should contain a summary to serve as an advance organizer or to help readers who want only an overview of the document.

Readers of progress reports want to know whether the project is proceeding according to schedule and (if applicable) on budget.

A brief statement of the context for the proposal. Note that the writers refer to the reader's having authorized the proposal.

An explanation of the problem: the current situation is inadequate because the medical center is not taking full advantage of tablets and because IT is spending a lot of time ensuring that the tablets in use are in compliance with federal requirements.

A formal statement of the task that Dr. Bremerton asked the writers to perform.

Most of the information in the introduction is taken directly from the proposal. This reuse of text is ethical because the writers created it for that earlier document.

The introduction concludes with an advance organizer for the rest of the proposal.

The writers begin by describing the organization of the results section. For a progress report, a chronological organization—completed work, then future work—makes good sense.

The writers follow the task structure that they used in the proposal.

Memo to Jill Bremerton, M.D. November 14, 2018 Page 2

Introduction

On October 8, 2018, Dr. Jill Bremerton, RRMC Chief Executive Officer, approved our proposal to determine the best course of action for integrating tablet computers into the RRMC clinical setting.

Currently, RRMC has no formal policy on tablet usage by clinical staff. By default, we are following a bring-your-own-device (BYOD) approach. More than half of our clinical staff use their personal tablets in their work. This situation is not ideal because not all clinical staff are taking advantage of the enormous potential for improving patient care and reducing costs by using tablets, and IT is struggling to keep up with the work needed to ensure that all the different tablets are working properly and that information-security protocols required by HIPAA and current health care laws are not being violated.

Dr. Bremerton approved our proposal to determine the best approach to making tablets available to all our clinical staff. Specifically, Dr. Bremerton asked us to perform five tasks:

- Determine how tablets are being used by clinical staff across the nation.
- Determine the RRMC clinical staff's knowledge of and attitudes toward tablet use.
- Determine how hospitals administer the use of tablets in a clinical setting.
- Establish criteria by which we might evaluate tablets for RRMC.
- Assess available tablets based on our criteria.

In the following sections, we present the results of our research to date, followed by an updated task schedule and references.

Results of Research

In this progress report, we present our completed work on Tasks 1–2 and our status on Task 3. Then we discuss our future work: Tasks 4–6.

Task 1. Acquire a basic understanding of tablet use by clinical staff across the nation

Since the introduction of the Apple iPad in 2010, the use of tablets by clinical staff in hospitals across the country has been growing steadily. Although there are no precise statistics on how many hospitals either distribute tablets to clinical staff or let them use their own devices in their work, the number of articles in trade magazines, exhibits at medical conferences, and discussions on discussion boards suggests that tablets are quickly becoming established in the clinical setting. And many hundreds of apps have already been written to enable users to carry out health-care-related tasks on tablets.

The most extensive set of data on tablets in hospitals relates to the use of the iPad, the first tablet on the market. Ottawa Hospital has distributed more than 1,000 iPads to clinical staff; California Hospital is piloting a program with more than 100 iPads for hospital use; Kaiser Permanente is testing the iPad for hospital and clinical workflow; and Cedars-Sinai Medical Center is testing the iPad in its hospital. The University of Chicago's Internal Medicine Residency Program uses the iPad; the iPad is also being distributed to first-year medical students at Stanford, University of California–Irvine, and University of San Francisco. In addition, there are reports of Windows-based and Android-based tablets being distributed at numerous other hospitals and medical schools (Husain, 2011).

The writers skillfully integrate their secondary research into their discussion of the task. By doing so, they enhance their credibility.

Today, tablets have five main clinical applications (Carr, 2011):

- *Monitoring patients and collecting data.* Clinical staff connect tablets to the hospital's monitoring instruments to collect patient information and transfer it to patients' health records without significant human intervention. In addition, staff access patient information on their tablets.
- *Ordering prescriptions, authorizations, and refills.* Clinical staff use tablets to communicate instantly with the hospital pharmacy and off-site pharmacies, as well as with other departments within the hospital, such as the Imaging Department.
- *Scheduling appointments.* Clinical staff use tablets to schedule doctor and nurse visits and laboratory tests, to send reminders, and to handle re-scheduling and cancellations.
- *Conducting research on the fly.* Clinical staff use tablets to access medication databases and numerous reference works.
- *Educating patients.* Clinical staff use videos and animations to educate patients on their conditions and treatment options.

Tablets provide clinical staff with significant advantages. Staff do not need to go back to their offices to connect to the Internet or to the hospital's own medical-record system. Staff save time, reduce paper usage, and reduce transcription errors by not having to enter nearly as much data by hand.

Task 2. Determine the RRMC clinical staff's knowledge of and attitudes toward tablet use

On October 14, 2018, we sent all 147 clinical staff members an email linking to a four-question Qualtrix survey. In the email, we said that we were seeking opinions about tablet use by clinical-staff members who already own tablets and made clear that the survey would take less than two minutes to complete. (The questionnaire, including the responses, appears in the Appendix, page 9.)

We received 96 responses, which represents 65 percent of the 147 staff members. We cannot be certain that all 96 respondents who indicated that they are tablet owners in fact own tablets. We also do not know whether all those staff members who own a tablet responded. However, given that some 75 percent of physicians in a 2013 poll own tablets, we suspect that the 96 respondents accurately represent the proportion of our clinical staff who own tablets (Drinkwater, 2013).

Here are the four main findings from the survey of tablet owners:

- Some 47 percent of respondents own an Apple iPad, and 47 percent own either a Samsung Galaxy or another tablet that uses the Android operating system. Only 6 percent use the Microsoft Surface, one of the several Windows-based tablets.
- Some 58 percent of the respondents strongly agree with the statement that they are expert users of their tablets. Overall, 90 percent agree more than they disagree with the statement.
- Some 63 percent of respondents use their tablets for at least one clinical application. They have either loaded apps on their tablets themselves or had IT do so for them.
- Some 27 percent of the respondents would prefer to continue to use their own tablets for clinical applications, whereas 38 percent would prefer to use a tablet supplied by RRMC. Some 35 percent had no strong feelings either way. None of the respondents indicated that they would prefer not to use a tablet at all for clinical applications.

Task 3. Assess the BYOD and hospital-owned tablet models

Currently, hospitals use one of two models for giving clinical staff access to tablets: the bring-your-own-device (BYOD) model and the purchase model, whereby the hospital purchases tablets to distribute to staff. In this section, we will present our findings on the relative advantages of each model.

The BYOD model is based on the fact that, nationally, some three-quarters of physicians already own tablets (with the Apple iPad the single most popular model) (Drinkwater, 2013). We could find no data on how many nurses own tablets.

The main advantage of the BYOD model is that clinical staff already know and like their tablets; therefore, they are motivated to use them and less likely to need extensive training. In addition, the hardware costs are eliminated (or almost eliminated, since some hospitals choose to purchase some tablets for staff who do not have their own). Todd Richardson, CIO with Deaconess Health System, Evansville, Indiana (Jackson, 2011), argues that staff members who own their own tablets use and maintain them carefully: they know how to charge, clean, store, and protect them. In addition, the hospital doesn't have to worry about the question of liability if staff members lose them during personal use. And if the staff member moves on to a new position at a different hospital, there is no dispute about who owns the information on the tablet. All the hospital has to do is disable the staff member's account.

However, there are three main disadvantages to the BYOD model:

- Some clinical staff do not have their own tablets, and some who do don't want to use them at work; to make the advantages of tablet use available to all the clinical staff, therefore, the hospital needs to decide whether to purchase tablets and distribute them to these staff members.
- Labor costs are high because each tablet needs to be examined carefully by the hospital IT department to ensure that it contains no software that might interfere with or be incompatible with the health-care software that needs to be loaded onto it. This labor-intensive assessment by IT can seriously erode the cost savings from not having to buy the tablet itself.
- Chances of loss increase because the staff member is more likely to use the tablet at home as well as in the hospital.

Currently, we are studying the advantages and disadvantages of the other model for making tablets available to clinical staff: for the hospital to purchase the same tablet for each staff member.

We are now completing Task 3 and beginning work on Task 4.

The writers explain that they are in the process of completing Task 3.

The Gantt chart shows the progress toward completing each of the project tasks. See the Tech Tip in Ch. 11, p. 306, for advice on how to create Gantt charts.

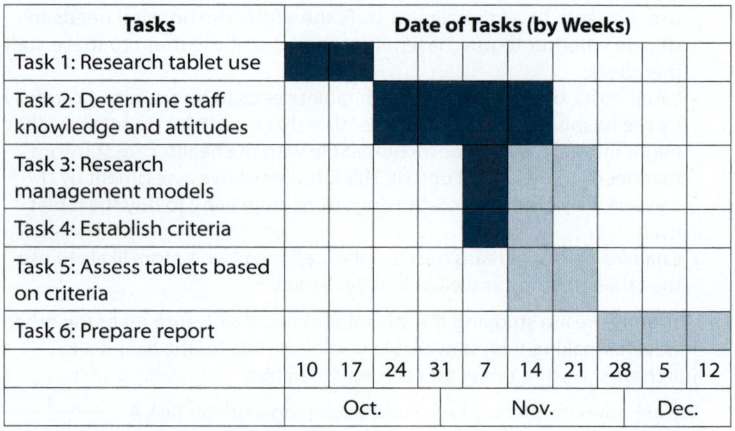

Memo to Jill Bremerton, M.D. November 14, 2018 Page 6

Task 4. Establish criteria for evaluating tablets

We will study the voluminous literature on tablets. We have already determined our first criterion: a cost of no more than $800 per device, fully configured with any commercial software needed to operate it. We will then determine the additional criteria to use in our study.

Task 5. Assess available tablets based on our criteria

We will study reviews in trade magazines and, if necessary, carry out on-site evaluations at RRMC.

Task 6. Analyze our data and prepare a recommendation report

We will draft our recommendation report and edit it in response to suggestions from interested readers among the clinical staff. Then, we will solicit one more round of edits and revise the report. Finally, we will present the report to Dr. Bremerton on December 14, 2018.

Updated Schedule

Figure 1 is an updated task schedule. The light blue bars represent tasks yet to be completed.

Tasks	Date of Tasks (by Weeks)									
Task 1: Research tablet use										
Task 2: Determine staff knowledge and attitudes										
Task 3: Research management models										
Task 4: Establish criteria										
Task 5: Assess tablets based on criteria										
Task 6: Prepare report										
	10	17	24	31	7	14	21	28	5	12
	Oct.				Nov.				Dec.	

FIGURE 1. Schedule of Project Tasks

Memo to Jill Bremerton, M.D. November 14, 2013 Page 7

Conclusion

We have successfully completed Tasks 1–2 (and part of 3) and begun Tasks 4 and 5. We are on schedule to complete all tasks by the December 14 deadline. We have a good understanding of how tablets are used nationwide, and we have completed our survey of tablet users among the RRMC clinical staff, as well as half of our study of the two administrative models used by hospitals. We are currently completing that task and are about to begin to establish criteria and analyze tablets based on them. In the report we will present on December 14, we will include our recommendation on how we think RRMC should proceed to take better advantage of the potential for clinical use of tablets.

Please contact Jeremy Elkins, at jelkins@rrmc.org or at 444-3967, or Eloise Carruthers, at ecarruthers@rrmc.org or at 444-3982, if you have questions or comments or would like to discuss this project further.

The conclusion summarizes the status of the project.

The writers end with a polite offer to provide additional information.

Memo to Jill Bremerton, M.D. November 14, 2018 Page 8

References

Carr, D. F. (2011, May 21). Healthcare puts tablets to the test. *InformationWeek Healthcare*. Retrieved October 1, 2018, from www.informationweek.com/healthcare/mobile-wireless/healthcare-puts-tablets-to-the-test/229503387.

Drinkwater, D. (2013, April 19). Three in 4 physicians are using tablets; some are even prescribing apps. *TabTimes*. Retrieved October 12, 2013, from http://tabtimes.com/news/ittech-stats-research/2013/04/19/3-4-physicians-are-using-tablets-some-are-even-prescribing.

Husain, I. (2011, March 10). Why Apple's iPad will beat Android tablets for hospital use. *iMedicalApps*. Retrieved October 15, 2013, from www.imedicalapps.com/2011/03/ipad-beat-android-tablets-hospital-medical-use/.

Jackson, S. (2011, August 3). Why physicians should buy their own mobile devices. *FierceMobileHealthcare*. Retrieved October 4, 2013, from www.fiercemobilehealthcare.com/story/why-physicians-should-buy-their-own-mobile-devices/2011-08-03.

This list of references follows the APA documentation style, which is discussed in Appendix, Part A. The APA documentation system calls for References to begin on a new page. Check with your instructor.

Presenting the percentage data in boldface after each question is a clear way to communicate how the respondents replied. Although most readers will not be interested in the raw data, some will.

Appendix: Clinical-Staff Questionnaire

This is the questionnaire we distributed to the 147 RRMC clinical staff members. We received 96 responses. The numbers in boldface below represent the percentage of respondents who chose each response.

Questionnaire on Tablet Use at RRMC

Directions: As you may know, Dr. Bremerton is conducting a study to determine whether to institute a formal policy on tablet use by clinical staff.

If you own a tablet device, please respond to the following four questions. Your opinions can help us decide whether and how to develop a policy for tablet use at RRMC. We greatly appreciate your answering the following four questions.

1. Which brand of tablet do you own?
   47% Apple iPad
   28% Samsung Galaxy
   9% Amazon Kindle Fire
   6% Microsoft Surface
   10% Other (please name the brand) **(Respondents named the Asus, Google Nexus, and a Toshiba model.)**

2. "I consider myself an expert user of my tablet."
   Strongly disagree_____ **8%** **2%** **13%** **19%** **58%** Strongly agree

3. Do you currently use your tablet for a clinical application, such as monitoring patients or ordering procedures?
   63% Yes
   37% No

4. If RRMC were to adopt a policy of using tablets for clinical applications (and to supply the appropriate software and training), which response best describes your attitude?
   27% I would prefer to use my own tablet.
   38% I would prefer to use a hospital-supplied tablet.
   35% I don't have strong feelings either way about using my own or a hospital-supplied tablet.
   0% I would prefer not to use any tablet at all for clinical applications.

Thank you!

Writing Incident Reports

An incident report describes an event such as a workplace accident, a health or safety emergency, or an equipment problem. (Specialized kinds of incident reports go by other names, such as *accident reports* or *trouble reports*.) The purpose of an incident report is to explain what happened, why it happened, and what the organization did (or is going to do) to follow up on the incident. Incident reports often contain a variety of graphics, including tables, drawings, diagrams, and photographs, as well as videos.

Incident reports can range from single-page forms that are filled out on paper or online to reports hundreds of pages long. Figure 12.4 shows an accident report form used at a university.

FIGURE 12.4 An Accident Report Form
Information from University of North Carolina-Chapel Hill Environment, Health, and Safety.

UNC – Chapel Hill – Facilities Services – Safety Plan

APPENDIX A

Employee's Accident Report Form
University of North Carolina at Chapel Hill

This form is to be completed by the employee and forwarded to the Health and Safety office as soon as practicable after the injury. (See Human Resources Manual)

Accident Date:

1. Name of employee:
2. Date and time of injury:
3. Describe how the injury occurred:
4. Describe what job duty you were doing at the time of your injury:
5. Describe what part of your body was injured:
6. Describe what you would recommend to prevent a reoccurrence:
7. Further information you would like to include regarding your injury:

Employee signature Date

HTTP://www.fac.unc.edu

Figure 12.5 is the executive summary of a National Transportation Safety Board accident report on a 2012 head-on collision between two freight trains in Oklahoma. Investigators spent many months researching and writing the full report.

FIGURE 12.5 Executive Summary of a Complex Accident Report
Information from National Transportation Safety Board, 2013.

The summary—two pages near the beginning of a 65-page report—begins with the basic facts about the accident.

On Sunday, June 24, 2012, at 10:02 a.m. central daylight time, eastbound Union Pacific Railroad (UP) freight train ZLAAH-22 and westbound UP freight train AAMMLX-22 collided head-on while operating on straight track on the UP Pratt subdivision near Goodwell, Oklahoma. Skies were clear, the temperature was 89°F, and visibility was 10 miles.

The collision derailed 3 locomotives and 24 cars of the eastbound train and 2 locomotives and 8 cars of the westbound train. The engineer and the conductor of the eastbound train and the engineer of the westbound train were killed. The conductor of the westbound train jumped to safety. During the collision and derailment, several fuel tanks from the derailed locomotives ruptured, releasing diesel fuel that ignited and burned. Damage was estimated at $14.8 million.

The writers discuss the probable cause of the accident and the resulting damage to the trains.

The National Transportation Safety Board determines that the probable cause of this accident was the eastbound Union Pacific Railroad train crew's lack of response to wayside signals because of the engineer's inability to see and correctly interpret the signals; the conductor's disengagement from his duties; and the lack of positive train control, which would have stopped the train and prevented the collision regardless of the crew's inaction. Contributing to the accident was a medical examination process that failed to decertify the engineer before his deteriorating vision adversely affected his ability to operate a train safely.

Information from National Transportation Safety Board, 2013.

The accident investigation focused on the following safety issues:

- The actions and responsibilities of the train crews: Crew conversations in the locomotive cab concerning signal aspects, radio transmissions, or any condition that can affect the safe operation of the train are important crew activities. In this accident, as the train passed signals for advance approach, approach, and stop, the engineer actively adjusted the throttle and dynamic brake as if all three signals were clear. The fact that the conductor was disengaged from his duties and did not appropriately intervene as the train proceeded through the signals demonstrates . . .
- The medical examination process for railroad engineer certification: The UP's medical records for the engineer of the eastbound train indicated that the engineer had passed his required vision test in 2009. However, the medical records from the engineer's personal physician, his ophthalmologist, and his optometrist documented . . .
- The survivability of event recorder data: The lead and trailing locomotives of both trains in this accident had event recorders to capture and preserve operational data that is important to accident investigation. However, most of the data could not be retrieved after the severe damage to the lead locomotives from the postaccident fire. . . .

(*continued*)

- The need for implementation of positive train control: Before reaching the Goodwell siding, the eastbound train crew had passed three signals without appropriately responding by slowing and then stopping their train. Regardless of the reason for the crew's nonresponse, had a positive train control system been in place in the area of the accident, it would have slowed and stopped the train, avoiding the collision.

As a result of this investigation, the National Transportation Safety Board makes safety recommendations to the Federal Railroad Administration, the Brotherhood of Locomotive Engineers and Trainmen, ... The National Transportation Safety Board also reiterates recommendations to the Federal Railroad Administration and the Association of American Railroads and reclassifies three recommendations to the Federal Railroad Administration.

FIGURE 12.5 Executive Summary of a Complex Accident Report *(continued)*

Finally, the writers list the results of the investigation.

Writing Meeting Minutes

Minutes, an organization's official record of a meeting, are distributed to all those who belong to the committee or group represented at the meeting. Sometimes, minutes are written by administrative assistants; other times they are written by technical professionals or technical communicators.

For more about conducting meetings, see Ch. 3, "Conducting Meetings."

In writing minutes, be clear, comprehensive, objective, and diplomatic. Do not interpret what happened; simply report it. Because meetings rarely follow the agenda perfectly, you might find it challenging to provide an accurate record of the meeting. If necessary, interrupt the discussion to request a clarification.

Do not record emotional exchanges between participants. Because minutes are the official record of the meeting, you want them to reflect positively on the participants and the organization. For example, in a meeting a person might say, undiplomatically, that another person's idea is stupid, a comment that might lead to an argument. Don't record the argument. Instead, describe the outcome: "After a discussion of the merits of the two approaches, the chair asked the Facilities Committee to consider the approaches and report back to membership at the next meeting."

Figure 12.6, an example of an effective set of minutes, was written using a Microsoft template. Many organizations today use templates like this one, which has three advantages:

- Because it is a word-processing template, the note taker can enter information on his or her computer or tablet during the meeting, reducing the time it takes to publish the minutes.
- Because the template is a form, it prompts the note taker to fill in the appropriate information, thus reducing the chances that he or she will overlook something important.
- Because the template is a table, readers quickly become accustomed to reading it and thereby learn where to look for the information they seek.

The first section of this template calls for information about the logistics of the meeting. You can modify the template to make it appropriate for your organization.

The second section of this template is devoted to the agenda items for the meeting.

Note that for each agenda item, the note taker is prompted to state how long the discussion took, the subject of the discussion, and the name of the person leading the discussion.

For each agenda item, the note taker records the main points of the discussion and the action items. Because the template calls for the action item (such as a vote or a task to be done), the name of the person responsible for doing the task, and the deadline for the task, there should be no confusion about who is to do which task and when it is due.

Weekly Planning Committee Meeting

MINUTES February 14, 2018 3:40 p.m. conference room

meeting called by	Principal Robert Barson
type of meeting	regular weekly
note taker	Zenda Hill
attendees	William Sipe, Patty Leahy, George Zaerr, Herbert Simon, Robert Barson, Zenda Hill. Absent: Heather Evett

Agenda topics

2 minutes | approval of minutes | Zenda Hill

discussion	The minutes of the February 7, 2016, meeting were read.	
action items	**person responsible**	**deadline**
One correction was made: In paragraph 2, "800 hours" was replaced with "80 hours." The minutes were then unanimously approved.	Zenda Hill	N/A

30 minutes | authorization for antidrug presentation by Alan Winston | Principal Barson

discussion	Principal Barson reported on his discussion with Peggy Giles of the School District, who offered positive comments about Winston's presentations at other schools in the district last year. Mr. Zaerr expressed concern about the effect of the visit on the teaching schedule. Principal Barson acknowledged that the visit would disrupt one whole day but said that the chairs unanimously approved of the visit. Student participation would be voluntary, and the chairs offered to give review sessions to those students who elected not to attend. Ms. Hill asked if there was any new business. There was none.	
action items	**person responsible**	**deadline**
Ms. Hill called for a vote on the motion. The motion carried 5–0, with one abstention.	Ms. Hill will arrange the Winston visit.	February 23, 2018
There being no new business, Ms. Hill moved that the committee adjourn. Motion passed. The committee adjourned at 4:12 p.m.	N/A	N/A

FIGURE 12.6 **A Set of Meeting Minutes**

WRITER'S CHECKLIST

- ☐ Did you choose an appropriate application for the informational report? *(p. 317)*

Does the directive

- ☐ clearly and politely explain your message? *(p. 318)*
- ☐ explain your reasoning, if appropriate? *(p. 318)*

Does the field report

- ☐ clearly explain the important information? *(p. 321)*
- ☐ use, if appropriate, a problem-methods-results-conclusion-recommendations organization? *(p. 321)*

Does the progress or status report

- ☐ clearly announce that it is a progress or status report? *(p. 323)*
- ☐ use an appropriate organization? *(p. 323)*
- ☐ clearly and honestly report on the subject and forecast the problems and possibilities of the future work? *(p. 323)*
- ☐ include, if appropriate, an appendix containing supporting materials that substantiate the discussion? *(p. 324)*

Does the incident report

- ☐ explain what happened? *(p. 333)*
- ☐ explain why it happened? *(p. 333)*
- ☐ explain what the organization did about it or will do about it? *(p. 333)*

Do the minutes

- ☐ provide the necessary logistics about the meeting? *(p. 336)*
- ☐ explain the events of the meeting accurately? *(p. 335)*
- ☐ reflect positively on the participants and the organization? *(p. 335)*

EXERCISES

For more about memos, see Ch. 9.

1. As the manager of Lewis, Lewis, and Wollensky Law, LPC, you have been informed by some clients that tattoos on the arms and necks of your employees are creating a negative impression. Write a directive in the form of a memo defining a new policy: employees are required to wear clothing that covers any tattoos on their arms and necks.
2. Write a progress report about the research project you are working on in response to Exercise 3 on page 315 in Chapter 11. If the proposal was a collaborative effort, collaborate with the same group members on the progress report.

CASE 12: Writing a Directive

You work for a company that is implementing a new waste-reduction initiative, with the goal of saving money and benefiting the environment. Your supervisor has given you an outline of the new policy and asked you to draft a directive that she will distribute to department heads. The directive should explain the purpose and importance of the policy and outline ways department heads and their employees will be expected to participate. To get started drafting the directive, go to LaunchPad.

CHAPTER 13 Writing Recommendation Reports

Rawlings
750
Ra

Date: December 14, 2018
To: Jill Bremerton, M.D.
Chief Executive Officer

CHAPTER 12 DISCUSSED informational reports: those in which the writer's main purpose is to present information. This chapter discusses recommendation reports. A *recommendation report* also presents information but goes one step further by offering suggestions about what the readers ought to do next.

Here are examples of the kinds of questions a recommendation report might address:

- **What should we do about Problem X?** What should we do about the increased cost of copper, which we use in manufacturing our line of electronic components?
- **Should we do Function X?** Although we cannot afford to pay tuition for all college courses our employees wish to take, can we reimburse them for classes directly related to their work?
- **Should we use Technology A or Technology B to do Function X?** Should we continue to supply our employees with laptops, or should we switch to tablets?
- **We currently use Method A to do Function X. Should we be using Method B?** We sort our bar-coded mail by hand; should we buy an automatic sorter?

Each of these questions can lead to a wide variety of recommendations, ranging from "do nothing" to "study this some more" to "take the following actions immediately."

Understanding the Role of Recommendation Reports

A recommendation report can be the final link in a chain of documents that begins with a proposal and continues with one or more progress reports. This last, formal report is often called a *final report*, a *project report*, a *recommendation report*, a *completion report*, or simply a *report*. The sample report beginning on page 357 is the recommendation report continuing the series about tablet computers at Rawlings Regional Medical Center presented in Chapters 11 and 12.

For more about proposals and progress reports, see Ch. 11 and Ch. 12.

A recommendation report can also be a freestanding document, one that was not preceded by a proposal or by progress reports. For instance, you might be asked for a recommendation on whether your company should offer employees comp time (compensating those who work overtime with time off) instead of overtime pay. This task would call for you to research the subject and write a single recommendation report.

Most recommendation reports discuss questions of feasibility. *Feasibility* is a measure of the practicality of a course of action. For instance, a company might conduct a *feasibility study* of whether it should acquire a competing company. In this case, the two courses of action are to acquire the competing company or not to acquire it. Or a company might do a study to determine which make and model of truck to buy for its fleet.

A feasibility report is a report that answers three kinds of questions:

- **Questions of possibility.** We would like to build a new rail line to link our warehouse and our retail outlet, but if we cannot raise the money, the project is not possible. Even if we can find the money, do we have government authorization? If we do, are the soil conditions appropriate for the rail link?
- **Questions of economic wisdom.** Even if we can afford to build the rail link, should we do so? If we use all our resources on this project, what other projects will have to be postponed or canceled? Is there a less expensive or a less financially risky way to achieve the same goals?
- **Questions of perception.** Because our company's workers have recently accepted a temporary wage freeze, they might view the rail link as an inappropriate use of funds. The truckers' union might see it as a threat to truckers' job security. Some members of the public might also be interested parties, because any large-scale construction might affect the environment.

Using a Problem-Solving Model for Preparing Recommendation Reports

The writing process for a recommendation report is similar to that for any other technical communication:

- **Planning.** Analyze your audience, determine your purpose, and visualize the deliverable: the report you will submit. Conduct appropriate secondary and primary research.

- **Drafting.** Write a draft of the report. Large projects often call for many writers and therefore benefit from shared document spaces and wikis.
- **Revising.** Think again about your audience and purpose, and then make appropriate changes to your draft.
- **Editing.** Improve the writing in the report, starting with the largest issues of development and emphasis and working down to the sections, paragraphs, sentences, and individual words.
- **Proofreading.** Go through the draft slowly, making sure you have written what you wanted to write. Get help from others.

For more about collaboration, see Ch. 3.

In addition to this model of the writing process, you need a problem-solving model for conducting the analysis that will enable you to write the recommendation report. The following discussion explains in more detail the problem-solving model shown in Figure 13.1.

FIGURE 13.1
A Problem-Solving Model for Recommendation Reports

As you work through this process, you might find that you need to go back to a previous step—or even to the first step—as you think more about your subject, audience, and purpose.

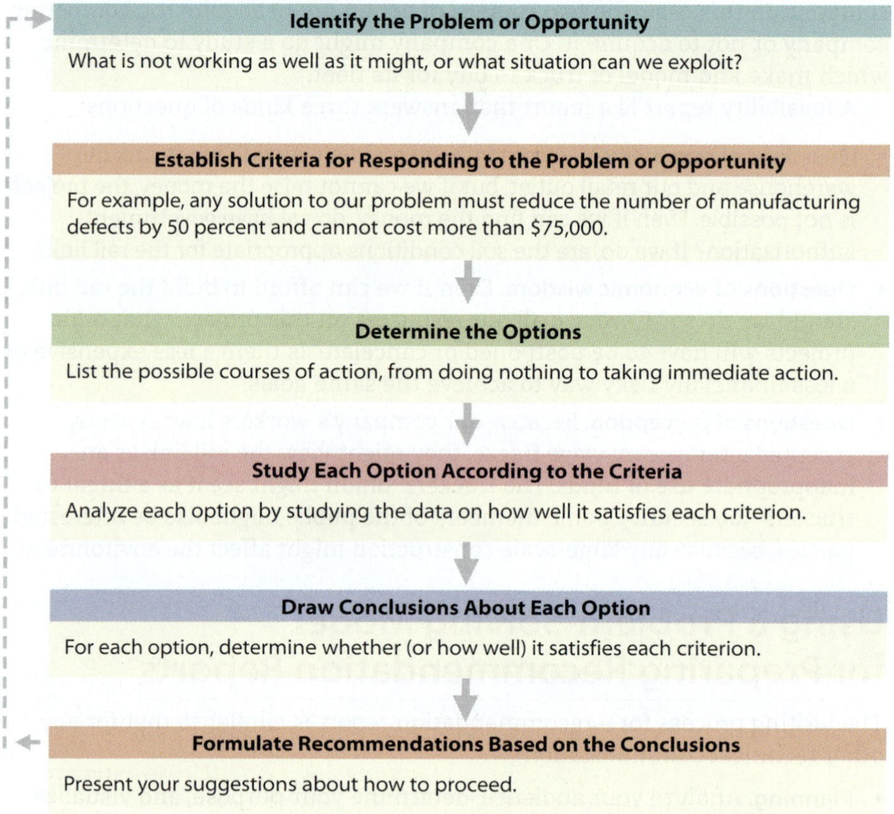

Identify the Problem or Opportunity
What is not working as well as it might, or what situation can we exploit?

Establish Criteria for Responding to the Problem or Opportunity
For example, any solution to our problem must reduce the number of manufacturing defects by 50 percent and cannot cost more than $75,000.

Determine the Options
List the possible courses of action, from doing nothing to taking immediate action.

Study Each Option According to the Criteria
Analyze each option by studying the data on how well it satisfies each criterion.

Draw Conclusions About Each Option
For each option, determine whether (or how well) it satisfies each criterion.

Formulate Recommendations Based on the Conclusions
Present your suggestions about how to proceed.

IDENTIFY THE PROBLEM OR OPPORTUNITY

What is not working or is not working as well as it might? What situation presents an opportunity to decrease costs or improve the quality of a product or service? Without a clear statement of your problem or opportunity, you cannot plan your research.

For example, your company has found that employees who smoke are absent and ill more often than those who don't smoke. Your supervisor has asked you to investigate whether the company should offer a free smoking-cessation program. The company can offer the program only if the company's insurance carrier will pay for it. The first thing you need to do is talk with the insurance agent; if the insurance carrier will pay for the program, you can proceed with your investigation. If the agent says no, you have to determine whether another insurance carrier offers better coverage or whether there is some other way to encourage employees to stop smoking.

ESTABLISH CRITERIA FOR RESPONDING TO THE PROBLEM OR OPPORTUNITY

Criteria are standards against which you measure your options. Criteria can be classified into two categories: *necessary* and *desirable*. For example, if you want to buy a photocopier, necessary criteria might be that each copy cost less than two cents to produce and that the photocopier be able to handle oversized documents. If the photocopier doesn't fulfill those two criteria, you will not consider it further. By contrast, desirable criteria might include that the photocopier be capable of double-sided copying and stapling. Desirable criteria let you make distinctions among a variety of similar objects, objectives, actions, or effects. If a photocopier does not fulfill a desirable criterion, you will still consider it, although it will be less attractive.

Until you establish your criteria, you don't know what your options are. Sometimes you are given your criteria: your supervisor tells you how much money you can spend, for instance, and that figure becomes one of your necessary criteria. Other times, you derive your criteria from your research.

DETERMINE THE OPTIONS

After you establish your criteria, you determine your options. *Options* are potential courses of action you can take in responding to a problem or opportunity. Determining your options might be simple or complicated.

Sometimes your options are presented to you. For instance, your supervisor asks you to study two vendors for accounting services and recommend one of them. The options are Vendor A or Vendor B. That's simple.

In other cases, you have to consider a series of options. For example, your department's photocopier is old and breaking down. Your first decision is

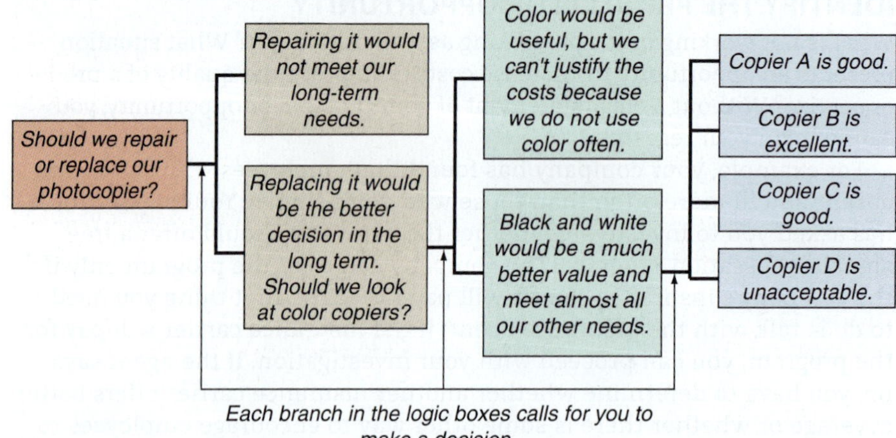

FIGURE 13.2 **Using Logic Boxes To Plot a Series of Options**

whether to repair it or replace it. Once you have answered that question, you might have to make more decisions. If you are going to replace it, what features should you look for in a new one? Each time you make a decision, you have to answer more questions until, eventually, you arrive at a recommendation. For a complicated scenario like this, you might find it helpful to use logic boxes or flowcharts to sketch the logic of your options, as shown in Figure 13.2.

As you research your topic, your understanding of your options will likely change. At this point, however, it is useful to understand the basic logic of your options or series of options.

STUDY EACH OPTION ACCORDING TO THE CRITERIA

For more about research techniques, see Ch. 5.

Once you have identified your options (or series of options), study each one according to the criteria. For the photocopier project, secondary research would include studying articles about photocopiers in technical journals and specification sheets from the different manufacturers. Primary research might include observing product demonstrations as well as interviewing representatives from different manufacturers and managers who have purchased different brands.

To make the analysis of the options as objective as possible, professionals sometimes create a *decision matrix*, a tool for evaluating each option according to each criterion. A decision matrix is a table (or a spreadsheet), as shown in Figure 13.3. Here the writer is nearly at the end of his series of options: he is evaluating three similar photocopiers according to three criteria. Each criterion has its own weight, which suggests how important it is. The greater the weight, the more important the criterion.

Criteria and Weight		Options					
		Ricoh		Xerox		Sharp	
Criterion	Weight	Rating	Score[1]	Rating	Score[1]	Rating	Score[1]
Pages/min.	1	9	9	6	6	3	3
Duplex	3	1	3	3	9	10	30
Color	4	10	40	1	4	10	40
Total Score			52		19		73

[1]Score = weight × rating.

FIGURE 13.3 A Decision Matrix
Spreadsheet programs often contain templates for creating decision matrices.

As shown in Figure 13.3, the criterion of pages per minute is relatively unimportant: it receives a weight of 1. For this reason, the Ricoh copier, even though it receives a high rating for pages per minute (9), receives only a modest score of 9 (1 × 9 = 9) on this criterion. However, the criterion of color copying is quite important, with a weight of 4. On this criterion, the Ricoh, with its rating of 10, achieves a very high score (4 × 10 = 40).

But a decision matrix cannot stand on its own. You need to explain your methods. That is, in the discussion or in footnotes to the matrix, you need to explain the following three decisions:

- **Why you chose each criterion—or didn't choose a criterion the reader might have expected to see included.** For instance, why did you choose duplexing (double-sided printing) but not image scanning?
- **Why you assigned a particular weight to each criterion.** For example, why is the copier's ability to make color copies four times more important than its speed?
- **Why you assigned a particular rating to each option.** For example, why does one copier receive a rating of only 1 on duplexing, whereas another receives a 3 and a third receives a 10?

A decision matrix is helpful only if your readers understand your methods and agree with the logic you used in choosing the criteria and assigning the weight and ratings for each option.

Although a decision matrix has its limitations, it is useful for both you and your readers. For you as the writer, the main advantage is that it helps you do a methodical analysis. For your readers, it makes your analysis easier to follow because it clearly presents your methods and results.

DRAW CONCLUSIONS ABOUT EACH OPTION

Whether you use a decision matrix or a less-formal means of recording your evaluations, the next step is to draw conclusions about the options you studied—by interpreting your results and writing evaluative statements about the options.

For the study of photocopiers, your conclusion might be that the Sharp model is the best copier: it meets all your necessary criteria and the greatest number of desirable criteria, or it scores highest on your matrix. Depending on your readers' preferences, you can present your conclusions in any one of three ways.

- **Rank all the options:** the Sharp copier is the best option, the Ricoh copier is second best, and so forth.
- **Classify all the options in one of two categories:** acceptable and unacceptable.
- **Present a compound conclusion:** the Sharp offers the most technical capabilities; the Ricoh is the best value.

FORMULATE RECOMMENDATIONS BASED ON THE CONCLUSIONS

If you conclude that Option A is better than Option B—and you see no obvious problems with Option A—recommend Option A. But if the problem has changed or your company's priorities or resources have changed, you might decide to recommend a course of action that is inconsistent with the conclusions you derived. Your responsibility is to use your judgment and recommend the best course of action.

ETHICS NOTE

PRESENTING HONEST RECOMMENDATIONS

As you formulate your recommendations, you might know what your readers want you to say. For example, they might want you to recommend the cheapest option, or one that uses a certain kind of technology, or one that is supplied by a certain vendor. Naturally, you want to be able to recommend what they want, but sometimes the facts won't let you. Your responsibility is to tell the truth—to do the research honestly and competently and then present the findings honestly. Your name goes on the report. You want to be able to defend your recommendations based on the evidence and your reasoning.

One worrisome situation that arises frequently is that none of the options would be a complete success or none would work at all. What should you do? You should tell the truth about the options, warts and all. Give the best advice you can, even if that advice is to do nothing.

Writing Recommendation Reports

The following discussion presents a basic structure for a recommendation report. Remember that every document you write should reflect its audience, purpose, and subject. Therefore, you might need to modify, add to, or delete some of the elements discussed here.

For more on collaboration, see Ch. 3.

Reports that are lengthy and complex are often written collaboratively. As you begin the project that will culminate in the report, consider whether it would make sense to set up a shared writing space, a wiki, or some other method for you and your team members to write and edit the report collaboratively.

The easiest way to draft a report is to think of it as consisting of three sections: the front matter, the body, and the back matter. Table 13.1 shows the purposes of and typical elements in these three sections.

You will probably draft the body before the front and the back matter. This sequence is easiest because you think through what you want to say in the body and then draft the front and back matter based on it.

If you are writing your recommendation report for readers from other cultures, keep in mind that conventions differ from one culture to another. In the United States, reports are commonly organized from general to specific. That is, the most general information (the abstract and the executive summary) appears early in the report. In many cultures, however, reports are organized from specific to general. Detailed discussions of methods and results precede discussions of the important findings.

Similarly, elements of the front and back matter are rooted in culture. For instance, in some cultures—or in some organizations—writers do not create executive summaries, or their executive summaries differ in length or organization from those discussed here. According to interface designer Pia Honold (1999), German users of high-tech products rely on the table of contents in a manual because they like to understand the scope and organization of the manual. Therefore, writers of manuals for German readers should include comprehensive, detailed tables of contents.

TABLE 13.1 Elements of a Typical Report

SECTION OF THE REPORT	PURPOSES OF THE SECTION	TYPICAL ELEMENTS IN THE SECTION
Front matter	• to orient the reader to the subject • to provide summaries for technical and managerial readers • to help readers navigate the report • to help readers decide whether to read the document	• letter of transmittal (p. 348) • cover (p. 348) • title page (p. 349) • abstract (p. 349) • table of contents (p. 350) • list of illustrations (p. 351) • executive summary (p. 351)
Body	• to provide the most comprehensive account of the project, from the problem or opportunity that motivated it to the methods and the most important findings	• introduction (p. 346) • methods (p. 347) • results (p. 347) • conclusions (p. 347) • recommendations (p. 347)
Back matter	• to present supplementary information, such as more-detailed explanations than are provided in the body • to enable readers to consult the secondary sources the writers used	• glossary (p. 354) • list of symbols (p. 354) • references (p. 356) • appendixes (p. 356)

Study samples of writing produced by people from the culture you are addressing to see how they organize their reports and use front and back matter.

WRITING THE BODY OF THE REPORT

The elements that make up the body of a report are discussed here in the order in which they usually appear in a report. However, you should draft the elements in whatever order you prefer. The sample recommendation report on pages 357–80 includes these elements.

Introduction The introduction helps readers understand the technical discussion that follows. Start by analyzing who your readers are. Then consider these questions:

- **What is the subject of the report?** If the report follows a proposal and a progress report, you can probably copy this information from one of those documents, modifying it as necessary. Reusing this information is efficient and ethical.
- **What is the purpose of the report?** The purpose of the report is not the purpose of the project. The purpose of the report is to explain a project from beginning (identifying a problem or an opportunity) to end (presenting recommendations).

For more about purpose statements, see Ch. 4, p. 50.

- **What is the background of the report?** Include this information, even if you have presented it before; some of your readers might not have read your previous documents or might have forgotten them.
- **What are your sources of information?** Briefly describe your primary and secondary research, to prepare your readers for a more detailed discussion of your sources in subsequent sections of the report.
- **What is the scope of the report?** Indicate the topics you are including, as well as those you are not.
- **What are the most significant findings?** Summarize the most significant findings of the project.
- **What are your recommendations?** In a short report containing a few simple recommendations, include those recommendations in the introduction. In a lengthy report containing many complex recommendations, briefly summarize them in the introduction, then refer readers to the more detailed discussion in the recommendations section.
- **What is the organization of the report?** Indicate your organizational pattern so that readers can understand where you are going and why.
- **What key terms are you using in the report?** The introduction is an appropriate place to define new terms. If you need to define many terms, place the definitions in a glossary and refer readers to it in the introduction.

Methods The methods section answers the question "What did you do?" In drafting the methods section, consider your readers' knowledge of the field, their perception of you, and the uniqueness of the project, as well as their reasons for reading the report and their attitudes toward the project. Provide enough information to enable readers to understand what you did and why you did it that way. If others will be using the report to duplicate your methods, include sufficient detail.

Results Whereas the methods section answers the question "What did you do?" the results section answers the question "What did you see or determine?"

Results are the data you discovered or compiled. Present the results objectively, without comment. Save the interpretation of the results—your conclusions—for later. If you combine results and conclusions, your readers might be unable to follow your reasoning and might not be able to tell whether the evidence justifies your conclusions.

Your audience's needs will help you decide how to structure the results. How much they know about the subject, what they plan to do with the report, what they expect your recommendation(s) to be—these and many other factors will affect how you present the results. For instance, suppose that your company is considering installing a VoIP phone system that will enable employees to make telephone calls over the Internet, and you conducted the research on the available systems. In the introduction, you explain the disadvantages of the company's current phone system. In the methods section, you describe how you established the criteria you applied to the available phone systems, as well as your research procedures. In the results section, you provide the details of each phone system you are considering, as well as the results of your evaluation of each system.

Conclusions Conclusions answer the question "What does it mean?" They are the implications of the results. To draw conclusions, you need to think carefully about your results, weighing whether they point clearly to a single meaning.

For more about evaluating evidence and drawing conclusions, see Ch. 5, pp. 86–88.

Recommendations Recommendations answer the question "What should we do?" As discussed earlier in this chapter, recommendations do not always flow directly from conclusions. Always consider recommending that the organization take no action or no action at this time.

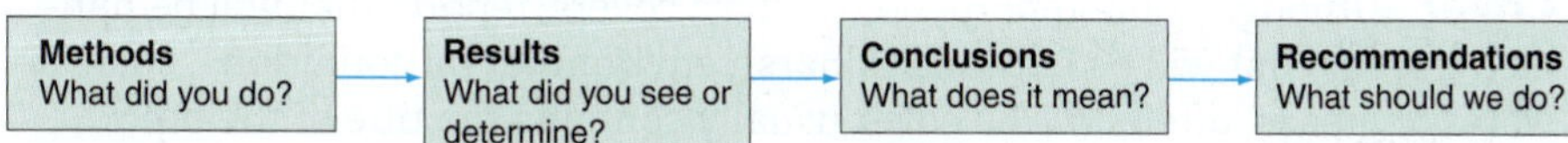

GUIDELINES Writing Recommendations

As you draft your recommendations, consider the following four factors:

- **Content.** Be clear and specific. If the project has been unsuccessful, don't simply recommend that your readers "try some other alternatives." What alternatives do you recommend and why?
- **Tone.** When you recommend a new course of action, be careful not to offend whoever formulated the earlier course. Do not write that following your recommendations will "correct the mistakes" that have been made. Instead, your recommendations should "offer great promise for success." A restrained, understated tone is more persuasive because it shows that you are interested only in the good of your company, not personal rivalries.
- **Form.** If the report leads to only one recommendation, use traditional paragraphs. If the report leads to more than one recommendation, consider a numbered list.
- **Location.** Consider including a summary of the recommendations—or, if they are brief, the full list—after the executive summary or in the introduction as well as at the end of the body of the report.

WRITING THE FRONT MATTER

Front matter is common in reports, proposals, and manuals. As indicated in Table 13.1 on page 345, front matter helps readers understand the whole report and find the information they seek. Most organizations have established formats for front matter. Study the style guide used in your company or, if there isn't one, examples from the files to see how other writers have assembled their reports.

For more about formatting a letter, see Ch. 9, p. 240.

Letter of Transmittal In the letter of transmittal, which can take the form of a letter or a memo, the writer introduces the primary reader to the purpose and content of the report. In addition, the writer often states who authorized or commissioned the report and acknowledges any assistance he or she received in carrying out the project. The letter of transmittal is attached to the report, bound in with it, or simply placed on top of it. Even though the letter likely contains little information that is not included elsewhere in the report, it is important because it is the first thing the reader sees. It establishes a courteous and professional tone. Letters of transmittal are customary even when the writer and the reader both work for the same organization. See the sample recommendation report (p. 357) for an example of a transmittal letter in the form of a memo.

Cover Although some reports do not have covers, reports that will be handled a lot or that will be exposed to harsh environmental conditions, such as water or grease, often do. The cover usually contains the title of the report, the name and position of the writer, the date of submission, and the name or logo of the writer's company. Sometimes the cover also includes a security notice or a statement of proprietary information.

Title Page A title page includes at least the title of the report, the name of the writer, and the date of submission. A more complex title page might also include a project number, a list of additional personnel who contributed to the report, and a distribution list. See the sample recommendation report (p. 359) for an example of a title page.

Abstract An abstract is a brief technical summary of the report, usually no more than 200 words. It addresses readers who are familiar with the technical subject and who need to decide whether they want to read the full report. In an abstract, you can use technical terminology and refer to advanced concepts in the field. Abstracts are sometimes published by abstract services, which are useful resources for researchers.

For more about abstract services, see Ch. 5, p. 82.

Abstracts often contain a list of half a dozen or so keywords, which are entered into electronic databases. As the writer, one of your tasks is to think of the various keywords that will lead people to the information in your report.

There are two types of abstracts: descriptive and informative. A *descriptive abstract*—sometimes called a *topical, indicative,* or *table-of-contents abstract*—describes the kinds of information contained in the report. It does not provide the major findings (important results, conclusions, or recommendations). It simply lists the topics covered, giving equal emphasis to each. Figure 13.4 is a descriptive abstract from a report by a utility company about its pilot program for measuring how much electricity its customers are using. A descriptive abstract is used most often when space is at a premium. Some government proposals, for example, call for a descriptive abstract to be placed at the bottom of the title page.

Abstract
"Design of a Radio-Based System for Distribution Automation"
by Brian D. Raven
A new survey by the Maryland Public Service Commission suggests that utilities have not effectively explained to consumers the benefits of smart meters. The two-year study of 86,000 consumers concludes that the long-term benefits of smart meters will not be realized until consumers understand the benefits of shifting some of their power usage to off-peak hours in response to the data they receive from their meters. The study presents recommendations for utilities and municipal governments to improve customer understanding of how to use the smart meters effectively.

Keywords: smart meters, distribution systems, load, customer attitudes, power consumption, utilities, Maryland Public Utilities Commission

This abstract is descriptive rather than informative because it does not present any of the major data from the survey or present the recommendations that are mentioned in the final sentence.

FIGURE 13.4 Descriptive Abstract

An *informative abstract* presents the major findings. If you don't know which kind of abstract the reader wants, write an informative one.

The distinction between descriptive and informative abstracts is not clear-cut. Sometimes you might have to combine elements of both in a single abstract. For instance, if there are 15 recommendations—far too many to list—you might simply note that the report includes numerous recommendations.

See the sample recommendation report (p. 360) for an example of an informative abstract.

Table of Contents The table of contents, the most important guide to navigating the report, has two main functions: to help readers find the information they want and to help them understand the scope and organization of the report.

A table of contents uses the same headings as the report itself. Therefore, to create an effective table of contents, you must first make sure that the headings are clear and that you have provided enough of them. If the table of contents shows no entry for five or six pages, you probably need to partition that section of the report into additional subsections. In fact, some tables of contents have one entry, or even several, for every report page.

The following table of contents, which relies exclusively on generic headings (those that describe an entire class of items), is too general to be useful.

Table of Contents

Introduction..1
Materials..3
Methods...4
Results...19
Recommendations...............................23
References..26
Appendixes.......................................28

This methods section, which goes from page 4 to page 18, should have subentries to break up the text and to help readers find the information they seek.

For more-informative headings, combine the generic and the specific:

Recommendations: Five Ways To Improve Information-Retrieval Materials Used in the Calcification Study

Results of the Commuting-Time Analysis

Then build more subheadings into the report itself. For instance, for the "Recommendations" example above, you could create a subheading for each of the five recommendations. Once you have established a clear system of headings within the report, use the same text attributes—capitalization, boldface, italics, and outline style (traditional or decimal)—in the table of contents.

For more about text attributes, see Ch. 7, p. 165.

When adding page numbers to your report, remember two points:

- The table of contents page does not contain an entry for itself.
- Front matter is numbered using lowercase Roman numerals (i, ii, and so forth), often centered at the bottom of the page. The title page of a report is not numbered, although it represents page i. The abstract is usually numbered page ii. The table of contents is usually not numbered, although it represents page iii. The body of the report is numbered with Arabic numerals (1, 2, and so on), typically in the upper outside corner of the page.

See the sample recommendation report (p. 361) for an example of a table of contents.

List of Illustrations A *list of illustrations* is a table of contents for the figures and tables. List the figures first, then the tables. (If the report contains only figures, call it a *list of figures*. If it contains only tables, call it a *list of tables*.) You may begin the list of illustrations on the same page as the table of contents, or you may begin the list on a separate page and include it in the table of contents. Figure 13.5 shows a list of illustrations.

Executive Summary The executive summary (sometimes called the *epitome*, *executive overview*, *management summary*, or *management overview*) is a brief condensation of the report addressed to managers. Most managers need only a broad understanding of the projects that an organization undertakes and how they fit together into a coherent whole.

LIST OF ILLUSTRATIONS

Figures

Tables

FIGURE 13.5 List of Illustrations

TECH TIP

Why To Make a Long Report Navigable

Whether your report is being read in print or online, your readers will appreciate page navigation guides to help them find the information they need. Headers, footers, and page numbers are useful, especially in print documents and PDFs, to help readers know where they are in a document. A table of contents can direct or link readers to the right location as well as provide them with a sense of the scope and organization of the report.

How To Make a Long Report Navigable

With programs such as Microsoft Word or Google Docs, you can easily format a report to include page navigation tools. To insert headers, footers, and page numbers, choose the function you want from the **Insert** tab. Both Word and Google Docs allow you to set up the header differently on the first page of the document. Word offers additional design options, allowing you to insert **section breaks** (in the **Layout** tab) so that you can have different headers and footers in different sections of a long report.

Microsoft Word **Google Docs**

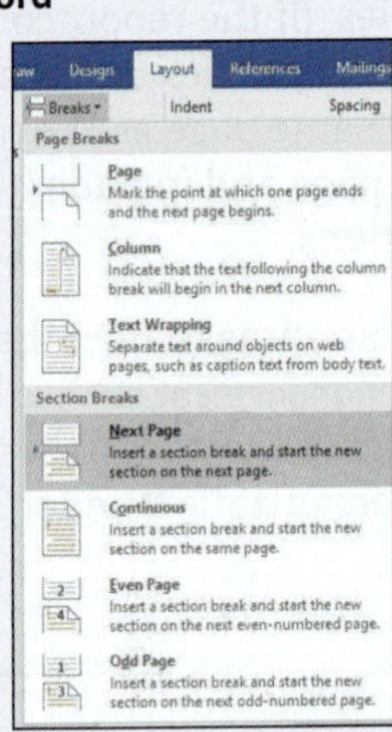

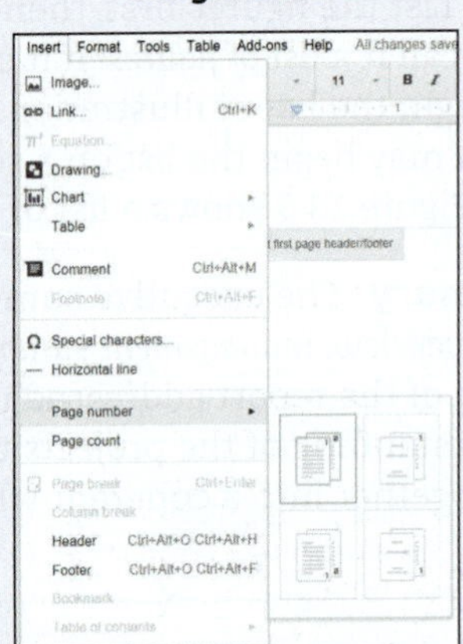

To create a preformatted table of contents, use Word's **Table of Contents** drop-down menu in the **References** tab; in Google Docs, choose the **Table of Contents** feature from the **Insert** tab. For online documents, you can choose to use hyperlinks instead of page numbers in your table of contents. As with headers, footers, and page numbers, Word offers additional customization tools for the table of contents.

Microsoft Word **Google Docs**

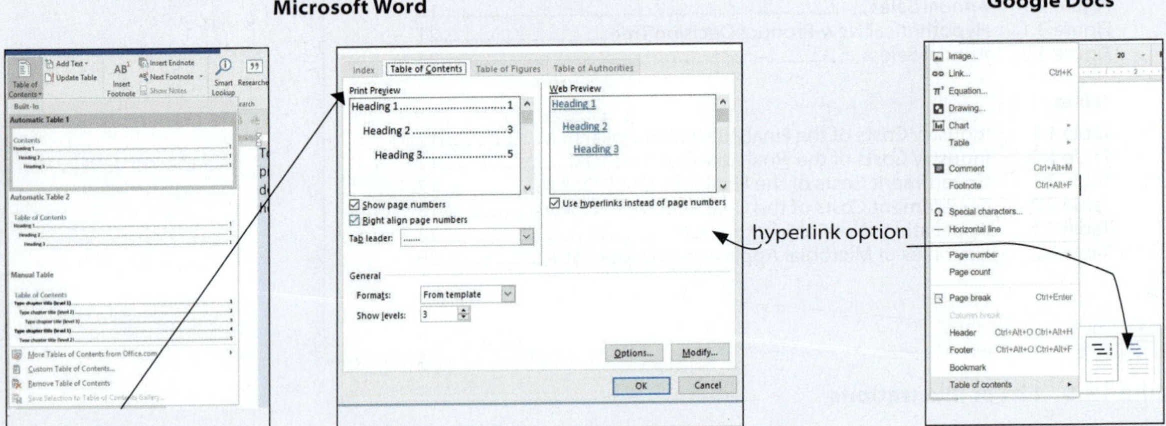

An executive summary for a report of under 20 pages is typically one page (double-spaced). For longer reports, the maximum length is often calculated as a percentage of the report, such as 5 percent.

The executive summary presents information to managers in two parts:

- **Background.** This section explains the problem or opportunity: what was not working or was not working effectively or efficiently, or what potential modification of a procedure or product had to be analyzed.
- **Major findings and implications.** This section might include a brief description—only one or two sentences—of the methods, followed by a full paragraph about the conclusions and recommendations.

An executive summary differs from an informative abstract. Whereas an abstract focuses on the technical subject (such as whether the public is taking advantage of the data from smart electric meters), an executive summary concentrates on the managerial implications of the subject for a particular company (such as whether PECO, the Philadelphia utility company, should carry out a public-information campaign to educate customers about how to use their smart meters).

GUIDELINES Writing an Executive Summary

Follow these five suggestions in writing executive summaries.

- **Use specific evidence in describing the background.** For most managers, the best evidence is in the form of costs and savings. Instead of writing that the equipment you are now using to cut metal foil is ineffective, write that the equipment jams once every 72 hours on average, costing $400 in materials and $2,000 in productivity each time. Then add up these figures for a monthly or an annual total.
- **Be specific in describing research.** For instance, research suggests that a computerized energy-management system could cut your company's energy costs by 20 to 25 percent. If the company's energy costs last year were $300,000, it could save $60,000 to $75,000.
- **Describe the methods briefly.** If you think your readers are interested, include a brief description of your methods—no more than a sentence or two.
- **Describe the findings according to your readers' needs.** If your readers want to know your results, provide them. If your readers are unable to understand the technical data or are uninterested, go directly to the conclusions and recommendations.
- **Ask an outside reader to review your draft.** Give the summary to someone who has no connection to the project. That person should be able to read your summary and understand what the project means to the organization.

See the sample recommendation report (p. 362) for an example of an executive summary.

Glossary

Applicant: A state agency, local government, or eligible private nonprofit organization that submits a request to the grantee for disaster assistance under the state's grant.

Case Management: A systems approach to providing equitable and fast service to applicants for disaster assistance. Organized around the needs of the applicant, the system consists of a single point of coordination, a team of on-site specialists, and a centralized, automated filing system.

Cost Estimating Format (CEF): A method for estimating the total cost of repair for large, permanent projects by use of construction industry standards. The format uses a base cost estimate and design and construction contingency factors, applied as a percentage of the base cost.

Declaration: The President's decision that a major disaster qualifies for federal assistance under the Stafford Act.

Hazard Mitigation: Any cost-effective measure that will reduce the potential for damage to a facility from a disaster event.

FIGURE 13.6 Glossary

WRITING THE BACK MATTER

The back matter of a recommendation report might include the following items: glossary, list of symbols, references, and appendixes.

Glossary and List of Symbols A *glossary*, an alphabetical list of definitions, is particularly useful if some of your readers are unfamiliar with the technical vocabulary in your report. Instead of slowing down your discussion by defining technical terms as they appear, you can use boldface, or some similar method of highlighting words, to indicate that the term is defined in the glossary. The first time a boldfaced term appears, explain this system in a footnote. For example, the body of the report might say, "Thus the positron* acts as the . . . ," while a note at the bottom of the page explains:

> *This and all subsequent terms in boldface are defined in the Glossary, page 26.

Although a glossary is usually placed near the end of the report, before the appendixes, it can also be placed immediately after the table of contents if the glossary is brief (less than a page) and if it defines essential terms. Figure 13.6 shows an excerpt from a glossary.

A list of symbols is formatted like a glossary, but it defines symbols and abbreviations rather than terms. It, too, may be placed before the appendixes or after the table of contents. Figure 13.7 shows a list of symbols.

List of Symbols

β	beta
CRT	cathode-ray tube
γ	gamma
Hz	hertz
rcvr	receiver
SNR	signal-to-noise ratio
uhf	ultra-high frequency
vhf	very high frequency

FIGURE 13.7 List of Symbols

DOCUMENT ANALYSIS ACTIVITY

Analyzing an Executive Summary

Executive Summary

On May 11, we received approval to study whether Android smartphones could help our 20 engineers send and receive email, monitor their schedules, take notes, and access reference sources they need in the field. In our study, we addressed these problems experienced by many of our engineers:

- They have missed deadlines and meetings and lost client information.
- They have been unable to access important files and reference materials from the field.
- They have complained about the weight—sometimes more than 40 pounds—of the binders and other materials that they have to carry.
- They have to spend time keyboarding notes that they take in the field.

In 2016, missed meetings and other schedule problems cost the company over $400,000 in lost business. And our insurance carrier settled a claim for $50,000 from an engineer who experienced back and shoulder problems due to the weight of his pack.

We researched the capabilities of Android smartphones, then established these criteria for our analysis:

- The device must weigh less than 5 ounces.
- It must run on Android 8.0 or higher.
- It must have at least 2GB RAM.
- It must have at least a 1.9-GHz Quad Core.
- It must have a microSD slot.
- It must have at least a 4-inch screen.
- It must have a camera with a resolution of 10.0MP or better.
- It must have on-device encryption.
- It must be Microsoft Office compatible.
- It must cost $700 or less.

On the basis of our analysis, we recommend that the company purchase 20 Samsung Galaxy 4 smartphones, for a total cost of $12,980. These devices best meet all our technical and cost criteria.

This executive summary comes from a corporate report on purchasing Android smartphones for employees. The questions below ask you to think about the discussion of executive summaries (beginning on p. 351).

1. How clearly do the writers explain the background? Identify the problem or opportunity they describe in this executive summary.
2. Do the writers discuss the methods? If so, identify the discussion.
3. Identify the findings: the results, conclusions, and recommendations. How clearly have the writers explained the benefits to the company?

DOCUMENT ANALYSIS ACTIVITY

To analyze a recommendation report presented as a video, go to LaunchPad.

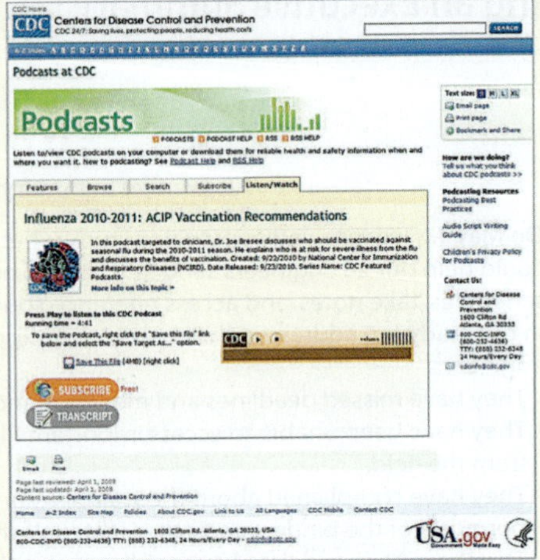

Check Your Steps! Make Every Injection Safe
Information from Centers for Disease Control and Prevention, 2014: www.oneandonlycampaign.org/content/audio-video.

References Many reports contain a list of references (sometimes called a *bibliography* or *list of works cited*) as part of the back matter. References and the accompanying textual citations throughout the report are called *documentation*. Documentation acknowledges your debt to your sources, establishes your credibility as a writer, and helps readers locate and review your sources. See Appendix, Part A, for a detailed discussion of documentation. See the sample recommendation report (p. 379) for an example of a reference list.

Appendixes An *appendix* is any section that follows the body of the report (and the glossary, list of symbols, or reference list). Appendixes (or *appendices*) convey information that is too bulky for the body of the report or that will interest only a few readers. Appendixes might include maps, large technical diagrams or charts, computations, computer printouts, test data, and texts of supporting documents.

Appendixes, usually labeled with letters rather than numbers (Appendix A, Appendix B, and so on), are listed in the table of contents and are referred to at appropriate points in the body of the report. Therefore, they are accessible to any reader who wants to consult them. See the sample recommendation report (p. 380) for an example of an appendix.

Sample Recommendation Report

The following example is the recommendation report on the tablet project proposed in Chapter 11 on page 308. The progress report for this project appears in Chapter 12 on page 325.

Rawlings Regional Medical Center
7500 Bannock Avenue
Rawlings, MT 59211

Date: December 14, 2018
To: Jill Bremerton, M.D.
Chief Executive Officer
Rawlings Regional Medical Center
From: Jeremy Elkins, Director of Information Technology
Eloise Carruthers, Director of Nursing
Rawlings Regional Medical Center
Subject: Recommendation Report for the Tablet Study at RRMC

Attached is the report for our study, "Selecting a Tablet Computer for the Clinical Staff at Rawlings Regional Medical Center: A Recommendation Report." We completed the tasks described in our proposal of October 6, 2018: familiarizing ourselves with tablet use by clinical staff in hospitals across the country, assessing RRMC clinical staff's knowledge of and attitudes toward tablet use, studying different models for administering tablet use, determining the criteria by which we might evaluate tablets, and performing the evaluations.

To carry out these tasks, we performed secondary and primary research. We studied the literature on tablet use, distributed a questionnaire to RRMC clinical staff who own tablets, and interviewed Dr. Bremerton. Then, we collected and analyzed our data and wrote the report.

Our main findings are that the clinical staff who already own tablets are very receptive to the idea of using tablets in a clinical setting and slightly prefer having the hospital supply the tablets. We, too, think the hospital-supplied model is preferable to the bring-your-own-device (BYOD) model. Although the best tablets for our needs would be those designed and built for health-care applications, those are too expensive for our budget. Because reports on the technical characteristics of computer products are notoriously unreliable, we cannot be sure whether the many

Transmittal "letters" can be presented as memos.

The writers include their titles and that of their primary reader. This way, future readers will be able to readily identify the reader and writers.

The subject heading indicates the subject of the report (the tablet study at RRMC) and the purpose of the report (recommendation report).

The purpose of the study. Notice that the writers link the recommendation report to the proposal, giving them an opportunity to state the main tasks they carried out in the study.

The methods the writers used to carry out the research.

The principal findings: the results and conclusions of the study. Notice that the writers state that they cannot be sure whether the technical information they have found is accurate. Is it okay to state that you are unsure about something? Yes, as long as you then propose a way to become sure about it.

Letter to Jill Bremerton, M.D.
December 14, 2018
page 2

general-purpose tablets can meet our standards for ease of disinfection or durability, and we are not sure whether they have sufficient battery life.

We recommend one of two courses of action: reconsidering the cost criterion or testing a representative sample of general-purpose tablets for disinfection and the other technical characteristics and letting the clinical staff try them out.

We appreciate the trust you have shown in inviting us to participate in this phase of the feasibility study, and we would look forward to working with you on any follow-up activities. If you have any questions or comments, please contact Jeremy Elkins, at jelkins@rrmc.org or at 444-3967, or Eloise Carruthers, at ecarruthers@rrmc.org or at 444-3982.

The major recommendation. The writers ask their supervisor if she will reconsider whether the hospital can afford tablets specifically designed for health-care environments. That's not insubordination. Just be polite about it.

A polite offer to participate further or to provide more information.

Selecting a Tablet Computer for the Clinical Staff at Rawlings Regional Medical Center: A Recommendation Report

A good title indicates the subject and purpose of the document. One way to indicate the purpose is to use a generic term—such as *analysis, recommendation, summary,* or *instructions*—in a phrase following a colon. For more about titles, see Ch. 6.

Prepared for: Jill Bremerton, M.D.
Chief Executive Officer
Rawlings Regional Medical Center

Prepared by: Jeremy Elkins, Director of Information Technology
Eloise Carruthers, Director of Nursing

The names and positions of the principal reader and the writers of the document.

December 14, 2018

The date the document was submitted.

Rawlings Regional Medical Center
7500 Bannock Avenue
Rawlings, MT 59211

The name or logo of the writers' organization often is presented at the bottom of the title page.

In the abstract, the title of the report is often enclosed in quotation marks because the abstract might be reproduced in another context (such as in a database), in which case the report title would be the title of a separate document.

Abstracts are often formatted as a single paragraph.

The background and purpose of the report.

The methods.

The major findings.

Note that the writers provide some technical information about tablet use, clinical staff attitudes, and technical characteristics of tablets.

The major recommendations.

A keywords list ensures that the report will appear in the list of results of an electronic search on any of the terms listed.

Abstract

"Selecting a Tablet Computer for the Clinical Staff
at Rawlings Regional Medical Center:
A Recommendation Report"

Prepared by: Jeremy Elkins, Director of Information Technology
Eloise Carruthers, Director of Nursing

On October 8, 2018, Dr. Jill Bremerton, Chief Executive Officer of Rawlings Regional Medical Center (RRMC), approved a proposal by Jeremy Elkins (Director of Information Technology) and Eloise Carruthers (Director of Nursing) to carry out a feasibility study on integrating tablet computers into the RRMC clinical setting and to report their findings. The authors began by performing research to better understand how tablets are being used by clinical staff in hospitals across the country. Then, they assessed RRMC clinical staff attitudes toward tablet use, studied two models for administering use of tablets in hospitals, determined the criteria by which tablets might be evaluated, and performed the evaluations. RRMC clinical staff who already own tablets are very receptive to the idea of using tablets in a clinical setting and slightly prefer having the hospital supply the tablets. The best tablets for RRMC needs are those designed and built for health-care applications because they meet hospital standards for disinfection, are durable, and offer numerous hardware and software options, such as barcode scanners, RFID readers, speech input, and smart-card readers. Unfortunately, they are too expensive for our budget. Because there are numerous health-care apps available for not only the iPad but also the many Android tablets and Windows-based tablets, any of these that meet our other needs would be acceptable. However, we are not sure whether the many general-purpose tablets can meet our standards for ease of disinfection or ruggedness, and we are not sure whether they have sufficient battery life. We recommend that, if we cannot reconsider the cost criterion, we test a representative sample of general-purpose tablets for disinfection and the other technical characteristics that would affect their usefulness in the clinical setting.

Keywords: tablets, health care, HIPAA, disinfection, iPad, Android, Windows, rugged, durability

ii

Table of Contents

Note that the typeface and design of the headings in the table of contents reflect the typeface and design of the headings in the report itself.

In this table of contents, the two levels of headings are distinguished by type style (boldface versus italic) and indentation. (Although the color blue is used in the report to distinguish the headings from the text, it would not be effective here, where there is no text to set it off.)

The executive summary describes the project with a focus on the managerial aspects, particularly the recommendation. Note the writers' emphasis on the problem at RRMC.

Here the writers present a brief statement of the subject of their report.

The background of the feasibility study that Dr. Bremerton is funding.

A brief statement of the methods the writers used to carry out their research. Note that throughout this report the writers use the active voice ("We studied the literature . . ."). See Ch. 6, p. 134 for more on the active voice. Note, too, that the discussion of the methods is brief: most managers are less interested in the details of the methods you used than in your findings.

Findings are the important results and conclusions of a study.

Note that the writers use the word *recommend*. Using key generic terms such as *problem*, *methods*, *results*, *conclusions*, and *recommendations* helps readers understand the role that each section plays in the document.

Because the executive summary is the report element addressed most directly to management, the writers make clear why they prefer looking again at whether the hospital can afford to purchase health-care-specific tablets.

1

Executive Summary

To determine the best way to integrate tablet computers into the RRMC clinical setting, Dr. Jill Bremerton, Chief Executive Officer, asked us to study national trends, determine clinical-staff attitudes, examine management models, devise criteria for assessing tablets, and present our findings and recommendations.

Currently, RRMC has no formal policy on tablet usage by clinical staff. By default, we are following a bring-your-own-device (BYOD) approach. More than half of our clinical staff already use their personal tablets in their work. Dr. Bremerton wanted us to determine the best way to make tablets available to all our clinical staff. This charge included assessing the available tablets and recommending which tablet we should make available to our clinical staff.

To carry out this study, we completed the tasks described in our proposal of October 6, 2018: we studied the literature on tablet use; distributed a questionnaire to every member of the RRMC clinical staff, requesting responses from those who own tablets; and interviewed Dr. Bremerton. Then, we collected and analyzed our data and wrote the report.

Our main finding is that the clinical staff who already own tablets are very receptive to the idea of using tablets in a clinical setting. Not one of these staff members thought we should not use tablets in the clinical setting. By a slight margin, these staff members prefer having the hospital supply the tablets. We, too, think the hospital-supplied model is preferable to the BYOD model because it will reduce the chances of privacy violations and streamline the work of the IT department. We concluded, too, that the best tablets for our needs are those designed and built for health-care applications. Unfortunately, they are too expensive for our budget. And we cannot be sure, simply from reading the literature, whether the many general-purpose tablets can meet our standards for ease of disinfection or durability. Nor can we be sure whether they have sufficient battery life. Any of the general-purpose tablets, regardless of operating system or brand, would be adequate if it met these standards.

We recommend one of two courses of action: reconsidering our cost criterion or testing a representative sample of general-purpose tablets for disinfection and the other technical characteristics and making suitable tablets available for clinical staff to demo. We believe reconsidering the cost criterion is the better approach for our needs because the health-care-specific tablets offer significant advantages over the general-purpose tablets.

2

Introduction

To determine the best course of action for integrating tablet computers into the RRMC clinical setting, Dr. Jill Bremerton, RRMC Chief Executive Officer, asked us to study national trends, determine clinical-staff knowledge of and attitudes toward tablets, examine administrative models for tablet use, devise criteria for assessing tablets, and present our findings and recommendations.

Currently, RRMC has no formal policy on tablet usage by clinical staff. By default, we are following a bring-your-own-device (BYOD) approach. More than half of our clinical staff use their personal tablets in their work. This situation is not ideal because not all clinical staff are taking advantage of the enormous potential for improving patient care and reducing costs by using tablets, and IT is struggling to keep up with the work needed to ensure that all the different tablets are working properly and that any information-security protocols required by HIPAA and the current health care laws are not being violated.

Therefore, Dr. Bremerton wanted us to determine the best way to make tablets available to all our clinical staff. Specifically, Dr. Bremerton asked us to perform five tasks:

- Determine how tablets are being used by clinical staff across the nation. We performed secondary research to complete this task.
- Determine the RRMC clinical staff's current knowledge of and attitudes toward tablet use. To complete this task, we wrote and distributed a survey to clinical staff who already own tablets.
- Determine how hospitals administer the use of tablets in a clinical setting. We performed secondary research to complete this task.
- Establish criteria by which we might evaluate tablets for RRMC. We performed secondary research to complete this task. In addition, we interviewed Dr. Bremerton for her suggestions about the most-important criterion.
- Assess available tablets based on our criteria. We performed secondary research to complete this task.

We found that tablet use in clinical settings is increasing quickly, and that clinical staff are finding many ways to use tablets to improve care and save time and money. Among the clinical staff at RRMC who already own

Some organizations require that each first-level heading begin on a new page.

A brief statement of the context for the report.

Note that the word *currently* is used to introduce the background of the study: the current situation is unsatisfactory for several reasons.

A formal statement of the task the committee was asked to perform. The writers paraphrase from the memo Dr. Bremerton gave them. Often in technical communication, you will quote or paraphrase words your reader wrote to you. This practice reminds the reader of the context and shows that you are carrying out your tasks professionally.

The writers incorporate a brief overview of their methods into the list of tasks.

The writers devote two paragraphs to their principal findings. The introduction can present the major findings of a report; technical communication is not about drama and suspense.

3

tablets, nearly half own an iPad and nearly half an Android tablet, most consider themselves expert users of their tablets, and more than two-thirds already use them in the clinical setting; by a slim margin, they would prefer a hospital-supplied model for tablet use to a BYOD model. Our research on the two models for making tablets available also found more advantages and fewer disadvantages to the hospital-supplied model.

Our principal finding regarding tablets themselves is that the best tablets for our use would be those designed and built for health-care applications. These tablets are rugged and easy to disinfect, and they offer a wealth of hardware and software options that would streamline our daily tasks without introducing any risks either to patient care or to data privacy. Unfortunately, purchasing enough of these tablets for all clinical staff would exceed our budget. To determine whether any of the general-purpose tablets meet all our needs, we would need to conduct hands-on testing regarding disinfection, battery life, durability, and several other technical criteria.

Notice the writers' use of the phrase "we recommend." Repeating key terms in this way helps readers understand the logic of a report and concentrate on the technical information it contains.

We recommend, first, that we reassess whether the budget will permit consideration of any of the health-care-specific tablets. If that is not possible, we recommend that we ask manufacturers of a small set of general-purpose tablets to let us test their products and invite our clinical staff to demo them. This option would yield data that would help us decide how to proceed.

An advance organizer for the rest of the report.

In the following sections, we provide additional details about our research methods, the results we obtained, the conclusions we drew from those results, and our recommendation.

4

Research Methods

We began our research by interviewing Dr. Bremerton, who emphasized that we need to maintain our focus on our priorities—patient care and service to the community—and not let technical questions about the tablets distract us from the needs of our clinical staff. "We're not going to do anything without the approval of the doctors and nurses," she said.

Early on in our research, we discovered an article that corroborated what Dr. Bremerton had told us (Narisi, 2013). Two keys to doing the research were to focus on security features—data-privacy issues mandated in HIPAA and in current health care laws—and to get the clinical staff's input.

To perform the analysis requested by Dr. Bremerton, we broke the project into six tasks:

1. acquire a basic understanding of tablet use by clinical staff across the nation
2. determine the RRMC clinical staff's knowledge of and attitudes toward tablet use
3. assess the BYOD and hospital-owned tablet models
4. establish criteria for evaluating tablets
5. assess available tablets based on our criteria
6. analyze our data and prepare this recommendation report

In the following discussion of how we performed each task, we explain the reasoning that guided our research.

Task 1. Acquire a basic understanding of tablet use by clinical staff across the nation

The writers use the same task organization as in the proposal and the progress report.

Dr. Bremerton pointed us to a number of resources on tablet use in clinical settings. In addition, we conducted our own literature review. Most of the research we studied fell into one of four categories:

- general introductions to tablet use in trade magazines and general-interest periodicals
- more-focused articles about tablets used in health care
- technical specifications of tablets provided in trade magazines and on manufacturers' websites
- trade-magazine articles about best practices for managing the use of tablets in clinical settings

By stating that they know that their sources are a mixture of different kinds of information, not all of which is equally useful for every kind of question that needs to be answered, the writers suggest that they are careful analysts.

The writers carefully explain the logic of their methods. Do not assume that your readers will automatically understand why you did what you did. Sometimes it is best to explain your thinking. Although technical communication contains a lot of facts and figures, like other kinds of writing it relies on clear, logical arguments.

As discussed in Ch. 5, some questions will misfire. Therefore, it is smart to field-test a questionnaire before you distribute it.

Including a page number in the cross-reference to the appendix is a convenience to the reader. When you do so, remember to add the correct page number after you determine where the appendix (or the several appendixes) will appear in the report.

5

As we expected, the information we acquired was a mix of user opinions, benchmark-test results, and marketing. We relied most heavily on case studies from hospital administrators and technical specialists in health IT. Because of the unreliability of information on manufacturers' websites, we were hesitant to rely on claims about product performance.

Task 2. Determine the RRMC clinical staff's knowledge of and attitudes toward tablet use

On October 14, 2018, we sent all 147 clinical staff members an email linking to a four-question Qualtrics survey. The email indicated that we were seeking opinions about tablet use by clinical-staff members who already own tablets and made clear that the survey would take less than two minutes to complete.

Initially, we considered collecting data from all 147 clinical staff members. However, as we constructed that survey, we realized that it would be cumbersome to gather and track information from three different populations: those who didn't own a tablet, those who owned one but didn't use it in the clinic, and those who owned one and did use it in the clinic. Eliciting information from these different groups would require a long, complex questionnaire and some statistical analysis to separate out the attitudes.

For this reason, we decided to address only the tablet owners, since we assumed that this group would constitute approximately two-thirds of the clinical staff (Drinkwater, 2013). With this streamlined focus, we were able to create a very brief survey, one that would likely yield a high return rate. We assumed, too, that the opinions expressed by current tablet owners would likely be of more value in helping us plan a formal program of tablet use than those of clinicians who were less likely to be experienced tablet users.

We field-tested the questionnaire with six clinical staff members, revised one of the questions, and then, with the authorization of Dr. Bremerton, uploaded the questionnaire to Qualtrics and sent an email to the clinical staff.

The questionnaire (including the responses) appears in the Appendix, page 19.

6

Task 3. Assess the BYOD and hospital-owned tablet models

Our research revealed that hospitals use one of two administrative models for giving clinical staff access to tablets: the bring-your-own-device (BYOD) model and purchasing tablets to distribute to staff. To present the advantages and disadvantages of each of these models, we relied on reports from hospital administrators who pioneered these models (Jackson, 2011a, 2011b). For statistics on the popularity of each of these administrative models, we relied on a survey (Terry, 2011).

Note that the writers present their references to their sources throughout the report.

Here, again, the writers explain the logic of their methods. They decided to rely on the experiences of hospital administrators. This approach will likely appeal to Dr. Bremerton.

Task 4. Establish criteria for evaluating tablets

We both studied the voluminous literature on tablets. Jeremy Elkins met informally with his five IT colleagues to discuss the data, and Eloise Carruthers met informally with her nursing staff and with selected physicians, including several who had responded to the survey.

We began with the first criterion: cost. Dr. Bremerton had told us in our interview that the budget for the project (assuming that RRMC would supply a tablet to each of the 147 members of the clinical staff) would be $800 per device, fully configured with any commercial software needed to operate it. For this reason, we did not conduct a thorough examination of the health-care-specific tablets, each of which costs $2,500–3,000.

Often you will begin your project with a cost criterion: your recommended solution must not cost more than a certain amount.

In addition, we paid particular attention to the complexities of the current tablet market, focusing on whether the various devices would work seamlessly with our health-records system and other security features ("Top Five," 2018), on the need to be able to disinfect the tablets (Carr, 2011), and on durability (Narisi, 2013). We knew from experience with all kinds of portable information technology that the question of battery life would be problematic because it can vary so much depending on load and other factors.

We concluded that two factors that might seem critically important were not: operating system and availability of relevant apps. Although the Apple iPad is the single most popular tablet, the Samsung Galaxy and other Android tablets are currently outselling iPads. As a result, all the important apps are being created for the Apple operating system (iOS), for Android, and for Windows (the OS used by the Microsoft Surface and the major health-care-specific tablets).

The writers present just enough information about the technologies to help the reader understand their logic. Writers sometimes present too much information; include only as much as your readers need to be able to follow your report.

Task 5. Assess available tablets based on our criteria

Because our budget did not permit us to recommend the tablets designed for the health-care industry, we decided not to study them in detail.

To study the general-purpose tablets, we relied on trade magazines. We soon realized, however, that several of our necessary criteria—namely, the ability to disinfect the tablet, as well as durability and battery life—were not adequately addressed in the literature because our needs as a hospital are so specific.

For instance, battery life is typically reported as the mean number of hours the battery will last. But that figure varies significantly, depending on the applications the device is running. In addition, there is the question of whether the device is hot-swappable (that is, whether the battery can be replaced without shutting down the device). However, some tablets boot very quickly, making this characteristic less important. Finally, there is an administrative question: will the clinical staff member use the same tablet every day or check one out at the start of each shift (or perhaps check out a second or even a third one during a 14-hour shift)? Will fresh batteries be available only in one location, or can they be checked out at several locations? Can a staff member grab a handful of fresh batteries at the start of a shift? All these questions bear on how we would need to think about the importance of battery life.

For this reason, we recommend that a representative sample of general-purpose tablets be evaluated for disinfection, durability, and battery life and that this study include a substantial on-site evaluation period at RRMC.

Task 6. Analyze our data and prepare this recommendation report

We drafted this report and uploaded it to a wiki that we created to make it convenient for the other IT staff members and interested clinical staff members to help us revise it. We incorporated most of our colleagues' suggestions and then presented a final draft of this report on the wiki to gather any final editing suggestions.

Because analyzing their data and writing this report is part of the study, it is appropriate to include it as one of the steps. In some organizations, however, this task is assumed to be part of the study and is therefore not presented in the report.

8

Results

In this section, we present the results of our research. For each of the tasks we carried out, we present the most important data we acquired.

The writers present an advance organizer for the results section.

Task 1. Acquire a basic understanding of tablet use by clinical staff across the nation

Since the introduction of the Apple iPad in 2010, the use of tablets by clinical staff in hospitals across the country has been growing steadily. Although there are no precise statistics on how many hospitals either distribute tablets to clinical staff or let them use their own devices in their work, the number of articles in trade magazines, exhibits at medical conferences, and discussions on discussion boards suggests that tablets are quickly becoming established in the clinical setting. And many hundreds of apps have already been written to enable users to carry out health-care-related tasks on tablets.

The most extensive set of data on tablets in hospitals relates to the use of the iPad, the first tablet on the market. Ottawa Hospital has distributed more than 1,000 iPads to clinical staff; California Hospital is piloting a program with more than 100 iPads for hospital use; Kaiser Permanente is testing the iPad for hospital and clinical workflow; and Cedars-Sinai Medical Center is testing the iPad in its hospital. The University of Chicago's Internal Medicine Residency Program uses the iPad; the iPad is also being distributed to first-year medical students at Stanford, University of California–Irvine, and University of San Francisco. In addition, there are reports of Windows-based and Android-based tablets being distributed at numerous other hospitals and medical schools (Husain, 2011).

Today, tablets have five main clinical applications (Carr, 2011):

- *Monitoring patients and collecting data.* Clinical staff connect tablets to the hospital's monitoring instruments to collect patient information and transfer it to patients' health records without significant human intervention. In addition, staff access patient information on their tablets.
- *Ordering prescriptions, authorizations, and refills.* Clinical staff use tablets to communicate instantly with the hospital pharmacy and off-site pharmacies, as well as with other departments within the hospital, such as the Imaging Department.

The writers continue to use the task structure that they used in the methods section.

- *Scheduling appointments.* Clinical staff use tablets to schedule doctor and nurse visits and laboratory tests, to send reminders, and to handle re-scheduling and cancellations.
- *Conducting research on the fly.* Clinical staff use tablets to access medication databases and numerous reference works.
- *Educating patients.* Clinical staff use videos and animations to educate patients on their conditions and treatment options.

Tablets provide clinical staff with significant advantages. Staff do not need to go back to their offices to connect to the Internet or to the hospital's own medical-record system. Staff save time, reduce paper usage, and reduce transcription errors by not having to enter nearly as much data by hand.

Task 2. Determine the RRMC clinical staff's knowledge of and attitudes toward tablet use

On October 14, 2018, we sent all 147 clinical staff members an email linking to a four-question Qualtrix survey. In the email, we said that we were seeking opinions about tablet use by clinical-staff members who already own tablets and made clear that the survey would take less than two minutes to complete.

We received 96 responses, which represents 65 percent of the 147 staff members. We cannot be certain that all 96 respondents who indicated that they are tablet owners in fact own tablets. We also do not know whether all those staff members who own a tablet responded. However, given that some 75 percent of physicians in a 2013 poll own tablets, we suspect that the 96 respondents reasonably accurately represent the proportion of our clinical staff who own tablets (Drinkwater, 2013).

Here are the four main findings from the survey of tablet owners:

- Some 47 percent of respondents own an Apple iPad, and 47 percent own either a Samsung Galaxy or another tablet that uses the Android operating system. Only 6 percent use the Microsoft Surface, one of the several Windows-based tablets.
- Some 58 percent of the respondents strongly agree with the statement that they are expert users of their tablets. Overall, 90 percent agree more than they disagree with the statement.

- Some 63 percent of respondents use their tablets for at least one clinical application. They have either loaded apps on their tablets themselves or had IT do so for them.
- Some 27 percent of the respondents would prefer to continue to use their own tablets for clinical applications, whereas 38 percent would prefer to use a tablet supplied by RRMC. Some 35 percent had no strong feelings either way. None of the respondents indicated that they would prefer not to use a tablet at all.

Task 3. Assess the BYOD and hospital-owned tablet models

Currently, hospitals use one of two models for giving clinical staff access to tablets: the bring-your-own-device (BYOD) model and the purchase model, whereby the hospital purchases tablets to distribute to staff. In this section, we will present our findings on the relative advantages of each model.

The BYOD model is based on the fact that, nationally, some three-quarters of physicians already own tablets (with the Apple iPad the single most popular model) (Drinkwater, 2013). We could find no data on how many nurses own tablets.

The main advantage of the BYOD model is that clinical staff already know and like their tablets; therefore, they are motivated to use them and less likely to need extensive training. In addition, the hardware costs are eliminated (or almost eliminated, since some hospitals choose to purchase some tablets for staff who do not own their own). Todd Richardson, CIO with Deaconess Health System, Evansville, Indiana (Jackson, 2011b), argues that staff members who own their own tablets use and maintain them carefully: they know how to charge, clean, store, and protect them. In addition, the hospital doesn't have to worry about the question of liability if staff members lose them during personal use. And if the staff member moves on to a new position at a different hospital, there is no dispute about who owns the information on the tablet. All the hospital has to do is disable the staff member's account.

However, there are three main disadvantages to the BYOD model:

- Some clinical staff do not have their own tablets, and some who do don't want to use them at work; to make the advantages of tablet use available to all the clinical staff, therefore, the hospital needs to decide whether to purchase tablets and distribute them to these staff members.

- Labor costs are high because each tablet needs to be examined carefully by the hospital IT department to ensure that it contains no software that might interfere with or be incompatible with the health-care software that needs to be loaded onto it. This labor-intensive assessment by IT can seriously erode the cost savings from not having to buy the tablet itself.
- Chances of loss increase because the staff member is more likely to use the tablet at home as well as in the hospital.

The other model for making tablets available to clinical staff is for the hospital to purchase the same tablet for each staff member.

The purchase model offers two distinct advantages, as described by Dale Potter, CIO of 1,300-bed Ottawa Hospital in Ontario, Canada. Potter has purchased more than 2,000 iPads for his staff (Jackson, 2011a):

- The hospital controls the software and apps loaded on the tablets and can even create its own apps. For instance, Potter hired 120 developers to create apps.
- The hospital reduces labor costs because IT can load exactly the same set of apps and other software on each machine. Updates and upgrades also are far simpler to manage when all the devices are the same.

The main disadvantages of the purchase model are the following:

- Staff members might not like the tablet that the hospital chooses.
- Staff members might need to be trained to use the tablet.
- Liability issues related to loss of the tablets must be addressed officially in the employment contract between the hospital and the staff member.

Beginning in 2010 with the introduction of the iPad, most hospitals used the BYOD model; clinical staff brought their own tablets to work and started to think about ways to use them in a clinical setting. As time passed, however, and more people began to acquire tablets from different manufacturers, hospitals began to see the advantages of standardizing tablet use. A 2011 survey (Terry, 2011) showed that 40 percent of hospitals use the BYOD model, whereas 55 percent of hospitals support only devices provided or owned by the institution.

12

Task 4. Establish criteria for evaluating tablets

Our interview with Dr. Bremerton, as well as our research in the available literature, yielded four *necessary* criteria and three *desirable* criteria for tablet use at RRMC.

A tablet that does not meet a necessary criterion would be eliminated from consideration. The four necessary criteria are the following:

- *Cost.* Each device must cost less than $800, fully configured for use.
- *HIPAA compliance.* We cannot use any device or technology that would jeopardize our compliance with the HIPAA and current health care privacy standards for medical information. What this means, in essence, is that a tablet must operate seamlessly with our electronic health-records system, Cerner, which we access through our current utility, Citrix. The tablet must duplicate our current desktop and remote-access capabilities.
- *Other security features.* The tablet must support basic security features, including encryption, remote wipe (IT's ability to remove data on a lost tablet), auto-lock (that cannot be turned off except by IT), and perimeter settings (IT's ability to prevent use of a tablet that has strayed a certain distance from the server) ("Top Five," 2018).
- *Ease of disinfection.* Because the tablets would be used in a clinical setting, they would be subject to the same standards of disinfection as any other equipment or device. A tablet that cannot be disinfected effectively and easily would not be an appropriate choice (Carr, 2011).

In addition, we have established two other criteria against which to evaluate tablets. We have deemed these criteria desirable; a tablet that does not meet a desirable criterion would not be eliminated from consideration.

- *Durability.* Because the tablet would be carried around within the hospital and sometimes would be used in close quarters, it likely would be dropped or bumped. The more durable, the more desirable the tablet would be.
- *Long battery life.* Our clinical staff routinely work shifts as long as 14 hours. The longer the battery life, the better. Related to battery life is the hot-swap feature. In a hot-swappable device, the battery can be removed and replaced with a fresh one without having to shut down the tablet.

Task 5. Assess available tablets based on our criteria

One challenge we faced is that it was impractical to do a comprehensive assessment of all of the approximately 100 tablets available on the market. A second challenge is that tablets fall into a dizzying variety of sometimes overlapping categories. For example, the single most popular tablet is the Apple iPad, which runs on Apple's proprietary operating system. Many other tablets run on the Android system. Some, like the Amazon Kindle Fire, run on specialized configurations of Android. And Microsoft's Surface runs on Windows. There are general-purpose tablets, like the Samsung, made for the consumer market, and there are specialized tablets, like the Motion, designed for the health-care industry (Phillips, 2013). There are rugged tablets designed to meet military specifications for durability. Some tablets have USB ports for easy connection to existing clinical instruments and devices, as well as specialized features such as barcode-reading and RFID capabilities. Some tablets come with dozens of native (pre-installed) apps related to health care.

In short, although many brands of tablet are similar, no two are identical. As a result, we decided to review a small set of tablets that are mentioned frequently in health-care magazines and journals. We were sure to include the iPad, the Microsoft Surface, and several Android tablets. We also looked briefly at several tablets specially designed and marketed for health-care applications.

We present our basic findings by returning to our set of necessary and desired criteria.

The four necessary criteria are the following:

- *Cost.* Unfortunately, our cost criterion of $800 per unit eliminates all the tablets designed for health-care applications. These tablets, which cost between $2,500 and $3,000, are easy to disinfect and highly durable and include many desirable features such as barcode scanners, RFID readers, speech, and smart-card readers. The leading tablets in this category are the ProScribe Medical Tablet PC, the Sahara Slate PC, the MediSlate MCA, the Motion C5v Medical Tablet PC, the Teguar TA-10 Medical Tablet PC, the Advantech Medical PC, and the Arbor Antibacterial Medical Tablet.
- *HIPAA compliance.* All of the tablets we reviewed would enable us to remain in compliance with HIPAA and the current health care laws, and all would enable us to use our current record system, Cerner, through our Citrix utility. Some of the tablets have native apps to support this access, whereas others use virtual private networks.

- *Other security features.* All of the tablets we reviewed would support the basic security features we identified, although some tablets would be easier than others for IT to configure.
- *Ease of disinfection.* This proved to be a very challenging criterion to assess. On the one hand, all of the so-called ruggedized tablets, as well as all those designed for health care, met this criterion because all the surfaces are sealed. On the other hand, the general-purpose tablets were not designed to withstand medical-grade disinfectants. However, this does not necessarily mean that they would not withstand clinical disinfection. According to John Curin, head of the health-care practice at Burwood, an IT consulting firm, "I couldn't disinfect [an iPad] if I wanted to." By contrast, Dr. John Halamka, the CIO of Beth Israel Deaconess Medical Center in Boston, argues that the iPad is "completely disinfectable." Even though Apple advises against using a disinfecting solution on the iPad, Halamka says he does so and has experienced no problems (Carr, 2011).

On the basis of the cost criterion, we must eliminate all the health-care-specific tablets and study only the general-purpose tablets. We need to devise a method to determine whether any of the general-purpose tablets can be adequately disinfected. For the purposes of this study, however, we decided to assess a representative subset of the general-purpose tablets based on our desirable criteria.

We will now present our major findings regarding the desirable criteria: durability and long battery life. We have decided against presenting a decision matrix because some of the technical characteristics are either imprecise or unknowable. For instance, battery life is a notoriously difficult characteristic to measure, because it can be affected greatly by so many factors.

Our tentative findings are that the Apple iPad, the Samsung Galaxy and the many other Android tablets, and the Microsoft Surface *appear* to meet our two desirable criteria. We would need to carry out our own tests of these tablets, configured with our software and appropriate apps, to determine their battery life. And we would need to determine whether any of the aftermarket protective cases would provide adequate durability. Rugged and waterproof iPad cases are available from Hard Candy Cases, OtterBox, and others. There are also waterproof solutions that are said to sterilize the iPad. In addition, there are some protective cases said to be designed specifically for clinical environments.

The function of a conclusion is to explain what the data mean. Here the writers explain how their results can help their readers determine how to proceed with the tablet study. Notice that a conclusion is not the same as a recommendation (which explains what writers think should be done next).

The writers present an advance organizer for the results section.

At this point in the report, the writers have decided to abandon the "task" labels. Their thinking is that they are focusing less on what they did and more on the meaning of the information they gathered. However, they retain the headings that help readers understand the topic they are discussing.

Conclusions

In this section, we present our conclusions based on our research related to the four questions we were asked to answer.

Tablet use by clinical staff

On the basis of our research, we conclude that increasing numbers of our clinical staff will begin to use tablets for more applications in a clinical setting. This increase will spur the creation of more health-care apps for all tablets.

The RRMC clinical staff's knowledge of and attitudes toward tablet use

On the basis of our survey of clinical staff who already own tablets, we conclude that they consider themselves proficient in using the tablets, and most already use them for at least one clinical application. Because they prefer the hospital-supplied model for tablet use, we conclude that they would welcome a formal plan to supply tablets.

The BYOD and hospital-owned tablet models

We conclude that the hospital-owned tablet model offers more advantages and fewer disadvantages than the BYOD model. Having all clinical staff use the same model of tablet saves money by streamlining the process of loading software and installing updates and upgrades. The medical center IT department can even create hospital-specific apps.

Criteria for evaluating tablets

Our first necessary criterion, cost, eliminated all the health-care-specific tablets from consideration, leaving us with only the general-purpose tablets. All the general-purpose tablets we evaluated met our HIPAA and health care law compliance and other security criteria. Unfortunately, because the general-purpose tablets are not designed for clinical settings, we could not determine from our research whether any of them are easy to disinfect. We would need to conduct our own tests to answer that question.

We would need to conduct our own testing to determine whether battery life and durability of any of the general-purpose tablets are acceptable.

16

Assessing available tablets based on our criteria

We drew two main conclusions from our study of available tablets:

- The best tablets for our use—the ones designed and intended for health-care settings—are out of our price range. Because they meet all our criteria, we should reassess whether our budget will permit us to assess them more carefully.
- Any one of the available general-purpose tablets is potentially acceptable (provided it meets the disinfection criterion). We see no compelling reason to favor one operating system over another. Apple iPads, Android tablets, and Windows tablets all come with acceptable power, and there are plenty of health-care apps available for each type. As Android machines establish their dominance over the iPad, health-care apps for Android devices will surpass those for the iPad in variety and number.

Recommendation

We recommend that the RRMC administration pursue one of two options:

Option 1: Reconsider the cost criterion

Although a health-care-specific tablet, at $2,500–3,000, is some three times the cost of a general-purpose tablet, it offers distinct advantages in terms of disinfection properties, ruggedness, and availability of specialized hardware and software for better integration with our other devices and equipment.

If it is not possible to provide health-care-specific tablets to all clinical staff, we might consider a two-tiered system (some staff members receive a health-care-specific tablet, others a general-purpose one) or a phased-implementation system.

Option 2: Test a representative sample of general-purpose tablets

This recommendation states explicitly what the writers think the reader should do next. Note that they are sketching in ideas that they have not discussed in detail but that might interest their readers.

If RRMC wishes to continue to assess the general-purpose tablets, we recommend that we contact manufacturers of the Apple iPad, the Samsung Galaxy, the Microsoft Surface, and several other Android tablets. We could request that they supply their most powerful products, equipped with the best set of medical apps, for our internal testing and evaluation.

We would then have IT test each tablet in a controlled environment for such technical characteristics as battery life and durability. We would also test each tablet for disinfection. Next, we would invite the manufacturers' representatives to attend a one- or two-day tablet fair, where we would make the products available to our clinical staff for demos and hands-on assessment. On the basis of these tests and follow-up questionnaires submitted by clinical staff, we would be in a good position to know how to proceed.

18

References

Carr, D. F. (2011, May 21). Healthcare puts tablets to the test. *InformationWeek Healthcare.* Retrieved October 1, 2018, from www.informationweek.com/healthcare/mobile-wireless/healthcare-puts-tablets-to-the-test/229503387.

Drinkwater, D. (2013, April 19). Three in 4 physicians are using tablets; some are even prescribing apps. *TabTimes.* Retrieved October 12, 2018, from tabtimes.com/news/ittech-stats-research/2013/04/19/3-4-physicians-are-using-tablets-some-are-even-prescribing.

Husain, I. (2011, March 10). Why Apple's iPad will beat Android tablets for hospital use. *iMedicalApps.* Retrieved October 15, 2018, from www.imedicalapps.com/2011/03/ipad-beat-android-tablets-hospital-medical-use/.

Jackson, S. (2011a, August 15). Five reasons hospitals should buy tablets for physicians. *FierceMobileHealthcare.* Retrieved October 2, 2018, from www.fiercemobilehealthcare.com/story/5-reasons-why-hospitals-should-buy-tablets-physicians/2011-08-15.

Jackson, S. (2011b, August 3). Why physicians should buy their own mobile devices. *FierceMobileHealthcare.* Retrieved October 4, 2018, from www.fiercemobilehealthcare.com/story/why-physicians-should-buy-their-own-mobile-devices/2011-08-03.

Narisi, S. (2013, April 24). Choosing the best tablets for doctors: 3 keys. *Healthcare Business & Technology.* Retrieved October 3, 2018, from www.healthcarebusinesstech.com/best-tablets-for-doctors.

Phillips, A. (2013, 10 April). Top 10 mobile tablets for healthcare professionals. *Healthcare Global.* Retrieved October 13, 2018, from www.healthcareglobal.com/top_ten/top-10-business/top-10-mobile-tablets-for-healthcare-professionals.

Terry, K. (2011, Dec. 9). Apple capitalizes on doctors' iPad romance. *InformationWeek Healthcare.* Retrieved October 2, 2018, from www.informationweek.com/healthcare/mobile-wireless/apple-capitalizes-on-doctors-ipad-roman/232300218.

Top Five Tablet Security Features. (2018). *Healthcare Data Solutions.* Retrieved October 2, 2018, from www.healthcaredatasolutions.com/top-five-tablet-security-features.html.

This list of references is written according to the APA documentation style, which is discussed in Appendix, Part A.

Presenting the percentage data in boldface after each question is a clear way to communicate how the respondents replied. Although most readers will not be interested in the raw data, some will.

Appendix: Clinical-Staff Questionnaire

This is the questionnaire we distributed to the 147 RRMC clinical staff members. We received 96 responses. The numbers in boldface below represent the percentage of respondents who chose each response.

Questionnaire on Tablet Use at RRMC

Directions: As you may know, Dr. Bremerton is conducting a study to determine whether to institute a formal policy on tablet use by clinical staff.

If you own a tablet device, please respond to the following four questions. Your opinions can help us decide whether and how to develop a policy for tablet use at RRMC. We greatly appreciate your answering the following four questions.

1. Which brand of tablet do you own?
   47% Apple iPad
   28% Samsung Galaxy
   9% Amazon Kindle Fire
   6% Microsoft Surface
   10% Other (please name the brand) **(Respondents named the Asus, Google Nexus, and a Toshiba model.)**
2. "I consider myself an expert user of my tablet."
   Strongly disagree ____ **8%** **2%** **13%** **19%** **58%** Strongly agree
3. Do you currently use your tablet for a clinical application, such as monitoring patients or ordering procedures?
   63% Yes
   37% No
4. If RRMC were to adopt a policy of using tablets for clinical applications (and to supply the appropriate software and training), which response best describes your attitude?
   27% I would prefer to use my own tablet.
   38% I would prefer to use a hospital-supplied tablet.
   35% I don't have strong feelings either way about using my own or a hospital-supplied tablet.
   0% I would prefer not to use any tablet at all for clinical applications.

Thank you!

WRITER'S CHECKLIST

In planning your recommendation report, did you

- ☐ analyze your audience? *(p. 339)*
- ☐ determine your purpose? *(p. 341)*
- ☐ identify the questions that need to be answered? *(p. 341)*
- ☐ carry out appropriate research? *(p. 342)*
- ☐ draw valid conclusions about the results (if appropriate)? *(p. 343)*
- ☐ formulate recommendations based on the conclusions (if appropriate)? *(p. 344)*

Does the transmittal letter

- ☐ clearly state the title and, if necessary, the subject and purpose of the report? *(p. 348)*
- ☐ clearly state who authorized or commissioned the report? *(p. 348)*
- ☐ acknowledge any assistance you received? *(p. 348)*
- ☐ establish a courteous and professional tone? *(p. 348)*

Does the cover include

- ☐ the title of the report? *(p. 348)*
- ☐ your name and position? *(p. 348)*
- ☐ the date of submission? *(p. 348)*
- ☐ the company name or logo? *(pp. 348–49)*

Does the title page

- ☐ include a title that clearly states the subject and purpose of the report? *(p. 349)*
- ☐ list your name and position and those of your principal reader(s)? *(p. 349)*
- ☐ include the date of submission of the report and any other identifying information? *(p. 349)*

Does the abstract

- ☐ list the report title, your name, and any other identifying information? *(p. 349)*
- ☐ clearly define the problem or opportunity that led to the project? *(p. 349)*
- ☐ briefly describe (if appropriate) the research methods? *(p. 349)*
- ☐ summarize the major results, conclusions, and recommendations? *(p. 349)*

Does the table of contents

- ☐ contain a sufficiently detailed breakdown of the major sections of the body of the report? *(p. 350)*
- ☐ reproduce the headings as they appear in your report? *(p. 350)*
- ☐ include page numbers? *(p. 351)*

☐ Does the list of illustrations (or list of tables or list of figures) include all the graphics found in the body of the report? *(p. 351)*

Does the executive summary

- ☐ clearly state the problem or opportunity that led to the project? *(p. 353)*
- ☐ explain the major results, conclusions, recommendations, and managerial implications of your report? *(p. 353)*
- ☐ avoid technical vocabulary and concepts that a managerial audience is not likely to be interested in? *(p. 353)*

Does the introduction

- ☐ explain the subject of the report? *(p. 346)*
- ☐ explain the purpose of the report? *(p. 346)*
- ☐ explain the background of the report? *(p. 346)*
- ☐ describe your sources of information? *(p. 346)*
- ☐ indicate the scope of the report? *(p. 346)*
- ☐ briefly summarize the most significant findings of the project? *(p. 346)*
- ☐ briefly summarize your recommendations? *(p. 346)*
- ☐ explain the organization of the report? *(p. 346)*
- ☐ define key terms used in the report? *(p. 346)*

Does the methods section

- ☐ describe your methods in sufficient detail? *(p. 347)*
- ☐ justify your methods where necessary, explaining, for instance, why you chose one method over another? *(p. 347)*

Are the results presented

- ☐ clearly? *(p. 347)*
- ☐ objectively? *(p. 347)*
- ☐ without interpretation? *(p. 347)*

Are the conclusions

- ☐ presented clearly? *(p. 347)*
- ☐ drawn logically from the results? *(p. 347)*

Are the recommendations

- ☐ clear? *(p. 348)*
- ☐ objective? *(p. 348)*
- ☐ stated politely? *(p. 348)*
- ☐ in an appropriate form (list or paragraph)? *(p. 348)*
- ☐ in an appropriate location? *(p. 348)*

- ☐ Does the glossary include definitions of all the technical terms your readers might not know? *(p. 354)*
- ☐ Does the list of symbols include all the symbols and abbreviations your readers might not know? *(p. 354)*
- ☐ Does the list of references include all your sources and adhere to an appropriate documentation system? *(p. 356)*
- ☐ Do the appendixes include supporting materials that are too bulky to present in the report body or that are of interest to only a small number of your readers? *(p. 356)*

EXERCISES

For more about memos, see Ch. 9, "Writing Memos."

1. An important element in carrying out a feasibility study is determining the criteria by which to judge each option. For each of the following topics, list five necessary criteria and five desirable criteria you might apply in assessing the options.
   - **a.** buying a smartphone
   - **b.** selecting a major
   - **c.** choosing a company to work for
   - **d.** buying a car
   - **e.** choosing a place to live while you attend college
2. Go to www.usa.gov and select the "Government Agencies" tab. Choose an agency and navigate to its website. Then find a recommendation report on a subject that interests you. In what ways does the structure of the report differ from the structure described in this chapter? In other words, does the report lack some of the elements described in this chapter, or does it have additional elements? Are the elements arranged in the order in which they are described in this chapter? In what ways do the differences reflect the audience, purpose, and subject of the report?
3. **TEAM EXERCISE** Write the recommendation report for the research project you proposed in response to Exercise 3 on page 315 in Chapter 11. Your instructor will tell you whether the report is to be written individually or collaboratively, but either way, work closely with a partner to review and revise your report. A partner can be very helpful during the planning phase, too, as you choose a topic, refine it, and plan your research.
4. Secure a recommendation report for a project subsidized by a city or federal agency, a private organization, or a university committee or task force. (Be sure to check your university's website; universities routinely publish strategic planning reports and other sorts of self-study reports. Also check www.nas.edu, which is the site for the National Academy of Sciences, the National Academy of Engineering, the Institute of Medicine, and the National Research Council, all of which publish reports on the web.) In a memo to your instructor, analyze the report. Overall, how effective is the report? How could the writers have improved the report? If possible, submit a copy of the report along with your memo.

CASE 13: Analyzing Decision Matrices

As president of a music-instruction company, you are exploring the possibility of purchasing an electric guitar that can be used for digital recording. You ask your director of guitar instruction to investigate options for you, but you find some problems with her methodology. To get started helping her research the matter more successfully and provide more useful recommendations, go to LaunchPad.

CHAPTER 14 Writing Definitions, Descriptions, and Instructions

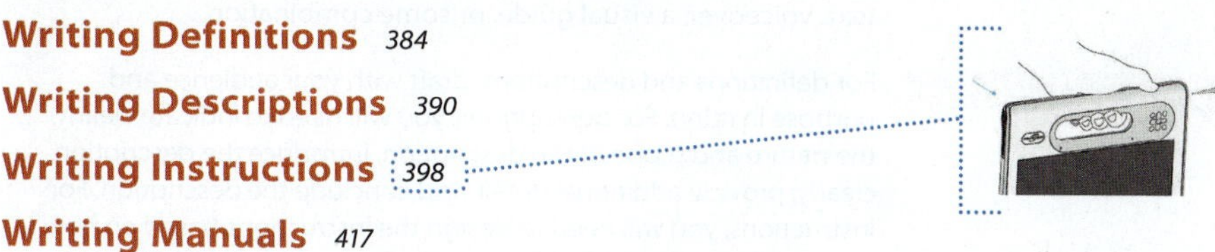

THIS CHAPTER DISCUSSES definitions, descriptions, and instructions. The first step is to define these three terms:

- A definition is typically a brief explanation, using words and sometimes graphics, of what an item is or what a concept means. You could write a definition of file format or of regenerative braking.
- A description is typically a longer explanation—usually accompanied by graphics—of the physical or operational features of an object, mechanism, or process. You could write a description of a wind turbine, of global warming, or of shale-oil extraction.
- A set of instructions is a kind of process description, almost always accompanied by graphics, intended to enable a person to carry out a task. You could write a set of instructions for installing a new roof or for using an app on your tablet.

Although each can appear independently, definitions, descriptions, and instructions are often presented together in a set of product information. For instance, a store that sells building materials for homeowners might create a product-information set about how to lay a brick patio. In this set might be definitions of tools (such as a mason's line), descriptions of objects (such as different types of edging materials, including plastic, metal, masonry, and wood), and step-by-step instructions for planning, laying, and maintaining the patio.

Regardless of your field, you will write definitions, descriptions, and instructions frequently. Whether you are communicating with other technical professionals, with managers, or with the public, you must be able to define key concepts, describe processes, and explain how to carry out tasks.

Photo: Courtesy of Xplore.

FOCUS ON PROCESS
Writing Definitions, Descriptions, and Instructions

In writing definitions, descriptions, and instructions, pay special attention to these steps in the writing process. For a complete process for writing technical documents, see page 11.

PLANNING For definitions, you will need to decide where to place the definition. Parenthetical and sentence definitions can be placed in the text, in a marginal gloss, in a separate hyperlinked file, in a footnote, or in a glossary (an alphabetized list of definitions). An extended definition can be a section in the body of the larger document or can be placed in an appendix. For descriptions and instructions, consider whether they would be most effective in text, voiceover, a visual guide, or some combination.

DRAFTING For definitions and descriptions, draft with your audience and purpose in mind. For descriptions, you will need to indicate clearly the nature and scope of the description, introduce the description clearly, provide additional detail, and conclude the description. For instructions, you will need to design the instructions based on how your readers will be using them, and you should include appropriate graphics. Do everything you can to ensure readers' safety.

REVISING Remember to keep audience and purpose in mind for definitions and descriptions. If you can, carry out usability testing on any instructions you write.

Writing Definitions

The world of business and industry depends on clear definitions. Suppose you learn at a job interview that the prospective employer pays tuition and expenses for employees' job-related education. You would need to study the employee-benefits manual to understand just what the company would pay for. Who, for instance, is an *employee*? Is it anyone who works for the company, or is it someone who has worked for the company full-time (40 hours per week) for at least six uninterrupted months? What is *tuition*? Does it include incidental laboratory or student fees? What is *job-related education*? Does a course about time management qualify? What, in fact, constitutes *education*?

Definitions are common in communicating policies and standards "for the record." Definitions also have many uses outside legal or contractual contexts. Two such uses occur frequently:

- **Definitions clarify a description of a new development or a new technology in a technical field.** For instance, a zoologist who has discovered a new animal species names and defines it.
- **Definitions help specialists communicate with less-knowledgeable readers.** A manual explaining how to tune up a car includes definitions of parts and tools.

Definitions, then, are crucial in many kinds of technical communication. All readers, from the general reader to the expert, need effective definitions to carry out their jobs.

ANALYZING THE WRITING SITUATION FOR DEFINITIONS

For more about audience and purpose, see Ch. 4.

The first step in writing effective definitions is to analyze the writing situation: the audience and the purpose of the document. Physicists wouldn't need a definition of *entropy*, but lawyers might. Builders know what a molly bolt is, but many insurance agents don't. When you write for people whose first language is not English, consider adding a *glossary* (a list of definitions), using Simplified English and easily recognizable terms in your definitions, and using graphics to help readers understand a term or concept.

Think, too, about your purpose. For readers who need only a basic understanding of a concept—say, *time-sharing vacation resorts*—a brief, informal definition is usually sufficient. However, readers who need to understand an object, process, or concept thoroughly and be able to carry out related tasks need a more formal and detailed definition. For example, the definition of a "Class 2 Alert" written for operators at a nuclear power plant must be comprehensive, specific, and precise.

The appropriate type of definition depends on your audience and purpose. You have three major choices: parenthetical, sentence, and extended.

CHOICES AND STRATEGIES Choosing the Appropriate Type of Definition

IF THE TERM IS . . .	AND THE DOCUMENT IS . . .	TRY THIS TYPE OF DEFINITION
Relatively simple	**Relatively informal**	**Parenthetical.** A *parenthetical definition* is a brief clarification within an existing sentence. Sometimes, a parenthetical definition is simply a word or phrase enclosed in parentheses or commas or introduced by a colon or a dash: "The computers were infected by a *Trojan horse* (a destructive program that appears to be benign)."
Relatively simple	**Relatively formal**	**Sentence.** A *sentence definition* usually follows a standard pattern in which the item to be defined is placed in a category of similar items and then distinguished from them: "Crippleware is shareware in which some features of the program are disabled until the user buys a license to use the program." Here, "shareware" is the category and the words that follow "shareware" are the distinguishing information. Writers often use sentence definitions to present a working definition for a particular document: "In this report, *electron microscope* refers to any microscope that uses electrons rather than visible light to produce magnified images." Such a definition is called a *stipulative definition.* See the Guidelines box for more about writing sentence definitions.
Relatively complex	**Informal or formal**	**Extended.** An *extended definition* is a detailed explanation—usually one or more paragraphs—of an object, process, or idea. Often an extended definition begins with a sentence definition, which is then elaborated. For instance, the sentence definition "An electrophorus is a laboratory instrument used to generate static electricity" tells you the basic function of the device, but it doesn't explain how it works, what it is used for, and its strengths and limitations. An extended definition would address these and other topics. See pages 386–90 for more about extended definitions.

WRITING SENTENCE DEFINITIONS

The following Guidelines box presents more advice on sentence definitions.

GUIDELINES Writing Effective Sentence Definitions

The following five suggestions can help you write effective sentence definitions.

- **Be specific in stating the category and the distinguishing characteristics.** If you write, "A Bunsen burner is a burner that consists of a vertical metal tube connected to a gas source," the imprecise category—"a burner"—ruins the definition: many types of large-scale burners use vertical metal tubes connected to gas sources.
- **Don't describe a specific item if you are defining a general class of items.** If you wish to define *catamaran*, don't describe a particular catamaran. The one you see on the beach in front of you might be made by Hobie and have a white hull and blue sails, but those characteristics are not essential to all catamarans.
- **Avoid writing circular definitions—that is, definitions that merely repeat the key words or the distinguishing characteristics of the item being defined in the category.** The definition "A required course is a course that is required" is useless: required of whom, by whom? However, in defining *electron microscope*, you can repeat *microscope* because *microscope* is not the difficult part of the term. The purpose of defining *electron microscope* is to clarify *electron* as it applies to a particular type of microscope.
- **Be sure the category contains a noun or a noun phrase rather than a phrase beginning with *when*, *what*, or *where*.**

  INCORRECT A brazier is what is used to . . .

  CORRECT A brazier is a metal pan used to . . .

  INCORRECT Hypnoanalysis is when hypnosis is used to . . .

  CORRECT Hypnoanalysis is a psychoanalytical technique in which . . .
- **Consider including a graphic.** A definition of an electron microscope would probably include a photograph, diagram, or drawing, for example.

WRITING EXTENDED DEFINITIONS

An *extended definition* is a more-detailed explanation—usually one or more paragraphs—of an object, process, or idea. Often an extended definition begins with a sentence definition, which is then elaborated. For instance, the sentence definition "An electrophorus is a laboratory instrument used to generate static electricity" tells you the basic function of the device, but it doesn't explain how it works, what it is used for, or its strengths and limitations. An extended definition would address these and other topics.

There is no one way to "extend" a definition. Your analysis of your audience and the purpose of your communication will help you decide which method to use. In fact, an extended definition sometimes employs several of the eight techniques discussed here.

Graphics Perhaps the most common way to present an extended definition in technical communication is to include a graphic and then explain it. Graphics are useful in defining not only physical objects but also concepts and ideas. A definition of *temperature inversion*, for instance, might include a diagram showing the forces that create temperature inversions.

The following passage from an extended definition of *additive color* shows how graphics can complement words in an extended definition.

> Additive color is the type of color that results from mixing colored light, as opposed to mixing pigments such as dyes or paints. When any two colored lights are mixed, they produce a third color that is lighter than either of the two original colors, as shown in this diagram. And when green, red, and blue lights are mixed together in equal parts, they form white light.
>
> We are all familiar with the concept of additive color from watching TV monitors. A TV monitor projects three beams of electrons—one each for red, blue, and green—onto a fluorescent screen. Depending on the combinations of the three colors, we see different colors on the screen.

The graphic effectively and economically clarifies the concept of additive color.

Examples Examples are particularly useful in making an abstract term easier to understand. The following excerpt is an extended definition of marine protected areas from the website of the National Oceanic and Atmospheric Administration (NOAA.gov, 2016).

> Marine protected areas (MPAs) in the U.S. come in a variety of forms and are established and managed by all levels of government. There are marine sanctuaries, estuarine research reserves, ocean parks, and marine wildlife refuges. Each of these sites differ. MPAs may be established to protect ecosystems, preserve cultural resources such as shipwrecks and archaeological sites, or sustain fisheries production.
>
> There is often confusion and debate regarding what the term "marine protected area" really means. Some people interpret MPAs to mean areas closed to all human activities, while others interpret them as special areas set aside for recreation (e.g., national parks) or to sustain commercial use (e.g., fishery management areas). These are just a few examples of the many types of MPAs.
>
> In reality, "marine protected area" is a term that encompasses a variety of conservation and management methods in the United States. If you have been fishing in central California, diving near a shipwreck in the Florida Keys, camping in Acadia, snorkeling in the Virgin Islands, or hiking along the Olympic Coast, you were probably one of thousands of visitors to an MPA.
>
> In the U.S., MPAs span a range of habitats, including the open ocean, coastal areas, intertidal zones, estuaries, and the Great Lakes. They also vary widely in purpose, legal authorities, agencies, management approaches, level of protection, and restrictions on human uses.

Partition Partitioning is the process of dividing a thing or an idea into smaller parts so that readers can understand it more easily. The following example (Brain, 2005) uses partition to define *computer infection*.

Types of Infection

When you listen to the news, you hear about many different forms of electronic infection. The most common are:

- Viruses—A virus is a small piece of software that piggybacks on real programs. For example, a virus might attach itself to a program such as a spreadsheet program. Each time the spreadsheet program runs, the virus runs, too, and it has the chance to reproduce (by attaching to other programs) or wreak havoc.
- Email viruses—An email virus moves around in email messages, and usually replicates itself by automatically mailing itself to dozens of people in the victim's email address book.
- Worms—A worm is a small piece of software that uses computer networks and security holes to replicate itself. A copy of the worm scans the network for another machine that has a specific security hole. It copies itself to the new machine using the security hole, and then starts replicating from there, as well.
- Trojan horses—A Trojan horse is simply a computer program. The program claims to do one thing (it may claim to be a game) but instead does damage when you run it (it may erase your hard disk). Trojan horses have no way to replicate automatically.

Principle of Operation Describing the principle of operation—the way something works—is an effective way to develop an extended definition, especially for an object or a process. The following excerpt from an extended definition of *adaptive cruise control* is based on the mechanism's principle of operation.

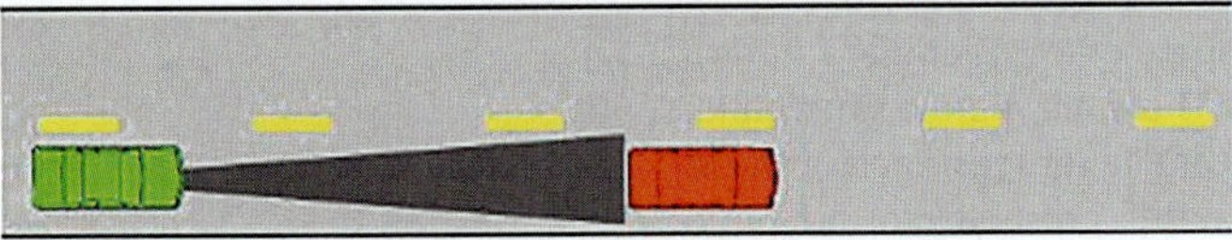

The ACC system in the green car detects the slower-moving red car.

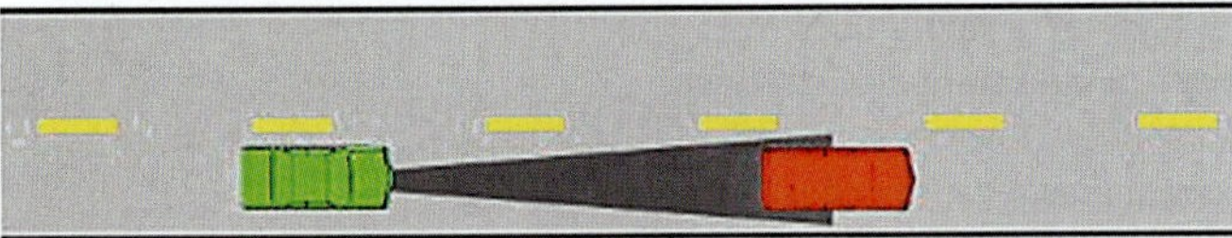

ACC reduces the green car's speed and maintains the preset headway.

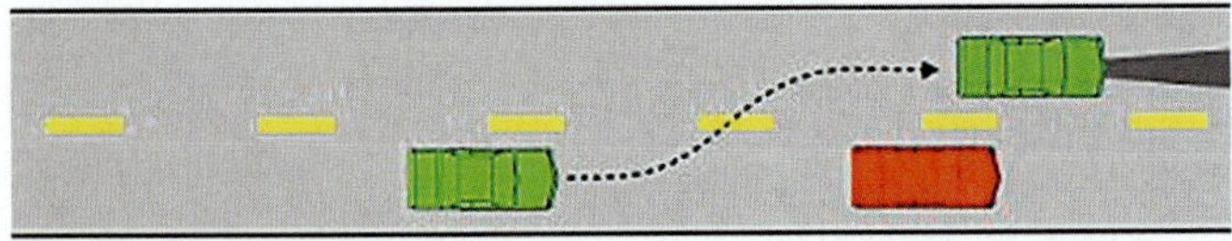

The green car overtakes the red car. The road ahead is clear. The ACC system resumes the vehicle's preset cruising speed.

Information from Canadian Association of Road Safety Professionals.

Adaptive cruise control (ACC) employs sensing and control systems to monitor the vehicle's position with respect to any vehicle ahead. When a vehicle equipped with ACC approaches a slower-moving vehicle, the ACC system reduces the vehicle speed in order to maintain a preset following distance (headway). However, when the traffic ahead clears, the ACC system automatically accelerates the vehicle back to the preset travel speed.

ACC systems use forward-looking radar or laser detection (lidar) systems to monitor the vehicle's position with respect to any vehicle in front and change the speed in order to maintain a preset following distance (headway).

The system typically allows the driver to preset a "following time," for example a two-second gap between vehicles. The ACC computer makes calculations of speed, distance and time based on the sensor inputs and makes appropriate adjustments to the vehicle's speed to maintain the desired headway.

Comparison and Contrast Using comparison and contrast, a writer discusses the similarities or differences between the item being defined and an item with which readers are more familiar. The following definition of VoIP (Voice over Internet Protocol) contrasts this new form of phone service to the form we all know.

> Voice over Internet Protocol is a form of phone service that lets you connect to the Internet through your cable or DSL modem. VoIP service uses a device called a telephony adapter, which attaches to the broadband modem, transforming phone pulses into IP packets sent over the Internet.
>
> VoIP is considerably cheaper than traditional phone service: for as little as $20 per month, users get unlimited local and domestic long-distance service. For international calls, VoIP service is only about three cents per minute, about a third the rate of traditional phone service. In addition, any calls from one person to another person with the same VoIP service provider are free.
>
> However, sound quality on VoIP cannot match that of a traditional land-based phone. On a good day, the sound is fine on VoIP, but frequent users comment on clipping and dropouts that can last up to a second. In addition, sometimes the sound has the distant, tinny quality of some of today's cell phones.

In this excerpt, the second and third paragraphs briefly compare VoIP and traditional phone service.

Analogy An *analogy* is a specialized kind of comparison. In a traditional comparison, the writer compares one item to another, similar item: an electron microscope to a light microscope, for example. In an analogy, however, the item being defined is compared to an item that is in some ways completely different but that shares some essential characteristic. For instance, the central processing unit of a computer is often compared to a brain. Obviously, these two items are very different, except that the relationship of the central processing unit to the computer is similar to that of the brain to the body.

The following example from a definition of *decellularization* (Falco, 2008) shows an effective use of an analogy.

> Researchers at the University of Minnesota were able to create a beating [rat] heart using the outer structure of one heart and injecting heart cells from another rat. Their findings are reported in the journal Nature Medicine. Rather than building a heart from scratch, which has often been mentioned as a possible use for stem cells, this procedure takes a heart and breaks it down to the outermost shell. It's similar to taking a house and gutting it, then rebuilding everything inside. In the human version, the patient's own cells would be used.

The writer of this passage uses the analogy of gutting a house to clarify the meaning of *decellularization*.

Negation A special kind of contrast is *negation*, sometimes called *negative statement*. Negation clarifies a term by distinguishing it from a different term with which readers might confuse it. The following example uses negation to distinguish the term *ambulatory* from *ambulance*.

> An ambulatory patient is not a patient who must be moved by ambulance. On the contrary, an ambulatory patient is one who can walk without assistance from another person.

Negation is rarely the only technique used in an extended definition; in fact, it is used most often in a sentence or two at the start. Once you have explained what something is not, you still have to explain what it is.

Etymology Citing a word's *etymology*, or derivation, is often a useful and interesting way to develop a definition. The following example uses the etymology of *spam*—unsolicited junk email—to define it.

> For many decades, Hormel Foods has manufactured a luncheon meat called Spam, which stands for "Shoulder Pork and hAM"/"SPiced hAM." Then, in the 1970s, the English comedy team Monty Python's Flying Circus broadcast a skit about a restaurant that served Spam with every dish. In describing each dish, the waitress repeats the word Spam over and over, and several Vikings standing in the corner chant the word repeatedly. In the mid-1990s, two businessmen hired a programmer to write a program that would send unsolicited ads to thousands of electronic newsgroups. Just as Monty Python's chanting Vikings drowned out other conversation in the restaurant, the ads began drowning out regular communication online. As a result, people started calling unsolicited junk email spam.

Etymology is a popular way to begin definitions of *acronyms*, which are abbreviations pronounced as words:

> RAID, which stands for redundant array of independent (or inexpensive) disks, refers to a computer storage system that can withstand a single (or, in some cases, even double) disk failure.

Etymology, like negation, is rarely used alone in technical communication, but it provides an effective way to introduce an extended definition.

A Sample Extended Definition Figure 14.1 is an example of an extended definition addressed to a general audience.

Writing Descriptions

Technical communication often requires descriptions: verbal and visual representations of the physical or operational features of objects, mechanisms, and processes.

- **Objects.** An object is anything from a natural physical site, such as a volcano, to a synthetic artifact, such as a battery. A tomato plant is an object, as is an automobile tire or a book.
- **Mechanisms.** A mechanism is a synthetic object consisting of a number of identifiable parts that work together. A cell phone is a mechanism, as is a voltmeter, a lawnmower, or a submarine.
- **Processes.** A process is an activity that takes place over time. Species evolve; steel is made; plants perform photosynthesis. *Descriptions of processes*, which explain how something happens, differ from *instructions*, which explain how to do something. Readers of a process description want to *understand* the process; readers of instructions want a step-by-step guide to help them *perform* the process.

FIGURE 14.1 An Extended Definition

Source: GPS.gov, 2013: www.gps.gov/systems/gps/.

What is GPS?

The Global Positioning System (GPS) is a U.S.-owned utility that provides users with positioning, navigation, and timing (PNT) services. This system consists of three segments: the space segment, the control segment, and the user segment. The U.S. Air Force develops, maintains, and operates the space and control segments.

Space Segment

The space segment consists of a nominal constellation of 24 operating satellites that transmit one-way signals that give the current GPS satellite position and time. *LEARN MORE...* ➡

Control Segment

The control segment consists of worldwide monitor and control stations that maintain the satellites in their proper orbits through occasional command maneuvers, and adjust the satellite clocks. It tracks the GPS satellites, uploads updated navigational data, and maintains health and status of the satellite constellation. *LEARN MORE...* ➡

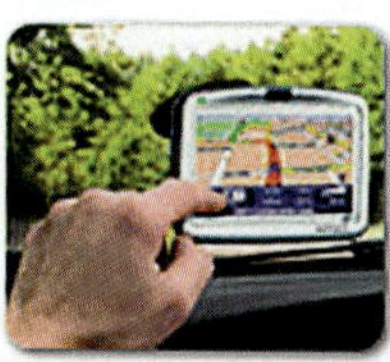

User Segment

The user segment consists of the GPS receiver equipment, which receives the signals from the GPS satellites and uses the transmitted information to calculate the user's three-dimensional position and time. *LEARN HOW GPS IS USED...* ➡

The first paragraph of this extended definition of GPS begins with a sentence definition.

The second sentence serves as an advance organizer for the definition, explaining that the definition will be extended by partition: GPS consists of three segments.

The body of this extended definition consists of three sections, each of which is introduced by a graphic and a topic sentence explaining the segment.

Links lead the reader to much more information about each segment, including more text, diagrams, and videos.

Descriptions of objects, mechanisms, and processes appear in virtually every kind of technical communication. For example, an employee who wants to persuade management to buy some equipment includes a description of the equipment in the proposal to buy it. A company manufacturing a consumer product provides a description and a photograph of the product on its website to attract buyers. A developer who wants to build a housing project includes in his proposal to municipal authorities descriptions of the geographical area and of the process he will use in developing that area. A maintenance manual for an air-conditioning system might begin with a description of the system to help the reader understand first how it operates and then how to fix or maintain it.

ANALYZING THE WRITING SITUATION FOR DESCRIPTIONS

Before you begin to write a description, consider carefully how the audience and the purpose of the document will affect what you write.

Your sense of your audience will determine not only how technical your vocabulary should be but also how long your sentences and paragraphs should be. Another audience-related factor is your use of graphics. Less-knowledgeable readers need simple graphics; they might have trouble understanding sophisticated schematics or decision charts. As you consider your audience, think about whether any of your readers are from other cultures and might therefore expect different topics, organization, or writing style in the description.

Consider, too, your purpose. What are you trying to accomplish with this description? If you want your readers to understand how a personal computer works, write a *general description* that applies to several brands and sizes of computers. If you want your readers to understand how a specific computer works, write a *particular description*. Your purpose will determine every aspect of the description, including its length, the amount of detail, and the number and type of graphics.

DRAFTING EFFECTIVE DESCRIPTIONS

There is no single organization or format used for descriptions. Because descriptions are written for different audiences and different purposes, they can take many shapes and forms. However, the following four suggestions will guide you in most situations:

- Indicate clearly the nature and scope of the description.
- Introduce the description clearly.
- Provide appropriate detail.
- End the description with a brief conclusion.

For more about titles and headings, see Ch. 6, p. 105.

Indicate Clearly the Nature and Scope of the Description If the description is to be a separate document, give it a title. If the description is to be part of a longer document, give it a section heading. In either case, clearly state the subject and indicate whether the description is general or particular. For instance, a general description of an object might be entitled "Description of a Minivan," and a particular description, "Description of the 2018 Honda Odyssey." A general description of a process might be called "Description of the Process of Designing a New Production Car," and a particular description, "Description of the Process of Designing the Chevrolet Malibu."

Introduce the Description Clearly Start with a general overview: you want to give readers a broad understanding of the object, mechanism, or process. Consider adding a graphic that introduces the overall concept. For example, in describing a process, you might include a flowchart summarizing the steps in the body of the description; in describing an object, such as a bicycle, you might include a photograph or a drawing showing the major components you will describe in detail in the body.

Table 14.1 shows some of the basic kinds of questions you might want to answer in introducing object, mechanism, and process descriptions. Figure 14.2 shows the introductory graphic accompanying a description of an electric bicycle.

TABLE 14.1 Questions To Answer in Introducing a Description

FOR OBJECT AND MECHANISM DESCRIPTIONS	FOR PROCESS DESCRIPTIONS
• **What is the item?** You might start with a sentence definition.	• **What is the process?** You might start with a sentence definition.
• **What is the function of the item?** If the function is not implicit in the sentence definition, state it: "Electron microscopes magnify objects that are smaller than the wavelengths of visible light."	• **What is the function of the process?** Unless the function is obvious, state it: "The main purpose of performing a census is to obtain current population figures, which government agencies use to revise legislative districts and determine revenue sharing."
• **What does the item look like?** Sometimes an object is best pictured with both graphics and words. Include a photograph or drawing if possible. (See Chapter 8 for more about incorporating graphics into your text.) If you cannot use a graphic, use an analogy or comparison: "The USB drive is a plastic- or metal-covered device, about the size of a pack of gum, with a removable cap that covers the type-A USB connection." Mention the material, texture, color, and other physical characteristics, if relevant.	• **Where and when does the process take place?** "Each year the stream is stocked with hatchery fish in the first week of March." Omit these facts only if your readers already know them.
• **How does the item work?** In a few sentences, define the operating principle. Sometimes objects do not "work"; they merely exist. For instance, a ship model has no operating principle.	• **Who or what performs the process?** If there is any doubt about who or what performs the process, state that information.
• **What are the principal parts of the item?** Limit your description to the principal parts. A description of a bicycle, for instance, would not mention the nuts and bolts that hold the mechanism together; it would focus on the chain, gears, pedals, wheels, and frame.	• **How does the process work?** "The four-treatment lawn-spray plan is based on the theory that the most effective way to promote a healthy lawn is to apply different treatments at crucial times during the growing season. The first two treatments—in spring and early summer—consist of . . . ."
	• **What are the principal steps in the process?** Name the steps in the order in which you will describe them. The principal steps in changing an automobile tire, for instance, are jacking up the car, replacing the old tire with the new one, and lowering the car back to the ground. Changing a tire also includes secondary steps, such as placing chocks against the tires to prevent the car from moving once it is jacked up. Explain or refer to these secondary steps at the appropriate points in the description.

Clicking on any of the red dots opens a window with a one-sentence description and graphic, as well as a link to an extended description and more-detailed graphic.

This web-based description includes six labeled systems. Here we see the expanded description of one of them: the regenerative braking system.

FIGURE 14.2 Graphic with Linked Graphics and Descriptions
Information from Trek Bicycle Corporation, 2013: www.trekbikes.com/us/en/collections/electric_assist/feature_tour.

Provide Appropriate Detail In the body of a description—the part-by-part or step-by-step section—treat each major part or step as a separate item. In describing an object or a mechanism, define each part and then, if applicable, describe its function, operating principle, and appearance. In discussing the appearance, include shape, dimensions, material, and physical details such as texture and color (if essential). In describing a process, treat each major step as if it were a separate process.

A description can have not only parts or steps but also subparts or substeps. A description of a computer system includes a keyboard as one of its main parts, and the description of the keyboard includes the numeric keypad as one of its subparts. And the description of the numeric keypad includes the arrow keys as one of its subparts. The same principle applies in describing processes: if a step has substeps, you need to describe who or what performs each substep.

Conclude the Description A typical description has a brief conclusion that summarizes it and prevents readers from overemphasizing the part or step discussed last.

A common technique for concluding descriptions of mechanisms and of some objects is to state briefly how the parts function together. A process description usually has a brief conclusion: a short paragraph summarizing the principal steps or discussing the importance or implications of the process.

A LOOK AT SEVERAL SAMPLE DESCRIPTIONS

A look at some sample descriptions will give you an idea of how different writers adapt basic approaches for a particular audience and purpose.

GUIDELINES Providing Appropriate Detail in Descriptions

Use the following techniques to flesh out your descriptions.

FOR MECHANISM AND OBJECT DESCRIPTIONS

- **Choose an appropriate organizational principle.** Descriptions can be organized in various ways. Two organizational principles are common:
  - Functional: how the item works or is used. In a radio, the sound begins at the receiver, travels into the amplifier, and then flows out through the speakers.
  - Spatial: based on the physical structure of the item (from top to bottom, east to west, outside to inside, and so forth).

  The description of a house, for instance, could be organized functionally (covering the different electrical and mechanical systems) or spatially (top to bottom, inside to outside, east to west, and so on). A complex description can use different patterns at different levels.

- **Use graphics.** Present a graphic for each major part. Use photographs to show external surfaces, drawings to emphasize particular items on the surface, and cutaways and exploded diagrams to show details beneath the surface. Other kinds of graphics, such as graphs and charts, are often useful supplements (see Chapter 8).

FOR PROCESS DESCRIPTIONS

- **Structure the step-by-step description chronologically.** If the process is a closed system—such as the cycle of evaporation and condensation—and therefore has no first step, begin with any principal step.
- **Explain causal relationships among steps.** Don't present the steps as if they had nothing to do with one another. In many cases, one step causes another. In the operation of a four-stroke gasoline engine, for instance, each step creates the conditions for the next step.
- **Use the present tense.** Discuss steps in the present tense unless you are writing about a process that occurred in the historical past. For example, use the past tense in describing how the Snake River aquifer was formed: "The molten material condensed . . . ." However, use the present tense in describing how steel is made: "The molten material is then poured into . . . ." The present tense helps readers understand that, in general, steel is made this way.
- **Use graphics.** Whenever possible, use graphics to clarify each point. Consider flowcharts or other kinds of graphics, such as photographs, drawings, and graphs. For example, in a description of how a four-stroke gasoline engine operates, use diagrams to illustrate the position of the valves and the activity occurring during each step.

Figure 14.3 shows the extent to which a process description can be based on a graphic. The topic is a household solar array. The audience is the general reader.

Figure 14.4 on page 397 shows an excerpt from a mechanism description of three different types of hybrid drivetrains used in automobiles: series, parallel, and series/parallel.

FIGURE 14.3
A Process Description Based on a Graphic
Source: Vanguard Energy Partners LLC, 2014

This description begins with an informal definition of "direct grid-tie system."

In the next section, the writer presents the steps of the process in chronological order.

The description uses boldfaced text to emphasize key terms, most of which appear in the graphic.

The description focuses on the operating principle of the system. It does not seek to explain the details of how the system works. Accordingly, the graphic focuses on the logic of the process, not on the particulars of what the components look like or where they are located in the house.

How Our Solar Electric System Works

Your solar electric system is most likely to be what is called a **direct grid-tie system**. This means it is connected into the electricity system provided by your utility company.

Here's how it works . . .

The sun strikes the panels of your **solar array** and a flow of **direct current (DC) electricity** is produced. This is the only type of current produced by solar cells.

Appliances and machinery, however, are run on higher voltage **alternating current (AC) electricity** as supplied by your utility.

The lower voltage DC is fed into an **inverter** that transforms it into alternating current. The AC feeds into the **main electrical panel** from which it powers your household's or your business's electrical needs.

Your electrical panel is also connected to a **specially installed bi-directional utility meter**. This is **connected to the electrical grid**, which is the utility's means of delivering electricity. This set up allows AC electricity to flow both into, and out of, your home or business.

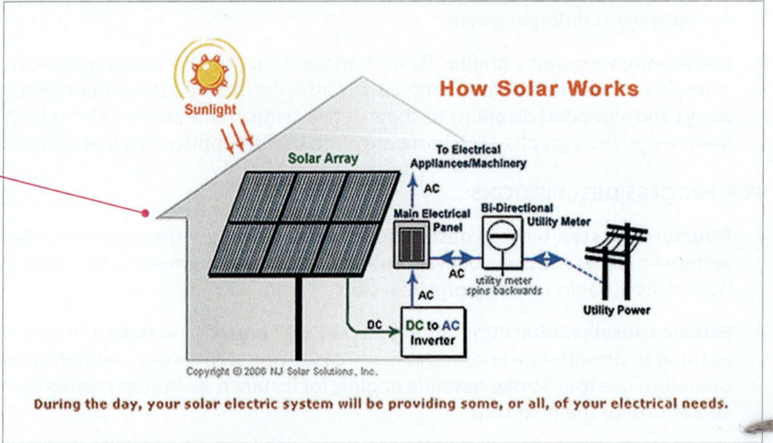

How much will depend, firstly, on the intensity of the sunlight; the system produces less power on cloudy days and during the winter months. It will also depend on the appliances or machinery you are running at the time.

If your solar system is not providing all the power you need at any time, the **balance is automatically provided by your utility.**

On days when sunlight is intense, **your system may well produce more than you need**. The excess is automatically fed into the grid. This is registered on your bi-directional meter which will spin backwards, **giving you credit for the electricity you are providing**. (This is known as net metering.)

At night, your utility company automatically provides your electrical needs.

If there is a utility power outage, your grid-tie system will shut down immediately for safety reasons. Your power will be reinstated moments after grid power is restored.

A grid-tie solar electric system does not provide power during outages **unless it incorporates a battery storage system**. If your home or business has critical needs that require an uninterrupted power supply, we'll be happy to take you through the various alternatives available to you.

Off-grid, or stand-alone, solar systems produce power independently of the utility grid. They are most appropriate for remote or environmentally sensitive areas; stand-alone systems may effectively provide farm lighting, fence charging or solar water pumps. Most of these systems rely on battery storage so that power produced during the day can be used at night.

FIGURE 14.4 Excerpt from a Mechanism Description

Information from Hybrids Under the Hood (Part 2): Drivetrains, 2013: www .ucsusa.org/.

Drivetrains

Now that we've covered the basic technology that defines hybrid vehicles, let's take a look at how they are put together to move the vehicle. The drivetrain of a vehicle is composed of the components that are responsible for transferring power to the drive wheels of your vehicle. With hybrids there are three possible setups for the drivetrain: the series drivetrain, the parallel drivetrain, and the series/parallel drivetrain.

This excerpt from a mechanism description begins with an advance organizer that helps the reader make the transition from the previous section (a basic description of hybrid technology) to the present section, which discusses the three types of drivetrains.

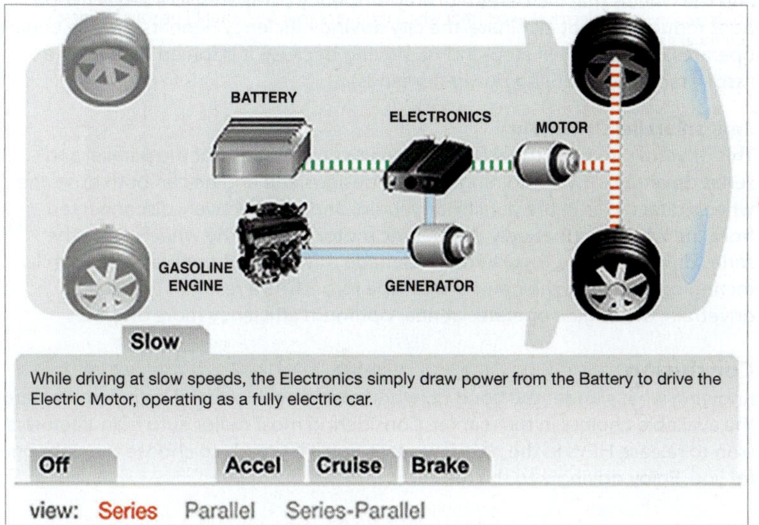

The interactive graphic enables the reader to view the principle of operation of three types of hybrid drivetrains, each during five modes. Here, the graphic shows how the components of the series drivetrain work together as the car operates at slow speeds. On the UCS website, the red and green striped lines are animated to highlight the components operating during the mode the reader has selected.

Series Drivetrain

This is the simplest hybrid configuration. In a series hybrid, the electric motor is the only means of providing power to get your wheels turning. The motor receives electric power from either the battery pack or from a generator run by a gasoline engine. A computer determines how much of the power comes from the battery or the engine/generator set. Both the engine/generator and regenerative braking recharge the battery pack. . . . While the engine in a conventional vehicle is forced to operate inefficiently in order to satisfy varying power demands of stop-and-go driving, series hybrids perform at their best in such conditions. This is because the gasoline engine in a series hybrid is not coupled to the wheels. This means the engine is no longer subject to the widely varying power demands experienced in stop-and-go driving and can instead operate in a narrow power range at near optimum efficiency. . . .

For each of the three types of drivetrains, the text begins with a description of the operating principle, followed by explanations of the strengths and weaknesses of the drivetrain. Notice that the audience and purpose of this description determine the kind of information it contains. Because this description seeks to explain the operating principle behind each of the three types of drivetrains, it focuses on the drivetrains' functions, not on the materials they are made of or on their technical specifications.

Parallel Drivetrain

Some up-and-coming hybrid models use a second electric motor to drive the rear wheels, providing electronic all-wheel drive that can improve handling and driving in bad weather conditions.

(continued)

FIGURE 14.4 Excerpt from a Mechanism Description (*continued*)

With a parallel hybrid electric vehicle, both the engine and the electric motor generate the power that drives the wheels. The addition of computer controls and a transmission allow these components to work together. This is the technology in the Insight, Civic, and Accord hybrids from Honda. Honda calls it their Integrated Motor Assist (IMA) technology. . . . Since the engine is connected directly to the wheels in this setup, it eliminates the inefficiency of converting mechanical power to electricity and back, which makes these hybrids quite efficient on the highway. Yet the same direct connection between the engine and the wheels that increases highway efficiency compared to a series hybrid does reduce, but not eliminate, the city driving efficiency benefits (i.e., the engine operates inefficiently in stop-and-go driving because it is forced to meet the associated widely varying power demands).

Series/Parallel Drivetrain

This drivetrain merges the advantages and complications of the parallel and series drivetrains. By combining the two designs, the engine can both drive the wheels directly (as in the parallel drivetrain) and be effectively disconnected from the wheels so that only the electric motor powers the wheels (as in the series drivetrain). The Toyota Prius has made this concept popular, and a similar technology is also in the new Ford Escape Hybrid. As a result of this dual drivetrain, the engine operates at near optimum efficiency more often. . . .

This section of the description ends with a brief conclusion.

Conclusion

Knowing what's under the hood of hybrid electric vehicles will help you evaluate the available choices in the market. Considering most major auto manufacturers plan to release HEVs in the next few years, you'll be ready to choose the right one for you. Enjoy driving into the future.

Figure 14.5 shows an excerpt from a set of specifications for Logitech headphones.

Figure 14.6 is a description of the process of turning biomass into useful fuels and other products.

Writing Instructions

This section discusses *instructions*, which are process descriptions written to help readers perform a specific task—for instance, installing a water heater in a house.

Although written instructions are still produced today, the growth of social media has radically changed how organizations instruct people on how to use their products and services. Now that technology has made it easy for people to participate in writing instructions and to view and make videos, most organizations try to present instructional material not only as formal written text but also through discussion forums, wikis, and videos. And users do not rely exclusively on the organizations themselves to create and present instructions. Rather, they create their own texts and videos.

FIGURE 14.5
Specifications
Information from Logitech, 2012: http://support.logitech.com/en_us/product/ue9000/specs.

Logitech UE 9000 Wireless Headphones Technical Specifications

Logitech UE 9000 Wireless Headphones

	Model Number (M/N)	PID - M/N - S/N Location
Headphones	A-00041	M/N, P/N: On headband behind left ear cup PID: On headband behind right ear cup

NOTE: Information is for reference only and may be subject to change.

General Product Information	[Compliance Certification (CE) Link]
Warranty / Self Help	Please see the product support page for warranty information and frequently asked questions.
Category	Headphones
Intended Usage	iPod, iPhone, iPad, Smartphones, Tablet PC, Portable MP3/media players
Software Support (at release)	No software needed.

Headset/Headphones Specifications	
Available Image(s)	
Device Type	Stereo
Connection Type	Wireless (USB for charging only)
Headband Design	Over-the-head, Over-the-ear
Frequency Response	20 Hz to 20 kHz
Input Impedance	55 Ohms; 2K Ohms (with active noise cancellation enabled)
Indicator Lights (LED)	Power and battery
Adjustable Headband	Yes
Inline Audio Controls	Play/Pause, Volume control, Multi-function
Earcup Audio Controls	Volume control, Power On/Off, Multi-function, Push to talk
Battery Details	Rechargeable, Non-accessible Size: Proprietary Quantity: 1 Type: Li-Ion
Battery Life	Recharge time: 3.5 hours (estimated) Discharge time: Up to 10 hours
Power Adapter	Yes (5v, 500mA)
Cable Length	6 feet or 1.8 meters
Microphone	Yes
Microphone Specifications	
Microphone Type	Built-in
Connection Type	Wireless
Noise Canceling	No
Frequency Response	100 Hz to 3.5 kHz
Input Sensitivity	Not available

Product Dimensions				
Product component	**Width**	**Depth/Length**	**Height**	**Weight**
Headphones	170 mm (6.7 inch)	90 mm (3.5 inch)	190 mm (7.5 inch)	330 g (11.6 ounce)

Package Contents	
Available Image(s)	
What is in the box	Headphones, 3.5mm audio cable, Case, User documentation, 1/4-inch adapter jack, USB AC adapter, Micro-fiber polishing cloth

Additional Product Information

- Wireless Bluetooth headphones with microphone.
- Optional 3.5mm connection available with microphone and remote (Apple-specific) included.
- Controls on right ear cup = volume, multi-function, power/bluetooth.
- Controls on left ear cup = push-to-talk, allows user to have face to face conversations without removing headphones using external microphones.
- Control functions designed for Apple products but other devices may have limited functionality.
- Can connect to PC/Mac w/ bluetooth radios.

An important kind of description is called a *specification*. A typical specification (or spec) consists of a graphic and a set of statistics about the device and its performance characteristics. Specifications help readers understand the capabilities of an item. You will see specifications on devices as small as transistors and as large as aircraft carriers.

Because this web-based spec sheet accompanies a consumer product, the arrangement of the specs is geared toward the interests of the likely purchasers. The audio specs are presented before the power specs because potential purchasers of high-end headphones will be most interested in the sound quality.

FIGURE 14.6
An Effective Process Description
Source: U.S. Department of Energy, 2013: www1.eere.energy.gov/bioenergy/pdfs/biochemical_four_pager.pdf.

This description begins with an overview of the process of biochemical conversion: the process of using fermentation and catalysis to make fuels and products.

The description includes a flowchart explaining the major steps in the process. The designers included photographs to add visual interest to the flowchart.

The lettered steps in the flowchart correspond to the textual descriptions of the steps in the process.

Most of the description is written in the passive voice (such as "Feedstocks for biochemical processes are selected . . ."). The passive voice is appropriate because the focus of this process description is on what happens to the materials, not on what a person does. By contrast, in a set of instructions the focus is on what a person does.

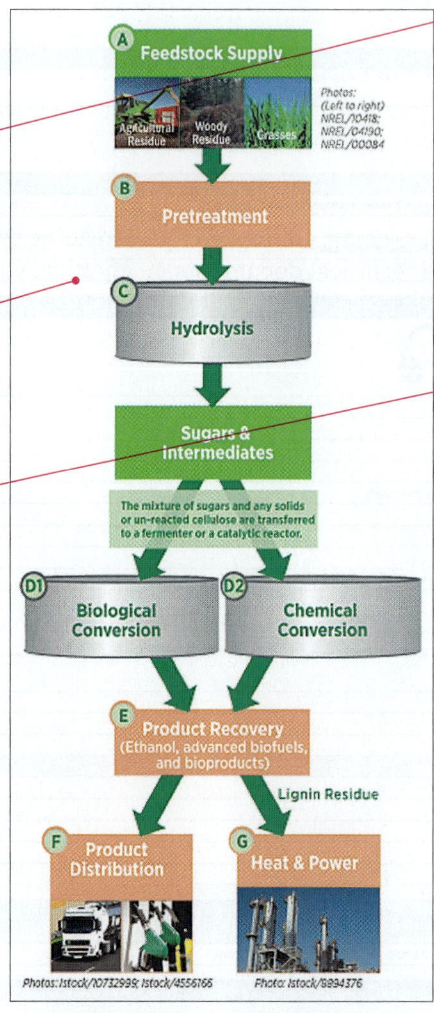

Biochemical conversion uses biocatalysts, such as enzymes, in addition to heat and other chemicals, to convert the carbohydrate portion of the biomass (hemicellulose and cellulose) into an intermediate sugar stream. These sugars are intermediate building blocks that can then be fermented or chemically catalyzed into ethanol, other advanced biofuels, and value-added chemicals. The overall process can be broken into the following essential steps:

A. Feedstock Supply: Feedstocks for biochemical processes are selected for optimum composition, quality, and size. Feedstock handling systems tailored to biochemical processing are essential to cost-effective, high-yield operations.

B. Pretreatment: Biomass is heated (often combined with an acid or base) to break the tough, fibrous cell walls down and make the cellulose easier to hydrolyze (see next step).

C. Hydrolysis: Enzymes (or other catalysts) enable the sugars in the pretreated material to be separated and released over a period of several days.

D1. Biological Conversion: Microorganisms are added, which then use the sugars to generate other molecules suitable for use as fuels or building-block chemicals.

D2. Chemical Conversion: Alternatively, the sugars can be converted to fuels or an entire suite of other useful products using chemical catalysis.

E. Product Recovery: Products are separated from water, solvents, and any residual solids.

F. Product Distribution: Fuels are transported to blending facilities, while other products and intermediates may be sent to traditional refineries or processing facilities for use in a diverse slate of consumer products.

G. Heat & Power: The remaining solids are mostly lignin, which can be burned for heat and power.

Your first job, then, in presenting instructions is to devise a strategy for incorporating user participation in the process and to choose the best mix of media for encouraging that participation. Written instructions (whether presented online or printed and put in the box with the product) will always have a role because they are portable and can include as much detail as possible for even the most complex tasks and systems. But consider whether your users will also benefit from accessing other people's ideas (in a discussion form or wiki) or from watching a video.

When you create instructions, whether written-only or using other media, you use the same planning, drafting, revising, editing, and proofreading process that you use with other kinds of technical documents. In addition, you might perform one other task—usability testing—to ensure that the instructions are accurate and easy to follow.

UNDERSTANDING THE ROLE OF INSTRUCTIONAL VIDEOS

The explosive growth of YouTube and other video-hosting sites has revolutionized how instructions are created and used. Product manufacturers and users alike make videos to help people understand how to perform tasks.

If you are producing instructional videos on behalf of your company, think about what style of video will be most effective for your audience, purpose, and subject. Companies often use simple, cartoon-style videos for basic conceptual information ("what can you do with a microblog?"), screenshot-based videos for computer-based tasks ("how to use master slides in PowerPoint"), and live-action videos for physical tasks ("how to install a ground-fault interrupter"). Video is particularly useful for communicating about physical tasks that call for subtle physical movements or that involve both sight and sound.

Although there are many software tools available for making videos, making professional-quality videos calls for professional skills, experience, and tools. If you are going to be making many videos, it makes sense to learn the process and acquire professional tools and equipment; otherwise, consult your company's media department or consider hiring freelance video producers.

Instructional videos tend to be brief. Whereas a reader of a document can navigate easily among various parts or steps, viewers of a video can only hit play, pause, and stop. For this reason, you should break long tasks into a series of brief videos: ideally 2–3 minutes, but no more than 12–15 minutes. Give each one a clear, specific title so viewers can easily tell whether they want to view it.

Similarly, you should make your instructional videos simple and uncluttered. Software makes it easy to add a lot of cinematic effects, but less is often more. The fewer distractions in the video, the easier it will be for viewers to see what to do. And remember that video is a warm medium. Connect with the viewer by being friendly, informal, and direct. But do not confuse being warm and informal with not needing to prepare. You *do* need to plan, write a script, and rehearse.

Be sure to build in the time and resources to revise the video. Good technical communication calls for reviewing, revising, and testing. This concept applies to video. Start testing even before production. Make sure your script and visuals are right for your audience and purpose. And repeat the process after you have created the rough cut of the video and after each major revision.

Use other sources responsibly. You need to obtain written permission to use any copyrighted text, images, videos, or music that will appear in your video. Because this process can be lengthy, difficult, and expensive, many organizations do not use any copyrighted material and instead generate their own images, text, and music, or they rely on material with Creative Commons licenses (see Chapter 2).

DESIGNING A SET OF INSTRUCTIONS

As you plan to write instructions, think about how readers will use them. Analyzing your audience and purpose and gathering and organizing your information will help you decide whether you should write a one-page set of instructions or a longer document that needs to be bound. You might realize that the information would work better as a web-based document that can include videos, be updated periodically, and provide readers with links to the information they need. Or you might decide to write several versions of the information: a brief paper-based set of instructions and a longer, web-based document with links.

As always in technical communication, imagining how readers will use what you write will help you plan your document. For example, having decided that your audience, purpose, and subject call for a printed set of instructions of perhaps 1,000 words and a dozen drawings and photographs, you can start to design the document. You will need to consider your resources, especially your budget: long documents cost more than short ones; color costs more than black and white; heavy paper costs more than light paper; secure bindings cost more than staples.

Designing a set of instructions is much like designing any other kind of technical document. As discussed in Chapter 7, you want to create a document that is attractive and easy to use. When you design a set of instructions, you need to consider a number of issues related to document design and page design:

- **What are your readers' expectations?** For a simple, inexpensive product, such as a light switch, readers will expect to find instructions written on the back of the package or printed in black and white on a small sheet of paper folded inside the package. For an expensive consumer product, such as a high-definition TV, readers will expect to find instructions in a more sophisticated full-color document printed on high-quality paper.

- **What are your readers' abilities?** If you're creating an instructional video on how to use a smartphone for banking transactions, will you need to include closed captioning for people with hearing impairments? In addition to closed captioning, you might also want to provide a stand-alone transcript of the video narration.
- **Do you need to create more than one set of instructions for different audiences?** If you are writing about a complex device such as an electronic thermostat, you might decide to create one set of instructions for electricians (who will install and maintain the device) and one set for homeowners (who will operate the device). In addition to producing paper copies of the documents, you might want to post them on the Internet, along with a brief video of the tasks you describe.
- **What languages should you use?** In most countries, several languages are spoken. You might decide to include instructions in two or more languages. Doing so will help you communicate better with more people, and it can help you avoid legal problems. In liability cases, U.S. courts sometimes find that if a company knows that many of its customers speak only Spanish, for example, the instructions should appear in Spanish as well as in English. You have two choices for presenting information in multiple languages: simultaneous presentation or sequential presentation. In a *simultaneous design*, you might use a multi-column page on which one column presents the graphics, another presents the text in English, and another presents the text in Spanish. Obviously, this won't work if you need to present information in more than two or three languages. But it is efficient because you present each graphic only once. In a *sequential design*, you present all the information in English (say, on pages 1–8), then all the information in Spanish (on pages 9–16). The sequential design is easier for readers to use because they are not distracted by text in other languages, but you will have to present the graphics more than once, which will make the instructions longer.

GUIDELINES Designing Clear, Attractive Pages

To design pages that are clear and attractive, follow these two guidelines:

- **Create an open, airy design.** Do not squeeze too much information onto the page. Build in space for wide margins and effective line spacing, use large type, and chunk the information effectively.
- **Clearly relate the graphics to the text.** In the step-by-step portion of a set of instructions, present graphics to accompany every step or almost every step. Create a design that makes it clear which graphics go with each text passage. One easy way to do this is to use a table, with the graphics in one column and the text in the other. A horizontal rule or extra line spacing separates the text and graphics for one step from the text and graphics for the next step.

For more about chunking, see Ch. 7, "Guidelines: Understanding Learning Theory and Page Design."

FIGURE 14.7 Cluttered and Attractive Page Designs in a Set of Instructions

The left-hand page is cluttered, containing far too much information. In addition, the page is not chunked effectively. As a result, the reader's eyes don't know where to focus. Would you look forward to using these instructions to assemble a cabinet?

The right-hand page is well designed, containing an appropriate amount of information presented in a simple two-column format. Notice the effective use of white space and the horizontal rules separating the steps.

Remember to follow the steps exactly as shown. Do not jump ahead, or you'll get tangled up!

Step 5. - Attach Second Side Panel and Installing Corner Blocks

Step 6. - Attaching Cabinet Legs

Step 7. - Fasten Cabinet to Wall

Step 8. - Hang Cabinet Doors

Step 9. - Installing Optional Adjustable Shelf

Step 10. - Congratulations!

CONGRATULATIONS! You have just finished assembling and installing a SLIDE-LOK cabinet.

SLIDE-LOK

NOTE: - Attaching Two or More Cabinets

Design Your Own SLIDE-LOK Storage Solution

Jack's Design Center

www.slide-lok.com | 800-835-1759

a. Cluttered design
Information from Slide-Lok.

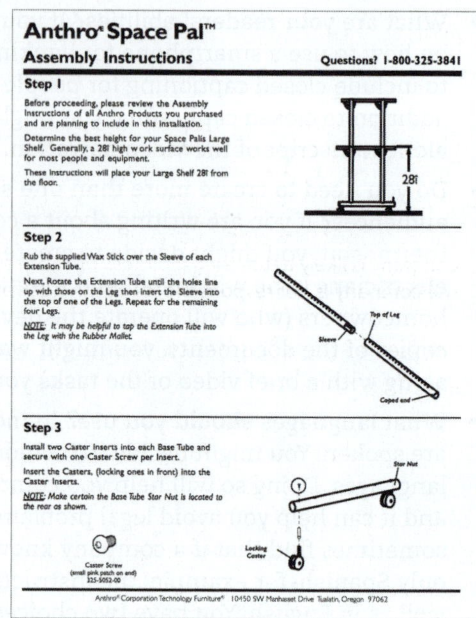

Anthro® Space Pal™

Assembly Instructions

Questions? 1-800-325-3841

Step 1

Before proceeding, please review the Assembly Instructions of all Anthro Products you purchased and are planning to include in this installation.

Determine the best height for your Space Pal's Large Shelf. Generally, a 28" high work surface works well for most people and equipment.

These instructions will place your Large Shelf 28" from the floor.

Step 2

Rub the supplied Wax Stick over the Sleeve of each Extension Tube.

Next, Rotate the Extension Tube until the holes line up with those on the Leg then insert the Sleeve into the top of one of the Legs. Repeat for the remaining four Legs.

NOTE: It may be helpful to tap the Extension Tube into the Leg with the Rubber Mallet.

Step 3

Install two Caster Inserts into each Base Tube and secure with one Caster Screw per Insert.

Insert the Casters, (locking ones in front) into the Caster Inserts.

NOTE: Make certain the Base Tube Star Nut is located to the rear as shown.

Anthro® Corporation Technology Furniture® 10450 SW Manhasset Drive Tualatin, Oregon 97062

b. Attractive design
Information from Anthro Corporation.

- **Will readers be anxious about the information?** If readers will find the information intimidating, make the design unintimidating. For instance, if you are writing for general readers about how to set up a wireless network for home computers, create open pages with a lot of white space and graphics. Use large type and narrow text columns so that each page contains a relatively small amount of information. Figure 14.7 illustrates the advantages of an open design.
- **Will the environment in which the instructions are read affect the document design?** If people will be using the instructions outdoors, you will need to use a coated paper that can tolerate moisture or dirt. If people will be reading the instructions while sitting in a small, enclosed area, you might select a small paper size and a binding that allows the reader to fold the pages over to save space. If people have a lot of room, you might decide to create poster-size instructions that can be taped to the wall and that are easy to read from across the room.

PLANNING FOR SAFETY

If the subject you are writing about involves safety risks, your most important responsibility is to do everything you can to ensure your readers' safety.

ETHICS NOTE

ENSURING YOUR READERS' SAFETY

To a large extent, the best way to keep your readers safe is to be honest and write clearly. If readers will encounter safety risks, explain what those risks are and how to minimize them. Doing so is a question of rights. Readers have a right to the best information they can get.

Ensuring your readers' safety is also a question of law. People who get hurt can sue the company that made the product or provided the service. This field of law is called liability law. Your company is likely to have legal professionals on staff or on retainer whose job is to ensure that the company is not responsible for putting people at unnecessary risk.

When you write safety information, be clear and concise. Avoid complicated sentences.

COMPLICATED It is required that safety glasses be worn when inside this laboratory.

SIMPLE You must wear safety glasses in this laboratory.

SIMPLE Wear safety glasses in this laboratory.

Sometimes a phrase works better than a sentence: "Safety Glasses Required."

Because a typical manual or set of instructions can contain dozens of comments—some related to safety and some not—experts have devised *signal words* to indicate the seriousness of the advice. Unfortunately, signal words are not used consistently. For instance, the American National Standards Institute (ANSI) and the U.S. military's MILSPEC publish definitions that differ significantly, and many private companies have their own definitions. Figure 14.8 presents the four most commonly used signal words. The first three signal words are accompanied by symbols showing the color combinations endorsed by ANSI in its standard Z535.4.

SIGNAL WORD	EXPLANATION	EXAMPLE
Danger ⚠ DANGER	*Danger* is used to alert readers about an immediate and serious hazard that will likely be fatal. Writers often use all uppercase letters for danger statements.	DANGER: EXTREMELY HIGH VOLTAGE. STAND BACK.
Warning ⚠WARNING	*Warning* is used to alert readers about the potential for serious injury or death or serious damage to equipment. Writers often use all uppercase letters for warning statements.	WARNING: TO PREVENT SERIOUS INJURY TO YOUR ARMS AND HANDS, YOU MUST MAKE SURE THE ARM RESTRAINTS ARE IN PLACE BEFORE OPERATING THIS MACHINE.
Caution ⚠CAUTION	*Caution* is used to alert readers about the potential for anything from moderate injury to serious equipment damage or destruction.	Caution: Do not use nonrechargeable batteries in this charging unit; they could damage the charging unit.
Note	*Note* is used for a tip or suggestion to help readers carry out a procedure successfully.	Note: Two kinds of washers are provided—regular washers and locking washers. Be sure to use the locking washers here.

FIGURE 14.8 Signal Words

FIGURE 14.9
A Typical Safety Label
Information from Clarion Safety Systems, 2013: www.clarionsafety.com/Electrical-Hazard-Safety-Labels.

The yellow triangle on the left is consistent with the ISO approach. Because ISO creates standards for international use, its safety labels use icons, not words, to represent safety hazards.

The signal word *Danger* and the text are consistent with the ANSI approach. The information is presented in English.

Whether safety information is printed in a document or on machinery or equipment, it should be prominent and easy to read. Many organizations use visual symbols to represent levels of danger, but these symbols are not standardized.

Organizations that create products that are used only in the United States design safety information to conform with standards published by ANSI and by the federal Occupational Safety and Health Administration (OSHA). Organizations that create products that are also used outside the United States design safety information to conform with standards published by the International Organization for Standardization (ISO). Figure 14.9 shows a safety label that incorporates both ANSI and ISO standards.

Part of planning for safety is determining the best location for safety information. This question has no easy answer because you cannot control how your audience reads your document. Be conservative: put safety information wherever you think the reader is likely to see it, and don't be afraid to repeat yourself. A reasonable amount of repetition—such as including the same safety comment at the top of each page—is effective. But don't repeat the same piece of advice in each of 20 steps, because readers will stop paying attention to it. If your company's format for instructions calls for a safety section near the beginning of the document, place the information there and repeat it just before the appropriate step in the step-by-step section.

Figure 14.10 shows one industry association's guidelines for placing safety information on conveyor belts.

DRAFTING EFFECTIVE INSTRUCTIONS

Instructions can be brief (a small sheet of paper) or extensive (20 pages or more). Brief instructions might be produced by a writer, a graphic artist, and a subject-matter expert. Longer instructions might call for the assistance of others, such as marketing and legal personnel.

Regardless of the size of the project, most instructions are organized like process descriptions. The main difference is that the conclusion of a set of instructions is not a summary but an explanation of how readers can make sure they have followed the instructions correctly. Most sets of instructions contain four elements: a title, a general introduction, step-by-step instructions, and a conclusion.

Drafting Titles A good title for instructions is simple and clear. Two forms are common:

- **How-to.** This is the simplest: "How To Install the J112 Shock Absorber."
- **Gerund.** The gerund form of a verb is the *-ing* form: "Installing the J112 Shock Absorber."

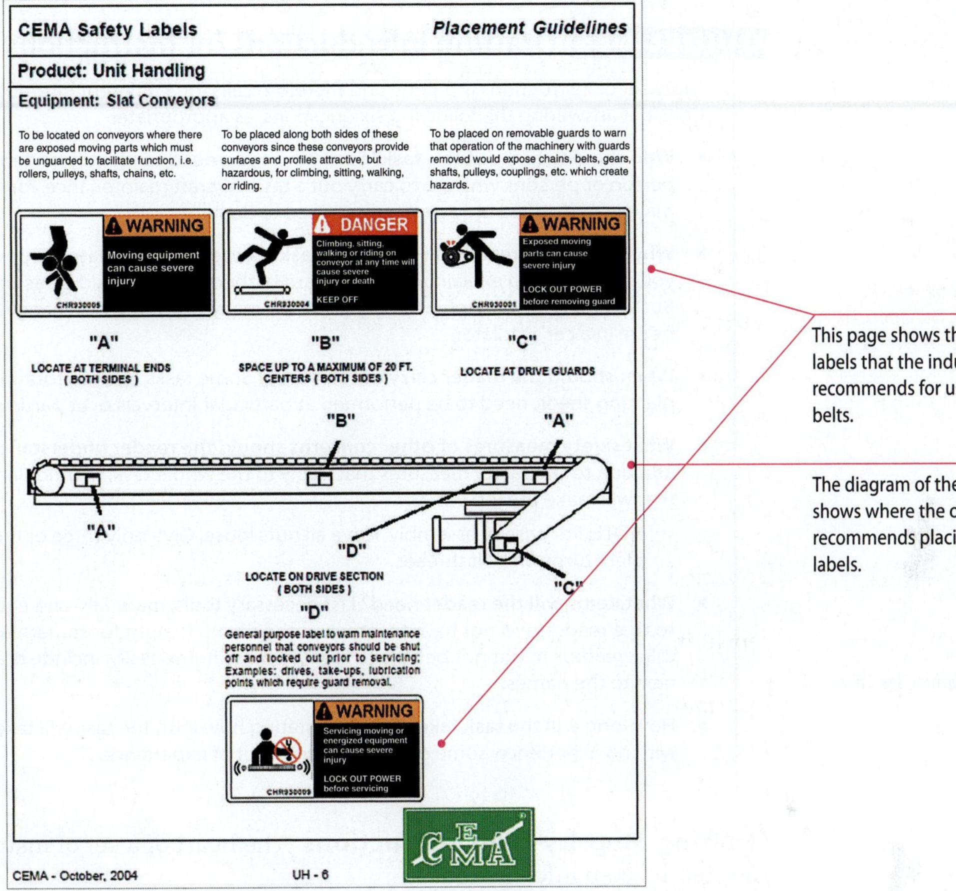
CEMA Safety Labels
Placement Guidelines
Product: Unit Handling
Equipment: Slat Conveyors
To be located on conveyors where there are exposed moving parts which must be unguarded to facilitate function, i.e. rollers, pulleys, shafts, chains, etc.
To be placed along both sides of these conveyors since these conveyors provide surfaces and profiles attractive, but hazardous, for climbing, sitting, walking, or riding.
To be placed on removable guards to warn that operation of the machinery with guards removed would expose chains, belts, gears, shafts, pulleys, couplings, etc. which create hazards.
WARNING
Moving equipment can cause severe injury
DANGER
Climbing, sitting, walking or riding on conveyor at any time will cause severe injury or death
KEEP OFF
WARNING
Exposed moving parts can cause severe injury
LOCK OUT POWER before removing guard
"A"
LOCATE AT TERMINAL ENDS (BOTH SIDES)
"B"
SPACE UP TO A MAXIMUM OF 20 FT. CENTERS (BOTH SIDES)
"C"
LOCATE AT DRIVE GUARDS
"B"
"A"
"A"
"D"
"C"
LOCATE ON DRIVE SECTION (BOTH SIDES)
"D"
General purpose label to warn maintenance personnel that conveyors should be shut off and locked out prior to servicing; Examples: drives, take-ups, lubrication points which require guard removal.
WARNING
Servicing moving or energized equipment can cause severe injury
LOCK OUT POWER before servicing
CEMA
CEMA - October, 2004
UH - 6

This page shows the four safety labels that the industry association recommends for use on conveyor belts.

The diagram of the conveyor belt shows where the organization recommends placing the safety labels.

FIGURE 14.10 **Placement of Safety Information on Equipment**
Information from Conveyor Equipment Manufacturers Association, 2004.

One form to avoid is the noun string, which is awkward and difficult for readers to understand: "J112 Shock Absorber Installation Instructions."

For more about noun strings, see Ch. 6, p. 138.

Drafting General Introductions The general introduction provides the preliminary information that readers will need to follow the instructions safely and easily.

GUIDELINES Drafting Introductions for Instructions

Every set of instructions is unique and therefore calls for a different introduction. Consider answering the following six questions, as appropriate:

- **Who should carry out this task?** Sometimes you need to identify or describe the person or persons who are to carry out a task. Aircraft maintenance, for example, may be performed only by those certified to do it.
- **Why should the reader carry out this task?** Sometimes the reason is obvious: you don't need to explain why a backyard barbecue grill should be assembled. But you do need to explain the rationale for many tasks, such as changing antifreeze in a car's radiator.
- **When should the reader carry out this task?** Some tasks, such as rotating tires or planting seeds, need to be performed at particular intervals or at particular times.
- **What safety measures or other concerns should the reader understand?** In addition to the safety measures that apply to the whole task, mention any tips that will make the job easier:

  NOTE: For ease of assembly, leave all nuts loose. Give only three or four complete turns on bolt threads.
- **What items will the reader need?** List necessary tools, materials, and equipment so that readers will not have to interrupt their work to hunt for something. If you think readers might not be able to identify these items easily, include drawings next to the names.
- **How long will the task take?** Consider stating how long the task will take readers with no experience, some experience, and a lot of experience.

For more about graphics, see Ch. 8.

Drafting Step-by-Step Instructions The heart of a set of instructions is the step-by-step information.

GUIDELINES Drafting Steps in Instructions

Follow these six suggestions for writing steps that are easy to understand.

- **Number the instructions.** For long, complex instructions, use two-level numbering, such as a decimal system:

  1
  - 1.1
  - 1.2

  2
  - 2.1
  - 2.2

  etc.

  If you need to present a long list of steps, group the steps logically into sets and begin each set with a clear heading. A list of 50 steps, for example, could be divided into 6 sets of 8 or 9 steps each.

(*continued*)

- **Present the right amount of information in each step.** Each step should define a single task the reader can carry out easily, without having to refer back to the instructions.

TOO MUCH INFORMATION	1. Mix one part cement with one part water, using the trowel. When the mixture is a thick consistency without any lumps bigger than a marble, place a strip of the mixture about 1″ high and 1″ wide along the face of the brick.
TOO LITTLE INFORMATION	1. Pick up the trowel.
RIGHT AMOUNT OF INFORMATION	1. Mix one part cement with one part water, using the trowel, until the mixture is a thick consistency without any lumps bigger than a marble.
	2. Place a strip of the mixture about 1″ high and 1″ wide along the face of the brick.

- **Use the imperative mood.** The imperative mood expresses a request or a command—for example, "Attach the red wire." The imperative is more direct and economical than the indicative mood ("You should attach the red wire" or "The operator should attach the red wire"). Avoid the passive voice ("The red wire is attached"), because it can be ambiguous: is the red wire already attached?

For more about the imperative mood and the passive voice, see Ch. 6, pp. 133 and 134.

- **Do not confuse steps and feedback statements.** A *step* is an action that the reader is to perform. A *feedback statement* describes an event that occurs in response to a step. For instance, a step might read "Upload your profile file." That step's feedback statement might read "The system will now update your user information." Do not present a feedback statement as a numbered step. Present it as part of the step to which it refers. Some writers give all feedback statements their own design.
- **Include graphics.** When appropriate, add a photograph or a drawing to show the reader what to do. Some activities—such as adding two drops of a reagent to a mixture—do not need an illustration, but they might be clarified by a chart or a table.
- **Do not omit articles (*a*, *an*, *the*) to save space.** Omitting articles can make the instructions unclear and hard to read. In the sentence "Locate midpoint and draw line," for example, the reader cannot tell if "draw line" is a noun (as in "locate the draw line") or a verb and its object (as in "draw a line").

Drafting Conclusions Instructions often conclude by stating that the reader has now completed the task or by describing what the reader should do next. For example:

> Now that you have replaced the glass and applied the glazing compound, let the window sit for at least five days so that the glazing can cure. Then, prime and paint the window.

Some conclusions end with *maintenance tips* or a *troubleshooting guide*. A troubleshooting guide, usually presented as a table, identifies common problems and explains how to solve them.

GUIDELINES Testing Instructions for Usability

The following seven guidelines will help you test instructions for usability.

- **Decide what you need to test.** All but the simplest of instructions should be tested for usability. You should also determine if you need to test your title, the introduction, the conclusion, or other specific aspects of the instructions.
- **Decide what you want the test to accomplish.** This means identifying the kinds of problems you want the test to uncover, whether it's of organization, navigation, visual design, language, or accessibility. You may need more than one type of test to help you identify different kinds of problems.
- **Decide on a testing approach.** Instructions can be tested in a variety of settings, ranging from specific, confined environments to broad contexts with many variables. In guided evaluations, instructions are tested by participants based on criteria determined by the creator of the document. In testing in a lab setting, instructions are evaluated by participants who are assigned to perform certain tasks. The test is performed in a controlled environment with limited variables. And in testing in context, instructions are also evaluated by participants who are assigned to perform certain tasks. In this case, however, the test is performed in an actual working environment, where many uncontrolled variables exist. As the setting expands from controlled environments to actual workplaces, the context more closely matches the real world and the testing approach grows more complex.
- **Decide when you need to conduct the test.** You might use a guided evaluation to test ideas for the design of your pages or screens, a lab test to evaluate a draft of the procedural steps themselves, or a test in context to see how well the final document is working for users. "Test early and often" is a good axiom for any technical communicator.
- **Decide who should be involved in the test.** Most usability tests should involve actual users, because they are the people most likely to help you improve your instructions. That said, there are cases in which it makes sense to enlist the help of user advocates or experts. For example, your test may be an informal evaluation of an early design of your document, conducted by your colleagues in a few hours. Or you may hire experts for their trained insights in a specialized area. But actual users should be the cornerstone of your testing.
- **Consider conducting a pilot test.** A pilot test is a usability test for the usability test. A pilot test can uncover problems with the equipment; the document, site, or software being tested; or the test design.
- **Decide what to do with the test results.** For small or informal tests, simply revise your instructions based on the feedback. For large or formal tests, write a clear, comprehensive report based on your findings. Writing a report will help you make sense of the test results and allow you to communicate them to relevant parties.

REVISING, EDITING, AND PROOFREADING INSTRUCTIONS

You know, of course, to revise, edit, and proofread all the documents you write to make sure they are honest, clear, accurate, comprehensive, accessible, concise, professional in appearance, and correct. When you write instructions, you should be extra careful, for two reasons.

First, your readers rely on your instructions to carry out a task. If they can't complete it—or they do complete it, but they don't achieve the expected outcome—they'll be unhappy. Nobody likes to spend a few hours assembling a garage-door opener, only to find half a dozen parts left over. Second, your readers rely on you to help them complete the task safely. To prevent injuries and liability actions, build time into the budget to revise, edit, and proofread the instructions carefully. Then, if you can, carry out usability testing on the instructions.

A LOOK AT SEVERAL SAMPLE SETS OF INSTRUCTIONS

Figure 14.11 is an excerpt from a set of instructions. Figure 14.12 on page 413 shows a list of tools and materials from a set of instructions. Figure 14.13 on page 414 is an excerpt from the safety information in a set of instructions. Figure 14.14 on page 415 is a portion of the troubleshooting guide in the instructions for a lawnmower. Figure 14.15 on page 415 is an excerpt from a thread in a discussion forum.

This page from the user's manual for a tablet computer used in health-care environments discusses how to use the barcode scanner.

Note that the writer uses a gerund (-*ing* phrase) in the major heading to describe the action ("Using the barcode scanner").

The writer explains why readers might want to scan barcodes.

The writer lists the types of barcodes the tablet can scan and then explains how to enable the tablet to scan additional types. Note that the more-conceptual information about the task precedes the instructional information. Why? Because readers want to understand the big picture before getting into the details.

The writer presents the steps. Note that the writer numbers the steps and uses the imperative mood for each one.

The drawing helps readers understand how to hold the tablet and aim it at the barcode. In cases such as this, simple drawings work better than photographs because they do not distract readers with unnecessary detail.

Using the barcode scanner

Your C5te/F5te is available with an optional integrated 1D and 2D barcode scanner that you can use to retrieve information from barcodes. Many applications use barcodes for asset tracking, identification, and process controls.

Supported barcode types

The C5te/F5te barcode scanner supports several different types of barcodes—a minimum set of barcodes is enabled at the factory.

The following symbologies are enabled by default:

- Aztec Code
- EAN-128
- Code 39
- UPC-A
- EAN-8
- Interleaved 2 of 5
- Micro PDF417
- RSS Limited
- Code 128
- EAN-UCC-CC-AB
- DataMatrix
- UPC-E
- EAN-13
- PDF417
- RSS-14

Other barcode types can be enabled by using EasySet® by Intermec. This application can be installed by running **setup**, which is located here: **C:\Motion\Software\EasySet.**

To use the barcode scanner:

1. Open the application that you want to receive the barcode data and place the insertion point in the appropriate field.
2. Hold the unit by the handle with the scanner lens in front of you.
3. Aim the scanner lens at the barcode.

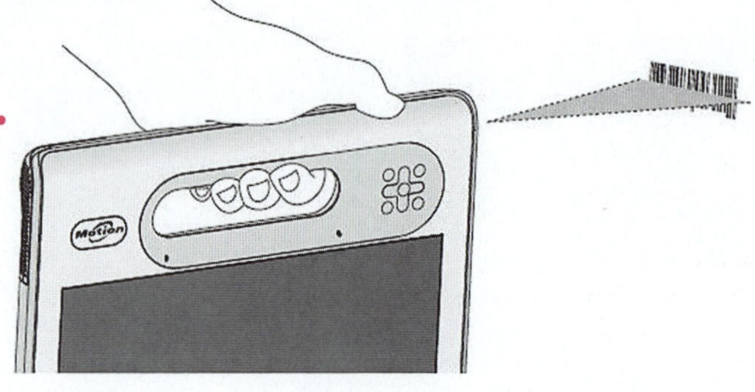

Chapter 2 Using your C5te/F5te   Using the barcode scanner 38

FIGURE 14.11 **Excerpt from a Set of Instructions**
Courtesy of Xplore.

Installation Instructions

PREPARE TO INSTALL THE RANGE

FOR YOUR SAFETY:

All rough-in and spacing dimensions must be met for safe use of your range. Electricity to the range can be disconnected at the outlet without moving the range if the outlet is in the preferred location (remove lower drawer).

To reduce the risk of burns or fire when reaching over hot surface elements, cabinet storage space above the cooktop should be avoided. If cabinet storage space is to be provided above the cooktop, the risk can be reduced by installing a range hood that sticks out at least 5" beyond the front of the cabinets. Cabinets installed above a cooktop must be no deeper than 13."

Be sure your appliance is properly installed and grounded by a qualified technician.

Make sure the cabinets and wall coverings around the range can withstand the temperatures (up to 200°F.) generated by the range.

TOOLS YOU WILL NEED

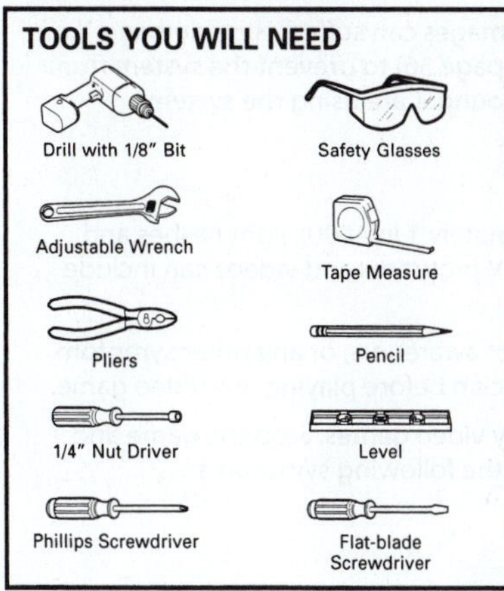

MATERIALS YOU MAY NEED

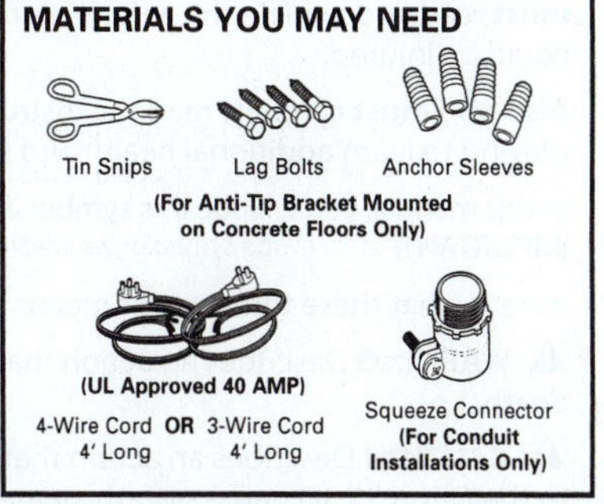

PARTS INCLUDED

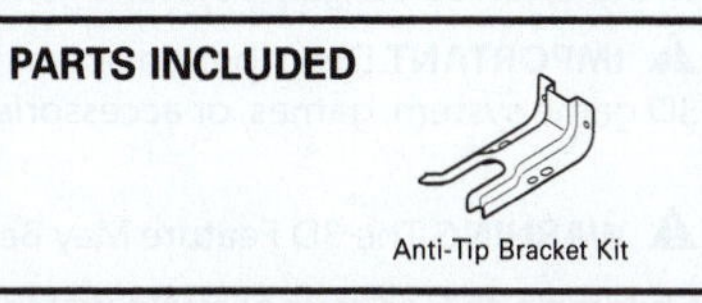

1 REMOVE SHIPPING MATERIALS

Remove packaging materials. Failure to remove packaging materials could result in damage to the appliance.

Drawings of tools, materials, and parts are more effective than lists.

FIGURE 14.12 List of Tools and Materials

Information from General Electric Corporation.

FIGURE 14.13 Excerpt from Safety Information

This excerpt from a user manual for a video-game player that displays 3D images describes two of the safety risks associated with playing video games.

Notice that the excerpt uses mandatory language: "You **must** . . ." Although politeness is desirable most of the time, you don't want to sound as if you are making suggestions or asking readers to do you favors. For instance, if a task calls for using safety goggles, do not write "You should consider wearing safety goggles." Instead, write "You must wear safety goggles when operating this equipment."

This set of safety information defines the keywords *warning*, *caution*, and *important*.

The safety information goes on to discuss eyestrain and motion sickness, repetitive motion injuries, and radio frequency interference.

HEALTH AND SAFETY INFORMATION

You **must** read the following warnings before you set up or use the Orion 35 3D game system. If young children will be using this product, a competent adult **must** read and explain this safety information to them. Otherwise, these children could be injured.

Also, you **must** carefully read the instruction booklet for the game you are playing to learn additional health and safety information.

In this manual, you will see this symbol ⚠ followed by WARNING, CAUTION, or IMPORTANT.

Here is what these three words mean:

⚠ **WARNING** Describes an action that could lead to a serious personal injury or death.

⚠ **CAUTION** Describes an action that could lead to personal injury or damage to the Orion 35 3D game system, games, or accessories.

⚠ **IMPORTANT** Describes an action that could lead to damage to the Orion 35 3D game system, games, or accessories.

⚠ **WARNING** The 3D Feature May Be Used Only by Children 7 and Older

Children age 6 or younger who watch 3D images can suffer vision damage. You must use the Parental Control feature (see page 36) to prevent the system from displaying 3D images when children 6 or younger are using the system.

⚠ **WARNING** Seizures

For a small percentage of people (approximately 1 in 4000), light flashes and patterns can cause seizures or blackouts. TV programs and videos can include these light flashes and patterns.

Anybody who has ever had a seizure, loss of awareness, or any other symptom linked to epilepsy **must** check with a physician before playing any video game.

Always watch your children when they play video games. Stop the game and consult a physician if your child has any of the following symptoms:

- Convulsions
- Eye or muscle twitching
- Loss of awareness
- Altered vision
- Involuntary movements
- Disorientation

To reduce the chance that you or a child will have a seizure while playing video games:

1. Sit or stand as far as possible from the screen.
2. Play the game on the smallest screen that is available.
3. Do not play any video game if you are tired.
4. Keep the room well-lit.
5. Every hour, take a break for 10 or 15 minutes.

PROBLEM	CAUSE	CORRECTION
The mower does not start.	1. The mower is out of gas. 2. The gas is stale. 3. The spark plug wire is disconnected from the spark plug	1. Fill the gas tank. 2. Drain the tank and refill it with fresh gas. 3. Connect the wire to the plug.
The mower loses power.	1. The grass is too high. 2. The air cleaner is dirty. 3. There is a buildup of grass, leaves, or trash in the underside of the mower housing.	1. Set the mower to a "higher cut" position. See page 10. 2. Replace the air cleaner. See page 11. 3. Disconnect the spark plug wire, attach it to the retainer post, and clean the underside of the mower housing. See page 8.

FIGURE 14.14 **Excerpt from a Troubleshooting Guide**

Original post from a user:

I am having difficulty printing TIFF files from a Microsoft program onto 11 x 17 sized paper to my WC 5335 [a Xerox printer]. When I print my file it will print only the 8-1/2 x 11 image on the 11 x 17 paper. I want the image to fill up the whole 11 x 17 page. When I print I select fit to size, but I still get the same thing. Has anyone ever experienced this situation? Does anybody have any tips on how I can resolve this issue? Thanks.

Reply from the Xerox representative monitoring the forum:

Thank you for using the Support Forum. Please take a look at this solution for printing to 11 x 17 sized paper and see if it relates to your issue. If this does not help please consider contacting your support centre for further assistance.

Thanks.

Response from a second user:

Yup indeed, I agree. And I also want to share you some information about TIFF: TIFF, originally called Tagged Image File Format, is a computer file format for storing images, including photographs, line art among graphic artists, the publishing industry, and both amateur and professional photographers in general.

TIFF format is supported widely in industry image processing applications, such as Photoshop (Adobe), GIMP (Jasc), PhotoImpact and Paint Shop Pro (Ulead) and Desktop publishing & Page Layout applications, like QuarkXPress and Adobe InDesign. Other applications, like scanning, faxing, word processing, OCR and more applications also support TIFF format. You can choose a TIFF processing SDK whose way of processing is simple and fast to process TIFF files. I am testing with the related SDKs. I hope we can have some communication later. Good luck.

Response from a third user:

Thanks for sharing, that's awesome but somewhat overpriced for me who will just use it only once. Do you have some cheaper or even free versions? Any suggestion will be appreciated!

Response from a fourth user:

I suggest you have a look at Gimp; it's open source and free.

This excerpt from a user-support group on the Xerox website shows the way many organizations collaborate with their customers to help solve customer problems.

In the original post, a customer explains a problem he or she is having printing a TIFF file with a Xerox printer.

The Xerox representative who monitors the user forum directs the customer to the company's statement that is intended to solve the user's problem.

A second user recommends the Xerox representative's reply and presents additional information about TIFF files. Notice in the last sentence that he or she offers to try to stay in touch with the customer who asked the original question.

A third user, who is experiencing the same problem that the first user wrote about, expresses appreciation for the second user's suggestions but raises a concern about the cost of the alternatives he or she describes.

A fourth user responds by pointing out that GIMP, one of the alternatives listed by the second user, is a free program.

FIGURE 14.15 **Excerpts from a Thread in a Customer-Support Forum**

Many companies use customer forums as an efficient and effective way to help their users solve problems. Although the forums can be messy—and sometimes users write nasty things about the company—companies realize that letting users participate increases the chances that they will find solutions to their problems.

Information from Xerox Corporation, 2013.

DOCUMENT ANALYSIS ACTIVITY

Presenting Clear Instructions

This page is from a set of instructions in an e-reader user's manual. The questions below ask you to think about the discussion of instructions on pages 398–410.

1. This page includes no graphics. Point out two or three passages on the page that might be easier to understand if they included graphics. Describe the graphics you would include.
2. Is the amount of information presented in each step appropriate?
3. How effectively has the designer used typography to distinguish the various kinds of information presented on this page?

Reading on Your NOOK

Reading a Book

To read a book on your NOOK, tap on its cover.

Turning Pages

To turn to the next page, tap along the right side of the screen. You can also turn to the next page by swiping to the left.

To turn to the previous page, tap along the left side of the screen. You can also turn to the previous page by swiping to the right.

Moving from One Part of a Book to Another

To move quickly through the book, do this:

1. Tap the center of the page to open the Reading Tools.
   Two panels of tools appear at the bottom of the page.
2. Use the Reading Tools to move through the book.
   - Drag your finger along the scrollbar to scroll through the book.
     A bubble appears above the scrollbar, telling you which chapter and page you have reached.
   - To go to a specific page, tap *Go to Page*. A dialog box appears. Type the page number you want, and press the Go button (an arrow).
3. When you have reached the page you want, tap the middle of the page to close the Reading Tools and continue reading.

Using the Table of Contents

To see the Table of Contents of the book, or to jump to a chapter or section listed in the Table of Contents, do this:

1. Tap the center of the page to open the Reading Tools.
2. Tap the Contents icon (a stack of horizontal lines).
   A panel opens with three tabs: *Table of Contents*, *Highlights and Notes*, and *Bookmarks*.
3. Tap the *Table of Contents* tab if it is not already highlighted.
   Your NOOK displays the Table of Contents for the book.
4. Tap on the chapter or section you want to read next.
   Your NOOK closes the Reading Tools and displays the page you selected.

Barnes & Noble NOOK HD+ User Guide 45

DOCUMENT ANALYSIS ACTIVITIES

To complete the activities below, go to LaunchPad.

Process Description Using Video Animation

Courtesy of the N.C. Department of Transportation.

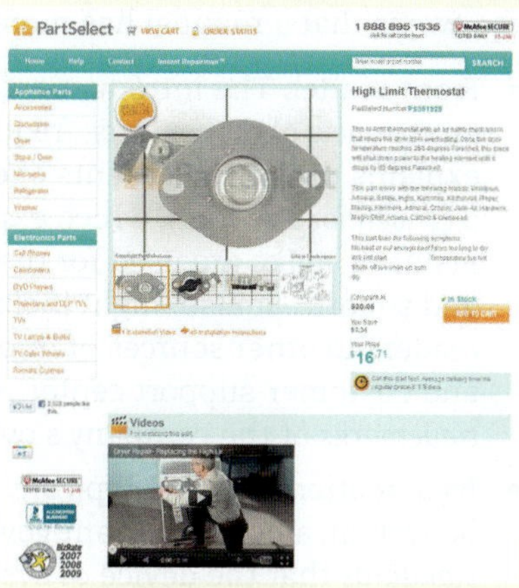

Instructions Using Video Demonstration

Courtesy Eldis Group.

Instructions Using a Combination of Video Demonstration and Screen Capture

Reprinted with permission of Texas Tech University.

Definition Using Video Animation

©ABC News.

Writing Manuals

There is no absolute distinction between a set of instructions and a manual. Typically, the two share a main purpose: to explain how to carry out a task safely, effectively, and efficiently. Both kinds of documents can include safety information. For example, a set of instructions on how to use an extension ladder explains how to avoid power lines and how to avoid falling off the ladder. A manual for a laptop explains how to avoid electrocution when you open the case. However, a set of instructions (which can range from 1 to 20 or more pages) is typically shorter than a manual and more limited in its subject. Obviously, using a laptop requires knowing about many more topics than does using a ladder.

A manual likely also includes some sections not found in a set of instructions. For instance, it typically has a title page. The main difference between the two is that a manual has more-elaborate front matter and back matter:

For more about typography, see Ch. 7.

- **Front matter.** The introduction, sometimes called a *preface*, often contains an *overview of the contents*, frequently in the form of a table, which explains the main contents of each section and chapter. It also contains a *conventions* section, which explains the typography of the manual. For instance, *italics* are used for the titles of books, **boldface** for keyboard keys, and so forth. It also might include a *where to get help* section, referring readers to other sources of information, such as the company's website and customer-support center. And it might contain a section listing the *trademarks* of the company's own products and those of other companies.
- **Back matter.** Manuals typically include a set of *specifications* of the device or system, a list of relevant government *safety regulations* and *industry standards* that the device or system supports, *tips on maintenance and servicing* the device, a *copyright page* listing bibliographic information about the manual, and an *index*. Many manuals also include *glossaries*.

Organizations also work hard to make their instructions and manuals appropriate for multicultural readers. Because important instructions and manuals might be read by readers from any number of cultures, you need to answer three important questions as you plan the documents:

- **In what language should the information be written?** You can either translate the document into readers' native languages or try to make the English easy to understand. Although translation is sometimes the best or only alternative, companies often use Simplified English or some other form of English with a limited grammar and vocabulary. Many organizations translate their manuals into various languages and post the translations on their websites as PDF documents for download.
- **Do the text and graphics need to be modified?** As discussed in Chapter 5, communicators need to be aware of cultural differences. For example, one printer manual aimed at an Italian audience presented nude models with strategically placed rectangles showing the various colors the machine could reproduce. Nudity would be inappropriate in almost all other countries. A software manual in the United States showed an illustration of a person's left hand. Because the left hand is considered unclean in many countries in the Middle East, the manual would need to be modified for those countries (Delio, 2002).
- **What is the readers' technological infrastructure?** If your readers don't have Internet access, there is no point in making a web version of the information. If your readers pay by the minute for Internet access, you will want to create web-based information that downloads quickly.

WRITER'S CHECKLIST

Parenthetical, Sentence, and Extended Definitions

- [ ] Are all necessary terms defined? *(p. 385)*

Are the parenthetical definitions

- [ ] appropriate for the audience? *(p. 385)*
- [ ] clear? *(p. 385)*
- [ ] smoothly integrated into the sentences? *(p. 385)*

Does each sentence definition

- [ ] contain a sufficiently specific category and distinguishing characteristics? *(p. 386)*
- [ ] avoid describing one particular item when a general class of items is intended? *(p. 386)*
- [ ] avoid circular definition? *(p. 386)*
- [ ] identify a category with a noun or a noun phrase? *(p. 386)*
- [ ] Are the extended definitions developed logically and clearly? *(p. 386)*

Descriptions of Objects and Mechanisms

- [ ] Are the nature and scope of the description clearly indicated? *(p. 392)*

In introducing the description, did you answer, if appropriate, the following questions:

- [ ] What is the item? *(p. 393)*
- [ ] What is its function? *(p. 393)*
- [ ] What does it look like? *(p. 393)*
- [ ] How does it work? *(p. 393)*
- [ ] What are its principal parts? *(p. 393)*
- [ ] Is there a graphic identifying all the principal parts? *(p. 393)*

In providing detailed information, did you

- [ ] answer, for each of the major components, the questions listed above in the section on introducing the description? *(p. 394)*
- [ ] choose an appropriate organizational principle? *(p. 395)*
- [ ] include graphics for each of the components? *(p. 395)*

In concluding the description, did you

- [ ] summarize the major points in the part-by-part description? *(p. 394)*
- [ ] include (where appropriate) a description of the item performing its function? *(p. 394)*

Process Descriptions

- [ ] Are the nature and scope of the description clearly indicated? *(p. 392)*

In introducing the description, did you answer, if appropriate, the following questions:

- [ ] What is the process? *(p. 393)*
- [ ] What is its function? *(p. 393)*
- [ ] Where and when does it take place? *(p. 393)*
- [ ] Who or what performs it? *(p. 393)*
- [ ] How does it work? *(p. 393)*
- [ ] What are its principal steps? *(p. 393)*
- [ ] Is there a graphic identifying all the principal steps? *(p. 395)*

In providing detailed information, did you

- [ ] answer, for each of the major steps, the questions listed above in the section on introducing the description? *(p. 393)*
- [ ] discuss the steps in chronological order or some other logical sequence? *(p. 395)*
- [ ] make clear the causal relationships among the steps? *(p. 395)*
- [ ] include graphics for each of the principal steps? *(p. 395)*
- [ ] In concluding the description, did you summarize the major points in the step-by-step description? *(p. 394)*

Instructions

- ☐ In planning the instructions, did you consider other media, such as wikis, discussion boards, and videos? *(p. 398)*
- ☐ Are the instructions designed effectively, with adequate white space and a clear relationship between the graphics and the accompanying text? *(p. 404)*
- ☐ Do the instructions have a clear title? *(p. 406)*

Does the introduction to the set of instructions

- ☐ state the purpose of the task? *(p. 408)*
- ☐ describe safety measures or other concerns that readers should understand? *(p. 408)*
- ☐ list necessary tools and materials? *(p. 408)*

Are the step-by-step instructions

- ☐ numbered? *(p. 408)*
- ☐ expressed in the imperative mood? *(p. 409)*
- ☐ simple and direct? *(p. 409)*
- ☐ accompanied by appropriate graphics? *(p. 409)*

Does the conclusion

- ☐ include any necessary follow-up advice? *(p. 409)*
- ☐ include, if appropriate, a troubleshooting guide? *(p. 410)*

EXERCISES

For more about memos, see Ch. 9, p. 248.

1. Add a parenthetical definition for the italicized term in each of the following sentences:

a. Reluctantly, he decided to *drop* the physics course.

b. The Anthropology Club decided to use *crowdfunding* to finance the semester's dig in Utah.

c. The department is using *shareware* in its drafting course.

2. Write a sentence definition for each of the following terms:

a. catalyst

b. job interview

c. website

3. Revise any of the following sentence definitions that need revision:

a. A thermometer measures temperature.

b. The spark plugs are the things that ignite the air-gas mixture in a cylinder.

c. Parallel parking is where you park next to the curb.

d. A strike is when the employees stop working.

e. Multitasking is when you do two things at once while you're on the computer.

4. Write a 500- to 1,000-word extended definition of one of the following terms or of a term used in your field of study. In a brief note at the start, indicate the audience and purpose for your definition. If you do secondary research, cite your sources clearly and accurately (see Appendix, Part A, for documentation systems). Check that any graphics you use are appropriate for your audience and purpose.

a. flextime

b. binding arbitration

c. robotics

d. an academic major (don't focus on any particular major; instead, define what a major is)

e. bioengineering

5. Write a 500- to 1,000-word description of one of the following items or of a piece of equipment used in your field. In a note preceding the description, specify your audience and indicate the type of description (general or particular) you are writing. Include appropriate graphics, and be sure to cite their sources correctly if you did not create them (see Appendix, Part A, for documentation systems).

a. GPS device

b. smartphone

c. waste electrical and electronic equipment

d. automobile jack

e. Bluetooth technology

6. Write a 500- to 1,000-word description of one of the following processes or a similar process with which you are familiar. In a note preceding the description, specify your audience and indicate the type of description (general or particular) you are writing. Include appropriate graphics. If you use secondary sources, cite them properly (see Appendix, Part A, for documentation systems).
   - **a.** how a wind turbine works
   - **b.** how a food co-op works
   - **c.** how a suspension bridge is constructed
   - **d.** how people see
   - **e.** how a baseball player becomes a free agent
7. Study a set of instructions from eHow: www.ehow.com. Write a memo to your instructor evaluating the quality of the instructions. Attach a screen shot or a printout of representative pages from the instructions.
8. You work in the customer-relations department of a company that makes plumbing supplies. The head of product development has just handed you a draft (see p. 422) of installation instructions for a sliding tub door. She has asked you to comment on their effectiveness. Write a memo to her, evaluating the instructions and suggesting improvements.
9. Write a brief manual for a process familiar to you. For example, you might write a procedures manual for a school activity or a part-time job, such as your work as the business manager of the school newspaper or as a tutor in the writing center.
10. **TEAM EXERCISE** Write instructions for one of the following activities or for a process used in your field. In a brief note preceding the instructions, indicate your audience and purpose. Include appropriate graphics.
    - **a.** how to change a bicycle tire
    - **b.** how to convert a WAV file to an MP3 file
    - **c.** how to find an online discussion board and subscribe to it
    - **d.** how to synchronize files and folders across two devices

    Exchange instructions with a partner. Each of you should observe the other person and take notes as he or she attempts to carry out the instructions. Then revise your instructions and share the revised version with your partner; discuss whether the revised instructions are easier to understand and apply and, if so, what made the difference. Submit the final version of your instructions to your instructor.

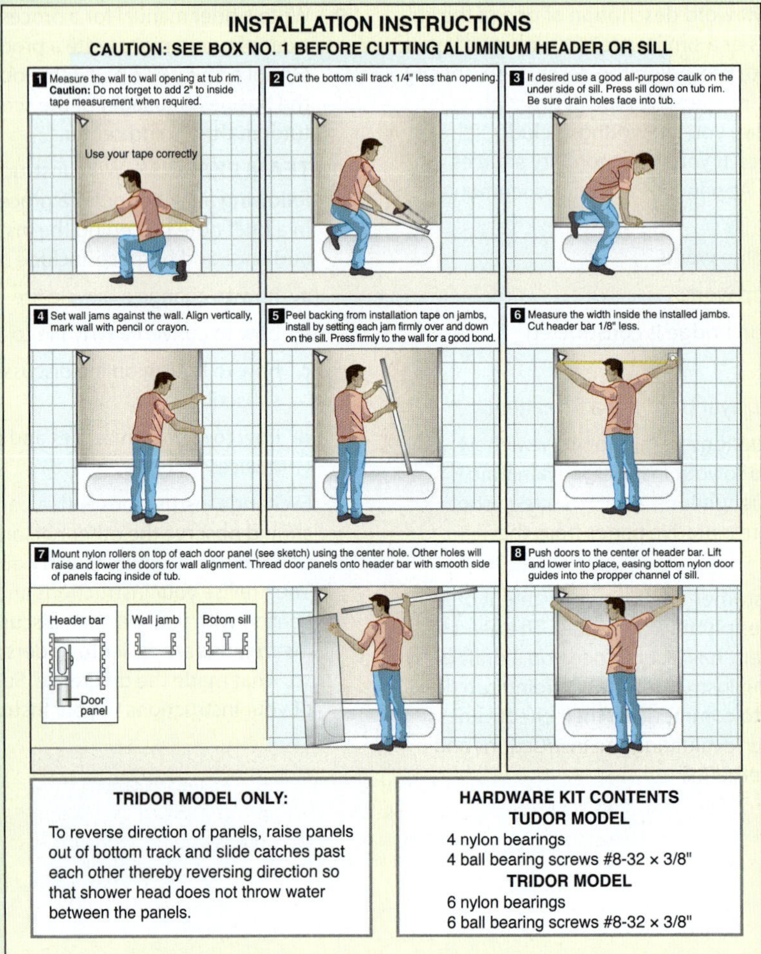

INSTALLATION INSTRUCTIONS
CAUTION: SEE BOX NO. 1 BEFORE CUTTING ALUMINUM HEADER OR SILL
1 Measure the wall to wall opening at tub rim. Caution: Do not forget to add 2" to inside tape measurement when required.
Use your tape correctly
2 Cut the bottom sill track 1/4" less than opening.
3 If desired use a good all-purpose caulk on the under side of sill. Press sill down on tub rim. Be sure drain holes face into tub.
4 Set wall jams against the wall. Align vertically, mark wall with pencil or crayon.
5 Peel backing from installation tape on jambs, install by setting each jam firmly over and down on the sill. Press firmly to the wall for a good bond.
6 Measure the width inside the installed jambs. Cut header bar 1/8" less.
7 Mount nylon rollers on top of each door panel (see sketch) using the center hole. Other holes will raise and lower the doors for wall alignment. Thread door panels onto header bar with smooth side of panels facing inside of tub.
Header bar
Door panel
Wall jamb
Botom sill
8 Push doors to the center of header bar. Lift and lower into place, easing bottom nylon door guides into the propper channel of sill.
TRIDOR MODEL ONLY:
To reverse direction of panels, raise panels out of bottom track and slide catches past each other thereby reversing direction so that shower head does not throw water between the panels.
HARDWARE KIT CONTENTS
TUDOR MODEL
4 nylon bearings
4 ball bearing screws #8-32 × 3/8"
TRIDOR MODEL
6 nylon bearings
6 ball bearing screws #8-32 × 3/8"

CASE 14: Choosing a Medium for Presenting Instructions

You work for the U.S. Department of Energy, which maintains an informational website called ENERGY STAR. One feature of the site is a series of videos on energy conservation produced for the general public. Your supervisor, concerned that few people are viewing most of the videos, has asked you to investigate the problem. Your assignment is to sample one of the videos and then do two tasks: (1) develop a revised script that addresses any problems you see in the original video and (2) write an outline for a print document that will deliver the same information. Once you've completed both tasks, you will need to determine what media are best for delivering this information to the public. To get started, go to LaunchPad.

CHAPTER

15 Making Oral Presentations

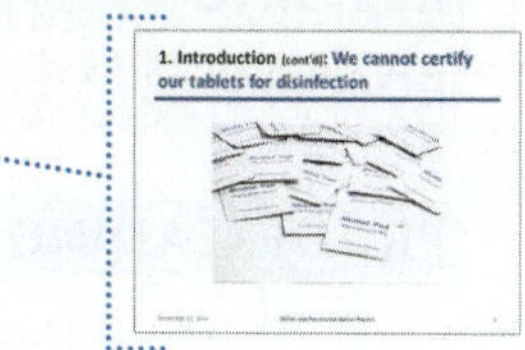

A RECENT SEARCH FOR "death by PowerPoint" on Google returned some 1,770,000 hits. Apparently, a lot of people have been on the receiving end of boring presentations built around bullet slides. But an oral presentation—with or without slides—doesn't have to be deadly dull.

And the process of creating and delivering a presentation doesn't have to be frightening. You might not have had much experience in public speaking, and perhaps your few attempts have been difficult. However, if you approach it logically, an oral presentation is simply another application you need to master in your role as a technical professional or technical communicator. Once you learn that the people in the room are there to hear what you have to say—not to stare at you or evaluate your clothing or catch you making a grammar mistake—you can calm down and deliver your information effectively while projecting your professionalism.

There are four basic types of presentations:

- **Impromptu presentations.** You deliver the presentation without advance notice. For instance, at a meeting, your supervisor calls on you to speak for a few minutes about a project you are working on.
- **Extemporaneous presentations.** You plan and rehearse the presentation, and you might refer to notes or an outline, but you create the sentences as you speak. At its best, an extemporaneous presentation is clear and sounds spontaneous.
- **Scripted presentations.** You read a text that was written out completely in advance (by you or someone else). You sacrifice naturalness for increased clarity and precision.
- **Memorized presentations.** You speak without notes or a script. Memorized presentations are not appropriate for most technical subjects because most people cannot memorize presentations longer than a few minutes.

This chapter discusses extemporaneous and scripted presentations.

Photo: BeeBright/Shutterstock.

FOCUS ON PROCESS Preparing an Oral Presentation

When preparing an oral presentation, pay special attention to these steps.

PLANNING	You will need to prepare effective presentation graphics that are visible, legible, simple, clear, and correct. Choose the appropriate technology based on the speaking situation and the available resources.
DRAFTING	Choose effective and memorable language. Your listeners will not be able to read your presentation to help them understand your message.
REVISING EDITING	Rehearse at least three times, making any necessary changes to your transitions, the order of your slides, or your graphics. Get feedback from others on ways to improve your presentation.
PROOFREADING	Make sure your spoken words and slides are well written. Make sure that your slides are free of errors.

Preparing the Presentation

When you see an excellent 20-minute presentation, you are seeing only the last 20 minutes of a process that took many hours. Experts recommend devoting 20 to 60 minutes of preparation time for each minute of the finished presentation (Nienow, 2013). That means that the average 20-minute presentation might take more than 13 hours to prepare. Obviously, there are many variables, including your knowledge of the subject and your experience creating graphics and giving presentations on that subject. But the point is that good presentations don't just happen.

As you start to prepare a presentation, think about ways to enlist others to help you prepare and deliver it. If possible, you should rehearse the presentation in front of others. You can also call on others to help you think about your audience and purpose, the organization of the information, the types of graphics to use, appropriate designs for slides, and so forth. The more extensively you work with other people as you plan, assemble, and rehearse, the more successful the presentation is likely to be.

Preparing an oral presentation requires five steps:

- analyzing the speaking situation
- organizing and developing the presentation
- preparing presentation graphics
- choosing effective language
- rehearsing the presentation

ANALYZING THE SPEAKING SITUATION

First, analyze your audience and purpose. Then determine how much information you can deliver in the allotted time.

Analyzing Your Audience and Purpose In planning an oral presentation, consider audience and purpose, just as you would in writing a document.

- **Audience.** What does the audience know about your subject? Your answer will help you determine the level of technical vocabulary and concepts you will use, as well as the types of graphics. Why are audience members listening to your presentation? Are they likely to be hostile, enthusiastic, or neutral? A presentation on the benefits of free trade, for instance, will be received one way by conservative economists and another way by U.S. steelworkers. Does your audience include nonnative speakers of English? If so, prepare to slow down the pace of the delivery and use simple vocabulary.
- **Purpose.** Are you attempting to inform or to both inform and persuade? If you are explaining how wind-turbine farms work, you will describe a process. If you are explaining why your company's wind turbines are an economical way to generate power, you will compare them with other power sources.

Your analysis of your audience and purpose will affect the content and the form of your presentation. For example, you might have to emphasize some aspects of your subject and ignore others altogether. Or you might have to arrange topics to accommodate an audience's needs.

TABLE 15.1 Time Allotment for a 20-Minute Presentation

TASK	TIME (MINUTES)
• Introduction	2
• Body	
– First Major Point	4
– Second Major Point	4
– Third Major Point	4
• Conclusion	2
• Questions	4

Budgeting Your Time At most professional meetings, each speaker is given a maximum time, such as 20 minutes. If the question-and-answer period is part of your allotted time, plan accordingly. Even for an informal presentation, you will probably have to work within an unstated time limit that you must determine from the speaking situation. If you take more than your time, eventually your listeners will resent you or simply stop paying attention.

For a 20-minute presentation, the time allotment shown in Table 15.1 is typical. For scripted presentations, most speakers need a little over a minute to deliver a double-spaced page of text effectively.

Considering Setting You may be delivering your presentation in person or remotely, with or without technical aids, to an audience ranging from 1 person to 1,000 or more people. All of these variables will affect the type of presentation you prepare and the way you prepare it. For example, if you will be speaking without technical aids to a small group of people, you may want to prepare handouts for them to take along. If you are preparing a remote online presentation on a platform such as WebEx or Join Me, keep in mind that viewers will be focused on the materials they see on their computer or tablet screens—so your presentation materials will be perhaps even more important than the words you speak.

ORGANIZING AND DEVELOPING THE PRESENTATION

The speaking situation will help you decide how to organize and develop the information you will present.

Start by considering the organizational patterns used typically in technical communication. For instance, if you are a quality-assurance engineer for a computer-chip manufacturer and must address your technical colleagues on why one of the company's products is experiencing a higher-than-normal failure rate, think in terms of cause and effect: the high failure rate is the effect, but what is the cause? Or think in terms of problem-method-solution: the high failure rate is the problem; the research you conducted to determine its cause is the method; your recommended action is the solution.

As you create an effective organizational pattern for your presentation, note the kinds of information you will need for each section of the presentation. Some of this information will be data; some of it will be graphics that you can use in your presentation; some might be objects that you want to pass around in the audience. Prepare an outline of your presentation.

This is also a good time to plan the introduction and the conclusion. Like an introduction to a written document, an introduction to an oral presentation helps your audience understand what you are going to say, why you are going to say it, and how you are going to say it. The conclusion reinforces what you have said and looks to the future.

GUIDELINES Introducing and Concluding the Presentation

In introducing a presentation, consider these five suggestions.

- **Introduce yourself.** Unless you are speaking to colleagues you work with every day, begin with an introduction: "Good morning. My name is Omar Castillo, and I'm the Director of Facilities here at United." If you are using slides, include your name and position on the title slide.
- **State the title of your presentation.** Like all titles, titles of presentations should name the subject and purpose, such as "Replacing the HVAC System in Building 3: Findings from the Feasibility Study." Include the title of your presentation on your title slide.
- **Explain the purpose of the presentation.** This explanation can be brief: "My purpose today is to present the results of the feasibility study carried out by the Facilities Group. As you may recall, last quarter we were charged with determining whether it would be wise to replace the HVAC system in Building 3."

(continued)

- **State your main point.** An explicit statement can help your audience understand the rest of the presentation: "Our main finding is that the HVAC system should be replaced as soon as possible. Replacing it would cost approximately $120,000. The payback period would be 2.5 years. We recommend that we start soliciting bids now, for an installation date in the third week of November."
- **Provide an advance organizer.** Listeners need an advance organizer that specifically states where you are going: "First, I'd like to describe our present system, highlighting the recent problems we have experienced. Next, I'd like to . . . . Then, I'd like to . . . . Finally, I'd like to invite your questions."

In concluding a presentation, consider these four suggestions.

- **Announce that you are concluding.** For example, "At this point, I'd like to conclude my talk with . . . ." This statement helps the audience focus on your conclusions.
- **Summarize the main points.** Because listeners cannot replay what you have said, you should briefly summarize your main points. If you are using slides, you should present a slide that lists each of your main points in one short phrase.
- **Look to the future.** If appropriate, speak briefly about what you think (or hope) will happen next: "If the president accepts our recommendation, you can expect the renovation to begin in late November. After a few hectic weeks, we'll have the ability to control our environment much more precisely than we can now—and start to reduce our expenses and our carbon footprint."
- **Invite questions politely.** You want to invite questions because they help you clarify what you said or communicate information that you did not present in the formal presentation. You want to ask politely to encourage people to speak up.

PREPARING PRESENTATION GRAPHICS

To watch a tutorial on creating presentation slides, go to LaunchPad.

Graphics clarify or highlight important ideas or facts. Statistical data, in particular, lend themselves to graphical presentation, as do abstract relationships and descriptions of equipment or processes. Researchers have known for decades that audiences remember information better if it is presented to them verbally and visually rather than only verbally (see, for instance, Fleming and Levie, 1978). Research reported by speaking coach Terry C. Smith (1991) indicates that presentations that include graphics are judged to be more professional, persuasive, and credible than those that do not. In addition, Smith notes, audiences remember the information better, with a retention rate of 65 percent three days later with graphics, as opposed to 10 percent without.*

For more about creating graphics, see Ch. 8.

Most speakers use presentation software to develop slides. By far the most-popular program is PowerPoint, but other software is becoming popular, too. One that has gained a lot of attention is Prezi, which takes a different approach than PowerPoint. Whereas PowerPoint uses a linear organization—the speaker presents each slide in sequence—Prezi uses a network or web pattern of organization. Figure 15.1 shows an example of a Prezi slide.

*Smith, Terry C. *Making Successful Presentations: A Self-Teaching Guide.* New York: Wiley, 1991. Copyright © 1991. Reprinted with permission from Dominick Abel Literary Agency, Inc.

Although the speaker can present each slide in sequence using the arrows at the bottom of the screen, Prezi makes it simple for the speaker to create a new sequence each time he or she delivers the presentation or to come back to a particular slide or set of slides during the question-and-answer period.

When the speaker clicks on any object on this overview slide, the software zooms in to reveal more objects within or next to that object. After discussing an object, the speaker can zoom out to the overview or zoom in to another object. This capability of Prezi not only makes it easy for the speaker to navigate within the presentation but also helps audience members remember the big picture because they see the overview graphic frequently as the speaker moves from point to point. Whereas PowerPoint is linear like a book—page 6 always precedes page 7—Prezi is more like a website in that the speaker can easily jump from one object to another.

The free version of Prezi is cloud-based; that is, you have to be connected to the Internet to use it. This feature means that you can deliver your Prezi presentation from any computer with an Internet connection, that your team members can collaborate from remote locations, and that your presentation is stored in the cloud. If you are using the basic version of Prezi, your presentation will be visible to all who visit the Prezi site; however, Prezi allows those with .edu email addresses to register for a free package that enables them to control privacy settings. If your presentation contains confidential information or information you do not wish to share with those beyond your original audience, you might want to use this option.

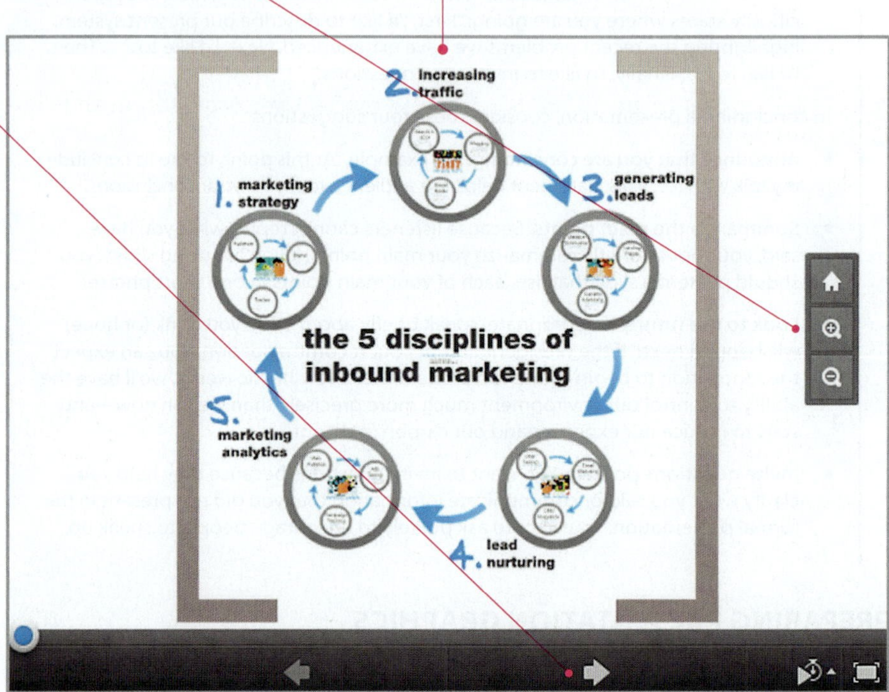

FIGURE 15.1 The Network Organization of Prezi
Shade Wilson, Scalability Project, LLC.

When preparing a presentation using a program such as PowerPoint or Prezi, it may be best to create your own simple design rather than rely on preexisting templates. In addition to templates, many presentation software programs contain animation effects. In PowerPoint, you can set the software so that when a new slide appears, it is accompanied by the sound of applause or of breaking glass, and the heading text spins around like a pinwheel. In Prezi, you can transition between two frames by rotating the canvas by as much as 90 degrees. But unless you have a good reason to use these animation effects, don't. Animation effects that are unrelated to your subject undercut your professionalism and quickly become tiresome.

However, one animation effect in PowerPoint, called *appear and dim*, is useful. When you create a bulleted list, you can set the software to show just the first bullet item and then make the next bullet item appear when you click the mouse. When you do so, the previous bullet item dims. This feature is useful because it focuses the audience's attention on the bullet item you are discussing. Regardless of whether you are using the appear-and-dim feature, set the software so that you use the mouse (or a colleague does) to advance from one graphic to the next. If you set the software so that the graphics advance automatically at a specified interval, such as 60 seconds, you might have to speed up or slow down your presentation to sync with the graphics.

Characteristics of an Effective Slide An effective presentation slide has five characteristics:

- **It presents a clear, well-supported claim.** In a presentation slide, the best way to present a claim and to support it is to put the claim in the headline section of the slide and the support in the body of the slide. For a Prezi presentation, include a claim and its support within a single frame, using design principles to clearly identify each item for the audience.
- **It is easy to see.** The most common problem with presentation text is that it is too small. In general, text has to be in 24-point type or larger to be visible on a screen. Figure 15.2 shows a slide containing so much information that most of it is too small to see easily.

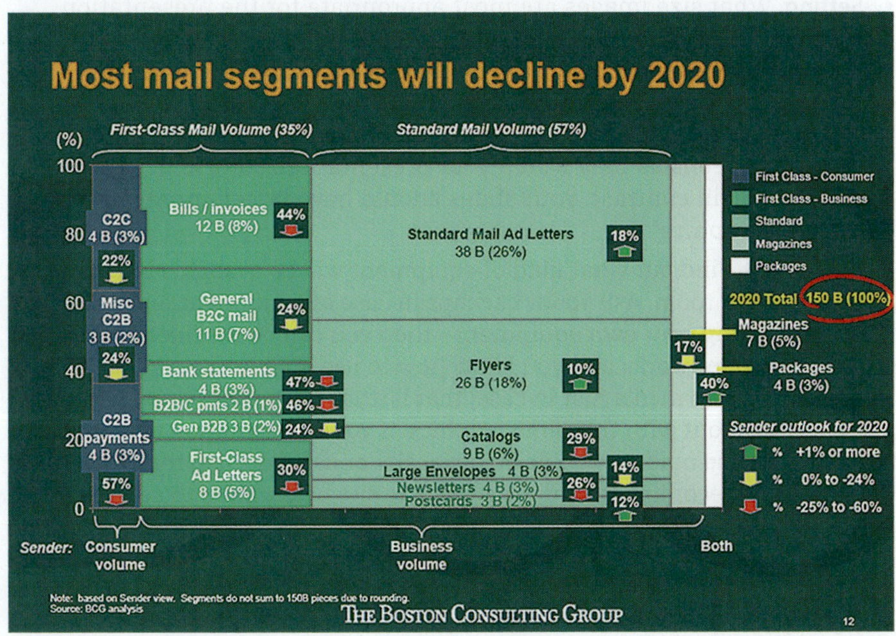

FIGURE 15.2 Too Much Information on a Slide
Information from Boston Group, 2010.

- **It is easy to read.** Use clear, legible lines for drawings and diagrams; black on white works best. Use legible typefaces for text; a boldface sans-serif typeface such as Arial or Helvetica is effective because it reproduces clearly on a screen. Avoid shadowed and outlined letters.
- **It is simple.** Each slide should present only one idea. Your listeners have not seen the graphic before and will not be able to linger over it.
- **It is correct.** Proofread your slides carefully. Everyone makes mistakes in grammar, punctuation, or spelling, but mistakes are particularly embarrassing when they are 10 inches tall on a screen.

To watch a tutorial on audio recording and editing, go to LaunchPad.

Graphics and the Speaking Situation To plan your graphics, analyze four aspects of the speaking situation:

- **Length of the presentation.** How many slides should you have? Smith (1991) suggests showing a different slide approximately every 30 seconds of the presentation. This figure is only a guideline; base your decision on your subject and audience. Still, the general point is valid: it is far better to have a series of simple slides than to have one complicated slide that stays on the screen for five minutes.
- **Audience aptitude and experience.** What kinds of graphics can your audience understand easily? You don't want to present scatter graphs, for example, if your audience does not know how to interpret them.
- **Setting.** What size images are most appropriate for the presentation context? Graphics suitable for a remote online WebEx presentation differ from those suitable for presentation in a small conference room or a 500-seat auditorium. In WebEx, you can provide more detail because viewers will most likely be sitting close to their computer or tablet screens as they watch your presentation. For use in a large auditorium, in contrast, your slides should have larger, more simplified text and graphics.
- **Equipment.** Find out what kind of equipment will be available in the presentation room. Ask about backups in case of equipment failure. If possible, bring your own equipment—then you can be confident that the equipment works and you know how to use it. Some speakers bring graphics in two media just in case; that is, they have slides, but they also bring a handout with the same graphics. If your presentation is going to be recorded to be made available on a website or as a podcast, try to arrange to have the recording technicians visit the site beforehand to see if there are any problems they will need to solve.

TECH TIP

Why To Create a Presentation Template

Master slides and templates allow you to maintain a consistent look throughout your presentation graphics without needing to design each individual slide. Although the templates provided by Prezi and PowerPoint can be useful, they often violate basic design principles, and they may not fit the needs of your presentation. When possible, create an original design that suits your topic and will appear fresh and original to your audience members, who have probably seen many of the predesigned templates used before.

How To Create a Master Page Design in Presentation Slides

PowerPoint

In PowerPoint, select **Slide Master** from the **Master Views** group on the **View** tab.

By selecting elements on the master slide and then using the commands on the **Slide Master** tab, you can add a background, choose a color scheme, and choose type styles and sizes.

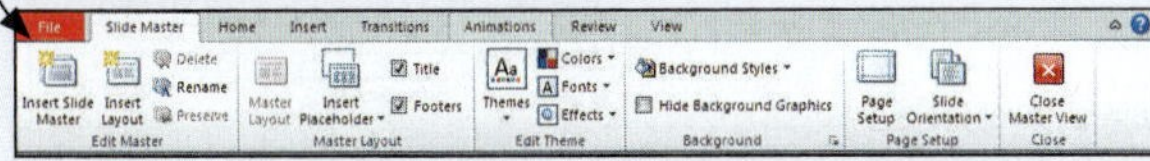

To add graphics to the master slide, use the **Images** and **Illustrations** groups on the **Insert** tab.

To save your page design so that you can use this design for another presentation, select **Save As** from the **File** tab, and then select **PowerPoint Template** from the drop-down menu.

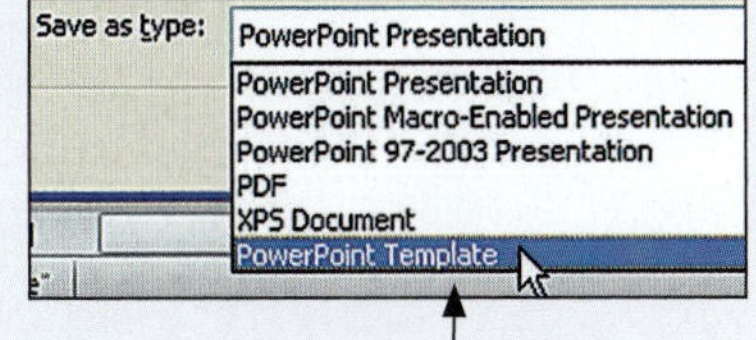

Prezi

In Prezi, select **Start blank Prezi** on the **Choose your template** page.

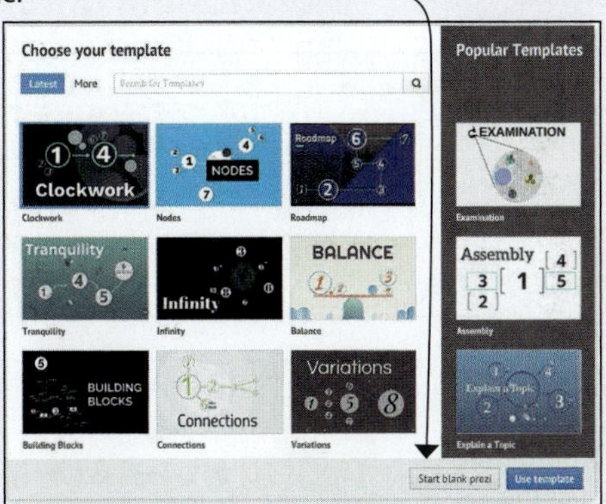

Click on the **Customize** option to launch the **Customize** panel, where you can choose the background colors and insert your own background images or content from your computer.

More background options are available from the **Advanced** button and the bottom of the **Customize** panel.

To save your customized background theme, choose **Save current theme**.

TECH TIP

Why To Set List Items To Appear and Dim During a Presentation

To help your audience focus on the point you are discussing, you can apply PowerPoint's custom animation feature to the **Master Page** so that a list item appears and then dims when the next item appears.

How To Set List Items To Appear and Dim During a Presentation

1. To apply a **custom animation**, select the **Title and Content Layout** slide in the **Slide Master** view, and then highlight the list on the slide.

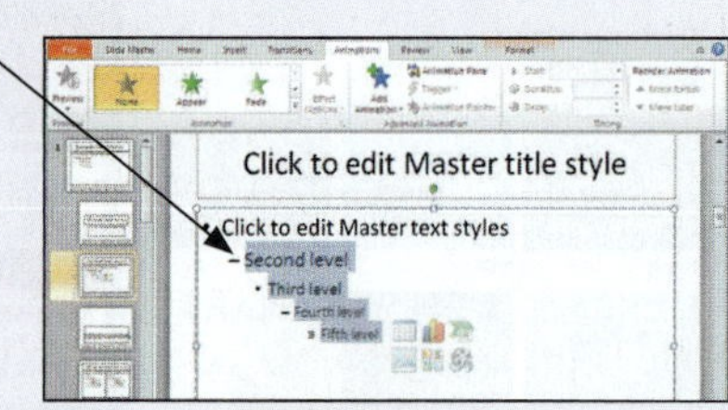

2. In the **Advanced Animation** group, select **Add Animation**, and then select the **Entrance** category and the **Appear** effect.

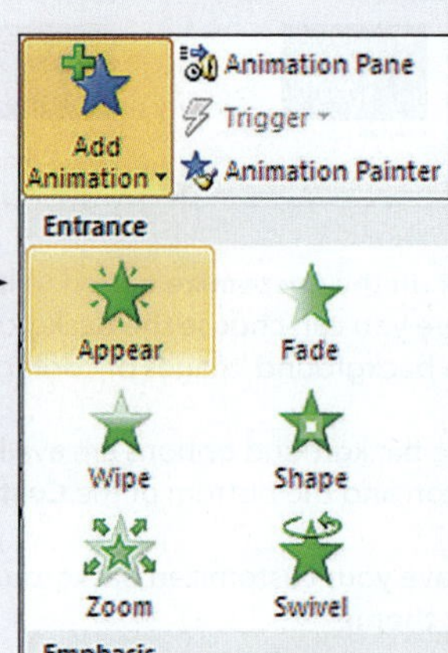

3. Select the **Animation Pane** button in the **Advanced Animation** group. In the **Animation Pane**, click the drop-down menu and select **Effect Options**.

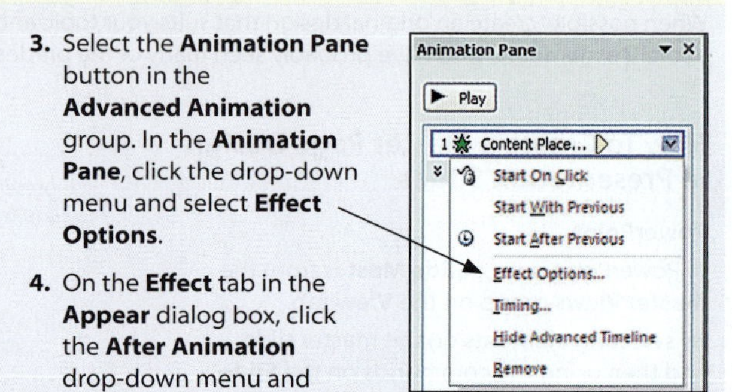

4. On the **Effect** tab in the **Appear** dialog box, click the **After Animation** drop-down menu and select a dim color.

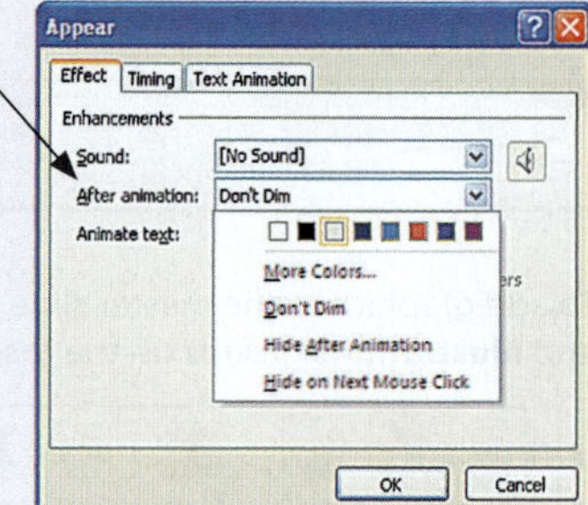

Using Graphics To Signal the Organization of a Presentation Used effectively, graphics can help you communicate how your presentation is organized. For example, you can use the transition from one graphic to the next to indicate the transition from one point to the next. Figure 15.3 shows the slides for a presentation that accompanied the report in Chapter 13 on tablet computer use at Rawlings Regional Medical Center (see p. 357).

Selecting a Tablet Computer for the Clinical Staff at Rawlings Regional Medical Center: A Recommendation Report

Prepared by:
Jeremy Elkins, Director of IT
Eloise Carruthers, Director of Nursing

December 15, 2013

The title slide shows the title of the presentation and the name and affiliation of each speaker. You might also want to include the date of the presentation.

Recommendation Report Outline

1. Introduction
2. Major Results
   2.1 Clinical-staff knowledge and attitudes
   2.2 BYOD v. hospital-owned model
   2.3 Determining criteria
   2.4 Assessing tablets
3. Conclusions
4. Recommendation

The next slide presents an overview, which outlines the presentation. The arrow identifies the point the speaker is addressing.

At the bottom of each slide in the body of the presentation is a footer with the date, the title of the presentation, and the number of the slide. The slide number gives audience members a way to refer to the slide when they ask questions.

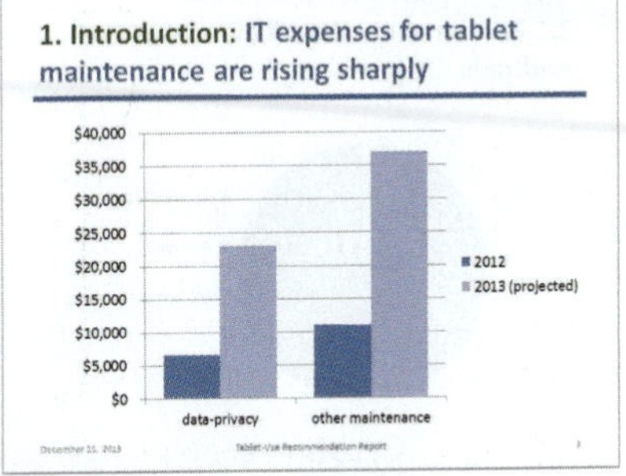

This slide uses a simple bar graph created in PowerPoint. Don't try to present a lot of information on a single slide.

The title of this slide uses the numbering system introduced in the previous slide. This cue helps the audience understand the structure of the presentation. Following the colon is an independent clause that presents the claim that will be supported in the slide.

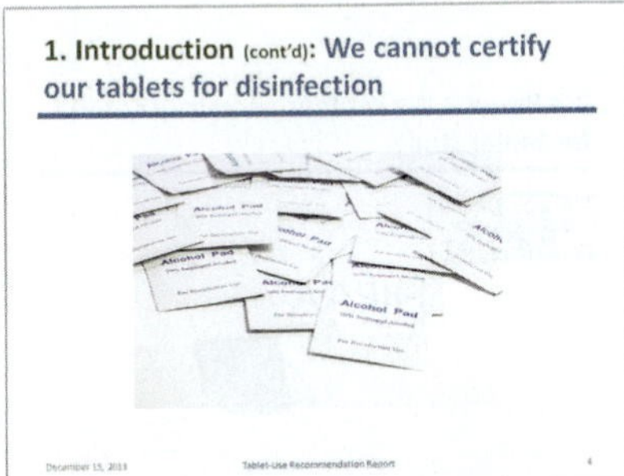

If the images in your presentation are your own intellectual property or are clip art that comes with the software, you can legally display them anywhere. If they are not, you need to cite their sources and obtain written permission. You have three choices for placing the source statements: at the bottom of the appropriate slides, in a sources slide that you show at the end of the presentation, or on a paper handout that you distribute at the end of the presentation.
BeeBright/Shutterstock.

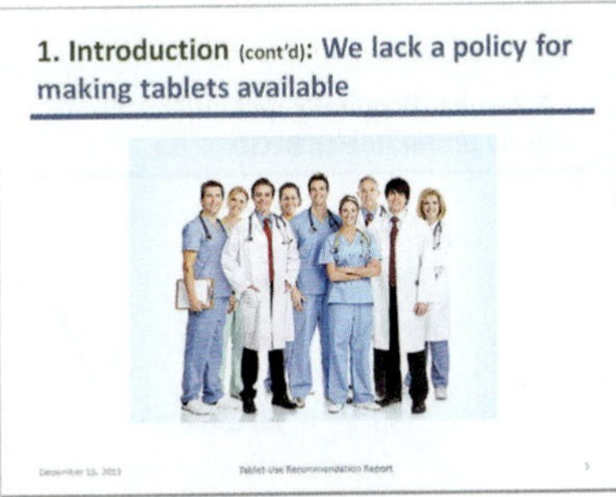

For second and subsequent slides within the same section of the presentation, use the "continued" abbreviation shown here. Notice that the speakers use one color for the generic term ("Introduction") and a different color for the claim.
kurhan/Shutterstock.

Recommendation Report Outline

1. Introduction
2. Major Results
   2.1 Clinical-staff knowledge and attitudes
   2.2 BYOD v. hospital-owned model
   2.3 Determining criteria
   2.4 Assessing tablets
3. Conclusions
4. Recommendation

This slide is identical to Slide 2, except that the arrow has moved. Use this organizing slide to help your audience remember the overall organization of your presentation. Don't overdo it; if you presented this organizing slide just a few slides ago, don't use it again until you make the transition to the next major unit in the presentation.

FIGURE 15.3 Sample PowerPoint Presentation

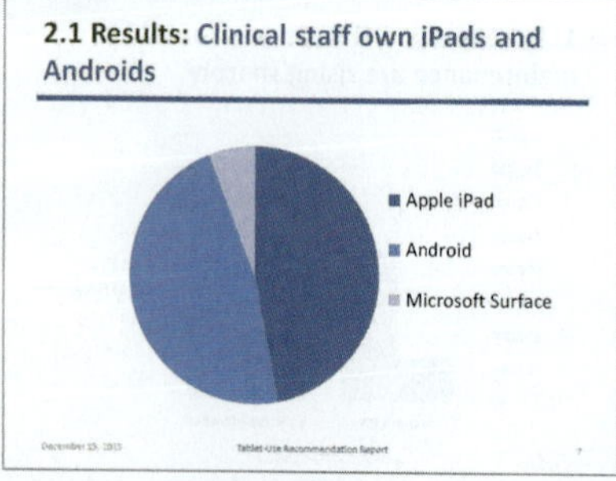

A pie chart is a logical choice for representing a small number of components (usually, seven or fewer) that add up to 100 percent.

Note that the speakers use conservative blues for all their graphics on the slide set. You don't need a rainbow full of colors. You need just enough difference so that the audience can distinguish between the different slices.

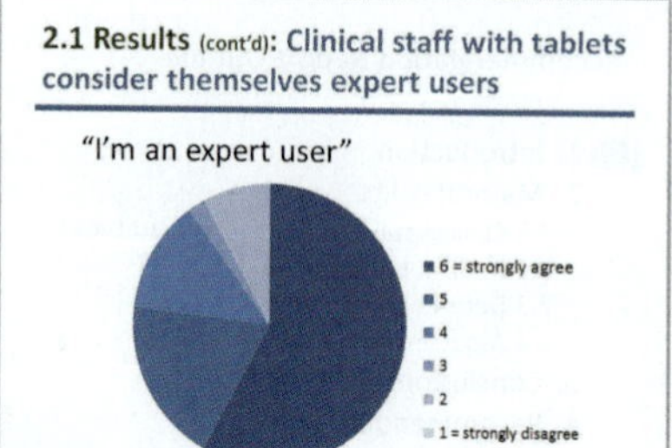

Slides work best if the text is brief. The speaker will explain that here are the responses to the statement "I consider myself an expert user of my tablet."

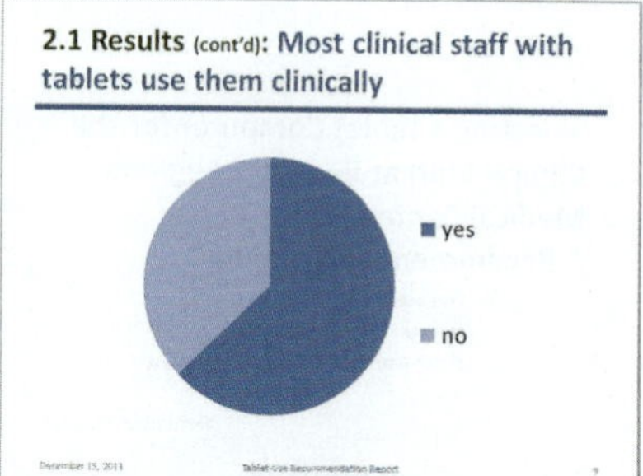

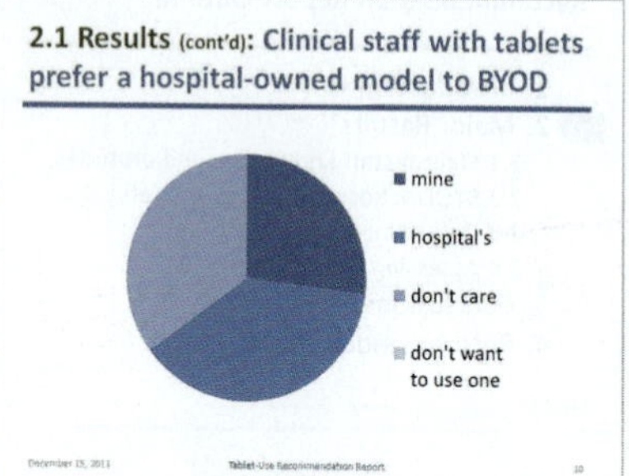

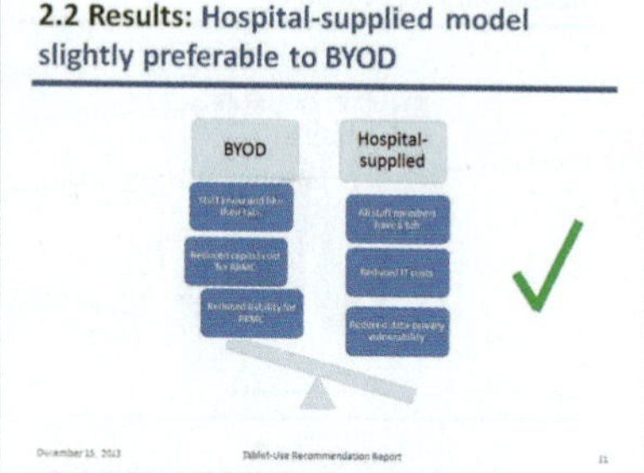

This slide was made using SmartArt graphics, which are part of PowerPoint. SmartArt graphics help you show logical relationships. Here, the relationship is that the hospital-supplied model is "heavier"—that is, has more to recommend it—than the BYOD model.

The speakers added the checkmark to emphasize that the hospital-supplied model is preferable to the BYOD model.

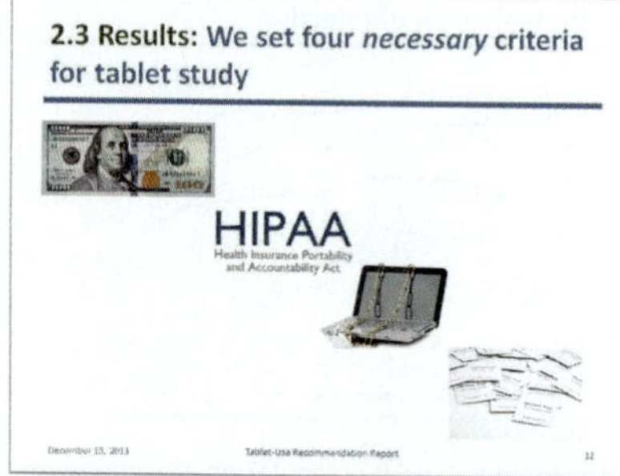

If the images you show are not your own intellectual property, you can legally display them in a college classroom because they are covered by the fair-use provisions of U.S. copyright law. However, if you display them in a business presentation, you would need formal written permission from the copyright holders. See Chapter 2, pp. 21–23, for more information.

Money: Bureau of Engraving and Printing. Chains on laptop: Elnur/Shutterstock. Disinfecting wipes: BeeBright/Shutterstock.

FIGURE 15.3 **Sample PowerPoint Presentation** *(continued)*

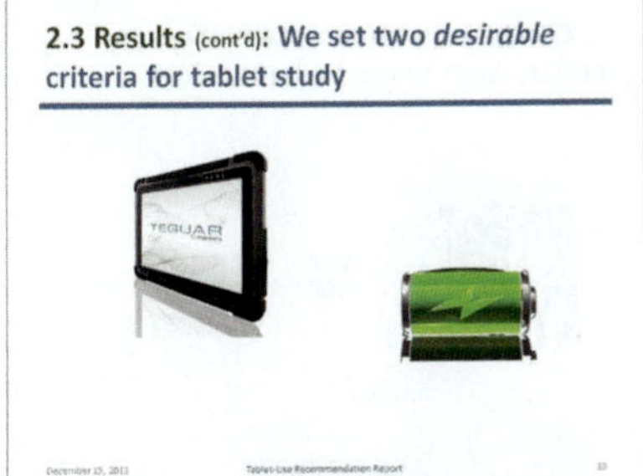

The speakers used the "appear" animation to display first the picture of the tablet and then the picture of the battery. This feature lets the speakers display on the screen only what they are discussing so that the audience is not distracted by other images.

TMT-1830-10 MEDICAL TABLET PC: Courtesy Teguar Corporation. Battery: oxanaart/Shutterstock.

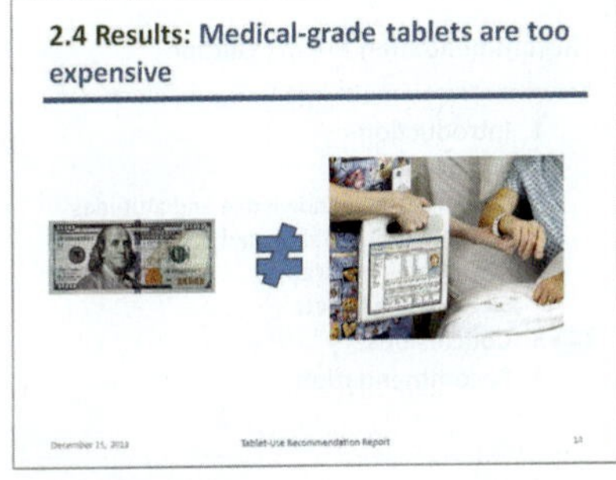

Money: Bureau of Engraving and Printing. Photo of Motion C5: Courtesy of Xplore.

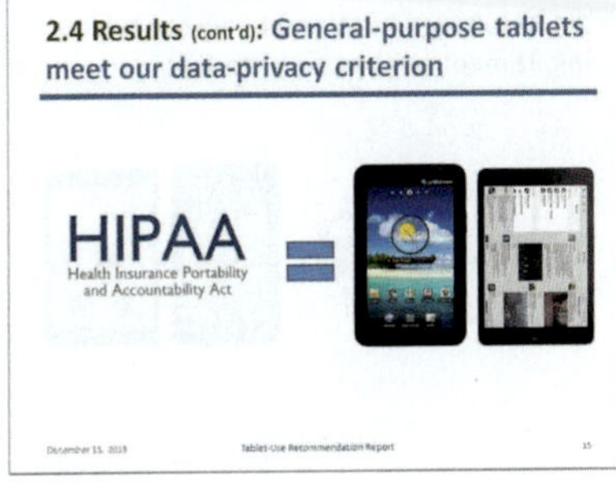

HootSuite App for iPod: HootSuite.

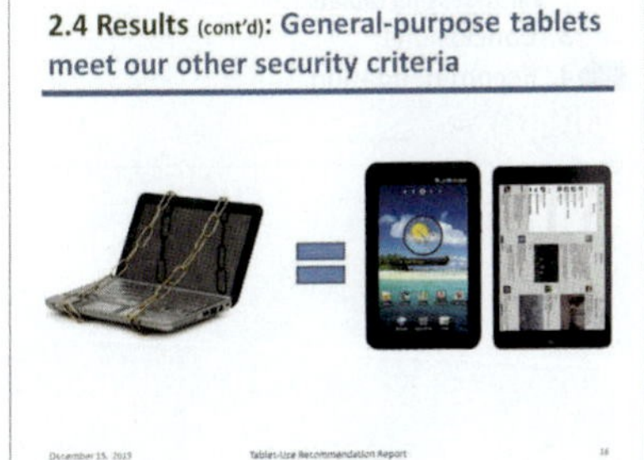

Chains on laptop: Elnur/Shutterstock. HootSuite App for iPad: HootSuite.

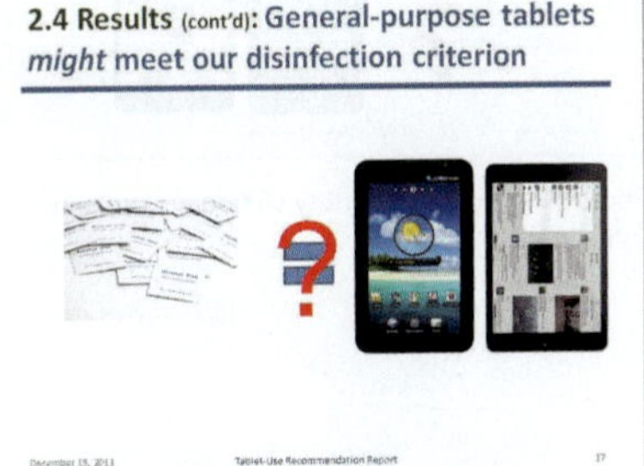

The formatting that appears throughout the slide set—the background color, the horizontal rule, and the footer—is created in the Slide-Master view. This formatting appears in every slide unless you modify or delete it for that slide. Note that the speakers use color—sparingly—for emphasis.

Disinfecting wipes: BeeBright/Shutterstock. HootSuite App for iPad: HootSuite.

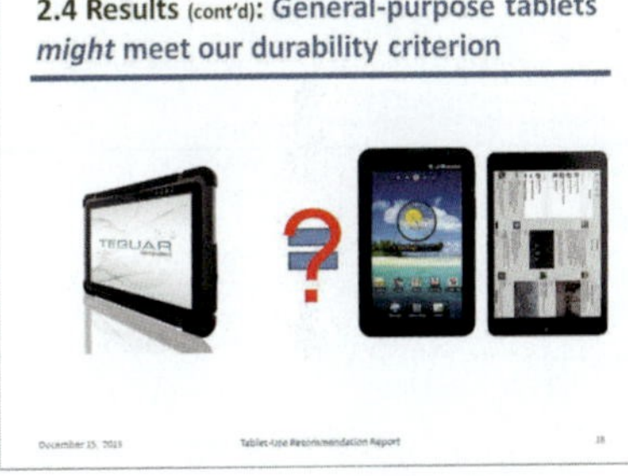

TMT-1830-10 MEDICAL TABLET PC: Courtesy Teguar Corporation. HootSuite App for iPad: HootSuite.

FIGURE 15.3 **Sample PowerPoint Presentation** *(continued)*

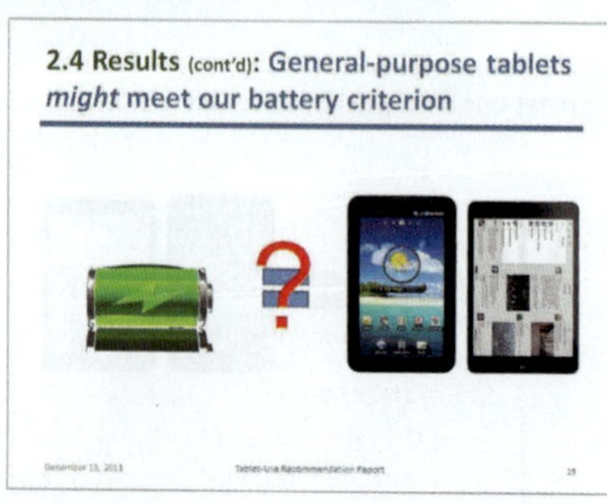

Battery: oxanaart/Shutterstock. HootSuite App for iPad: HootSuite.

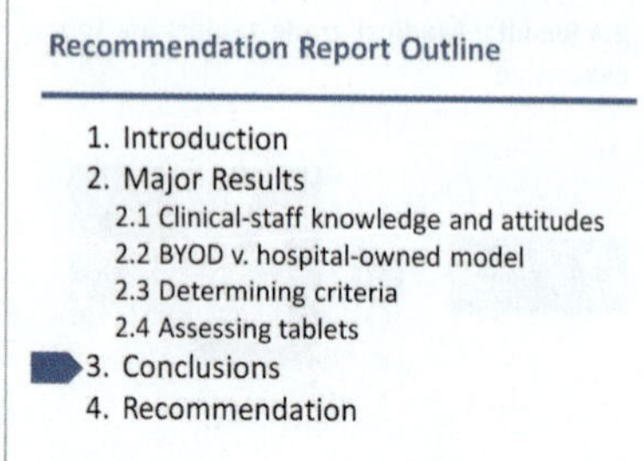

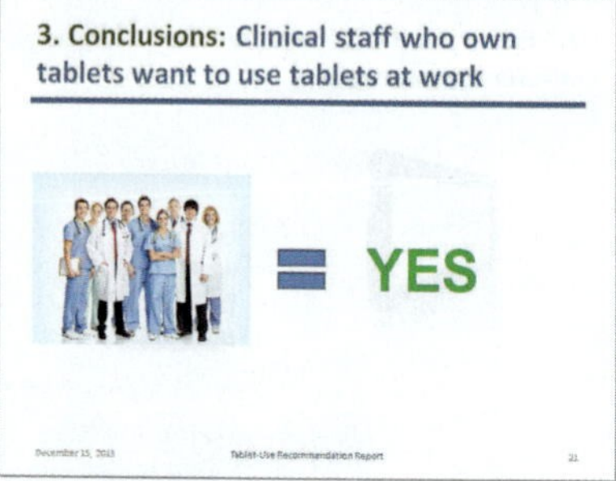

As discussed in Chapter 13, conclusions are inferences you draw from results.
kurhan/Shutterstock.

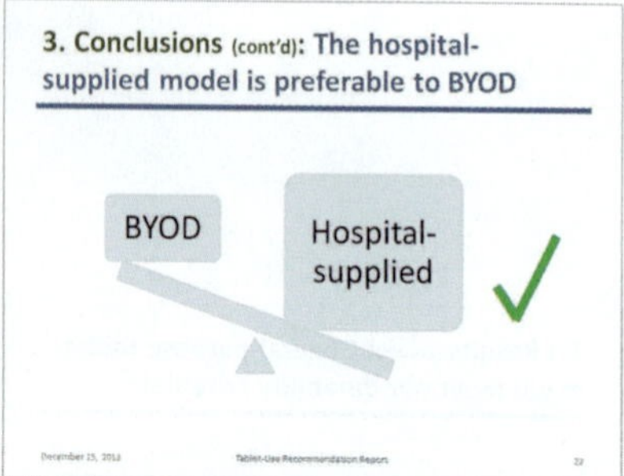

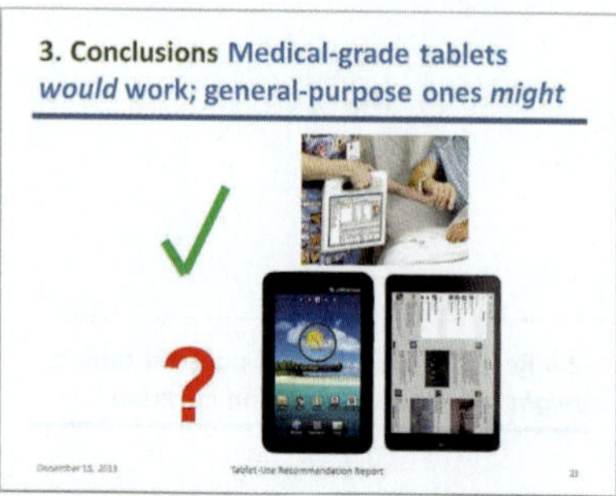

Photo of Motion C5: Courtesy of Xplore. HootSuite App for iPad: HootSuite.

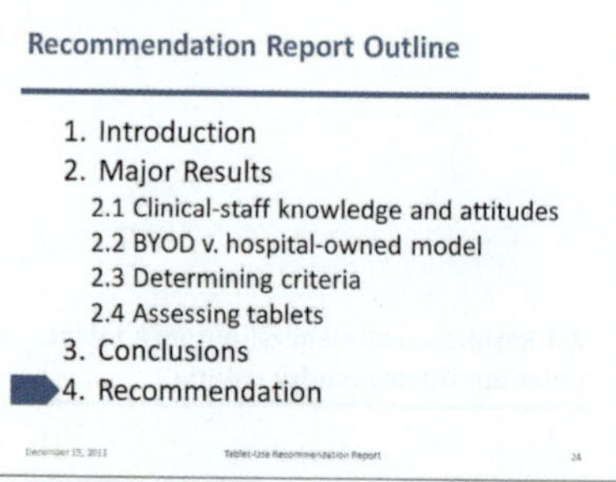

4. Recommendation: We recommend one of two options

1. Reconsider budget, study medical-grade tablets.
2. Test general-purpose tablets on-site.

As discussed in Chapter 13, recommendations are statements about what you think should be done next.

Some speakers like to make a final slide with the word "Questions?" on it to signal the end of the presentation. You can also display contact information (such as your email address) to encourage audience members to get in touch with you.

Photo of Motion C5: Courtesy of Xplore. HootSuite App for iPad: HootSuite.

FIGURE 15.3 **Sample PowerPoint Presentation** (*continued*)

Presentation software allows you to create two other kinds of documents—*speaking notes* and *handouts*—that can enhance a presentation. Figure 15.4 shows a page of speaking notes. Figure 15.5 shows a page from a handout created from PowerPoint slides.

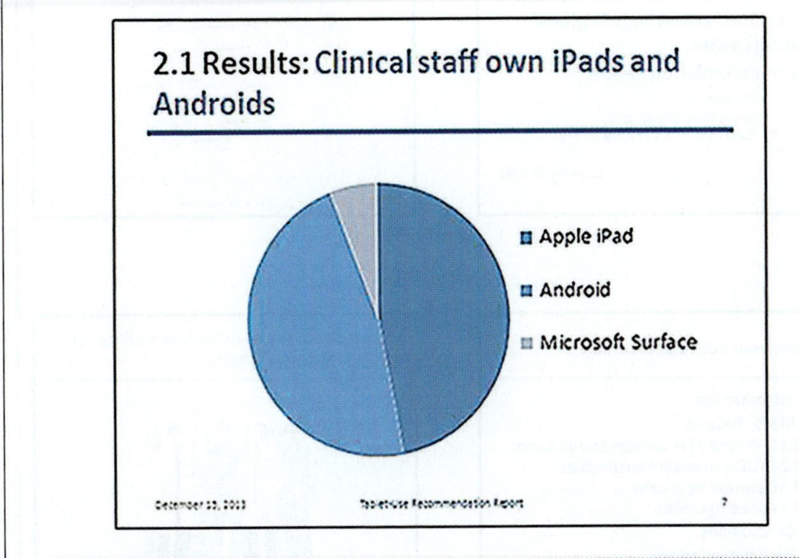

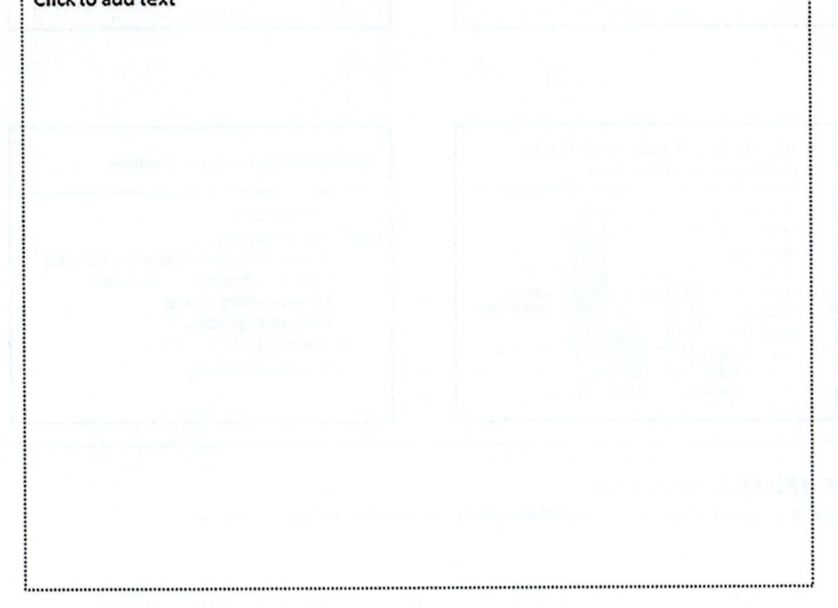

FIGURE 15.4
Speaking Notes

To create speaking notes for each slide, type the notes in the box under the picture of the slide, and then print the notes pages. You can print the slides on your notes pages in color or black and white.

The problem with using speaking notes is that you cannot read your notes and maintain eye contact at the same time.

In PowerPoint, you use the **Page Setup** tab to configure the file for printing. You can set the software to display from one to nine slides on a page.

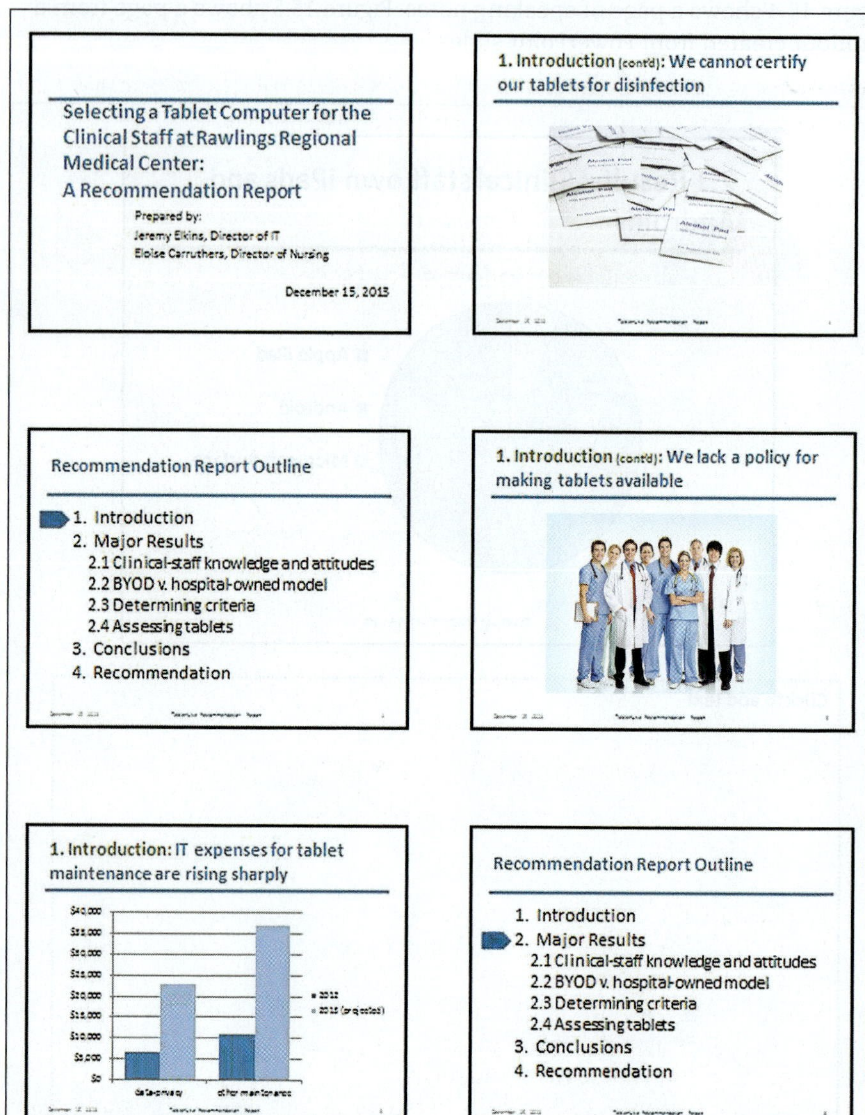

FIGURE 15.5 Handout

Disinfecting wipes: BeeBright/Shutterstock. Photo of health-care workers: kurhan/Shutterstock.

DOCUMENT ANALYSIS ACTIVITY

Integrating Graphics and Text on a Presentation Slide

The following slide is part of a presentation about the Human Genome Project. The questions in the margin ask you to think about the discussion of preparing presentation graphics.

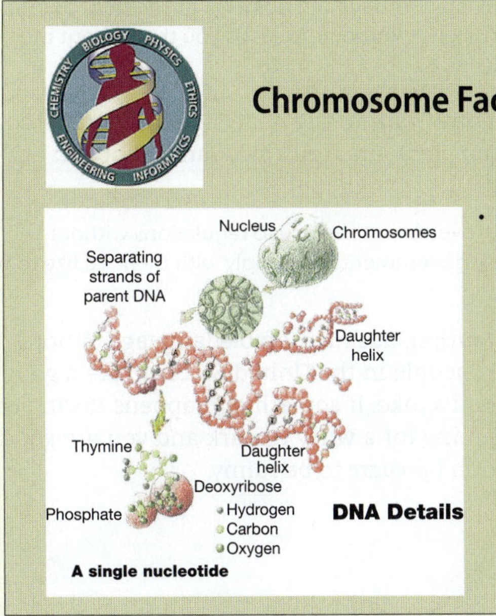

Chromosome Facts

- number of chromosomes: 22 pairs + 1 pair sex-determining chromosomes = 46
  - one chromosome of each pair donated from each parent's egg or sperm
  - sex chromosomes: X,Y for males; X,X for females
  - largest chromosome: #1 = ~263 million base pairs (bp)
  - smallest chromosome: Y = ~59 million bp

1. How effective is the Human Genome Project logo in the upper left-hand corner of the slide?
2. How well does the graphic of DNA support the accompanying text on chromosome facts?
3. Overall, how effective is the presentation graphic?

CHOOSING EFFECTIVE LANGUAGE

Delivering an oral presentation is more challenging than writing a document because listeners can't reread something they didn't understand. In addition, because you are speaking live, you must maintain your listeners' attention, even if they are hungry or tired or the room is too hot. Using language effectively helps you meet these two challenges.

Even if you use graphics effectively, listeners cannot "see" the organization of a presentation as well as readers can. For this reason, use language to alert your listeners to advance organizers, summaries, and transitions.

- **Advance organizers.** Use an advance organizer (a statement that tells the listener what you are about to say) in the introduction. In addition, use advance organizers when you introduce main ideas in the body of the presentation.

- **Summaries.** The major summary is in the conclusion, but you might also summarize at strategic points in the body of the presentation. For instance, after a three- to four-minute discussion of a major point, you might summarize it in one sentence before going on to the next major point. Here is a sample summary from a conclusion:

    Let me conclude by summarizing my three main points about the implications of the new RCRA regulations on the long-range waste-management strategy for Radnor Township. The first point is . . . . The second point is . . . . The third point is . . . . I hope this presentation will give you some ideas as you think about the challenges of implementing the RCRA.

- **Transitions.** As you move from one point to the next, signal the transition clearly. Summarize the previous point, and then announce that you are moving to the next point:

    It is clear, then, that the federal government has issued regulations without indicating how it expects county governments to comply with them. I'd like to turn now to my second main point. . . .

To maintain your listener's attention, use memorable language. A note about humor: only a few hundred people in the United States make a good living being funny. Don't plan to tell a joke. If something happens during the presentation that provides an opening for a witty remark and you are good at making witty remarks, fine. But don't *prepare* to be funny.

GUIDELINES Using Memorable Language in Oral Presentations

Draw on these three techniques to help make a lasting impression on your audience.

- **Involve the audience.** People are more interested in their own concerns than in yours. Talk to the audience about their problems and their solutions. In the introduction, establish a link between your topic and the audience's interests. For instance, a presentation to a city council about waste management might begin like this:

    Picture yourself on the Radnor Township Council two years from now. After exhaustive hearings, proposals, and feasibility studies, you still don't have a waste-management plan that meets federal regulations. What you do have is a mounting debt: the township is being fined $1,000 per day until you implement an acceptable plan.

(continued)

- **Refer to people, not to abstractions.** People remember specifics; they forget abstractions. To make a point memorable, describe it in human terms:

    What could you do with that $365,000 every year? In each computer lab in each school in the township, you could replace each laptop every three years instead of every four years. Or you could expand your school-lunch program to feed every needy child in the township. Or you could extend your after-school programs to cover an additional 3,000 students.

- **Use interesting facts, figures, and quotations.** Search the Internet for interesting information about your subject. For instance, you might find a brief quotation from an authoritative figure in the field or a famous person not generally associated with the field (for example, Theodore Roosevelt on waste management and the environment).

REHEARSING THE PRESENTATION

Even the most gifted speakers need to rehearse. It is a good idea to set aside enough time to rehearse your speech thoroughly.

- **First rehearsal.** Don't worry about posture or voice projection. Just deliver your presentation aloud with your presentation slides. Your goal is to see if the speech makes sense—if you can explain all the points and create effective transitions. If you have trouble, stop and try to figure out the problem. If you need more information, get it. If you need a better transition, create one. You are likely to learn that you need to revise the order of your slides. Pick up where you left off and continue the rehearsal, stopping again where necessary to revise.
- **Second rehearsal.** This time, the presentation should flow more easily. Make any necessary changes to the slides. When you have complete control over the organization and flow, check to see if you are within the time limit.
- **Third rehearsal.** After a satisfactory second rehearsal, try the presentation under more realistic circumstances—if possible, in front of others. The listeners might offer questions or constructive advice about your speaking style. If people aren't available, record a video of the presentation on your computer or phone, and then evaluate your own delivery. If you can visit the site of the presentation to rehearse there, you will find giving the actual speech a little easier.

Rehearse again until you are satisfied with your presentation, but don't try to memorize it.

THINKING VISUALLY

Delivering the Presentation

Calming Your Nerves

Apprehension about public speaking is common. Keep in mind three facts about nervousness:

- You are much more **AWARE** of your nervousness than the audience is.
- Nervousness gives you **ENERGY** and **ENTHUSIASM**.
- After a few minutes, your **NERVOUSNESS WILL PASS**.

Experienced speakers offer three tips for coping with nervousness:

- Realize that you are **PREPARED**.
- Realize that **THE AUDIENCE IS THERE TO HEAR YOU**, not to judge you.
- Realize that your audience is made up of **INDIVIDUAL PEOPLE** who happen to be sitting in the same room.

Releasing Nervous Energy

Experienced speakers suggest the following four strategies for dealing with nervousness before a presentation:

- **WALK AROUND BRISKLY** for a minute or two.
- **TALK WITH SOMEONE** for a few minutes.
- **GO OFF BY YOURSELF** for a few minutes and compose your thoughts.
- **TAKE SEVERAL DEEP BREATHS**, exhaling slowly.

Following these six recommendations can help you handle nervousness during a presentation:

- **WALK UP SLOWLY** and **ARRANGE YOUR NOTES** at the podium.
- **TAKE A SIP OF WATER** if it is available.
- **LOOK OUT AT THE AUDIENCE** for a few seconds before you begin.
- **BEGIN WITH "GOOD MORNING"** (or "Good afternoon" or "Good evening"), and refer to any officers and dignitaries present. If you have not been introduced, introduce yourself. In less-formal contexts, just begin.
- Use your voice and your body to **PROJECT AN ATTITUDE OF RESTRAINED SELF-CONFIDENCE.**
- **SHOW INTEREST** in your topic and knowledge about your subject.

From left to right: (1) Majivecka/Shutterstock; (2) Fiscus777/Shutterstock.

Using Your Voice Effectively

Inexperienced speakers often have problems with five aspects of vocalizing:

- **VOLUME.** Acoustics vary greatly from room to room, so you won't know how well your voice will carry in a particular setting. Ask if the people in the back of the room can hear you. Look at your audience's body language to see if you need to adjust the volume if you're using a microphone.
- **SPEED.** Nervousness makes people speak quickly. Even if you think you are speaking at the right rate, you might be going too fast. For particularly difficult points, slow down for emphasis. After finishing one major point, pause.
- **PITCH.** Try to let the pitch of your voice go up or down, as it would in a normal conversation, rather than flattening it.
- **ARTICULATION.** Nervousness can accentuate sloppy pronunciation. When a speaker uses a phrase over and over, it tends to get clipped.
- **NONFLUENCIES.** Avoid such meaningless fillers as *you know, like, okay, right, uh*, and *um*.

Using Your Body Effectively

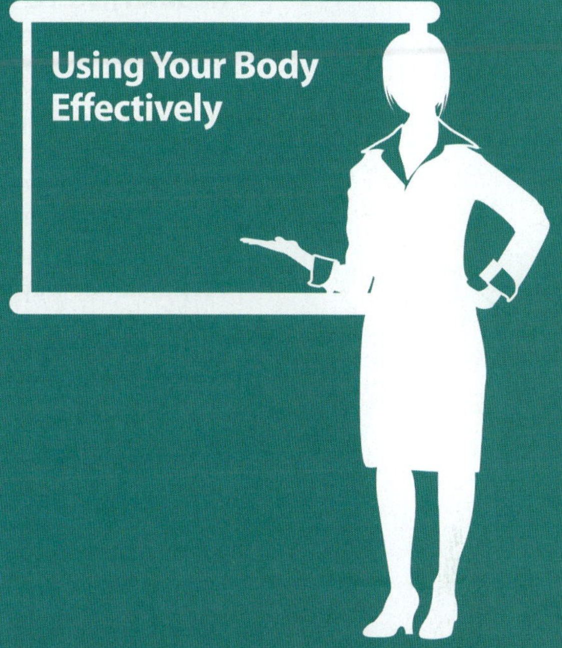

As you give a presentation, keep in mind these four guidelines about physical movement:

- **MAINTAIN EYE CONTACT** so you can see how the audience is receiving the presentation. With small groups, look at each listener randomly; with larger groups, look at each segment frequently.
- **USE NATURAL GESTURES.** Supplement your natural gestures by using your arms and hands to signal pauses and to emphasize points. When referring to graphics, walk toward the screen and point. Avoid physical gestures such as pacing that serve no useful purpose.
- Don't block the audience's view of the screen. **STAND OFF TO THE SIDE OF THE SCREEN** and refer to it with a pointer.
- **CONTROL** the audience's **ATTENTION.** When you want the audience to look at you and listen to you, remove the graphics from the screen or let the screen go blank.

From left to right: (1) NowikSylwia/Shutterstock; (2) Vectomart/Shutterstock.

PRESENTING TO ALL AUDIENCES

If your audience includes people of different cultures and native languages, keep in mind the following three suggestions:

- **Hire translators and interpreters if necessary.** If many people in the audience do not understand your language, hire interpreters (people who translate your words as you speak them) and translators (people who translate your written material in advance).
- **Use graphics effectively to reinforce your points for nonnative speakers.** Try to devise ways to present information using graphics—flowcharts, diagrams, and so forth—to help your listeners understand you. Putting more textual information on graphics will allow your listeners to see as well as hear your points.
- **Be aware that gestures can have culturally based meanings.** American hand gestures (such as the thumbs-up sign or the "okay" gesture) have different—and sometimes insulting—meanings in other cultures. Therefore, it's a good idea to avoid the use of these gestures. You can't go wrong with an arms-out, palms-up gesture that projects openness and inclusiveness.

If your audience includes people with varying physical abilities, keep in mind the following four suggestions:

- Provide handouts in accessible formats ahead of time for audience members to study and review.
- Hire a sign-language specialist who can translate your presentation for audience members with hearing impairments.
- Make a transcript of your talk available for audience members with hearing impairments.
- Provide a recording of the presentation for audience members who can't take notes.

Answering Questions After a Presentation

When you finish a presentation, thank the audience simply and directly: "Thank you for your attention." Then invite questions. Don't abruptly ask, "Any questions?" This phrasing suggests that you don't really want any questions. Instead, say something like this: "If you have any questions, I'll be happy to try to answer them now." If invited politely, people will be much more likely to ask questions, and you will be more likely to succeed in communicating your information effectively.

When you respond to questions, you might encounter any of these four situations:

- **You're not sure everyone heard the question.** Ask if people heard it. If they didn't, repeat or paraphrase it, perhaps as an introduction to your response: "Your question is about the efficiency of these three techniques." Some speakers always repeat the question, which gives them an extra moment to prepare an answer.
- **You don't understand the question.** Ask for clarification. After responding, ask if you have answered the question adequately.
- **You have already answered the question during the presentation.** Restate the answer politely. Begin your answer with a phrase such as the following: "I'm sorry I didn't make that point clear in my talk. I wanted to explain how . . . ." Never insult an audience member by pointing out that you already answered the question.
- **A belligerent member of the audience rejects your response and insists on restating his or her original point.** Politely offer to discuss the matter further after the presentation. If you are lucky, the person won't continue to bore or annoy the rest of the audience.

If it is appropriate to stay after the session to talk individually with members of the audience, offer to do so.

ETHICS NOTE

ANSWERING QUESTIONS HONESTLY

If an audience member asks a question to which you do not know the answer, admit it. Simply say, "I don't know" or "I'm not sure, but I think the answer is . . . ." Smart people know that they don't know everything. If you have some ideas about how to find out the answer—by checking a particular reference source, for example—share them. If the question is obviously important to the person who asked it, you might offer to meet with him or her to discuss ways for you to provide a more complete response later, perhaps by email.

SPEAKER'S CHECKLIST

- ☐ Did you analyze the speaking situation—the audience and purpose of the presentation? *(p. 425)*
- ☐ Did you determine how much information you can communicate in your allotted time? *(p. 425)*
- ☐ Did you choose an appropriate organizational pattern and determine what kinds of information to present? *(p. 426)*
- ☐ Did you create an outline? *(p. 426)*
- ☐ Did you plan your introduction and your conclusion? *(p. 426)*

Does each presentation graphic have these five characteristics?

- ☐ It presents a clear, well-supported claim. *(p. 429)*
- ☐ It is easy to see. *(p. 429)*
- ☐ It is easy to read. *(p. 430)*
- ☐ It is simple. *(p. 430)*
- ☐ It is correct. *(p. 430)*

- ☐ In planning your graphics, did you consider the length of your presentation, your audience's aptitude and experience, the setting for the presentation, and the equipment available? *(p. 430)*
- ☐ Did you plan your graphics to help the audience understand the organization of your presentation? *(p. 432)*
- ☐ Did you use language to signal advance organizers, summaries, and transitions? *(p. 439)*
- ☐ Did you choose language that is vivid and memorable? *(p. 440)*
- ☐ Did you rehearse your presentation several times with an audio recorder, video camera, or live audience? *(p. 441)*

EXERCISES

1. Learn some of the basic functions of a presentation software program. For instance, modify a template, create your own original design, add footer information to a master slide, insert a graphic on a slide, and set the animation feature to make just the first bullet item on a slide appear and then show the next bullet item after a mouse click.
2. Using PowerPoint, create a design to be used for the master slide for a presentation to be delivered in one of your classes. If you are unfamiliar with how to create a master slide, consult PowerPoint's help files. Be prepared to explain the design to your classmates.
3. Prepare a five-minute presentation, including graphics, on one of the following topics. The audience for your presentation will consist of the other students in your class, and your purpose will be to introduce them to an aspect of your academic field.
   a. Define a key term or concept in your field.
   b. Describe how a particular device or technology is used in your field.
   c. Describe how to carry out a procedure common in your field.

   Your instructor and the other students will evaluate the presentation by filling out the evaluation form located in LaunchPad.
4. **TEAM EXERCISE** Prepare a five-minute presentation based either on the proposal for a research project that you prepared in Chapter 11 or on your recommendation report for that project. Your audience will consist of the other students in your class, and your purpose will be to introduce them to your topic. The instructor and the other students will evaluate your presentation by filling out the evaluation form located in LaunchPad. Your instructor might have you work on this assignment collaboratively.

CASE 15: Understanding the Claim-and-Support Structure for Presentation Graphics

You have been invited to a student conference to make an oral presentation of a paper you wrote for a class. Because you do not have much experience delivering presentations, you run a few ideas by a friend, who offers great pointers. She's offered to look at your revised slides, so you need to get to work creating the slides and determining the best organization scheme for them. To get started on your presentation, go to LaunchPad.

Appendix: Reference Handbook

Part A: Documenting Your Sources

Documentation identifies the sources of the ideas and the quotations in your document. Documentation consists of the citations in the text throughout your document and the reference list (or list of works cited) at the end of your document. Documentation serves three basic functions:

- **It helps you acknowledge your debt to your sources.** Complete and accurate documentation is a professional obligation, a matter of ethics. Failure to document a source, whether intentional or unintentional, is plagiarism. At most colleges and universities, plagiarism can mean automatic failure of the course and, in some instances, suspension or expulsion. In many companies, it is grounds for immediate dismissal.
- **It helps you establish credibility.** Effective documentation helps you place your document within the general context of continuing research and helps you define it as a responsible contribution to knowledge in the field. Knowing how to use existing research is one mark of a professional.
- **It helps your readers find your source in case they want to read more about a particular subject.**

  Three kinds of material should always be documented:

- Any quotation from a written source or an interview, even if it is only a few words
- A paraphrased idea, concept, or opinion gathered from your reading
- Any graphic from a written or an electronic source

For more about using graphics from other sources, see Ch. 8, p. 200.

Just as organizations have their own rules for formatting and punctuation, many organizations also have their own documentation styles. For documents prepared in the workplace, find out your organization's style and abide by it. Check with your instructor to see which documentation system to use in the documents you write for class. The documentation systems included in this section of the appendix are based on the following style manuals:

- *Publication manual of the American Psychological Association* (6th ed.). (2010). Washington, DC: APA.
- *IEEE editorial style manual* [PDF]. (2016). Piscataway, NJ: IEEE.
- *MLA handbook for writers of research papers* (8th ed.). (2016). New York: Modern Language Association.

Note Taking

Most note taking involves three kinds of activities: paraphrasing, quoting, and summarizing.

PARAPHRASING

A paraphrase is a restatement, in your own words, of someone else's words. If you simply copy someone else's words—even a mere two or three in a row—you must use quotation marks.

In taking notes, what kind of material should you paraphrase? Any information that you think might be useful: background data, descriptions of mechanisms or processes, test results, and so forth.

GUIDELINES Paraphrasing Accurately

- **Study the original until you understand it thoroughly.**
- **Rewrite the relevant portions of the original.** Use complete sentences, fragments, or lists, but don't compress the material so much that you'll have trouble understanding it later.
- **Title the information so that you'll be able to identify its subject at a glance.** The title should include the general subject and the author's attitude or approach to it, such as "Criticism of open-sea pollution-control devices."
- **Include the author's last name, a short title of the article or book, and the page number of the original.** You will need this information later in citing your source.

Figure A.1 shows examples of paraphrasing based on the following discussion. The author is explaining the concept of performance-centered design.

Original Passage

In performance-centered design, the emphasis is on providing support for the structure of the work as well as the information needed to accomplish it. One of the best examples is TurboTax, which meets all the three main criteria of effective performance-centered design:

- *People can do their work with no training on how to use the system.* People trying to do their income taxes have no interest in taking any kind of training. They want to get their taxes filled out correctly and quickly, getting all the deductions they are entitled to. These packages, over the years, have moved the interface from a forms-based one, where the user had to know what forms were needed, to an interview-based one that fills out the forms automatically as you answer questions. The design of the interface assumes no particular computer expertise.
- *The system provides the right information at the right time to accomplish the work.* At each step in the process, the system asks only those questions that are relevant based on previous answers. The taxpayer is free to ask for more detail or may proceed through a dialog that asks more-detailed questions if the taxpayer doesn't know the answer to the higher-level question. If a taxpayer is married filing jointly, the system presents only those questions for that filing status.

- *Both tasks and systems change as the user understands the system.* When I first used TurboTax 6 years ago I found myself going to the forms themselves. Doing my taxes generally took about 2 days. Each year I found my need to go to the forms to be less and less. Last year, it took me about 2 hours to do my taxes, and I looked at the forms only when I printed out the final copy.

This paraphrase is inappropriate because the three bulleted points are taken word for word from the original. The fact that the student omitted the explanations from the original is irrelevant. These are direct quotes, not paraphrases.

> Lovgren, "Achieving Performance-Centered Design"
> www.reisman-consulting.com/pages/a-Perform.html
>
> example of performance-centered design:
> TurboTax® meets three main criteria:
>
> - People can do their work with no training on how to use the system.
> - The system provides the right information at the right time to accomplish the work.
> - Both tasks and systems change as the user understands the system.

a. Inappropriate paraphrase

This paraphrase is appropriate because the words are different from those used in the original.

When you turn your notes into a document, you are likely to reword your paraphrases. As you revise your document, check a copy of the original source document to be sure you haven't unintentionally reverted to the wording from the original source.

> Lovgren, "Achieving Performance-Centered Design"
> www.reisman-consulting.com/pages/a-Perform.html
>
> example of performance-centered design:
> TurboTax® meets three main criteria:
>
> - You don't have to learn how to use the system.
> - The system knows how to respond at the appropriate time to what the user is doing.
> - As the user gets smarter about using the system, the system gets smarter, making it faster to complete the task.

b. Appropriate paraphrase

FIGURE A.1 Inappropriate and Appropriate Paraphrased Notes

Information from Lovgren, 2000: www.reisman-consulting.com/pages/a-Perform.html.

QUOTING

For more about formatting quotations, see "Quotation Marks," "Ellipses," and "Square Brackets" in Appendix, Part B, pp. 491–93.

Sometimes you will want to quote a source, either to preserve the author's particularly well-expressed or emphatic phrasing or to lend authority to your discussion. Avoid quoting passages of more than two or three sentences, or your document will look like a mere compilation. Your job is to integrate an author's words and ideas into your own thinking, not merely to introduce a series of quotations.

Although you probably won't be quoting long passages in your document, recording a complete quotation in your notes will help you recall its meaning and context more accurately when you are ready to integrate it into your own work.

The simplest form of quotation is an author's exact statement:

> As Jones states, "Solar energy won't make much of a difference for at least a decade."

To add an explanatory word or phrase to a quotation, use brackets:

As Nelson states, "It [the oil glut] will disappear before we understand it."

Use ellipses (three spaced dots) to show that you are omitting part of an author's statement:

ORIGINAL STATEMENT	"The generator, which we purchased in May, has turned out to be one of our wisest investments."
ELLIPTICAL QUOTATION	"The generator . . . has turned out to be one of our wisest investments."

According to the documentation style recommended by the Modern Language Association (MLA), if the author's original statement has ellipses, you should add brackets around the ellipses that you introduce:

ORIGINAL STATEMENT	"I think reuse adoption offers . . . the promise to improve business in a number of ways."
ELLIPTICAL QUOTATION	"I think reuse adoption offers . . . the promise to improve business [. . .] ."

SUMMARIZING

Summarizing is the process of rewriting a passage in your own words to make it shorter while still retaining its essential message. Writers summarize to help them learn a body of information or create a draft of one or more of the summaries that will go into the document.

GUIDELINES Summarizing

The following advice focuses on extracting the essence of a passage by summarizing it.

- **Read the passage carefully several times.**
- **Underline key ideas.** Look for them in the titles, headings, topic sentences, transitional paragraphs, and concluding paragraphs.
- **Combine key ideas.** Study what you have underlined. Paraphrase the underlined ideas. Don't worry about your grammar, punctuation, or style at this point.
- **Check your draft against the original for accuracy and emphasis.** Check that you have recorded statistics and names correctly and that your version of a complicated concept faithfully represents the original. Check that you got the proportions right; if the original devotes 20 percent of its space to a particular point, your draft should not devote 5 percent or 50 percent to that point.
- **Record the bibliographic information carefully.** Even though a summary might contain all your own words, you still must cite it, because the main ideas are someone else's. If you don't have the bibliographic information in an electronic form, put it on a card.

APA Style

APA (American Psychological Association) style consists of two elements: citations in the text and a list of references at the end of the document.

APA Style for Textual Citations

APA Style for Reference List Entries

APA Style for Reference List Entries (*continued*)

APA TEXTUAL CITATIONS

In APA style, a textual citation typically includes the name of the source's author and the date of its publication. Textual citations vary depending on the type of information cited, the number of authors, and the context of the citation. The following models illustrate a variety of common textual citations; for additional examples, consult the *Publication Manual of the American Psychological Association*.

1. Summarized or Paraphrased Material For material or ideas that you have summarized or paraphrased, include the author's name and the publication date in parentheses immediately following the borrowed information.

> This phenomenon was identified more than 60 years ago (Wilkinson, 1948).

If your sentence already includes the source's name, do not repeat it in the parenthetical notation.

> Wilkinson (1948) identified this phenomenon more than 60 years ago.

2. Quoted Material or Specific Fact If the reference is to a specific fact, idea, or quotation, add the page number(s) from the source to your citation.

> This phenomenon was identified more than 60 years ago (Wilkinson, 1948, p. 36).
>
> Wilkinson (1948) identified this phenomenon more than 60 years ago (p. 36).

3. Source with Multiple Authors For a source written by two authors, cite both names. Use an ampersand (&) in the parenthetical citation itself, but use the word *and* in regular text.

> (Tyshenko & Paterson, 2014)
>
> Tyshenko and Paterson (2014) argued . . .

For a source written by three or more authors, include only the last name of the first author followed by *et al.*

Cashman et al. (2016) found . . .

4. Source Authored by an Organization If the author is an organization rather than a person, use the name of the organization.

There is currently ongoing discussion of the scope and practice of nursing informatics (American Nurses Association, 2015).

In a recent publication, the American Nurses Association (2015) discusses the scope and practice of nursing informatics.

If the organization name is commonly abbreviated, you may include the abbreviation in the first citation and use it in any subsequent citations.

First Text Citation
(International Business Machines [IBM], 2018)

Subsequent Citations
(IBM, 2018)

5. Source with an Unknown Author If the source does not identify an author, use a shortened version of the title in your parenthetical citation.

Hawking made the discovery that under precise conditions, thermal radiation could exit black holes ("World Scientists," 2009).

If the author is identified as anonymous—a rare occurrence—treat *Anonymous* as a real name.

(Anonymous, 2016)

6. Multiple Authors with the Same Last Name Use first initials if two or more sources have authors with the same last name.

B. Porter (2012) created a more stable platform for database transfers, while A. L. Porter (2012) focused primarily on latitudinal peer-to-peer outcome interference.

7. Multiple Sources in One Citation When you refer to two or more sources in one citation, present the sources in alphabetical order, separated by a semicolon.

This phenomenon has been well documented (Houlding, 2016; Jessen, 2015).

8. Personal Communication When you cite personal interviews, phone calls, letters, memos, and emails, include the words *personal communication* and the date of the communication. Cite personal communications that cannot be accessed by readers in the text only, not in the reference list.

> D. E. Walls (personal communication, April 3, 2016) provided the prior history of his . . .

9. Electronic Document Cite the author and date for an electronic source as you would for other kinds of documents. If the author is unknown, give a shortened version of the title in your parenthetical citation. If the date is unknown, use *n.d.* (for *no date*).

> Interpersonal relationships are complicated by differing goals (Hoffman, n.d.).

If the document is posted as a PDF file, include the page number in the citation. If a page number is not available but the source contains paragraph numbers, give the paragraph number.

> (Tong, 2010, para. 4)

If no paragraph or page number is available and the source has headings, cite the appropriate heading and paragraph.

> The CDC (2007) warns that babies born to women who smoke during pregnancy are 30% more likely to be born prematurely (The Reality section, para. 3).

THE APA REFERENCE LIST

A reference list provides the information your readers will need in order to find each source you have cited in the text. It should not include sources you read but did not use.

For a sample APA-style reference list, see p. 464.

Following are some guidelines for an APA-style reference list.

- **Arranging entries.** Arrange the entries alphabetically by author's last name. If two or more works are by the same author, arrange them by date, earliest to latest. If two or more works are by the same author in the same year, list them alphabetically by title and include a lowercase letter after the date: 2015a, 2015b, and so on. Alphabetize works by an organization by the first significant word in the name of the organization.
- **Book titles.** Italicize titles of books. Capitalize only the first word of the book's title, the first word of the subtitle, and any proper nouns.
- **Publication information.** For books, give the publisher's name in as brief a form as is intelligible; retain the words *Books* and *Press*.
- **Periodical titles.** Italicize titles of periodicals and capitalize all major words.

- **Article titles.** Do not italicize titles of articles or place them in quotation marks. Capitalize only the first word of the article's title and subtitle and any proper nouns.
- **Electronic sources.** Include as much information as you can about electronic sources, such as author, date of publication, identifying numbers, and retrieval information. Include the digital object identifier (DOI) when one exists, and format it as a link. Remember that electronic information changes frequently. If the content of an electronic source is likely to change, be sure to record the date you retrieved the information.
- **Indenting.** Use a hanging indent, with the first line of each entry flush with the left margin and all subsequent lines indented one-half inch:

  Sokolova, G. N. (2010). Economic stratification in Belarus and Russia: An experiment in comparative analysis. *Sociological Research, 49*(3), 25–26.

  Your instructor might prefer a paragraph indent, in which the first line of each entry is indented one-half inch:

  Sokolova, G. N. (2010). Economic stratification in Belarus and Russia: An experiment in comparative analysis. *Sociological Research, 49*(3), 25–26.

- **Spacing.** Double-space the entire reference list. Do not add extra space between entries.
- **Page numbers.** When citing a range of page numbers for an article, always give the complete numbers (for example, 121–124, *not* 121–24 or 121–4). If an article continues on subsequent pages after being interrupted by other articles or advertisements, use a comma to separate the page numbers. Use the abbreviation *p.* or *pp.* only with articles in newspapers, chapters in edited books, and articles from proceedings published as a book.
- **Dates.** Follow the format year, month, day, with a comma after only the year: (2016, October 31).

Following are models of reference list entries for a variety of sources. For further examples of APA-style citations, consult the *Publication Manual of the American Psychological Association*.

BOOKS

10. Book by One Author Begin with the author's last name, followed by the first initial or initials. Include a space between initials. Place the year of publication in parentheses, then give the title of the book, followed by the location and name of the publisher.

Power, G. A. (2010). *Dementia beyond drugs: Changing the culture of care.* Health Professions Press.

11. Book by Multiple Authors When citing a work by from two to seven authors, separate the authors' names with a comma or commas, and use an ampersand (&) instead of *and* before the final author's name.

Tyshenko, M. G., & Paterson, C. (2010). *SARS unmasked: Risk communication of pandemics and influenza in Canada*. McGill-Queen's University Press.

To cite more than seven authors, list only the first six, followed by three dots (an ellipsis) and the last author's name.

12. Multiple Books by the Same Author Arrange the entries by date, with the earliest date first.

Tabloski, P. A. (2007). *Clinical handbook for gerontological nursing*. Pearson/Prentice Hall.

Tabloski, P. A. (2010). *Gerontological nursing*. Pearson.

If you use multiple works by the same author written in the same year, arrange the books alphabetically by title and include *a*, *b*, and so forth after the year—both in your reference list and in your parenthetical citations.

Agger, B. (2007a). *Fast families, virtual children: A critical sociology of families and schooling*. Paradigm.

Agger, B. (2007b). *Public sociology: From social facts to literary acts*. Rowman & Littlefield.

13. Book Authored by an Organization Use the full name of the organization in place of an author's name. If the organization is also the publisher, use the word *Author* in place of the publisher's name.

American Nurses Association. (2010). *Nursing's social policy statement: The essence of the profession* (3rd ed.). Author.

14. Book by an Unknown Author If the author of the book is unknown, begin with the title in italics.

The PDR pocket guide to prescription drugs (9th ed.). (2010). Pocket Books.

15. Edited Book Place the abbreviation *Ed.* (singular) or *Eds.* (plural) in parentheses after the name(s), followed by a period.

Haugen, D., Musser, S., & Lovelace, K. (Eds.). (2010). *Global warming*. Greenhaven Press.

16. Chapter or Section in an Edited Book

Jyonouchi, H. (2010). Possible impact of innate immunity in autism. In A. Chauhan, V. Chauhan, & W. T. Brown (Eds.), *Autism: Oxidative stress, inflammation, and immune abnormalities* (pp. 245–276). CRC Press.

17. Book in an Edition Other Than the First Include the edition number in parentheses following the title.

Quinn, G. R. (2010). *Behavioral science* (2nd ed.). McGraw-Hill Medical.

18. Multivolume Work Include the number of volumes after the title.

Weiner, I. B., & Craighead, W. E. (Eds.). (2010). *The Corsini encyclopedia of psychology* (Vols. 1–4). Wiley.

19. Translated Book Name the translator after the title.

Bieler, A., & Gutmann, H.-M. (2010). *Embodying grace: Proclaiming justification in the real world* (L. M. Maloney, Trans.). Fortress Press.

20. Non-English Book Give the original title, then the English translation in brackets.

Hernandez, G. H., Moreno, A. M., Zaragoza, F. G., & Porras, A. C. (Eds.). (2010). *Tratado de medicina farmacéutica* [Treatise on pharmaceutical medicine]. Editorial Médica Panamericana.

21. Entry in a Reference Work Begin with the title of the entry if it has no author.

Kohlrabi. (2010). In R. T. Wood (Ed.), *The new whole foods encyclopedia: A comprehensive resource for healthy eating* (2nd ed., pp. 178–179). Penguin Books.

PERIODICALS

22. Journal Article Follow the author's name and the year of publication with the article title; then give the journal title, followed by a comma. For all journals, include the volume number (italicized). For journals that begin each issue with page 1, also include the issue number in parentheses (not italicized). Insert a comma and end with the page number(s).

Cumsille, P., Darling, N., & Martinez, M. L. (2010). Shading the truth: The pattern of adolescents' decisions to avoid issues, disclose, or lie to parents. *Journal of Adolescence, 33*, 285–296.

23. Magazine Article Include the month after the year. If it's a weekly magazine, include the day. Give the volume and issue numbers, if any, after the magazine title.

Stix, G. (2011, March). The neuroscience of true grit. *Scientific American, 304*(3), 28–33.

24. Newspaper Article Include the specific publication date following the year.

Seltz, J. (2010, December 26). Internet policies examined: Schools aim to clarify social rules. *Boston Globe*, p. 1.

25. Newsletter Article Cite a newsletter article as you would a magazine article. If the date is given as a season, insert a comma following the year and then include the season. If it's not necessarily clear that you are citing a newsletter, you may include the label "[Newsletter]" following the title.

Meyerhoff, M. K. (2010, September/October). Paying attention to attention [Newsletter]. *Pediatrics for Parents, 26*(9/10), 8–9.

ELECTRONIC SOURCES

Generally, include the same elements for electronic sources as you would for print sources. Include any information required to locate the item. Many scholarly publishers are now assigning a digital object identifier (DOI) to journal articles and other documents. A DOI is a unique alphanumeric string assigned by a registration agency. It provides a persistent link to unchanging content on the Internet. When available, substitute the DOI for a URL, using a link format. If the content is subject to change, include the retrieval date before the URL. Use the exact URL for open-source material; use the home-page or menu-page URL for subscription-only material or content presented in frames, which make exact URLs unworkable. Break URLs before a punctuation mark (or after http://), and avoid using punctuation after a URL or DOI so as not to confuse the reader. If you encounter a DOI in a different format, change the beginning to "https://doi.org/" and add the DOI number for the specific work.

26. Nonperiodical Web Document To cite a nonperiodical web document, provide as much of the following information as possible: author's name, date of publication or most recent update (use *n.d.* if there is no date), document title (in italics), and URL (or DOI, if available) for the document.

For a tutorial on citing websites in APA style, go to LaunchPad.

Centers for Disease Control and Prevention. (2012, October). *Teens drinking and driving.* Retrieved from https://www.cdc.gov/vitalsigns/teendrinkinganddriving/index.html

If the author of a document is not identified, begin the reference with the title of the document. If the document is from a university program's website, identify the host institution and the program or department, followed by a colon and the URL for the document.

Safety manual. (2011, March 18). Retrieved from Harvard University, Center for Nanoscale Systems website: http://www.cns.fas.harvard.edu/users/Forms/CNS_Safety_Manual.pdf

Journal Articles

27. Article with DOI Assigned

Iemolo, F., Cavallaro, T., & Rizzuto, N. (2010). Atypical Alzheimer's disease: A case report. *Neurological Sciences, 31*, 643–646. https://doi.org/10.1007/s10072-010-0334-1

28. Article with No DOI Assigned

Srivastava, R. K., & More, A. T. (2010). Some aesthetic considerations for over-the-counter (OTC) pharmaceutical products. *International Journal of Biotechnology, 11*(3–4), 267–283.

29. Preprint Version of Article

Wang, T. J., Larson, M. G., Vasan, R. S., Cheng, S., Rhee, E. P., McCabe, E., . . . Gerszten, R. E. (2011). Metabolite profiles and the risk of developing diabetes. *Nature Medicine.* Advance online publication. https://doi.org/10.1038/nm.2307

Electronic Books

30. Entire Book

Einstein, A. (n.d.). *Relativity: The special and general theory.* Retrieved from http://www.gutenberg.org/etext/5001

Dissertations and Theses

For a tutorial on citing databases in APA style, go to LaunchPad.

31. Dissertation Retrieved from Database For a commercial database, include the database name, followed by the accession number. For an institutional database, include the URL.

Siegel, R. S. (2010). *Mediators of the association between risk for mania and close relationship quality in adolescents* (Doctoral dissertation). http://scholarlyrepository.miami.edu/oa_dissertations/426

Reference Materials

Give the home-page or index-page URL for reference works.

32. Online Encyclopedia

Cross, M. S. (2011). Social history. In J. H. Marsh (Ed.), *The Canadian encyclopedia.* http://www.thecanadianencyclopedia.com

33. Online Dictionary

Conductance. (n.d.). In *Merriam-Webster's online dictionary*. http://www.merriam-webster.com/dictionary/conductance

34. Wiki

Tsunami. (n.d.). Retrieved March 20, 2016, from http://en.wikipedia.org/wiki/Tsunami

Raw Data

35. Data Set

Department of Health and Human Services. (2018). FDA peanut product recalls [Data set]. https://catalog.data.gov/dataset/fda-peanut-product-recalls-baef7

36. Graphic Representation of Data

U.S. Department of Labor, Bureau of Labor Statistics. (2011, April 4). *Civilian unemployment rate (UNRATE).* [Line graph]. http://research.stlouisfed.org/fred2/series/UNRATE

37. Qualitative Data

Jaques, C. (2010). *They called it slums but it was never a slum to me* [Audio stream]. https://storycorps.org/listen/carol-jacques/

Other Electronic Documents

38. Technical or Research Report

Moran, R., Rampey, B. D., Dion, G. S., & Donahue, P. L. (2008). *National Indian education study 2007, Part 1. Performance of American Indian and Alaska native students at grades 4 and 8 on NAEP 2007 reading and mathematics assessments* (Report No. NCES 2008–457). http://nces.ed.gov/nationsreportcard/pdf/studies/2008457.pdf

39. Presentation Slides

Wyominginspector. (2010). *Cell phone use in the mining industry* [PowerPoint slides]. http://www.slideshare.net/wyominginspector/cell-phone-use-in-the-mining-industry

General-Interest Media and Alternative Presses

40. Newspaper Article

Applebaum, A. (2011, February 14). Channeling Egypt's energy of the crowd into positive change. *The Washington Post.* http://www.washingtonpost.com

41. Audio Podcast Include the presenter, producer, or other authority, if known; date; episode title; any episode or show identifier in brackets, such as *[Show 13]*; show name; the words *Audio podcast* in brackets; and retrieval information.

du Sautoy, Marcus. (Presenter). (2010, June 25). Nicolas Bourbaki. *A Brief History of Mathematics* [Audio podcast]. https://www.bbc.co.uk/programmes/b00srz5b/episodes/downloads

42. Online Magazine Content Not Found in Print Version

Greenemeier, L. (2010, November 17). Buzz kill: FDA cracks down on caffeinated alcoholic beverages. *Scientific American.* https://www.scientificamerican.com/article/fda-caffeinated-alcohol/

Online Communities

43. Message Posted to an Electronic Mailing List, Online Forum, or Discussion Group If an online posting is not archived and therefore is not retrievable, cite it as a personal communication and do not include it in the reference list. If the posting can be retrieved from an archive, provide the author's name (or the author's screen name if the real name is not available), the exact date of the posting, the title or subject line or thread name, and a description of the type of post in brackets. Finish with the address.

Gomez, T. N. (2010, December 20). Food found in archaeological environments [Electronic mailing list message]. http://cool.conservation-us.org/byform/mailing-lists/cdl/2010/1297.html

44. Blog Post

Laden, Greg. (2017, October 10). Hacking The American Election System: Getting It Right [Web log post]. http://gregladen.com/blog/2017/10/10 /hacking-the-american-election-system-getting-it-right/

Telecom. (2011, February 22). Cellphone use tied to changes in brain activity [Web log comment]. http://well.blogs.nytimes.com/2011/02/22/cellphone-use-tied-to-changes-in-brain-activity/#comment-643942

45. Email Message or Real-Time Communication Do not cite email messages in the reference list. Instead, cite them in the text as personal communications. (See item 8 on page 455.)

OTHER SOURCES

46. Technical or Research Report Include an identifying number in parentheses after the report title. If appropriate, include the name of the service used to locate the item in parentheses after the publisher.

Arai, M., & Mazuka, R. (2010). *Linking syntactic priming to language development: A visual world eye-tracking study* (TL2010-18). Institute of Electronics, Information and Communication Engineers.

47. Government Document For most government agencies, use the abbreviation *U.S.* instead of spelling out *United States*. Include any identifying document number after the publication title.

U.S. Department of State. (2010, June). *Trafficking in persons report* (10th ed.). Government Printing Office.

48. Brochure or Pamphlet After the title of the document, include the word *Brochure* or *Pamphlet* in brackets.

U.S. Department of Health and Human Services, Centers for Disease Control and Prevention. (2010, October). *How to clean and disinfect schools to help slow the spread of flu* [Pamphlet]. Author.

49. Article from Conference Proceedings After the proceedings title, give the page numbers on which the article appears.

Sebastianelli, R., Tamimi, N., Gnanendran, K., & Stark, R. (2010). An examination of factors affecting perceived quality and satisfaction in online MBA courses. In *Proceedings of the 41st Annual Meeting of the Decision Sciences Institute* (pp. 1641–1646). Decision Sciences Institute.

50. Lecture or Speech

Culicover, P. W. (2010, March 3). *Grammar and complexity: Language at the intersection of competence and performance.* Lecture presented at the Ohio State University, Columbus, OH.

51. Audio Recording Give the role (narrator, producer, director, or the like) of the person whose name appears at the beginning of the entry in parentheses after the name. Give the medium in brackets after the title.

Young, J. K. (Lecturer). (2007). *The building blocks of human life: Understanding mature cells and stem cells* [CD]. Recorded Books.

52. Film Give the name of at least one primary contributor, such as the producer or director, and follow the film's title with the word *Film* in brackets. List the country in which the film was produced and the studio's name. If the film was not widely distributed, give instead the distributor's name and address in parentheses.

Fincher, D. (Director). (2010). *The social network* [Film]. Columbia Pictures.

53. Television Program Start with the director, producer, or other principal contributor and the date the program aired. Include the words *Television broadcast* or *Television series* in brackets after the program title.

Fine, S. (Executive Producer). (2011). *NOVA scienceNOW* [Television series]. WGBH.

For a single episode in a television series, start with the writer and director of the episode or other relevant editorial personnel. Include the words *Television series episode* in brackets after the episode title. Also include information about the series. End with the location and name of the station or network.

Dart, K., Evans, N., & Stubberfield, T. (Producers & Directors). (2010, October 26). Emergency mine rescue [Television series episode]. In H. Swartz (Executive Producer), *NOVA*. WGBH.

54. Published Interview If it is not clear from the title that the entry is an interview, or if there is no title, include the words *Interview with* and the subject's name in brackets.

Jackson, L. (2010, December 6). The EPA is not the villain [Interview with Daniel Stone]. *Newsweek, 156*(23), 14.

55. Personal Interview Consider interviews you conduct, whether in person or over the telephone, as personal communications and do not include them in the reference list. Instead, cite them in the text. (See item 8 on page 455.)

56. Personal Correspondence Like emails, personal letters and memos should not be included in the reference list. Instead, cite them in the text. (See item 8 on page 455.)

57. Unpublished Data Where the title would normally appear, include a description of the data in brackets.

Standifer, M. (2007). [Daily temperatures, 2007, Barton Springs municipal pool, Austin, TX]. Unpublished raw data.

SAMPLE APA REFERENCE LIST

Following is a sample reference list using the APA citation system.

Nonperiodical web document with no DOI

Journal article, paginated by volume

Online article with a DOI

Chapter in an edited book

Book in an edition other than the first

Online article, paginated by issue, with no DOI

References

Centers for Disease Control and Prevention. (2012, October). *Teen Drinking and Driving*. https://www.cdc.gov/vitalsigns/teendrinkinganddriving/index.html

Cumsille, P., Darling, N., & Martinez, M. L. (2010). Shading the truth: The pattern of adolescents' decisions to avoid issues, disclose, or lie to parents. *Journal of Adolescence, 33*, 285–296.

Iemolo, F., Cavallaro, T., & Rizzuto, N. (2010). Atypical Alzheimer's disease: A case report. *Neurological Sciences, 31*, 643–646. https://doi.org/10.1007/s10072-010-0334-1

Jyonouchi, H. (2010). Possible impact of innate immunity in autism. In A. Chauhan, V. Chauhan, & W. T. Brown (Eds.), *Autism: Oxidative stress, inflammation, and immune abnormalities* (pp. 245–276). CRC Press.

Quinn, G. R. (2010). *Behavioral science* (2nd ed.). McGraw-Hill Medical.

Srivastava, R. K., & More, A. T. (2010). Some aesthetic considerations for over-the-counter (OTC) pharmaceutical products. *International Journal of Biotechnology, 11*(3–4), 267–283. http://www.inderscience.com

IEEE Style

IEEE style consists of two elements: citations in the text and a reference list at the end of the document.

IEEE Style for Reference List Entries

IEEE TEXTUAL CITATIONS

In the IEEE (originally, Institute of Electrical and Electronics Engineers) documentation system, citations in the text are bracketed numbers, keyed to a numbered list of references that appears at the end of the document. Entries in the list are arranged in the order in which they are cited in the text and are numbered sequentially. Once a reference has been listed, the same number is used in all subsequent citations of that source.

To cite references in the text, place the reference number or numbers immediately after the author's name, in square brackets, before any punctuation. Use *et al.* if there are three or more author names.

> A recent study by Goldfinkel [5] shows that this is not an efficient solution. Murphy [8]–[10] comes to a different conclusion.

You can also use the bracketed citation number or numbers as a noun.

> In addition, [5] shows that this is not an efficient solution; however, [8]–[10] come to a different conclusion.

NOTE: Because references are listed *in the order in which they first appear in the text*, if you add a new citation within the text while rewriting or editing, you will need to renumber the reference list as well as the citations in the text. For example, if in rewriting you were to add a new reference between the first citations of the Murphy references originally numbered [8] and [9], the previous example would then read:

> . . . [8], [10], [11] come to a different conclusion.

To make a reference more precise, you can provide extra information.

> A recent study by Goldfinkel [5, pp. 12–19] shows that this is not an efficient solution.

THE IEEE REFERENCE LIST

For a sample IEEE-style reference list, see p. 469.

The following guidelines will help you prepare IEEE-style references. For additional information on formatting entries, consult the latest edition of *The IEEE Editorial Style Manual*.

- **Arranging entries.** Arrange the entries in the order in which they first are cited in the text, and then number them sequentially. Place the numbers in square brackets and set them flush left in a column of their own, separate from the body of the references. Place the entries in their own column, with no indents for turnovers.
- **Authors.** List the author's first initial (or first and middle initials, separated by spaces), followed by the last name. In the case of multiple authors, use all names; use *et al.* after the first author's name only if the other names are not given. If an entry has an editor or translator in place of an author, add the abbreviation *Ed.* (or *Eds.* for *editors*) or *Trans.* following the name.
- **Book titles.** Italicize titles of books. In English, capitalize the first word and all major words. In foreign languages, capitalize the first word of the title and subtitle, as well as any words that would be capitalized in that language.
- **Publication information.** For books, give the city of publication, the country (if other than the United States), the publisher's name (abbreviated), and the year of publication. When two or more cities are given on a book's copyright page, include only the first. If the city is not well known or could be confused with another city, add the abbreviation for the name of the state or province (in Canada). If the publisher's name indicates the state, no state abbreviation is necessary.
- **Periodical titles.** Italicize and abbreviate titles of periodicals. Capitalize all major words in the title.
- **Article titles.** Place titles of print articles in quotation marks; do not use quotation marks for titles of articles found in electronic sources. Capitalize the first word of the title and subtitle. Do not capitalize the remaining words unless they are proper nouns.
- **Electronic sources.** Follow the special style for electronic sources in which, most notably, the sequence of information is different from that for print material (the date follows the author, and the year comes before the month). Do not place article titles in quotation marks, and use periods rather than commas to separate sections. In addition to the basic information, give the medium and provide a way to locate the source by including, for example, a URL.

- **Spacing.** Single-space the reference list, and do not add extra space between entries.
- **Page numbers.** To give a page or a range of pages for a specific article in a book or periodical, use the abbreviation *p.* or *pp.* Write numbers in full (152–159, *not* 152–59 or 152–9).
- **Dates.** For print sources, follow the format month (abbreviated), day, year (for example, Apr. 3, 2016 or Feb. 22–23, 2016). Do not abbreviate May, June, or July. For electronic sources, follow the format year, month (abbreviated), day (for example, 2016, Oct. 14).

BOOKS

1. Book by One Author Include the author's first initial and middle initial (if available), the author's last name, the book title (in italics), the place of publication, the publisher, the year of publication, and the page range of the material referenced.

[1] B. Mehlenbacher, *Instruction and Technology: Designs for Everyday Learning.* Cambridge, MA: MIT Press, 2010, pp. 22–28.

2. Book by Multiple Authors List all the authors' names. Use *et al.* after the first author's name only if the other names are not given. Do not invert names, and include a comma before *and* only if there are three or more names.

[2] S.-T. Yau and S. J. Nadis, *The Shape of Inner Space: String Theory and the Geometry of the Universe's Hidden Dimensions.* New York: Basic Books, 2010, pp. 254–255.

3. Book Authored by an Organization The organization takes the place of the author.

[3] World Bank, *World Development Report 2011: Conflict, Security, and Development.* Washington, DC: World Bank, 2011, pp. 25–31.

4. Edited Book Include the abbreviation *Ed.* (singular) or *Eds.* (plural) after the name(s).

[4] J. Dibbell, Ed., *The Best Technology Writing 2010.* New Haven: Yale University Press, 2010, pp. 157–162.

5. Chapter or Section in an Edited Book Give the author and the title of the chapter or section first (enclosed in quotation marks and with only the first word capitalized), followed by the word *in*, the book title, and the book editor(s). Then give the publication information for the book and the page numbers where the chapter or section appears.

[5] E. Castronova, "The changing meaning of play," in *Online Communication and Collaboration: A Reader,* H. M. Donelan, K. L. Kear, and M. Ramage, Eds. New York: Routledge, 2010, pp. 184–189.

6. Book in an Edition Other Than the First The edition number follows the title of the book and is preceded by a comma.

[6] L. Xinju, *Laser Technology*, 2nd ed. Boca Raton, FL: CRC Press, 2010, pp. 203–205.

PRINT PERIODICALS

7. Journal Article Include the author, the article title, and the journal title (abbreviated where possible), followed by the volume number, issue number, page number(s), abbreviated month, and year (or abbreviated month, day, and year for weekly periodicals).

[7] R. C. Weber, P.-Y. Lin, E. J. Garnero, Q. Williams, and P. Lognonne, "Seismic detection of the lunar core," *Science*, vol. 331, no. 6015, pp. 309–312, Jan. 21, 2011.

8. Magazine Article List the author, the article title, and the magazine title (abbreviated where possible), followed by the page number(s) and the issue date.

[8] J. Villasenor, "The hacker in your hardware," *Scientific Amer.*, pp. 82–87, Aug. 2010.

9. Newspaper Article List the author, the article title, and the newspaper name, followed by the section and the date.

[9] M. Woolhouse, "For many, snow day is business as usual," *Boston Globe*, sec. B, Jan. 13, 2011.

ELECTRONIC SOURCES

10. Article in an Online Journal or Magazine

[10] R. Marani and A. G. Perri. (2010). An electronic medical device for preventing and improving the assisted ventilation of intensive care unit patients. *Open Elect. Electron. Eng. J.* [Online]. *4*, pp. 16–20. Available: https://benthamopen.com/ABSTRACT/TOEEJ-4-16

11. Website

[11] American Institute of Physics. (2011). American Institute of Physics [Online]. Available: http://www.aip.org

12. Document on a Government Website

[12] U.S. Department of Health and Human Services, Centers for Disease Control and Prevention (1999). Bioterrorism Readiness Plan: A Template for Healthcare Facilities [Online]. Available: https://emergency.cdc.gov/bioterrorism/pdf/13apr99APIC-CDCBioterrorism.pdf

OTHER SOURCES

13. Thesis or Dissertation

[13] J. L. Beutler, "Frequency response and gain enhancement of solid-state impact-ionization multipliers (SIMs)," Ph.D. dissertation, Dept. Elect. Eng., Brigham Young Univ., Provo, UT, 2010.

14. Standard For standards, include the title in italics, the standard number, and the date.

[14] *Testing and Evaluation Protocol for Spectroscopic Personal Radiation Detectors (SPRDs) for Homeland Security*, ANSI Standard T&E Protocol N42.48, 2010.

15. Scientific or Technical Report

[15] E. G. Fernando, "Investigation of rainfall and regional factors for maintenance cost allocation," Texas Transportation Inst. Texas A&M, College Station, TX, Report 5-4519-01-1, Aug. 2010.

16. Paper Published in Conference Proceedings

[16] T. O'Brien, A. Ritz, B. J. Raphael, and D. H. Laidlaw, "Gremlin: An interactive visualization model for analyzing genomic rearrangements," in *Proc. IEEE Information Visualization Conf.*, 2010, vol. 16, no. 6, pp. 918–926.

17. Government Document

[17] W. R. Selbig and R. T. Bannerman, "Characterizing the size distribution of particles in urban stormwater by use of fixed-point sample-collection methods," U.S. Geological Survey, Open-File Report 2011-1052, 2011.

18. Unpublished Document

[18] S. Reed, "An approach to evaluating the autistic spectrum in uncooperative adolescents," unpublished.

SAMPLE IEEE REFERENCE LIST

Following is a sample reference list using the IEEE numbered reference system. The references are listed in the order in which they might appear in a fictional document.

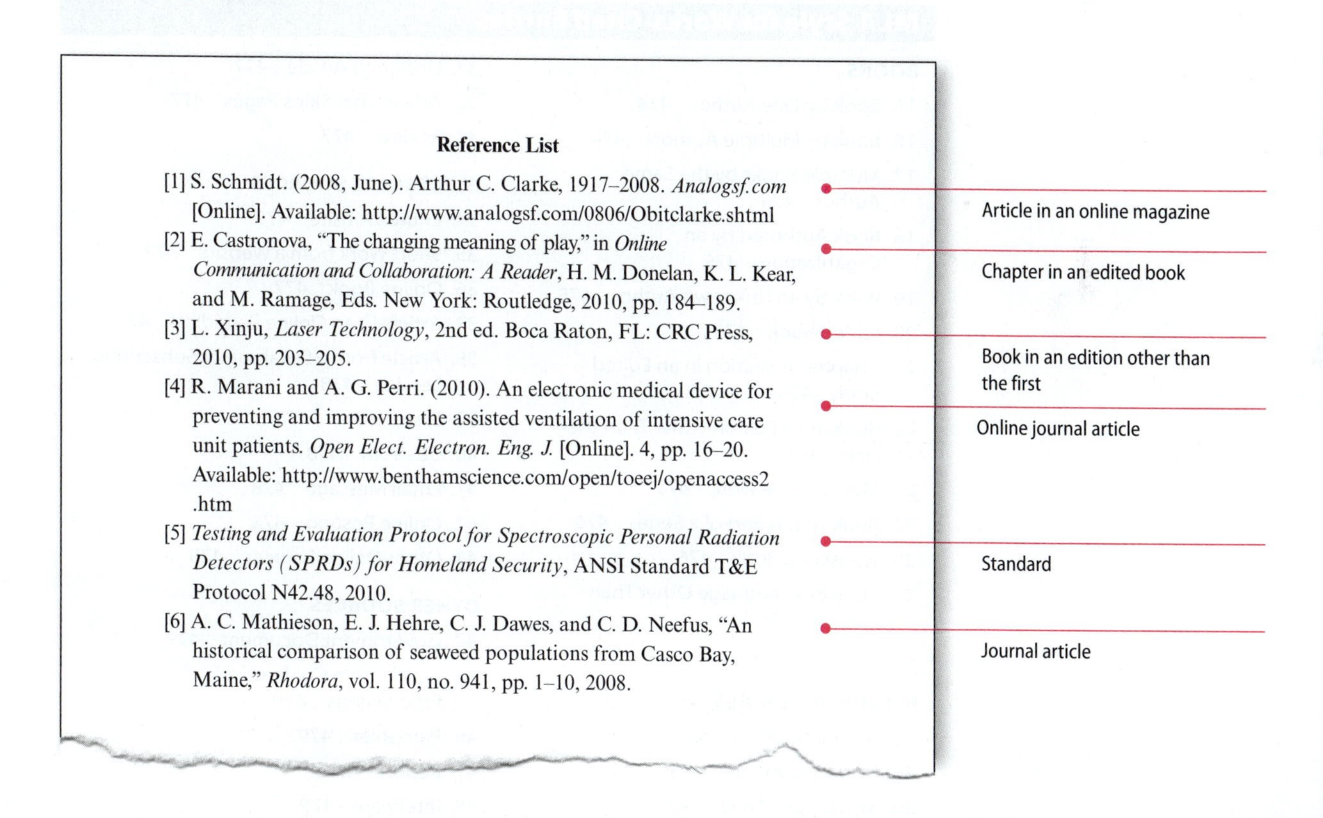

Reference List

[1] S. Schmidt. (2008, June). Arthur C. Clarke, 1917–2008. *Analogsf.com* [Online]. Available: http://www.analogsf.com/0806/Obitclarke.shtml

[2] E. Castronova, "The changing meaning of play," in *Online Communication and Collaboration: A Reader*, H. M. Donelan, K. L. Kear, and M. Ramage, Eds. New York: Routledge, 2010, pp. 184–189.

[3] L. Xinju, *Laser Technology*, 2nd ed. Boca Raton, FL: CRC Press, 2010, pp. 203–205.

[4] R. Marani and A. G. Perri. (2010). An electronic medical device for preventing and improving the assisted ventilation of intensive care unit patients. *Open Elect. Electron. Eng. J.* [Online]. 4, pp. 16–20. Available: http://www.benthamscience.com/open/toeej/openaccess2.htm

[5] *Testing and Evaluation Protocol for Spectroscopic Personal Radiation Detectors (SPRDs) for Homeland Security*, ANSI Standard T&E Protocol N42.48, 2010.

[6] A. C. Mathieson, E. J. Hehre, C. J. Dawes, and C. D. Neefus, "An historical comparison of seaweed populations from Casco Bay, Maine," *Rhodora*, vol. 110, no. 941, pp. 1–10, 2008.

Article in an online magazine

Chapter in an edited book

Book in an edition other than the first

Online journal article

Standard

Journal article

MLA Style

MLA (Modern Language Association) style consists of two elements: citations in the text and a list of works cited at the end of the document.

MLA Style for Textual Citations

MLA Style for Works-Cited Entries

MLA Style for Works-Cited Entries (*continued*)

MLA TEXTUAL CITATIONS

In MLA style, the textual citation typically includes the name of the source's author and the number of the page being referred to. Textual citations vary depending on the type of source cited and the context of the citation. The following models illustrate a variety of common situations; for additional examples, consult the *MLA Handbook*.

1. Entire Work If you are referring to the whole source, not to a particular page or pages, use only the author's name.

> Harwood's work gives us a careful framework for understanding the aging process and how it affects communication.

2. Specific Page(s) Immediately following the material you are quoting or paraphrasing, include a parenthetical reference with the author's name and the page number(s) being referred to. Do not add a comma between the name and the page number, and do not use the abbreviation *p.* or *pp.*

> Each feature evolves independently, so there can't be a steady progression of fossils representing change (Prothero 27).

If your sentence already includes the author's name, put only the page number in the parenthetical citation.

> Prothero explains why we won't find a steady progression of human fossils approaching modern humans, as each feature evolves independently (27).

3. Work Without Page Numbers Give a paragraph, section, or screen number, if provided. Use *par.* (singular) or *pars.* (plural) to indicate paragraph numbers. Either spell out or use standard abbreviations (such as *col.*, *fig.*, *pt.*, *ch.*, or *l.*) for other identifying words. Use a comma after the author's name if it appears in the parenthetical citation.

> Under the right conditions, humanitarian aid forestalls health epidemics in the aftermath of natural disasters (Bourmah, pars. 3–6).
>
> Maternal leave of at least three months has a significantly positive effect on the development of attachment in the infant (Ling, screen 2).

4. Multiple Sources by the Same Author If you cite two or more sources by the same author, either include the full source title in the text or add a shortened title after the author's name in the parenthetical citation to prevent confusion.

Chatterjee believes that diversification in investments can take many forms (*Diversification* 13).

Risk is a necessary component of a successful investment strategy (Chatterjee, *Failsafe* 25).

5. Source with Multiple Authors For a source written by two or three authors, cite all the names.

Grendel and Chang assert that . . .

This phenomenon was verified in the late 1970s (Grendel and Chang 281).

For a source written by four or more authors, either list all the authors or give only the first author, followed by the abbreviation *et al.* Follow the same format as in the works-cited list.

Studies show that incidences of type 2 diabetes are widespread and rising quickly (Gianarikas et al.).

6. Source Quoted Within Another Source Give the source of the quotation in the text. In the parenthetical citation, give the author and page number(s) of the source in which you found the quotation, preceded by *qtd. in.*

Freud describes the change in men's egos as science proved that the earth was not the center of the universe and that man was descended from animals (qtd. in Prothero 89–90).

Only the source by Prothero will appear in the list of works cited.

7. Source Authored by an Organization If the author is an organization rather than a person, use the name of the organization. When giving the organization's name in parentheses, abbreviate common words.

In a recent booklet, the Association of Sleep Disorders discusses the causes of narcolepsy (2–3).

The causes of narcolepsy are discussed in a recent booklet (Assn. of Sleep Disorders 2–3).

8. Source with an Unknown Author If the source does not identify an author, shorten the title to the first noun phrase in your parenthetical citation.

Multidisciplinary study in academia is becoming increasingly common ("Interdisciplinary Programs" 23).

In a web document, the author's name is often at the end of the document or in small print on the home page. Do some research before assuming that a website does not have an author. Remember that an organization might be the author. (See item 7.)

9. Multiple Sources in One Citation When you refer to two or more sources at the same point, separate the sources with a semicolon.

Much speculation exists about the origin of this theory (Brady 42; Yao 388).

10. Multiple Authors with the Same Last Name If the authors of two or more sources have the same last name, spell out the first names of those authors in the text and use the authors' first initials in parenthetical citations.

> In contrast, Albert Martinez has a radically different explanation (29).
>
> The economy's strength may be derived from its growing bond market (J. Martinez 87).

11. Chapter or Section in an Edited Book Cite the author of the work, not the editor of the anthology.

> Wolburg and Treise note that college binge drinkers include students with both high and low GPAs (4).

12. Multivolume Work If you use only one volume of a multivolume work, list the volume number in the works-cited list only. If you use more than one volume of a multivolume work, indicate the specific volume you are referring to, followed by a colon and the page number, in your parenthetical citation.

> Many religious organizations opposed the Revolutionary War (Hazlitt 2: 423).

13. Entry in a Reference Work If the entry does not have an author, use the word or term you looked up. You do not need to cite page numbers for entries in encyclopedias and dictionaries because they are arranged alphabetically.

> The term *groupism* is important to understand when preparing to communicate with Japanese business counterparts ("Groupism").

14. Electronic Source When citing electronic sources, follow the same rules as for print sources, providing author names and page numbers, if available. If an author's name is not given, use either the full title of the source in the text or the first noun phrase of the title in one parenthetical citation. (See item 8 on page 472.) If no page numbers appear, include other identifying numbers, such as paragraph or section numbers, only if they are provided in the source.

> Twenty million books were in print by the early sixteenth century (Rawlins).

THE MLA LIST OF WORKS CITED

A list of works cited provides the information your readers will need to find each source you have cited in the text. It should not include sources you consulted for background reading. Following are some guidelines for an MLA-style list of works cited.

For a sample MLA-style list of works, see p. 481.

- **Arranging entries.** Arrange the entries alphabetically by the author's last name. If two or more works are by the same author, arrange them alphabetically by title. Alphabetize works by an organization by the first significant word in the name of the organization.
- **Book titles.** Italicize titles of books and capitalize all major words. Note that in MLA style, prepositions are not capitalized.

- **Publishers.** Give the full name of the publisher, including any words such as *Books* or *Publisher*. Omit terms such as *Inc.* or *Company*. If the publisher is a university press, you may use the abbreviation *UP*. If a source lists two equal publishers, use both names separated by a slash (/). For imprints, use the name of the parent company. For divisions, use the division name.
- **Periodical titles.** Italicize titles of periodicals and capitalize all major words. Omit any initial article.
- **Article titles.** Place titles of articles and other short works in quotation marks and capitalize all major words.
- **Electronic sources.** Include as much information as you can about electronic sources, such as author, date of publication, identifying numbers, and retrieval information. If no date of publication is provided, include the date you accessed the information at the end of the citation. If no author is known, start with the title of the website. Italicize titles of entire websites; treat titles of works within websites, such as articles and video clips, as you would for print sources. Include the URL for any sources you access on the Internet. If you access a source through a database, include the DOI (digital object identifier), if provided, or a stable URL.
- **Indenting.** Use a hanging indent, with the first line of each entry flush with the left margin and all subsequent lines indented one-half inch.
- **Spacing.** Double-space the entire works-cited list. Do not add extra space between entries.
- **Page numbers.** Use the abbreviation *p.* or *pp.* when giving page numbers. For a range of pages, give only the last two digits of the second number if the previous digits are identical (for example, 243–47, *not* 243–247 or 243–7). Use a plus sign (+) to indicate that an article continues on subsequent pages, interrupted by other articles or advertisements.
- **Dates.** Follow the format day month year, with no commas (for example, 20 Feb. 2009). Spell out *May*, *June*, and *July*; abbreviate all other months (except *Sept.*) using the first three letters followed by a period. For journals, give the season or month and year in addition to volume and issue numbers.

Following are models of works-cited-list entries for a variety of sources. For further examples of MLA-style citations, consult the *MLA Handbook*.

BOOKS

15. Book by One Author Include the author's full name, in reverse order, followed by the book title. Next give the location and name of the publisher, followed by the year of publication.

Gleick, James. *The Information: A History, a Theory, a Flood.* Pantheon, 2011.

16. Book by Multiple Authors For a book by two or three authors, present the names in the sequence in which they appear on the title page. Use reverse order for the name of the first author only. Use a comma to separate the names of the authors.

Burt, Stephen, and David Mikics. *The Art of the Sonnet*. Belknap-Harvard UP, 2010.

For a book by four or more authors, either name all the authors or use the abbreviation *et al.* after the first author's name.

Thomas, David N., et al. *The Biology of Polar Regions*. Oxford UP, 2008.

17. Multiple Books by the Same Author For the second and subsequent entries by the same author, use three hyphens followed by a period in place of the name. Arrange the entries alphabetically by title, ignoring *A*, *An*, or *The*.

Hassan, Robert. *Empires of Speed: Time and the Acceleration of Politics and Society*. Leiden, Brill, 2009.

—. *The Information Society: Cyber Dreams and Digital Nightmares*. Cambridge, Eng., Polity, 2008.

18. Book Authored by an Organization The organization takes the position of the author.

World Bank. *Atlas of Global Development: A Visual Guide to the World's Greatest Challenges*. Washington, World Bank, 2011.

19. Book by an Unknown Author If the author of the book is unknown, begin with the title.

The World Almanac Notebook Atlas. Union, Hammond, 2010.

Note that you would ignore *The* in alphabetizing this entry.

20. Edited Book List the book editor's name, followed by *editor* (or *editors* if more than one editor), in place of the author's name.

Levi, Scott Cameron, and Ron Sela, editors. *Islamic Central Asia: An Anthology of Historical Sources*. Indiana UP, 2010.

21. Chapter or Section in an Edited Book Give the author and title of the article first, followed by the book title and editor. Present the editor's name in normal order, preceded by *edited by* and followed by a comma. After the publication information, give the pages on which the material appears.

Marx, Karl. "Proletarians and Communists." *Marx Today: Selected Works and Recent Debates*, edited by John F. Sitton, Macmillan, 2010, pp. 51–56.

22. Book in an Edition Other Than the First List the edition number after the title of the book.

Geary, Patrick, editor. *Readings in Medieval History*. 4th ed., U of Toronto P, 2010.

23. Multivolume Work If you use two or more volumes from a multivolume work, indicate the total number of volumes (for example, *4 vols.*) before the place of publication. If you use only one volume, give the volume number before the publisher, and give the total number of volumes after the date.

Sophocles. *The Complete Sophocles.* Edited by Peter Burian and Alan Shapiro, vol. 1, Oxford UP, 2010. 2 vols.

24. Book That Is Part of a Series End the entry with the series name as it appears on the title page (but use common abbreviations, such as *Ser.*), followed by the series number, if any. Do not italicize the series name.

Aune, David Edward, editor. *The Blackwell Companion to the New Testament.* Wiley-Blackwell, 2010. Blackwell Companions to Religion.

25. Translated Book After the title, present the translator's name in normal order, preceded by *Translated by*.

Torre, Domingo de la, Romin Teratol, and Antzelmo Peres. *Travelers to the Other World: A Maya View of North America.* Translated by Robert M. Laughlin, edited by Carol Karasik, U of New Mexico P, 2010.

26. Book in a Language Other Than English You may give a translation of the book's title in brackets.

Moine, Fabienne. *Poésie et identité féminines en Angleterre: le genre en jeu, 1830–1900* [*Poetry and Female Identity in England: Genre/Gender at Play*]. L'Harmattan, 2010.

27. Entry in a Reference Work If the work is well known, you do not need to include the publisher or place of publication. If entries are listed alphabetically, you do not need to include a page number.

"Desdemona." *Women in Shakespeare: A Dictionary*, edited by Alison Findlay, Bloomsbury, 2014. Arden Shakespeare Dictionary Series.

PRINT PERIODICALS

For a tutorial on citing articles in MLA style, go to LaunchPad.

28. Journal Article List the author's name, the article title (in quotation marks), and the journal title (italicized), followed by the volume number, issue number, year, and page number(s).

Mooney, William. "Sex, Booze, and the Code: Four Versions of the *Maltese Falcon.*" *Literature-Film Quarterly*, vol. 39, no. 1, Jan. 2011, pp. 54–72.

29. Magazine Article List the author's name, the article title (in quotation marks), and the magazine title (italicized), followed by the issue date and page number(s).

Seabrook, John. "Crush Point." *The New Yorker*, 7 Feb. 2011, pp. 32–38.

30. Newspaper Article List the author's name, the article title (in quotation marks), and the newspaper name (italicized), followed by the issue date and the page number(s) (which might include a section letter).

Robertson, Campbell. "Beyond the Oil Spill, the Tragedy of an Ailing Gulf." *The New York Times*, 21 Apr. 2011, p. A17.

31. Unsigned Article If the author of an article is not indicated, begin with the title. Alphabetize the work by title, ignoring any initial article.

"How Much Is Enough?" *The Economist*, 26 Feb. 2011, p. 5.

32. Article That Skips Pages Give the page on which the article starts, followed by a plus sign (+) and a period.

Kennicott, Philip. "Out-Vermeering Vermeer." *The Washington Post*, 10 Apr. 2011, pp. E1+.

33. Review For a book or film review, give the author of the review and the title of the review (in quotation marks), followed by the words *Review of* and the title of the work reviewed (italicized). Insert a comma and the word *by*, then give the name of the author of the work reviewed. (Instead of *by*, you might use *edited by*, *translated by*, or *directed by*, depending on the work.) End with the publication information for the periodical in which the review was published.

Wynne, Clive. "Our Conflicted Relationship with Animals." Review of *Some We Love, Some We Hate, Some We Eat*, by Hal Herzog. *Nature*, vol. 467, no. 7313, 16 Sept. 2010, pp. 275–76.

ELECTRONIC SOURCES

34. Entire Website If you are citing an entire website, begin with the name of the author or editor (if given) and the title of the site (italicized). Then give the name of the sponsoring institution or organization, the date of publication or most recent update, and the URL, followed by a period. If no publication date is available, include the date you accessed the site at the end of the entry.

For a tutorial on citing websites in MLA style, go to LaunchPad.

Poets.org. Academy of American Poets, www.poets.org. Accessed 12 Jan. 2017.

35. Short Work from a Website If you are citing a portion of a website, begin with the author, the title of the material (in quotation marks), and the title of the site (italicized). Then include the publisher, the date of publication, and the URL. If the publisher name is the same as the site title, you do not need to repeat it.

Ferenstein, Greg. "How Mobile Technology Is a Game Changer for Developing Africa." *Mashable*, 19 July 2010, mashable.com/2010/07/19/mobile-africa/#mahVbvylbaqO.

36. Online Book Begin with the author's name and the title of the work, along with publication information about the print source, if the book has been published in print. Then include the name of the site where you accessed the book and the URL.

Piketty, Thomas. *Capital in the Twenty-First Century*. Translated by Arthur Goldhammer, Harvard UP, 2014. *Google Books*, books.google.com/books?isbn=0674369556.

37. Article in an Online Periodical Begin with the author's name and include the title of the document, the name of the periodical, and the date of publication. If the periodical is a scholarly journal, include relevant identifying numbers, such as volume, issue, and page numbers before the date. End with the URL.

Maas, Korey D. "Natural Law, Lutheranism, and the Public Good." *Lutheran Witness*, vol. 130, no. 3, 2 Mar. 2011, blogs.lcms.org/2011/natural-law-lutheranism-and-the-public-good-3-2011.

For magazine and newspaper articles found online, give the author, the title of the article (in quotation marks), the title of the magazine or newspaper (italicized), the date of publication and the URL.

Crowell, Maddy. "How Computers Are Getting Better at Detecting Liars." *The Christian Science Monitor*, 12 Dec. 2015, www.csmonitor.com/Science/Science-Notebook/2015/1212/How-computers-are-getting-better-at-detecting-liars.

For a tutorial on citing databases in MLA style, go to LaunchPad.

38. Article from a Database or Subscription Service After giving the print article information, give the name of the database (italicized) and the DOI of the article. If no DOI is available and the database provides a stable URL or permalink, give the complete URL. For subscription databases such as EBSCO, you may use a truncated URL.

Coles, Kimberly Anne. "The Matter of Belief in John Donne's Holy Sonnets." *Renaissance Quarterly*, vol. 68, no. 3, Fall 2015, pp. 899–931. *JSTOR*, doi:10.1086/683855.

39. Dissertation The title appears in quotation marks if the dissertation is unpublished or in italics if it is published.

Zimmer, Kenyon. *The Whole World Is Our Country: Immigration and Anarchism in the United States, 1885–1940.* Dissertation, U of Pittsburgh, 2010.

40. CD-ROM Treat material on a CD-ROM as you would if it were in print form and include the descriptive term *CD-ROM* at the end of the entry. That is, if the source is an article in a database, include *CD-ROM* after the page numbers, followed by the database title, the vendor, and the publication date of the database. If it is a book, treat it as a book.

Greek-Cypriot Maritime Guide 2011. Alimos, Greece, Marine Information Services, 2011. CD-ROM.

41. Email Message Include the email's subject line as the title. Include the words *Received by* followed by the name of the recipient (if you were the recipient, use the phrase *the author*). End with the date the email was sent.

Lange, Frauke. "Data for Genealogical Project." Received by the author, 26 Dec. 2018.

42. Online Posting List the author's name, the title (or subject line) in quotation marks, the name of the discussion group or newsgroup in italics, the publisher, the posting date, and the URL. If the posting includes the time when it was posted, list the time along with the date.

@MarsCuriosity. "Can you see me waving? How to spot #Mars in the night sky: https://youtu.be/hv8hVvJlcJQ." *Twitter*, 5 Nov. 2015, 11:00 a.m., twitter.com/marscuriosity/status/672859022911889408.

43. Other Online Sources Follow the MLA guidelines already discussed, adapting them as appropriate to the electronic medium. The following examples are for a podcast and a blog, respectively.

McDougall, Christopher. "How Did Endurance Help Early Humans Survive?" *TED Radio Hour*, National Public Radio, 20 Nov. 2015, www.npr.org/2015/11/20/455904655/how-did-endurance-help-early-humans-survive.

Cimons, Marlene. "Why Cities Could Be the Key to Solving the Climate Crisis." *Thinkprogress.org*, Center for American Progress Action Fund, 10 Dec. 2015, thinkprogress.org/why-cities-could-be-the-key-to-solving-the-climate-crisis-2af518316bef.

OTHER SOURCES

44. Government Document Begin with the author. If the author is a government agency, begin with the name of the country and the agency. Follow with the document title, publisher, and date.

United States, National Commission on the Causes of the Financial and Economic Crisis. *The Financial Crisis Inquiry Report: Final Report of the National Commission on the Causes of the Financial and Economic Crisis in the United States.* US Government Publishing Office, 2011.

For an online source, include the URL.

Phelps, G. A., et al. *A Refined Characterization of the Alluvial Geology of Yucca Flat and Its Effect on Bulk Hydraulic Conductivity.* Open-File Report 2010-1307, US Department of the Interior/US Geological Survey, 2011, pubs.usgs.gov/of/2010/1307/of2010-1307.pdf.

45. Article from Conference Proceedings List the author's name, the article title, the proceedings title, and the editor's name, followed by the publication information.

Glicksman, Robert. "Climate Change Adaptation and the Federal Lands." *The Past, Present, and Future of Our Public Lands: Celebrating the 40th Anniversary of the Public Land Law Review Commission's Report*, edited by Gary C. Bryner, Natural Resources Law Center, 2010.

46. Pamphlet Cite a pamphlet as you would a book.

The Legendary Sleepy Hollow Cemetery. Friends of Sleepy Hollow Cemetery, 2008.

47. Report Cite a report as you would a book.

Liebreich, Michael, et al. *Green Investing 2010: Policy Mechanisms to Bridge the Financing Gap.* World Economic Forum, 2010.

48. Interview For a published interview, begin with the name of the person interviewed. If the interview has a title, enclose it in quotation marks. Insert the words *Interview by* and give the interviewer's name followed by the information of the work in which the interview was published.

Walcott, Derek. "Purple Prose." Interview by Alexander Newbauer, *Harper's Magazine*, Feb. 2010, pp. 24–26.

If you conducted the interview yourself, give the interviewee's name, the words *Personal interview*, and the date.

Youngblood, Adelaide. Personal interview, 5 Jan. 2018.

49. Letter or Memo If the letter or memo was addressed to you, give the writer's name, followed by the words *Letter* [or *Memo*] *to the author* and the date it was written.

Jakobiak, Ursula. Letter to the author, 27 Oct. 2018.

If the letter or memo was addressed to someone other than you, give the recipient's name in place of the words *the author*.

50. Lecture or Speech Give the speaker's name, the title of the lecture or speech (if known), the event and sponsoring organization (if applicable), and the place and date. End with a medium-descriptive term, such as *Lecture* or *Keynote Speech*.

Wang, Samuel. "Neuroscience and Everyday Life." Freshman Assembly, 12 Sept. 2010, Princeton University, Princeton, NJ. Lecture.

51. Map or Chart Give the author (if known), the title (in quotation marks), and the publication information. For an online source, include the name of the website (italicized), the name of the site's publisher, the date of publication, and the URL.

"Map of Sudan." *Global Citizen*, Citizens for Global Solutions, 2011, globalsolutions.org/blog/bashir#.VthzNMfi_FI.

52. Photograph or Work of Art Give the name of the artist; the title of the artwork, italicized; the date of composition; and the institution and city in which the artwork can be found. For artworks found online, include the title of the website on which you found the work, the publisher, and the URL.

Smedley, W. T. *On the Beach at Narragansett Pier.* 1900, *Cabinet of American Illustration*, Library of Congress, Prints and Photographs Division, www.loc.gov/pictures/collection/cai/item/2010718015/.

53. Legal Source For a legal case, give the name of the first plaintiff and first defendant, the law report number, the name of the court, and the year of the decision, followed by the publication information.

Utah v. Evans. 536 US 452. Supreme Court of the US. 2002. *Legal Information Institute*, Cornell U Law School, www.law.cornell.edu/supremecourt/text/536/452.

For a legislative act, give the name of the act, the Public Law number, the Statutes at Large volume and page numbers, and the date the law was enacted.

Protect America Act of 2007. Pub. L. 110-55. 5 Stat. 121–552. 5 Aug. 2007.

54. Radio or Television Program Give the title of the episode or segment, if applicable, and the title of the program. Include relevant information about the narrator, director, or performers. Then give the network and the broadcast date. If you accessed the program on the web, include the URL.

"Aircraft Safety." *Nightline*, narrated by Cynthia McFadden, ABC, 4 Apr. 2011.

"The Cathedral." *Reply All*, narrated by Sruthi Pinnamaneni, episode 50, Gimlet Media, 7 Jan. 2016, gimletmedia.com/episode/50-the-cathedral.

55. Film, Video, or DVD Give the title of the film and the name of the director. You may also give the names of major performers (*performances by*) or the narrator (*narrated by*). Give the distributor and the year of original release.

Lady Bird. Directed by Greta Gerwig, performances by Saoirse Ronan, Laurie Metcalf, and Beanie Feldstein, A24, 2017.

56. Advertisement Include the name of the product, organization, or service being advertised, and the publication information. At the end of the entry, add the descriptive term *Advertisement.*

NeutronicEar. *Smithsonian*, Mar. 2011, p. 89. Advertisement.

SAMPLE MLA LIST OF WORKS CITED

Following is a sample list of works cited using the MLA citation system.

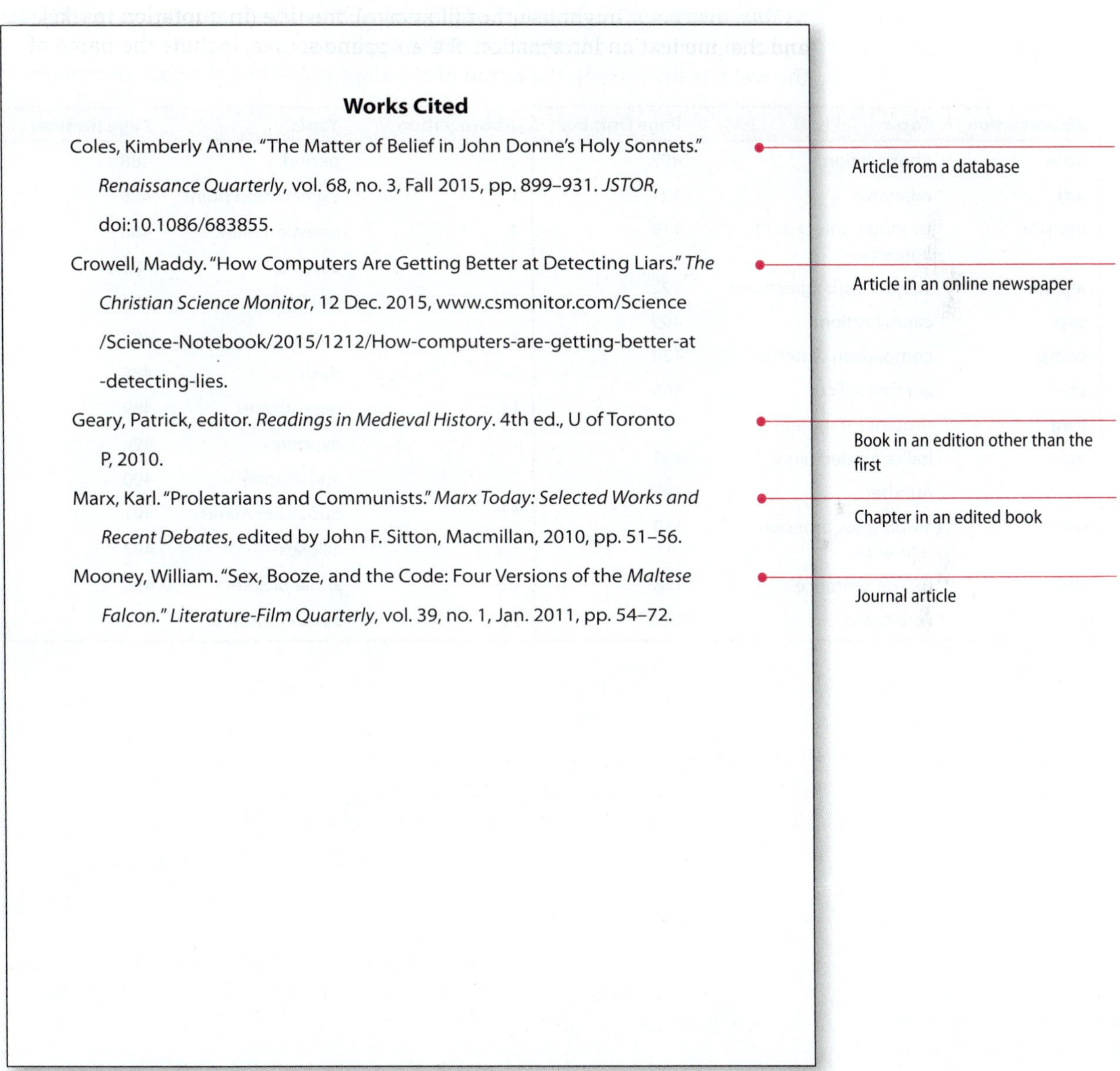

Works Cited

Coles, Kimberly Anne. "The Matter of Belief in John Donne's Holy Sonnets." *Renaissance Quarterly*, vol. 68, no. 3, Fall 2015, pp. 899–931. *JSTOR*, doi:10.1086/683855.

Crowell, Maddy. "How Computers Are Getting Better at Detecting Liars." *The Christian Science Monitor*, 12 Dec. 2015, www.csmonitor.com/Science/Science-Notebook/2015/1212/How-computers-are-getting-better-at-detecting-lies.

Geary, Patrick, editor. *Readings in Medieval History*. 4th ed., U of Toronto P, 2010.

Marx, Karl. "Proletarians and Communists." *Marx Today: Selected Works and Recent Debates*, edited by John F. Sitton, Macmillan, 2010, pp. 51–56.

Mooney, William. "Sex, Booze, and the Code: Four Versions of the *Maltese Falcon*." *Literature-Film Quarterly*, vol. 39, no. 1, Jan. 2011, pp. 54–72.

Part B: Editing and Proofreading Your Documents

This part of the Appendix contains advice on editing your documents for grammar, punctuation, and mechanics. If your organization or professional field has a style guide with different recommendations about grammar and usage, you should of course follow those guidelines.

Your instructor might use the following abbreviations to refer you to specific topics in the text and in this part of the Appendix.

Abbreviation	Topic	Page Number
abbr	abbreviation	497
adj	adjective	121
agr p/a	pronoun-antecedent agreement	119
agr s/v	subject-verb agreement	122
cap	capitalization	499
comp	comparison of items	120
cs	comma splice	485
frag	sentence fragment	117
ital	italics (underlining)	493
num	number	495
ref	ambiguous pronoun reference	119
run	run-on sentence	118
t	verb tense	122

Abbreviation	Topic	Page Number
.	period	488
!	exclamation point	488
?	question mark	488
,	comma	483
;	semicolon	486
:	colon	487
—	dash	488
( )	parentheses	489
-	hyphen	494
'	apostrophe	490
" "	quotation marks	491
. . .	ellipses	492
< >	angle brackets	494
[ ]	square brackets	493

Punctuation

, COMMAS

The comma is the most frequently used punctuation mark, as well as the one about whose usage writers most often disagree. Examples of common misuses of the comma accompany the following guidelines. The section concludes with advice about editing for unnecessary commas.

1. **Use a comma in a compound sentence to separate two independent clauses linked by a coordinating conjunction (*and, or, nor, but, so, for,* or *yet*).**

   INCORRECT The mixture was prepared from the two premixes and the remaining ingredients were then combined.

   CORRECT The mixture was prepared from the two premixes, and the remaining ingredients were then combined.

2. **Use a comma to separate items in a series composed of three or more elements.**

   The manager of spare parts is responsible for ordering, stocking, and disbursing all spare parts for the entire plant.

   Despite the presence of the conjunction *and*, most technical-communication style manuals require a comma after the second-to-last item. The comma clarifies the separation and prevents misreading.

   CONFUSING The report will be distributed to Operations, Research and Development and Accounting.

   CLEAR The report will be distributed to Operations, Research and Development, and Accounting.

3. **Use a comma to separate introductory words, phrases, and clauses from the main clause of the sentence.**

   However, we will have to calculate the effect of the wind.

   To facilitate trade, the government holds a yearly international conference.

   In the following example, the comma actually prevents misreading:

   Just as we finished eating, the rats discovered the treadmill.

   NOTE: Writers sometimes make errors by omitting commas following introductory words, phrases, or clauses. A comma is optional only if the introductory text is brief and cannot be misread.

   CORRECT First, let's take care of the introductions.

   CORRECT First let's take care of the introductions.

   INCORRECT As the researchers sat down to eat the laboratory rats awakened.

   CORRECT As the researchers sat down to eat, the laboratory rats awakened.

4. **Use a comma to separate a dependent clause from the main clause.**

   Although most of the executive council saw nothing wrong with it, the advertising campaign was canceled.

   Most tablet computers use green technology, even though it is relatively expensive.

5. **Use commas to separate nonrestrictive modifiers (parenthetical clarifications) from the rest of the sentence.**

   Jones, the temporary chairman, called the meeting to order.

For more about restrictive and nonrestrictive modifiers, see Ch. 6, p. 131.

   NOTE: Writers sometimes introduce an error by dropping one of the commas around a nonrestrictive modifier.

   INCORRECT The data line, which was installed two weeks ago had to be disconnected.

   CORRECT The data line, which was installed two weeks ago, had to be disconnected.

6. **Use a comma to separate interjections and transitional elements from the rest of the sentence.**

   Yes, I admit that your findings are correct.

   Their plans, however, have great potential.

   NOTE: Writers sometimes introduce an error by dropping one of the commas around an interjection or a transitional element.

   INCORRECT Our new statistician, however used to work for Konaire, Inc.

   CORRECT Our new statistician, however, used to work for Konaire, Inc.

7. **Use a comma to separate coordinate adjectives.**

   The finished product was a sleek, comfortable cruiser.

   The heavy, awkward trains are still being used.

   The commas in these examples take the place of the conjunction *and.*

For more about coordinate adjectives, see Ch. 6, p. 121.

   If the adjectives are not coordinate—that is, if one of the adjectives modifies the combined adjective and noun—do not use a comma:

   They decided to go to the first general meeting.

8. **Use a comma to signal that a word or phrase has been omitted from a sentence because it is implied.**

   Smithers is in charge of the accounting; Harlen, the data management; Demarest, the publicity.

   The commas after *Harlen* and *Demarest* show that the phrase *is in charge of* has not been repeated.

9. **Use a comma to separate a proper noun from the rest of the sentence in direct address.**

John, have you seen the purchase order from United?

What I'd like to know, Betty, is why we didn't see this problem coming.

10. **Use a comma to introduce most quotations.**

He asked, "What time were they expected?"

11. **Use a comma to separate cities or towns, states, and countries.**

Bethlehem, Pennsylvania, is the home of Lehigh University.

He attended Lehigh University in Bethlehem, Pennsylvania, and the University of California at Berkeley.

Note that a comma precedes and follows *Pennsylvania*.

12. **Use a comma to set off the year in a date.**

August 1, 2016, is the anticipated completion date.

If the month separates the date and the year, you do not need to use commas because the numbers are not next to each other:

The anticipated completion date is 1 August 2016.

13. **Use a comma to clarify numbers.**

12,013,104

NOTE: European practice is to reverse the use of commas and periods in writing numbers: periods separate hundreds and thousands, thousands and millions, and so on, while commas separate whole numbers from decimals.

12.013,4

14. **Use a comma to separate names from professional or academic titles.**

Harold Clayton, PhD
Marion Fewick, CLU
Joyce Carnone, PE

The comma also follows the title in a sentence:

Harold Clayton, PhD, is the featured speaker.

MISUSE OF COMMAS

Writers often introduce errors by using inappropriate commas. Commas are not used in the following situations:

For more about comma splices, see Ch. 6, p. 118.

- Commas are not used to link two independent clauses without a coordinating conjunction (an error known as a *comma splice*).

INCORRECT	All the motors were cleaned and dried after the water had entered, had they not been, additional damage would have occurred.
CORRECT	All the motors were cleaned and dried after the water had entered; had they not been, additional damage would have occurred.

CORRECT All the motors were cleaned and dried after the water had entered. Had they not been, additional damage would have occurred.

- Commas are not used to separate the subject from the verb in a sentence.

INCORRECT Another of the many possibilities, is to use a "first in, first out" sequence.

CORRECT Another of the many possibilities is to use a "first in, first out" sequence.

- Commas are not used to separate the verb from its complement.

INCORRECT The schedules that have to be updated every month are, numbers 14, 16, 21, 22, 27, and 31.

CORRECT The schedules that have to be updated every month are numbers 14, 16, 21, 22, 27, and 31.

- Commas are not used with a restrictive modifier.

INCORRECT New and old employees, who use the processed order form, do not completely understand the basis of the system.

The phrase *who use the processed order form* is a restrictive modifier necessary to the meaning: it defines which employees do not understand the system.

CORRECT New and old employees who use the processed order form do not completely understand the basis of the system.

INCORRECT A company, that has grown so big, no longer finds an informal evaluation procedure effective.

The clause *that has grown so big* is a restrictive modifier.

CORRECT A company that has grown so big no longer finds an informal evaluation procedure effective.

- Commas are not used to separate two elements in a compound subject.

INCORRECT Recent studies, and reports by other firms confirm our experience.

CORRECT Recent studies and reports by other firms confirm our experience.

; SEMICOLONS

Semicolons are used in the following instances:

1. **Use a semicolon to separate independent clauses not linked by a coordinating conjunction.**

   The second edition of the handbook is more up-to-date; however, it is also more expensive.

2. **Use a semicolon to separate items in a series that already contains commas.**

   The members elected three officers: Jack Resnick, president; Carol Wayshum, vice president; Ahmed Jamoogian, recording secretary.

Here the semicolon acts as a "supercomma," grouping each name with the correct title.

MISUSE OF SEMICOLONS

Sometimes writers incorrectly use a semicolon when a colon is called for:

INCORRECT We still need one ingredient; luck.

CORRECT We still need one ingredient: luck.

: COLONS

Colons are used in the following instances:

1. **Use a colon to introduce a word, phrase, or clause that amplifies, illustrates, or explains a general statement.**

   The project team lacked one crucial member: a project leader.

   Here is the client's request: we are to provide the preliminary proposal by November 13.

   We found three substances in excessive quantities: potassium, cyanide, and asbestos.

   The week was productive: 14 projects were completed, and another dozen were initiated.

   NOTE: The text preceding a colon should be able to stand on its own as a sentence:

   INCORRECT We found: potassium, cyanide, and asbestos.

   CORRECT We found the following: potassium, cyanide, and asbestos.

   CORRECT We found potassium, cyanide, and asbestos.

2. **Use a colon to introduce items in a vertical list if the sense of the introductory text would be incomplete without the list.**

   We found the following:

   - potassium
   - cyanide
   - asbestos

3. **Use a colon to introduce long or formal quotations.**

   The president began: "In the last year . . . ."

MISUSE OF COLONS

Writers sometimes incorrectly use a colon to separate a verb from its complement:

INCORRECT The tools we need are: a plane, a level, and a T square.

CORRECT The tools we need are a plane, a level, and a T square.

CORRECT We need three tools: a plane, a level, and a T square.

. PERIODS

Periods are used in the following instances:

1. **Use a period at the end of sentences that do not ask questions or express strong emotion.**

   The lateral stress still needs to be calculated.

For more about abbreviations, see page 497.

2. **Use a period after some abbreviations.**

   U.S.A.

   etc.

3. **Use a period with decimal fractions.**

   4.056

   $6.75

   75.6 percent

! EXCLAMATION POINTS

The exclamation point is used at the end of a sentence that expresses strong emotion, such as surprise.

> The nuclear plant, which was originally expected to cost $1.6 billion, eventually cost more than $8 billion!

In technical documents, which require objectivity and a calm, understated tone, exclamation points are rarely used.

? QUESTION MARKS

The question mark is used at the end of a sentence that asks a direct question.

> What did the commission say about effluents?

NOTE: When a question mark is used within quotation marks, no other end punctuation is required.

> She asked, "What did the commission say about effluents?"

MISUSE OF QUESTION MARKS

Do not use a question mark at the end of a sentence that asks an indirect question.

> He wanted to know whether the procedure had been approved for use.

— DASHES

To make a dash, use two uninterrupted hyphens (—). Do not add a space before or after the dash. Some word-processing programs turn two hyphens into a dash, but with others, you have to use a special combination of keys to make a dash; there is no dash key on the keyboard.

Dashes are used in the following instances:

1. **Use a dash to set off a sudden change in thought or tone.**

   The committee found—can you believe this?—that the company bore full responsibility for the accident.

   That's what she said—if I remember correctly.

2. **Use a dash to emphasize a parenthetical element.**

   The managers' reports—all 10 of them—recommend production cutbacks for the coming year.

   Arlene Kregman—the first woman elected to the board of directors—is the next scheduled speaker.

3. **Use a dash to set off an introductory series from its explanation.**

   Wet suits, weight belts, tanks—everything will have to be shipped in.

   NOTE: When a series follows the general statement, a colon replaces the dash.

   Everything will have to be shipped in: wet suits, weight belts, and tanks.

MISUSE OF DASHES

Sometimes writers incorrectly use a dash as a substitute for other punctuation marks:

INCORRECT	The regulations—which were issued yesterday—had been anticipated for months.
	There would be no reason to emphasize this parenthetical element.
CORRECT	The regulations, which were issued yesterday, had been anticipated for months.
INCORRECT	Many candidates applied—however, only one was chosen.
CORRECT	Many candidates applied; however, only one was chosen.

() PARENTHESES

Parentheses are used in the following instances:

1. **Use parentheses to set off incidental information.**

   Please call me (x3104) when you get the information.

   Galileo (1564–1642) is often considered the father of modern astronomy.

   The cure rate for lung cancer almost doubled in thirty years (Capron, 2016).

2. **Use parentheses to enclose numbers and letters that label items listed in a sentence.**

   To transfer a call within the office, (1) place the party on HOLD, (2) press TRANSFER, (3) press the extension number, and (4) hang up.

   Use both a left and a right parenthesis—not just a right parenthesis—in this situation.

MISUSE OF PARENTHESES

For more about square brackets, see p. 493.

Sometimes writers incorrectly use parentheses instead of brackets to enclose their insertion within a quotation:

INCORRECT He said, "The new manager (Farnham) is due in next week."

CORRECT He said, "The new manager [Farnham] is due in next week."

' APOSTROPHES

Apostrophes are used in the following instances:

1. **Use an apostrophe to indicate possession.**

   the manager's goals the employees' credit union

   the workers' lounge Charles's T square

   For joint possession, add an apostrophe and an s only to the last noun or proper noun:

   Watson and Crick's discovery

   For separate possession, add an apostrophe and an s to each of the nouns or pronouns:

   Newton's and Galileo's theories

   NOTE: Do not add an apostrophe or an s to possessive pronouns: *his*, *hers*, *its*, *ours*, *yours*, *theirs*.

2. **Use an apostrophe to indicate possession when a noun modifies a gerund.**

   We were all looking forward to Bill's joining the company.

   The gerund *joining* is modified by the proper noun *Bill*.

3. **Use an apostrophe to form contractions.**

   I've shouldn't

   can't it's

   The apostrophe usually indicates an omitted letter or letters:

   can(no)t = can't

   it (i)s = it's

   NOTE: Some organizations discourage the use of contractions. Find out what the policy of your organization is.

4. **Use an apostrophe to indicate special plurals.**

   three 9's

   two different JCL's

   the why's and how's of the problem

NOTE: For plurals of numbers and abbreviations, some style guides omit the apostrophe: *9s*, *JCLs*. Because usage varies considerably, check with your organization.

MISUSE OF APOSTROPHES

Writers sometimes incorrectly use the contraction *it's* in place of the possessive pronoun its.

INCORRECT The company's management does not believe that the problem is it's responsibility.

CORRECT The company's management does not believe that the problem is its responsibility.

“ ” QUOTATION MARKS

Quotation marks are used in the following instances:

1. **Use quotation marks to indicate titles of short works, such as articles, essays, or chapters.**

   Smith's essay "Solar Heating Alternatives" was short but informative.

2. **Use quotation marks to call attention to a word or phrase used in an unusual way or in an unusual context.**

   A proposal is "wired" if the sponsoring agency has already decided who will be granted the contract.

   NOTE: Do not use quotation marks to excuse poor word choice:

   INCORRECT The new director has been a real "pain."

3. **Use quotation marks to indicate a direct quotation.**

   "In the future," he said, "check with me before authorizing any large purchases."

   As Breyer wrote, "Morale *is* productivity."

   NOTE: Quotation marks are not used with indirect quotations:

   INCORRECT He said that "third-quarter profits will be up."

   CORRECT He said that third-quarter profits will be up.

   CORRECT He said, "Third-quarter profits will be up."

For more about quoting sources, see Appendix, Part A, p. 450.

Also note that quotation marks are not used with quotations that are longer than four lines; instead, set the quotation in block format. In a word-processed manuscript, a block quotation is usually introduced by a complete sentence followed by a colon and indented one-half inch from the left-hand margin.

Different style manuals recommend variations on the basic rules; the following example illustrates APA style.

McFarland (2011) writes:

> The extent to which organisms adapt to their environment is still being charted. Many animals, we have recently learned, respond to a dry winter

with an automatic birth control chemical that limits the number of young to be born that spring. This prevents mass starvation among the species in that locale. (p. 49)

Hollins (2012) concurs. She writes, "Biological adaptation will be a major research area during the next decade" (p. 2).

USING QUOTATION MARKS WITH OTHER PUNCTUATION

- If the sentence contains a *tag*—a phrase identifying the speaker or writer—a comma separates it from the quotation:

  Wilson replied, "I'll try to fly out there tomorrow."

  "I'll try to fly out there tomorrow," Wilson replied.

  Informal and brief quotations require no punctuation before a quotation mark:

  She asked herself "Why?" several times a day.

- In the United States (unlike most other nations where English is spoken), commas and periods following quotations are placed within the quotation marks:

  The project engineer reported, "A new factor has been added."

  "A new factor has been added," the project engineer reported.

- Question marks, dashes, and exclamation points are placed inside quotation marks when they are part of the quoted material:

  He asked, "Did the shipment come in yet?"

- When question marks, dashes, and exclamation points apply to the whole sentence, they are placed outside the quotation marks:

  Did he say, "This is the limit"?

- When a punctuation mark appears inside a quotation mark at the end of a sentence, do not add another punctuation mark:

  INCORRECT Did she say, "What time is it?"?

  CORRECT Did she say, "What time is it?"

... ELLIPSES

Ellipses (three spaced periods) indicate the omission of material from a direct quotation. (Word processors have a special character for ellipses.)

SOURCE My team will need three extra months for market research and quality-assurance testing to successfully complete the job.

QUOTE She responded, "My team will need three extra months . . . to successfully complete the job."

Insert an ellipsis after a period if you are omitting entire sentences that follow:

> Larkin refers to the project as "an attempt . . . to clarify the issue of compulsory arbitration. . . . We do not foresee an end to the legal wrangling . . . but perhaps the report can serve as a definition of the areas of contention."
>
> The writer has omitted words from the source after *attempt* and after *wrangling*. After *arbitration*, the writer has inserted an ellipsis after a period to indicate that a sentence has been omitted.

NOTE: If the author's original statement has ellipses, MLA style recommends that you insert brackets around an ellipsis that you introduce in a quotation.

> Sexton thinks "reuse adoption offers . . . the promise to improve business [. . .] worldwide."

[] SQUARE BRACKETS

Square brackets are used in the following instances:

1. **Use square brackets around words added to a quotation.**

   > As noted in the minutes of the meeting, "He [Pearson] spoke out against the proposal."

   A better approach would be to shorten the quotation:

   > The minutes of the meeting note that Pearson "spoke out against the proposal."

2. **Use square brackets to indicate parenthetical information within parentheses.**

   > (For further information, see Charles Houghton's *Civil Engineering Today* [1997].)

Mechanics

ital ITALICS

Although italics are generally preferred, you may use underlining in place of italics. Whichever method you choose, be consistent throughout your document. Italics (or underlining) are used in the following instances:

1. **Use italics for words used as words.**

   > In this report, the word *operator* will refer to any individual who is in charge of the equipment, regardless of that individual's certification.

2. **Use italics to indicate titles of long works (books, manuals, and so on), periodicals and newspapers, long films, long plays, and long musical works.**

See Houghton's *Civil Engineering Today*.

We subscribe to the *Wall Street Journal*.

Note that *the* is not italicized or capitalized when the title is used in a sentence.

NOTE: The MLA style guide recommends that the names of websites be italicized.

The Library of Congress maintains *Thomas*, an excellent site for legislative information.

3. **Use italics to indicate the names of ships, trains, and airplanes.**

The shipment is expected to arrive next week on the *Penguin*.

4. **Use italics to set off foreign expressions that have not become fully assimilated into English.**

Grace's *joie de vivre* makes her an engaging presenter.

The replacement part came from one of the *marchés aux puces* in Paris.

Check a dictionary to determine whether a foreign expression has become assimilated.

5. **Use italics to emphasize words or phrases.**

Do not press the red button.

< > ANGLE BRACKETS

Some style guides advocate using angle brackets around URLs in print documents to set them off from the text.

Our survey included a close look at three online news sites: the *New York Times* <www.nytimes.com>, the *Washington Post* <www.washingtonpost.com>, and CNN <www.cnn.com>.

You might want to check with your instructor or organization before following this recommendation.

- HYPHENS

Hyphens are used in the following instances:

1. **Use hyphens to form compound adjectives that precede nouns.**

general-purpose register
meat-eating dinosaur
chain-driven saw

NOTE: Hyphens are not used after adverbs that end in -ly.

newly acquired terminal

Also note that hyphens are not used when the compound adjective follows the noun:

For more about compound adjectives, see Ch. 6.

The Woodchuck saw is chain driven.

Many organizations have their own policy about hyphenating compound adjectives. Check to see if your organization has a policy.

2. **Use hyphens to form some compound nouns.**

   once-over

   go-between

   NOTE: There is a trend away from hyphenating compound nouns (*vice president*, *photomicroscope*, *drawbridge*); check your dictionary for proper spelling.

3. **Use hyphens to form fractions and compound numbers.**

   one-half

   fifty-six

4. **Use hyphens to attach some prefixes and suffixes.**

   post-1945

   president-elect

5. **Use hyphens to divide a word at the end of a line.**

   We will meet in the pavil-
   ion in one hour.

   Whenever possible, however, avoid such line breaks; they slow the reader down. Even when your word processor is determining the line breaks, you might have to check the dictionary occasionally to make sure a word has been divided between syllables. If you need to break a URL at the end of a line, do not add a hyphen. Instead, break the URL before a single slash or before a period:

   http://www.stc.org/
   ethical.asp

num NUMBERS

Ways of handling numbers vary considerably. Therefore, in choosing between words and numerals, consult your organization's style guide. Many organizations observe the following guidelines:

1. **Express technical quantities of any size in numerals, especially if a unit of measurement is included.**

3 feet	43,219 square miles
12 grams	36 hectares

2. **Express nontechnical quantities of fewer than 10 in words.**

    three persons

    six whales

3. **Express nontechnical quantities of 10 or more in numerals.**

    300 persons

    12 whales

4. **Write out approximations.**

    approximately ten thousand people

    about two million trees

5. **Express round numbers over nine million in a combination of words and numerals.**

    14 million light-years

    $64 billion

6. **Express decimals in numerals.**

    3.14

    1,013.065

    Decimals of less than one are preceded by a zero:

    0.146

    0.006

7. **Write out fractions, unless they are linked to units of measurement.**

    two-thirds of the members

    3½ hp

8. **Express time of day in numerals if A.M. or P.M. is used; otherwise, write it out.**

    6:10 A.M.

    six o'clock

    the nine-thirty train

9. **Express page numbers and figure and table numbers in numerals.**

    Figure 1

    Table 13

    page 261

10. **Write back-to-back numbers using a combination of words and numerals.**

    six 3-inch screws

    fourteen 12-foot ladders

    3,012 five-piece starter units

    In general, the quantity linked to a unit of measurement should be expressed with the numeral. If the nontechnical quantity would be cumbersome in words, however, use the numeral for it instead.

11. **Use both words and numerals to represent numbers in legal contracts or in documents intended for international readers.**

    thirty-seven thousand dollars ($37,000)

    five (5) relays

12. **Use both words and numerals in some street addresses.**

    3801 Fifteenth Street

NUMBERS: SPECIAL CASES

- A number at the beginning of a sentence should be spelled out:

  Thirty-seven acres was the size of the lot.

  Many writers would revise the sentence to avoid spelling out the number:

  The lot was 37 acres.

- Within a sentence, numbers with the same unit of measurement should be expressed consistently in either numerals or words:

  INCORRECT On Tuesday, the attendance was 13; on Wednesday, eight.

  CORRECT On Tuesday, the attendance was 13; on Wednesday, 8.

  CORRECT On Tuesday, the attendance was thirteen; on Wednesday, eight.

- In general, months should not be expressed as numbers. In the United States, 3/7/16 means March 7, 2016; in many other countries, it means July 3, 2016. The following forms, in which the months are written out, are preferable:

  March 7, 2016

  7 March 2016

abbr ABBREVIATIONS

Abbreviations save time and space, but you should use them carefully because your readers might not understand them. Many companies and professional organizations provide lists of approved abbreviations.

Analyze your audience to determine whether and how to abbreviate. If your readers include a general audience unfamiliar with your field, either write out the technical terms or attach a list of abbreviations. If you are new to an organization or are publishing in a field for the first time, find out which abbreviations are commonly used. If for any reason you are unsure about a term, write it out.

The following are general guidelines about abbreviations:

1. **When an unfamiliar abbreviation is introduced for the first time, give the full term, followed by the abbreviation in parentheses. In subsequent references, the abbreviation may be used alone. For long works, the full term and its abbreviation may be written out at the start of major units, such as chapters.**

   The heart of the new system is the self-loading cartridge (SLC).

   The liquid crystal display (LCD) is your control center.

2. **To form the plural of an abbreviation, add an s, either with or without an apostrophe, depending on the style used by your organization.**

   GNP's or GNPs

   PhD's or PhDs

   Abbreviations for most units of measurement do not take plurals:

   10 in.

   3 qt

3. **Do not use periods with most abbreviations in scientific writing.**

   lb

   cos

   dc

   If an abbreviation can be confused with another word, however, a period should be used:

   in.

   Fig.

4. **If no number is used with a unit of measurement, do not use an abbreviation.**

INCORRECT	How many sq meters is the site?
CORRECT	How many square meters is the site?

cap CAPITALIZATION

For the most part, the conventions of capitalization in general writing apply in technical communication:

1. **Capitalize proper nouns, titles, trade names, places, languages, religions, and organizations.**

   William Rusham

   Director of Personnel

   Quick-Fix Erasers

   Bethesda, Maryland

   Italian

   Methodism

   Society for Technical Communication

   In some organizations, job titles are not capitalized unless they refer to specific people.

   Alfred Loggins, Director of Personnel, is interested in being considered for vice president of marketing.

2. **Capitalize headings and labels.**

   A Proposal To Implement the Wilkins Conversion System

   Mitosis

   Table 3

   Section One

   The Problem

   Rate of Inflation, 2006–2016

   Figure 6

Proofreading Symbols and Their Meanings

Mark in margin	Instructions	Mark on manuscript	Corrected type
ℯ	Delete	$10 billion ~~dollars~~	$10 billion
^	Insert	envirogment (n inserted)	environment
(stet)	Let stand	let ~~it~~ stand	let it stand
(cap)	Capitalize	the english language	the English language
(lc)	Make lowercase	the English /Language	the English language
—	Italicize	Technical Communication (underlined)	*Technical Communication*
(tr)	Transpose	reicve	receive
⌢	Close up space	diagnostic ultra sound	diagnostic ultrasound
(sp)	Spell out	(Pres) Smithers	President Smithers
#	Insert space	3amp light	3 amp light
¶	Start paragraph	. . . the results.¶These results	. . . the results. These results
run in	No paragraph	. . . the results. For this reason,	. . . the results. For this reason,
(sc)	Set in small capitals	Needle-nose pliers (double underlined)	NEEDLE-NOSE PLIERS
(bf)	Set in boldface	Needle-nose pliers (wavy underlined)	**Needle-nose pliers**
⊙	Insert period	Fig 21	Fig. 21
^,	Insert comma	the plant which was built	the plant, which was built
=	Insert hyphen	menu driven software	menu-driven software
(:)	Insert colon	Add the following	Add the following:
^;	Insert semicolon	. . . the plan however, the committee	. . . the plan; however, the committee
’	Insert apostrophe	the users preference	the user’s preference
“/”	Insert quotation marks	Furthermore, she said . . .	“Furthermore,” she said . . .
(/)	Insert parentheses	Write to us at the Newark office	Write to us (at the Newark office)
[/]	Insert brackets	President [John] Smithers	President [John] Smithers
1/N	Insert en dash	1984 2001	1984–2001
1/M	Insert em dash	Our goal victory	Our goal—victory
2	Insert superscript	4,000 ft2	4,000 ft^2
2	Insert subscript	H2O	H_2O
//	Align	$123.05 $86.95	$123.05 $86.95
[	Move to the left	[PVC piping	PVC piping
]	Move to the right	PVC piping]	PVC piping
⊓	Move up	PVC piping	PVC piping
⊔	Move down	PVC piping	PVC piping

References

CHAPTER 1: Introduction to Technical Communication

Adecco Staffing. (2013). *Lack of soft skills negatively impacts today's workforce*. Retrieved July 9, 2015, from http://www.adeccousa.com/about/press/Pages/20130930-lack-of-soft-skills-negatively-impacts-todays-us-workforce.aspx

Grammarly. (2014). *Strong writing skills make you better at your job and are linked to higher pay (infographic)*. Retrieved June 5, 2015, from http://www.grammarly.com/blog/2015/strong-writing-skills-make-you-better-at-your-job-and-pay-more-infographic/

National Association of Colleges and Employers. (2014). *Job outlook 2014*, p. 8. Retrieved January 20, 2015, from http://www.naceweb.org/s10242012/skills-abilities-qualities-new-hires

TIME. (2013). *The real reason new college grads can't get hired*. Retrieved January 20, 2015, from http://business.time.com/2013/11/10/the-real-reason-new-college-grads-cant-get-hired/

CHAPTER 2: Understanding Ethical and Legal Obligations

Bettinger, B. (2010). *Social media implications for intellectual property law*. Retrieved May 23, 2013, from www.slideshare.net/blaine_5/social-media-implications-for-intellectual-property-law

Donaldson, T. (1991). *The ethics of international business*. New York, NY: Oxford University Press.

Ethics Resource Center. (2014). *2013 national business ethics survey: Workplace ethics in transition*. Retrieved January 20, 2015, from http://www.ethics.org/nbes/files/FinalNBES-web.pdf

Lipus, T. (2006). International consumer protection: Writing adequate instructions for global audiences. *Journal of Technical Writing and Communication, 36*(1), 75–91.

Malachowski, D. (2013). *Wasted time at work costing companies billions*. Retrieved June 12, 2013, from www.salary.com/wasted-time-at-work-still-costing-companies-billions-in-2006

Proskauer Rose LLP. (2014). *Social media in the workplace around the world 3.0*. Retrieved November 20, 2014, from http://www.proskauer.com/files/uploads/social-media-in-the-workplace-2014.pdf

Sigma Xi. (2000). *Honor in science*. New Haven, CT: Author.

Trillos-Decarie, J. (2012). Marketing + sales + communication + legal = social strategy (PowerPoint presentation). Retrieved May 23, 2013, from http://www.slideshare.net/JasmineTrillos-Decarie/jasmine-vegas-conf-final

U.S. Census Bureau. (2016). *ProQuest statistical abstract of the United States: 2015*. Washington, DC: U.S. Government Printing Office.

U.S. Consumer Product Safety Commission. (2016). *2015 annual report to the President and Congress*. Bethesda, MD: Author. Retrieved April 4, 2018, from https://www.cpsc.gov/s3fs-public/FY15AnnualReport.pdf

U.S. Federal Trade Commission. (2009, October 5). *FTC publishes final guides governing endorsements, testimonials*. Retrieved March 22, 2013, from www.ftc.gov/opa/2009/10/endortest.shtm

Velasquez, M. G. (2011). *Business ethics: Concepts and cases* (7th ed.). Upper Saddle River, NJ: Pearson.

CHAPTER 3: Writing Collaboratively

Cisco Systems, Inc. (2010). *Cisco 2010 midyear security report*. Retrieved July 22, 2010, from www.cisco.com/en/US/prod/collateral/vpndevc/security_annual_report_mid2010.pdf

Duin, A. H., Jorn, L. A., & DeBower, M. S. (1991). Collaborative writing—courseware and telecommunications. In M. M. Lay & W. M. Karis (Eds.), *Collaborative writing in industry: Investigations in theory and practice* (pp. 146–69). Amityville, NY: Baywood.

Karten, N. (2002). *Communication gaps and how to close them*. New York, NY: Dorset House.

Kaupins, G., & Park, S. (2010, June 2). Legal and ethical implications of corporate social networks. *Employee Responsibilities and Rights Journal*. Retrieved July 9, 2010, from www.springerlink.com/content/446x810tx0134588/fulltext.pdfDOI10.1007/s10672-010-9149-8

Lustig, M. W., & Koester, J. (2012). *Intercultural competence: Interpersonal communication across cultures* (7th ed.). Boston, MA: Allyn & Bacon.

SocialTimes. (2016, April 4). Here's how many people are on Facebook, Instagram, Twitter, and other big social media networks. *Adweek Blog Network*. Retrieved October 15, 2016, from www.adweek.com/socialtimes/heres-how-many-people-are-on-facebook-instagram-twitter-other-big-social-networks/637205

CHAPTER 4: Analyzing Your Audience and Purpose

Bell, A. H. (1992). *Business communication: Toward 2000*. Cincinnati, OH: South-Western.

Bosley, D. S. (1999). Visual elements in cross-cultural technical communication: Recognition and comprehension as a function of cultural conventions. In C. R. Lovitt & D. Goswami (Eds.), *Exploring the rhetoric of international professional communication: An agenda for teachers and researchers* (pp. 253–76). Amityville, NY: Baywood.

Bureau of Economic Analysis. (2014a). Table 1308: U.S. direct investment position abroad, capital outflows, and income by industry of foreign affiliates: 2000 to 2013 [selected years]. *ProQuest statistical abstract of the United States: 2014, online edition*. Retrieved November 21, 2014, from http://statabs.proquest.com.libproxy.boisestate.edu/sa/abstract.html?table-no=1313&acc-no=C7095-1.28&year=2014&z=B444DFDF3BB71CBD73720D6BDC7897016C7DFC5E

Bureau of Economic Analysis. (2014b). Table 1300: U.S. international transactions by type of transaction: 2000 to 2013 [selected years]. *ProQuest statistical abstract of the United States: 2014, online edition*. Retrieved November 21, 2014, from http://statabs.proquest.com.libproxy.boisestate.edu/sa/abstract.html?table-no=1313&acc-no=C7095-1.28&year=2014&z=B444DFDF3BB71CBD73720D6BDC7897016C7DFC5E

Centers for Disease Control and Prevention. (2013). *CDC.gov and social media metrics: April 2013*. Retrieved July 9, 2013, from www.cdc.gov/metrics/reports/2013/oadcmetricsreportapril2013.pdf

Hoft, N. L. (1995). *International technical communication: How to export information about high technology*. New York, NY: Wiley.

Lovitt, C. R. (1999). Introduction: Rethinking the role of culture in international professional communication. In C. R. Lovitt & D. Goswami (Eds.), *Exploring the rhetoric of international professional communication: An agenda for teachers and researchers* (pp. 1–13). Amityville, NY: Baywood.

Tebeaux, E., & Driskill, L. (1999). Culture and the shape of rhetoric: Protocols of international document design. In C. R. Lovitt & D. Goswami (Eds.), *Exploring the rhetoric of international professional communication: An agenda for teachers and researchers* (pp. 211–251). Amityville, NY: Baywood.

U.S. Census Bureau. (2014). Table 43. *ProQuest statistical abstract of the United States: 2014, online edition*. Retrieved November 20, 2014, from http://statabs.proquest.com.libproxy.boisestate.edu/sa/abstract.html?table-no=43&acc-no=C7095-1.1&year=2014&z=AE0C3946262E8E270D0DFB6AFBFEFBA826334A9C

Yan, Sophia. (2015, September 24). Microsoft ditches Bing for Baidu in China. *CNN*. Retrieved September 8, 2016, from http://money.cnn.com/2015/09/24/technology/microsoft-baidu-xiaomi-china-deal/

CHAPTER 5: Researching Your Subject

DeVault, G. (2013). *Market research case study—Nielsen Twitter TV rating metric*. Retrieved June 26, 2013, from http://marketresearch.about.com/od/market.research.social.media/a/Market-Research-Case-Study-Nielsen-Twitter-Tv-Rating-Metric.htm

McCaney, K. (2013, June 18). Energy lab team explores new ways of analyzing social media. *GCN*. Retrieved June 26, 2013, from http://gcn.com/articles/2013/06/18/Energy-lab-new-ways-analyzing-social-media.aspx

CHAPTER 6: Writing for Your Readers

Benson, P. (1985). Writing visually: Design considerations in technical publications. *Technical Communication, 32*, 35–39.

Snow, K. (2009). *People first language*. Retrieved August 2, 2010, from www.disabilityisnatural.com/images/PDF/pfl-sh09.pdf

U.S. Census Bureau. (2012). *Americans with disabilities: 2010*. Retrieved July 23, 2013, from www.census.gov/prod/2012pubs/p70-131.pdf

Williams, J. M. (2007). *Style: Lessons in clarity and grace* (9th ed.). New York, NY: Pearson Longman.

CHAPTER 7: Designing Print and Online Documents

Biggs, J. R. (1980). *Basic typography*. New York, NY: Watson-Guptill.

Haley, A. (1991). All caps: A typographic oxymoron. *U&lc, 18*(3), 14–15.

Internet World Stats. (2016). *Internet world users by language*. Retrieved February 27, 2017, from www.internetworldstats.com/stats.htm

Keyes, E. (1993). Typography, color, and information structure. *Technical Communication, 40*, 638–654.

Poulton, E. (1968). Rate of comprehension of an existing teleprinter output and of possible alternatives. *Journal of Applied Psychology, 52*, 16–21.

U.S. Department of Agriculture. (2002, March 5). *Thermometer usage messages and delivery mechanisms for parents of young children*. Retrieved April 4, 2002, from www.fsis.usda.gov/oa/research/rti_thermy.pdf

WebAIM. (2016). *Introduction to web accessibility*. Retrieved October 15, 2016, from http://webaim.org/intro

Williams, R. (2015). *The non-designer's design book* (4th ed.). Berkeley, CA: Peachpit Press.

Williams, T., & Spyridakis, J. (1992). Visual discriminability of headings in text. *IEEE Transactions on Professional Communication, 35*, 64–70.

CHAPTER 8: Creating Graphics

Brockmann, R. J. (1990). *Writing better computer user documentation: From paper to hypertext*. New York, NY: Wiley.

Gatlin, P. L. (1988). Visuals and prose in manuals: The effective combination. In *Proceedings of the 35th International Technical Communication Conference* (pp. RET 113–15). Arlington, VA: Society for Technical Communication.

Grimstead, D. (1987). Quality graphics: Writers draw the line. In *Proceedings of the 34th International Technical Communication Conference* (pp. VC 66–69). Arlington, VA: Society for Technical Communication.

Horton, W. (1992). Pictures please—presenting information visually. In Carol Barnum and Saul Carliner (Eds.), *Techniques for technical communicators*. New York, NY: Longman.

Horton, W. (1993). The almost universal language: Graphics for international documents. *Technical Communication, 40*, 682-693.

Levie, W. H., & Lentz, R. (1982). Effects of text illustrations: A review of research. *Journal of Educational Psychology, 73*, 195–232.

Morrison, C., & Jimmerson, W. (1989, July). Business presentations for the 1990s. *Video Manager, 4*, 18.

ProZ.com. (2013). Retrieved July 12, 2013, from http://search.proz.com/employers/rates

U.S. Census Bureau. (2013). *ProQuest statistical abstract of the United States*. Washington, DC: U.S. Government Printing Office. Retrieved July 20, 2013, from http://statab.conquestsystems.com/sa/abstract.html?table-no=815&acc-no=C7095-1.16&year=2013&z=D221BAE892DDA2D7BFCBA95DB0B859487C40FCF2

White, J. V. (1984). *Using charts and graphs: 1000 ideas for visual persuasion*. New York, NY: R. R. Bowker.

White, J. V. (1990). *Color for the electronic age*. New York, NY: Watson-Guptill.

CHAPTER 9: Corresponding in Print and Online

Hall, S. (2009, August 21). Twitter fan companies say it helps them get closer to customers. *ITBusinessEdge*. Retrieved November 11, 2013, from http://www.itbusinessedge.com/cm/community/features/articles/blog/twitter-fan-companies-say-it-helps-them-get-closer-to-customers

Meyer, K. (2016, January 9). 2015: The year in social media disasters. *Medium.com*. Retrieved November 11, 2016, from https://medium.com/the-social-reader/2015-the-year-in-social-media-disasters-9cf0d53b60aa#.9xptx7afx

Sasaki, U. (2010). *Japanese business etiquette for email*. Retrieved September 3, 2010, from www.ehow.com/about_6523223_japanese-business-etiquette-email.html

CHAPTER 10: Applying for a Job

Auerbach, D. (2012, August 29). *How to get that computer to send your résumé to a real person*. Retrieved August 4, 2013, from www.careerbuilder.com/Article/CB-3185-Resumes-Cover-Letters-How-to-get-that-computer-to-send-your-resume-to-a-real-person

Bowers, T. (2013, March 18). Would you hire someone with poor grammar skills? *TechRepublic*. Retrieved August 6, 2013, from www.techrepublic.com/blog/career-management/would-you-hire-someone-with-poor-grammar-skills/

Fottrell, Q. (2014). *How job recruiters screen you on LinkedIn*. Retrieved December 3, 2014, from http://www.marketwatch.com/story/how-recruiters-screen-you-on-linkedin-2014-05-02

Halzack, S. (2013, August 4). Tips for using LinkedIn to find a job. *Washington Post*. Retrieved August 4, 2013, from www.washingtonpost.com/business/capitalbusiness/tips-for-using-linkedin-to-find-a-job/2013/08/01/7c50c418-e0ff-11e2-8ae9-5db15d3c0fca_story.html

Isaacs, K. (2012). *Lying on your résumé: What are the consequences?* Retrieved August 10, 2013, from http://college.monster.com/training/articles/51-lying-on-your-resume-what-are-the-consequences

Lorenz, M. (2012, May 31). *Six ways hiring managers are spotting résumé lies*. Retrieved August 5, 2013, from www.careerbuilder.com/Article/CB-3077-Resumes-Cover-Letters-6-ways-hiring-managers-are-spotting-resume-lies

Society for Human Resource Management. (2014). *SHRM survey findings: Resumes, cover letters and interviews, 2014*. Retrieved December 3, 2014, from http://www.shrm.org/research/surveyfindings/articles/pages/resume-cover-letter.aspx

Tarpey, M. (2015, May 14). More employers checking out candidates on social media. *TheHiringSite Blog*. Retrieved October 10, 2016, from http://thehiringsite.careerbuilder.com/employers-checking-candidates-social-media/

U.S. Department of Labor. (2015, March 31). Number of jobs held, labor market activity, and earnings growth among the youngest baby boomers: Results from a longitudinal survey summary. *Economic News Release* (Document USDL-15-0528), p. 1. Retrieved October 6, 2016, from www.bls.gov/news.release/pdf/nlsoy.nr0.htm

CHAPTER 11: Writing Proposals

Newman, L. (2011). *Proposal guide for business and technical professionals* (4th ed.). Farmington, UT: Shipleyn Associates.

Ohio Office of Criminal Justice Services. (2003). *Sample grant proposal*. Retrieved March 19, 2008, from www.graduate.appstate.edu/gwtoolbox/ocjs_sample_grant.pdf

Thrush, E. (2000, January 20). *Writing for an international audience: Part I. Communication skills*. Retrieved November 5, 2002, from www.suite101.com/article.cfm/5381/32233

USAspending.gov. (2016). *Advanced data search*. Retrieved October 4, 2016, from www.usaspending.gov/Pages/AdvancedSearch.aspx?sub=y&ST=C&FY=2015&A=0&SS=USA&AA=9700

CHAPTER 13: Writing Recommendation Reports

Honold, P. (1999). Learning how to use a cellular phone: Comparison between German and Chinese users. *Technical Communication, 46*(2), 195–205.

CHAPTER 14: Writing Definitions, Descriptions, and Instructions

Brain, M. (2005). *How computer viruses work*. Retrieved June 20, 2005, from http://computer.howstuffworks.com/virus1.htm

Delio, M. (2002, June 4). Read the f***ing story, then RTFM. *Wired News*. Retrieved June 6, 2002, from www.wired.com/culture/lifestyle/news/2002/06/52901

Falco, M. (2008, January 14). *New hope may lie in lab-created heart*. Retrieved March 21, 2008, from www.cnn.com/2008/HEALTH/01/14/rebuilt.heart

National Oceanic and Atmospheric Administration. (2016, October 19). *What is a marine protected area?* Retrieved November 30, 2016, from http://oceanservice.noaa.gov/facts/mpa.html

CHAPTER 15: Making Oral Presentations

Fleming, M., & Levie, W. H. (Eds.). (1978). *Instructional message design*. Englewood Cliffs, NJ: Educational Technology Publications.

Nienow, S. (2013, July 6). *How long does it take to create a really great presentation?* Retrieved August 28, 2013, from http://redzestdesign.com/2013/10/how-long-does-it-take-to-create-a-really-great-presentation

Smith, T. C. (1991). *Making successful presentations: A self-teaching guide*. New York, NY: Wiley.

Index

Note: *f* indicates a figure and *t* indicates a table.

B

C

E

F

G

Index of Features

FOCUS ON PROCESS

GUIDELINES

TECH TIPS